AF249014

Continued after index

George B. Benedek
Felix M.H. Villars

Physics With Illustrative Examples From Medicine and Biology

STATISTICAL PHYSICS

Second Edition

Foreword by Irving M. London

With 156 Illustrations

Springer

George B. Benedek
Department of Physics
Massachusetts Institute of Technology
Cambridge, MA 02139-4037
USA

Felix M.H. Villars
Department of Physics
Massachusetts Institute of Technology
Cambridge, MA 02139-4037
USA

Cover illustration: (Top right) a pair of dice; (bottom right) schematic of the glomerulus and Bowman's Capsule in the kidney; (bottom left) the eye as a light detector.

Library of Congress Cataloging-in-Publication Data
Benedek, George Bernard, 1928–
    Physics with illustrative examples from medicine and biology/George B. Benedek, Felix M.H. Villars. — 2nd ed.
        p. cm.
    Contents: v. 1. Mechanics — v. 2. Statistical physics — v. 3. Electricity and magnetism.
    ISBN 0-387-98769-X (hc.: alk. paper) — ISBN 0-387-98754-1 (hc)
    — ISBN 0-387-98770-3 (hc) — ISBN 0-387-98952-8 (hc)
    1. Physics.   2. Medicine.   3. Biology.   I. Villars, Felix, 1921– . II. Title.
    QC23 .B424 2000
    530—dc21   99-033040

Printed on acid-free paper.

Production managed by Frank M^CGuckin; manufacturing supervised by Joe Quatela.
Typeset by Integre Technical Publishing Co., Inc., Albuquerque, NM.
Printed and bound by Hamilton Printing Co., Rensselaer, NY.
Printed in the United States of America.

9 8 7 6 5 4 3 2 1

ISBN 0-387-98754-1 Springer-Verlag New York Berlin Heidelberg   SPIN 10708862

Foreword to the First Edition

During the past several decades the biological and medical sciences have flourished with the growth of knowledge of the chemistry of living matter. Ours is a biochemical era. Increasingly, however, biology and medicine are dependent on the physical sciences for a deeper and more quantitative understanding of important biomedical problems and for the development of successful solutions to these problems. Accordingly, students of the life sciences, and specifically of human biology and medicine, have a growing need to acquire an effective working knowledge of the physical sciences.

To help meet this need, Professors George Benedek and Felix Villars have developed a rigorous and, indeed, challenging introductory course in physics with illustrative content derived from biology and medicine. The course not only provides an excellent means for the study of physics, but it also demonstrates the value and utility of the quantitative approach of the physicist in affording illuminating insights into a wide range of important biological problems. This book and the course on which it is based have been sponsored by the Harvard–MIT Program in Health Sciences and Technology. One of the objectives of the Program is the advancement of the health sciences by promoting their productive interaction with the physical sciences and engineering. This course of study is designed to help meet this objective. The enthusiastic reception of the course at MIT prompted the recommendation that Professors Benedek and Villars make it available to a wider audience. They have responded, fortunately, by the writing of this book.

Gassin, France
July 1973

IRVING M. LONDON

Series Preface

The field of biological physics is a broad, multidisciplinary, and dynamic one, touching on many areas of research in physics, biology, chemistry and medicine. New findings are published in a large number of publications within these disciplines, making it difficult for students and scientists working in biological physics to keep up with advances occurring in disciplines other than their own. The Biological Physics Series is intended therefore to be a comprehensive one covering a broad range of topics important to the study of biological physics. Its goal is to provide scientists and engineers with text books, monographs and reference books to address the growing need for information.

Books in the Biological Physics Series will emphasize frontier areas of science including molecular, membrane, and mathematical biophysics; photosynthetic energy harvesting and conversion; information processing; physical principles of genetics; sensory communications; automata networks, neural networks, and cellular automata. Equally important will be coverage of current and potential applied aspects of biological physics such as biomolecular electronic components and devices, biosensors, medicine, imaging, physical principles of renewable energy production, and environmental control and engineering.

We are fortunate to have a distinguished roster of consulting editors on the Editorial Board, reflecting the breadth of biological physics. We believe that the Biological Physics Series can help advance the knowledge in the field by providing a home for publications in the field and that scientists and practitioners from many disciplines will find much to learn from the upcoming volumes.

Oak Ridge, Tennessee   ELIAS GREENBAUM
Series Editor-in-Chief

Preface to the Second Edition

Already in the early 1970's it was becoming evident that the biological and medical sciences were on the threshold of major new advances. It was also clear that the promotion of these advances required interdisciplinary interactions between the physical sciences and engineering on the one hand and the biological and medical sciences on the other. To facilitate such interactions, the Harvard–MIT Division of Health Sciences and Technology was created. Under the supportive auspices of this unique inter-university structure, we created, in the period 1973 to 1979, the first edition of this three-volume text. Its purpose was to teach students the principles of physics using examples and applications taken from biology and medicine. By this means we sought to help educate scientists who could work effectively at the fruitful interface between the physical and biological disciplines.

The first edition of these texts, which were soft cover photocopies of a type-written manuscript, illustrated by the authors, ultimately was allowed to go out of print in January of 1990. However, the extraordinary advances at the interface between the physical and biological sciences have continued to accelerate. Increasingly, science and engineering departments are recognizing the need to train students and faculty to be capable of research and teaching at this active interface. As a result, growing numbers of our colleagues, who had found these texts to be uniquely useful, have urgently requested that they be made available once again.

This increasing need was clearly understood by our editor, Maria Taylor. At her initiative, the American Institute of Physics and Springer-Verlag have under-taken to publish this second edition. We reviewed our original text and, even though our colleagues still accorded it high value, we decided to update various sections. We also have included supplementary references so as to help make connections with the current literature.

We hope that this beautifully printed, bound, and illustrated second edition will attract the careful attention and interest of yet a new generation of intellectually adventurous students and teachers.

Cambridge, Massachusetts   GEORGE B. BENEDEK
January 2000   FELIX M.H. VILLARS

Acknowledgments for the Second Edition

A renewed publication more than twenty years after the original issue can only take place with the aid of far-sighted supporters. We are grateful to Maria Taylor, our editor, whose vision and advocacy was essential to initiate the entire project. All the effort associated with the preparation of an updated, accurate second edition required reallocation of the authors' time from their normal academic teaching responsibilities. The authors wish to express their appreciation to Professors Martha Gray, and Joseph Bonventre, the Co-Directors of the Harvard–MIT Division of Health Sciences and Technology for their continued support of our teaching and writing efforts. Finally, we wish to give thanks to those many colleagues, both in the United States and abroad who have closely studied these books, understood their value and requested its renewed availability. Their urgent requests led to the reappearance of our work in the form of this Second Edition.

Cambridge, Massachusetts   GEORGE B. BENEDEK
January 2000   FELIX M.H. VILLARS

Preface to the First Edition

The structure of Classical Physics rests on three cornerstones: (1) Mechanics, (2) Electricity and Magnetism, and finally (3) Statistical Mechanics and Thermodynamics.

Newtonian Mechanics, the primary subject of Volume I of this series, provides a *deterministic* description of physical phenomena. If the forces acting on a body are known, and if the initial coordinates and momenta are known, then Newton's Laws of motion permit us to predict with great precision the subsequent motion of the body. Mechanics is a manifestly *causal* science. Newton's laws of motion provide the link between the causes (the forces) and the effects (the motion).

Electricity and Magnetism, the subject of Volume III of this series also provides a deterministic description of physical phenomena. In this case Maxwell's equations provide a deterministic framework which links the electric charges and their motion with the electric and magnetic fields.

The dominant role of these two disciplines and their customary location at the beginning of the physics curriculum naturally creates in the student's mind the view that physics is a rigidly deterministic discipline.

In the present volume we develop and present the ideas of statistical physics, of which statistical mechanics and thermodynamics are but one part. We seek to demonstrate to students, early in their career, the power, the broad range, and the astonishing usefulness of a probabilistic, non-deterministic view of the origin of a wide range of physical phenomena. By applying this approach analytically and quantitatively to problems such as: the size of random coil polymers; the diffusive flow of solutes across permeable membranes; the survival of bacteria after viral attack; the attachment of oxygen to the binding sites on the hemoglobin molecule; and the effect of solutes on the boiling point and vapor pressure of volatile solvents; we demonstrate that the probabilistic analysis of statistical physics provides a satisfying understanding of important phenomena in fields as diverse as physics, biology, medicine, physiology, and physical chemistry.

xiii

The development of the ideas of statistical physics does not follow a traditional and well-established pattern as in the case of mechanics or electromagnetism. It is quite impossible to summarize the essence of statistical physics with a few compact expressions like Maxwell's Equations or Newton's Equations. We have nevertheless endeavored to develop our views of statistical physics into a logically satisfying structure starting with the elements of the theory of probability and concluding, in Chapter 4, with the deep insights of statistical mechanics and thermodynamics.

In Chapter 1 we present, along with some historical vignettes, a development of the fundamental ideas of probability. By considering the coin-toss process, which in fact is isomorphic to a wide variety of biological and physical processes, we develop the very basic binomial probability distribution. This distribution is then applied to three problems: the sex distribution of children in families of various size; the mean square radius of random coil polymers; and the mean and mean square electric charge and solubility of the protein hemoglobin, in solution.

Chapter 2 starts with an analysis of the one dimensional random walk problem in terms of the Bernoulli probability distribution. The time space evolution of this distribution leads to an understanding of the fundamental phenomena of diffusion and transport of solute particles in solution and across biological and synthetic membranes. This subject in fact is carried up to the most current level of understanding of the process of passive transport. Applications are made to artificial kidney dialysis, to the physiological function of the kidney and to the passage of solute and solvent across capillary walls.

Chapter 3, entitled "Poisson Statistics," begins with a careful analysis of the general conditions under which the Poisson probability distribution is applicable. In fact this distribution appears in the analysis of a wide variety of subjects ranging from economics to epidemiology to sports and war. The Poisson distribution plays a central role in the understanding of the detection of light at both very low and high light illumination levels as the student will see from our discussion of the famous Hecht, Shlaer, Pirenne experiment and the threshold intensity discrimination experiments of H. Barlow. We also analyze, in this chapter, the Luria-Delbrück experiment which proved, with the aid of purely statistical reasoning, that bacterial immunity to viral attack is the result of mutation and is not the result of the transmission of an acquired characteristic.

This volume closes with Chapter 4 which deals with the Boltzmann probability distribution and the thermal equilibrium of many body systems. In this final chapter we have devoted particular attention to the clear development and physical understanding of the central ideas of statistical mechanics, e.g., the Boltzmann factor and the entropy. We recognize the entropy as the quantity whose maximization establishes the equilibrium state (Second Law of Thermodynamics), and use this to develop thermodynamics as a description of the relationship between equilibrium states of simple and composite systems. We believe that Chapter 4 contains many unique applications. For example we present the theory for the temperature dependence of the vapor pressure of a solid and show that this classical macroscopic phenomena permits a quite accurate determination of Planck's constant. We also present a combined microscopic-thermodynamic theory which describes, with good accuracy, the shape of the important oxygen dissociation curves of myoglobin and hemoglobin. We believe that Chapter 4 will prove useful not only to students of physics and the life sciences, but will be particularly valuable for students of physical chemistry and chemical thermodynamics.

At the end of of each chapter the reader will find many problems which clarify the physical, biological, or mathematical ideas presented in the text. The problems are graded in order of difficulty and are classified according to the textual topic where they apply.

The essential mathematical skills needed in the theory of probability are the ability to count possible patterns, and the ability to sum series, particularly the geometric series and the binomial series. These skills employ a knowledge of high school algebra and a student versed in this subject should encounter little difficulty with the mathematics of probability theory. Elementary differential and integral calculus are employed here on about the same level as in Volume I. In the present volume we frequently use the partial derivative of a function $f(x, y, z)$ of several variables. The student should understand that the partial derivative of f with respect to some variable, say z, $(\partial f / \partial z)$ is the same as the usual first derivative except that in calculating it one holds constant all the other independent variables $(x, y = \text{const.})$.

The experienced teacher will recognize that this volume contains more material than can normally be presented in a one semester course. The text has been written in such a way as to permit selective omissions without loss of information necessary for subsequent sections. For example, one need not present all the applications included in Chapter 1. Also, in Chapter 2, one might elect not to include section 2.5 or 2.6. Chapter 3 is largely self-contained and is not a prerequisite for Chapter 4. In Chapter 4 the instructor may wish, for example, to present the statistical mechanics portion and omit the thermodynamics which follow. In general we encourage the instructor to make use of the flexibility offered by the structure of this text, so as to employ that material most suited to his particular objectives.

We have presented the subject of Statistical Physics (Vol. 2) before that of Electricity and Magnetism (Vol. 3) because of our deeply felt belief in the overriding importance of probabilistic ideas in physics. These ides are central in the description of phenomena of crucial importance for students of physics, chemistry, and the life sciences. We recognize, of course, Electricity and Magnetism is taught conventionally immediately after Mechanics. Volume 3, *Electricity and Magnetism with Illustrative Examples from Medicine and Biology*, will be written in such a way as to permit the instructor to maintain that conventional pattern, if he desires.

Cambridge, Massachusetts   GEORGE B. BENEDEK
August 1973   FELIX M.H. VILLARS

Acknowledgments for the First Edition

We acknowledge, with gratitude, the continued encouragement, and farsighted support of Professor Irving London, M.D., and the Harvard-M.I.T. Program in Health Sciences and Technology.

We are deeply indebted to our exceptional secretary Ms. Mary Patton Fitzgerald. Despite the heavy press of many other duties she typed this book with remarkable accuracy and speed. She is responsible for the visual balance and clarity of this text.

Cambridge, Massachusetts
August 1974

GEORGE B. BENEDEK
FELIX M.H. VILLARS

Contents

Table of Important Constants   **621**

Table of Units and Conversion Factors   **623**

Index   **627**

CHAPTER 1

Elements of the Theory of Probability: The Binomial Distribution: Applications

1.1 Definition of Probability. The Two Rules. Illustrative Examples

The mathematical concept of probability, and the theory of probability built on it, grew out of the practice of gambling. Games of chance are as old as recorded history. In ancient times one aspect of such games was an appeal to fate, the search for an omen, which would give guidance before an impending decision. (This has survived in the practice of drawing lots.) However, the role of games of chance as entertainment was probably always predominant, and as such games were played for stakes, the desire to have a quantitative measure of one's chances of winning emerged.

While attempts at solving such problems began in fifteenth-century Italy, the decisive step toward a clear and rigorous mathematical formulation of questions of chance was made in France in the middle of the seventeenth century, a period when gambling for high stakes flourished.

The seventeenth century saw an unprecedented development of mathematics; let us just mention the names of Descartes, Fermat, Pascal, Leibnitz, and Newton. It so happened (by chance?) that Fermat and Pascal came to lay the foundations of what we now call the theory of probability.

Blaise Pascal (1623–1662) is known to posterity more for his literary and polemic work on the religious controversies of the time than for his scientific work. Yet he made important contributions to geometry, established the laws of hydro-static pressure, and built one of the first calculating machines. Later, in his thirties, he was overcome by a feeling of "sinfulness" of such endeavors, and abandoned them altogether.

When he was 28, his father died and left him in comfortable circumstances. For about two years Pascal enjoyed the pleasures of a man about town, and it is surmised that during this period he made the acquaintance of the Chevalier de Méré, an inveterate gambler. His friend consulted him on various occasions on questions of how to assess properly the odds in games he was playing, and Pascal's interest was aroused. Knowing of Pierre de Fermat's reputation as the foremost mathematician of his time, he got in touch with him through a common friend, and a correspondence resulted in 1654, which we may consider as the beginning of the mathematical theory of probability. Fermat, a lawyer by profession, was at that time King's councillor at the Parliament of Toulouse in the south of France. (Their correspondence is preserved, except for Pascal's first letter, and can be found in its entirety in the book, *Games, Gods and Gambling* by N. F. David [2].)

From then on, the topic was launched as a challenge to mathematicians, and saw a rapid development in the following 150 years. Landmarks in this progress were the works of Jacob Bernoulli (*Ars Conjectandi*, 1713) and of Pierre Simon de Laplace (*Théorie Analytique des Probabilités*, 1812).

We should not be astonished that the mathematical concept of probability was so late in coming into being. It is indeed a subtle notion, capable of many shades of interpretations, but generally based on the observed *regularities* in long *repetitions* of a process, which taken singly, has an *unpredictable* outcome.

The fact that the notion of probability may be used with different meanings (or shades of meaning) becomes quite apparent as soon as we cite some examples. In its most transparent application the concept has an *operational* definition; that is, it is based on verifiable experience: We can throw a die as often as our endurance will permit, and determine the frequency of occurrence of any given number, 1 to 6. The statement that an even numbered face occurs with the probability of 50% then expresses an *expectation*, based on past experience, that in repeated throws of that die an even numbered face will show up in about half the cases.

A similar but somewhat watered down concept of probability is used by life insurance companies in setting the price of an insurance policy depending on the age of the customer. No two customers of a given age are alike, and the company's policy must be based on the attributes of an "average person," which as nearly as possible represents the common characteristics of a large number of people of a given age.

Applied to "unpredictable" events like the outcome of a horse race, or an election, the word probability (e.g., the probability that a particular horse will win) lacks the attribute of verification by repetition under the same circumstances. The odds one is willing to give in betting on a horse may represent a "plausible inference" from past experience, but the "probability" of victory of that horse is of a kind we shall not have to deal with.

We now describe the setting in which the concept of probability, *as used here*, applies, and proceed to give a definition:

(a) We deal with an *operation* or *process*—subsequently called a "trial"—which has several distinct possible *outcomes*, none of which can be predicted with certainty in a single trial. Let n stand for the number of distinct outcomes which we assume to be mutually exclusive. We label them by an index i running from 1 to n.

(b) The trial can be repeated any number of times under *manifestly identical* conditions. Let $\mathcal{N}$ be the number of repetitions, and let $\mathcal{N}_i$ stand for the number of times that the outcome "i" is reached. Clearly

$$\mathcal{N}_1 + \mathcal{N}_2 + \cdots + \mathcal{N}_i + \mathcal{N}_n = \mathcal{N}. \tag{1-1}$$

(c) If we then observe that the ratio between the number $\mathcal{N}_i$ of outcomes "i" and the total number $\mathcal{N}$ of repetitions tends to a fixed value as we make $\mathcal{N}$ larger and larger, we *call* the limit

$$P(i) = \left(\frac{\mathcal{N}_i}{\mathcal{N}}\right)_{\mathcal{N}\to\infty} \tag{1-2}$$

the *probability* of the outcome "i" for the trial in question.

Two results follow immediately from this definition. First, the probability $P(i)$ for outcome "i" is a number between zero and one:

$$0 \le P(i) \le 1.$$

Second, the sum of the probabilities of *all* possible, mutually exclusive, outcomes is one:

$$\sum_{\text{all outcomes}} P(i) = \sum_i \left(\frac{\mathcal{N}_i}{\mathcal{N}}\right)_{\mathcal{N}\to\infty} = \left(\frac{\sum \mathcal{N}_i}{\mathcal{N}}\right)_{\mathcal{N}\to\infty} = \left(\frac{\mathcal{N}}{\mathcal{N}}\right)_{\mathcal{N}\to\infty} = 1. \tag{1-3}$$

Since we have emphasized the point of mutual exclusion, let us illustrate this. Consider a die, with faces labeled 1 to 6. Only one of them will be up in any trial. We have therefore six *mutually exclusive* outcomes, and these six also describe *all possible* outcomes. There is no other. If, and only if, these two conditions, namely mutual exclusion and completeness, are satisfied, does (1-3) hold. As a counterexample, consider a trial that consists in pulling a card from a shuffled deck, and consider the probabilities of:

(1) pulling a spade;

(2) pulling a diamond; and

(3) pulling a king.

Now these outcomes are *not* mutually exclusive; the king of diamonds belongs to both outcomes (2) and (3). Nor is the table of outcomes complete; the ten of hearts belongs to neither (1), (2), nor (3). Although we can define probabilities (by means of (1-2)) for the three stated outcomes, (1-3) does not hold in this case:

$$P(1) + P(2) + P(3) \neq 1.$$

The definition of probability given above appears to require that we *verify* the *existence* of the limit (1-2) as the number $\mathcal{N}$ of repetitions of the trial gets very large. Actually, in most applications, one will take the existence of this limit for granted. Our experience tells us that we are on pretty safe ground here as long as the key provision—repetition under manifestly identical conditions—remains satisfied. This will indeed generally be the case. As an example, if we ask for the probability that a molecule in a gas travels a certain distance between two consecutive collisions, we see at once that all the molecules are continuously carrying out trials under conditions that are manifestly the same.

Our main concern in the present section is to develop a set of *rules* which will enable us to determine the probabilities for the possible outcomes of trials that are the *combination* of more elementary operations, for which the probabilities are known. As an example, we may want to know the probability of the occurrence of at least one double six in four consecutive throws of a pair of dice. This probability can be determined from the knowledge of the probability that a six will show in one throw of a *single* die; we shall show how this is done. Clearly then, we need the "rules of the game." *Two* basic rules are all that we require, and we will take them up in turn.

The *first* rule is the so-called *addition rule*. Consider a trial with n mutually exclusive outcomes labeled $i = 1$ to $i = n$; let $P(i)$ be the probability for outcome "i." We ask for the probability for finding the outcome "either i or j" (i being distinct from j). What this means is that repeating the trial $\mathcal{N}$ times, and finding $\mathcal{N}_i$ outcomes "i," $\mathcal{N}_j$ outcomes "j," we will take the sum $(\mathcal{N}_i + \mathcal{N}_j)$ as the number of outcomes "either i or j." We shall designate this outcome by "$i + j$," and the probability for it by $P(i + j)$: This probability is defined, according to (1-2), as

$$P(i + j) = \left(\frac{N_i + N_j}{N}\right)_{N \to \infty}$$
$$= \left(\frac{N_i}{N}\right)_{N \to \infty} + \left(\frac{N_j}{N}\right)_{N \to \infty} = P(i) + P(j).$$

This is the substance of the addition rule, which we now state as follows:

The probability $P(i + j)$ for the occurrence of the outcome "either (i) or (j)," (i) and (j) being distinct and mutually exclusive, is

$$P(i + j) = P(i) + P(j), \tag{1-4}$$

$P(i)$ and $P(j)$ being the probabilities for the separate outcomes "i" and "j," respectively.

As an illustration, consider a die with faces labeled 1 to 6. Assume now that face 1 is painted green, faces 2 and 3 blue, and faces 4, 5, and 6 red. Throwing this die, we may want to know the probabilities for throwing red, blue, or green. If we know *already* the probabilities for throwing 1 to 6, $P(1)$ to $P(6)$, then it follows from the addition rule that:

$$P(\text{green}) = P(1); \quad P(\text{blue}) = P(2) + P(3); \quad P(\text{red}) = P(4) + P(5) + P(6).$$

Simple as it is, students often fail to take advantage of this rule. Consider, for instance, the question: What is the probability of *not* throwing a 6 with one die? The outcome in question here is "not 6," which means that it is "either 1 or 2 or 3 or 4 or 5." By the addition rule, then, the probability $P(\text{not } 6)$ is

$$P(\text{not } 6) = P(1) + P(2) + P(3) + P(4) + P(5)$$
$$= 1 - P(6),$$

since the sum of the probabilities for all outcomes adds up to 1.

The second rule is the multiplication rule for the joint occurrence of independent events. As an example, consider a trial which consists of throwing two dice. To distinguish the dice, let the faces of one of them be labeled 1 to 6, the faces of the other $A, B, \ldots, F$. This trial has 36 distinct, mutually exclusive outcomes, which we label as $1A, 2A, \ldots, 1B, 2B, \ldots, 1F, 2F, \ldots, 6F$. Now let this pair of dice be thrown a very large number $\mathcal{N}$ of times, and the number of outcomes $\mathcal{N}_{1A}, \mathcal{N}_{2A}, \ldots, \mathcal{N}_{6F}$ counted and tabulated. We can arrange this table (Figure 1.1) as a square of six rows and six columns: Outside this square, we have also put down the subtotals for each row and each column

$$\mathcal{N}_1 = \mathcal{N}_{1A} + \mathcal{N}_{1B} + \cdots + \mathcal{N}_{1F} = \text{column subtotal,}$$
$$\mathcal{N}_A = \mathcal{N}_{1A} + \mathcal{N}_{2A} + \cdots + \mathcal{N}_{6A} = \text{row subtotal,}$$

and in the bottom corner the total number of throws

$$\mathcal{N} = \mathcal{N}_1 + \mathcal{N}_2 + \cdots + \mathcal{N}_6 = \mathcal{N}_A + \mathcal{N}_B + \cdots + \mathcal{N}_F.$$

With the help of this set of numbers, we can now define the probabilities for various outcomes. Thus, the probability for the event "$1A$" is

$$P(1A) = \left(\frac{\mathcal{N}_{1A}}{\mathcal{N}}\right)_{\mathcal{N}\to\infty} \tag{1-5}$$

and a corresponding definition holds for any of the 36 outcomes $1A$ to $6F$.

$\mathcal{N}_{1A}$	$\mathcal{N}_{2A}$	$\cdots$	$\mathcal{N}_{6A}$	$\mathcal{N}_A$
$\mathcal{N}_{1B}$	$\mathcal{N}_{2B}$	$\cdots$	$\mathcal{N}_{6B}$	$\mathcal{N}_B$
$\vdots$	$\vdots$		$\vdots$	
$\mathcal{N}_{1F}$	$\mathcal{N}_{2F}$	$\cdots$	$\mathcal{N}_{6F}$	$\mathcal{N}_F$
$\mathcal{N}_1$	$\mathcal{N}_2$	$\cdots$	$\mathcal{N}_6$	$\mathcal{N}$

Figure 1.1. Tabulation of the possible outcomes of throwing two different dies $\mathcal{N}$ times.

Now assume we have also practiced trials with one die *alone* (these dice may be loaded!), and have established the probabilities

$P(1), P(2), \ldots, P(6)$ for the outcomes 1 to 6 on the numbered die, and the probabilities

$$P(A), P(B), \ldots, P(F) \qquad \text{for the outcomes } A \text{ to } F \text{ on the lettered die.}$$

The *question* now is: Can we use the knowledge of the *single die* probabilities listed above to *infer* what the 36 probabilities $P(1A)$ to $P(6F)$ for the outcomes of throwing a pair of dice are? The answer is yes, provided we are justified in making the assumption of independence of the two dice. We have an intuitive understanding of this in the case of two dice thrown from a tumbler: The way the first die falls does not affect the second, on the average. This would not be true, however, if the dice had been "fixed" by putting a magnet bar inside each.

What conclusions can we draw from this independence assumption? The first, and rather obvious conclusion, is that

$$P(1) = \left(\frac{\mathcal{N}_1}{\mathcal{N}}\right)_{\mathcal{N}\to\infty}. \tag{1-6}$$

Let us see why this is so: $\mathcal{N}_1$ is the number of outcomes in which the numbered die shows "1," *irrespective* of what the letter on the other die is; $\mathcal{N}$ is the total number of throws. The ratio $(\mathcal{N}_1/\mathcal{N})$ is therefore the relative frequency of outcomes "1," obtained by simply *ignoring* the reading on the second (lettered) die. Now the meaning of independence is exactly this: That although this second die is in the tumbler, and thrown along with the first, its mere presence cannot have an effect on the frequency with which a "1" or "2" shows up on the first die. These frequencies must be the same as if this die were not there at all.

By the same argument we infer that

$$P(A) = \left(\frac{\mathcal{N}_A}{\mathcal{N}}\right)_{\mathcal{N}\to\infty}, \tag{1-7}$$

where $\mathcal{N}_A$ is the number of throws in which "A" shows up on the second, lettered die, irrespective of the number of the first die.

We have, however, not yet squeezed the last drop of contents from the independence assumption. This assumption also leads to the somewhat less obvious conclusion that

$$P(1) = \left(\frac{\mathcal{N}_{1A}}{\mathcal{N}_A}\right)_{\mathcal{N}\to\infty} = \left(\frac{\mathcal{N}_{1B}}{\mathcal{N}_B}\right)_{\mathcal{N}\to\infty} = \cdots. \tag{1-8}$$

To understand this point, let us look at the meaning of the two ratios on the right-hand side of (1-8): $\mathcal{N}_A$ is the number of all throws in which the second die shows "A." A certain fraction of those throws, namely $\mathcal{N}_{1A}$, also show "1" on the first die. So $(\mathcal{N}_{1A}/\mathcal{N}_A)$ is the frequency of the occurrence of a "1" among those throws, in which the lettered die shows "A." Similarly, $(\mathcal{N}_{1B}/\mathcal{N}_B)$ is the frequency of the occurrence of a "1" among the throws, in which the lettered die showed "B." Equation (1-8) claims that these frequencies must be the *same*. Indeed, assume they were different: that would mean that the frequency of the outcome "1" on the first die did in fact *depend* on what face the second die showed; this is contrary to the independence assumption we made.

One therefore concludes that

$$\left(\frac{\mathcal{N}_{1A}}{\mathcal{N}_A}\right) = \left(\frac{\mathcal{N}_{1B}}{\mathcal{N}_B}\right) = \cdots \left(\frac{\mathcal{N}_{1F}}{\mathcal{N}_F}\right). \tag{1-9}$$

But if that is so, these ratios are also equal to* $P(1)$, i.e.,

$$\frac{\mathcal{N}_{1A} + \mathcal{N}_{1B} + \cdots + \mathcal{N}_{1F}}{\mathcal{N}_A + \mathcal{N}_B + \cdots + \mathcal{N}_F} = \frac{\mathcal{N}_1}{\mathcal{N}} = P(1). \tag{1-10}$$

This result ((1-9)) is of crucial importance. Combined with (1-7), it leads to the result

$$P(1) = \left(\frac{\mathcal{N}_{1A}}{\mathcal{N}_A}\right)_{\mathcal{N}\to\infty} = \left(\frac{\mathcal{N}_{1A}}{P(A)\mathcal{N}}\right)_{\mathcal{N}\to\infty} = \frac{1}{P(A)}\left(\frac{\mathcal{N}_{1A}}{\mathcal{N}}\right)_{\mathcal{N}\to\infty}.$$

But the ratio $(\mathcal{N}_{1A}/\mathcal{N})_{\mathcal{N}\to\infty}$ is just the probability $P(1A)$ defined in (1-5). We have therefore the result

$$P(1A) = P(1)P(A). \tag{1-11}$$

The probability of the outcome "$1A$" in a throw of two independent dice is equal to the product of the probability $P(1)$ for the outcome "1" on the first die, and the probability $P(A)$ for the outcome "A" on the second die. Clearly, the same rule holds for any of the 36 outcomes "$1A$" to "$6F$."

This result is the multiplication rule. Let us state it in a general form:

---

*Elementary algebra shows that if $\frac{a}{b} = \frac{c}{d}$, then $\frac{a}{b} = \frac{c}{d} = \left(\frac{a+c}{b+d}\right)$. Since $\left(\frac{a}{b}\right) = \left(\frac{c}{d}\right)$ we may write $c = \lambda a$ and $d = \lambda b$ where λ is some constant: Thus $\left(\frac{a+c}{b+d}\right) = \frac{a(1+\lambda)}{b(1+\lambda)} = \frac{a}{b}$. Q.E.D.

If a trial T_1 has n distinct, mutually exclusive outcomes, "i" ($i = 1, 2, \ldots, n$), that occur with probability $P(i)$, and a trial T_2 has m outcomes "α" ($\alpha = 1, 2, \ldots, m$) occurring with probability $P(\alpha)$, then the probability $P(i\alpha)$ of the outcome "$i\alpha$" meaning "both i and α" in the combined trial $T_1 T_2$ is given by

$$P(i\alpha) = P(i)P(\alpha). \tag{1-12}$$

Provided that the two trials T_1 and T_2 are independent.

An important, but straightforward extension of the multiplication rule is the application to a trial consisting of the N-fold *repetition* of a given operation. Let this trial have outcomes "i" (independent, mutually exclusive) occurring with probability $P(i)$. Then any *specific outcome* of the N-fold repetition can be labeled by the *set*

$$\text{"}i_1, i_2, \ldots, i_N.\text{"}$$

This means that the first outcome is i_1, that of the first repetition is i_2, and that of the Nth repetition is i_N. For example, let a single trial consist of throwing a die five times. In this case, N is equal to 5. A specific outcome of this single trial requires the specification of the set of five face numbers, e.g., "2, 5, 3, 1, 6." Alternatively, we can envision the single trial as consisting of simultaneously casting five independent dice into some ordered slots. The outcome "2, 5, 3, 1, 6" in this case represents the spatial sequence of face numbers occurring in the slots.

It is important to recognize that the symbol N is introduced here to specify the number of elements in a single trial. This symbol must not be confused with $\mathcal{N}$, which we have used, and will continue to use, to specify the total number of test trials.

Under the independence assumption, the probability $P(i_1, i_2, \ldots, i_N)$ for the specific outcome "$i_1, i_2, \ldots, i_N$" is given by

$$P(i_1, i_2, \ldots, i_N) = P(i_1)P(i_2)\ldots P(i_N) \tag{1-13}$$

The proof of this is left as a problem assignment.

We shall now present a few illustrative examples of the application of these rules. (Further examples are given in the problem set.) In all these examples we consider an ideal die, for which the probability $P(i)$, for any of the six faces "i" showing, is equal to one-sixth.

Example 1. What is the probability of throwing an even number with one die? This is a simple application of the addition rule: Three of the six faces carry an even number, so

$$P(\text{even}) = \left(\frac{\mathcal{N}_2 + \mathcal{N}_4 + \mathcal{N}_6}{\mathcal{N}}\right)_{\mathcal{N}\to\infty} = P(2) + P(4) + P(6) = \tfrac{1}{6} + \tfrac{1}{6} + \tfrac{1}{6} = \tfrac{1}{2}.$$

Example 2. What is the probability of *not* throwing a 6? Again the addition rule enters. "Not 6" means "either 1 or 2 or 3 or 4 or 5." So

$$P(\text{not } 6) = \sum_{1}^{5} P(i) = \tfrac{5}{6}.$$

Example 3. What is the probability that two dice thrown show the same face? By the multiplication rule any specific combination (i, j) of outcomes has the probability $\tfrac{1}{6} \times \tfrac{1}{6} = \tfrac{1}{36}$. This holds, in particular, for $(1, 1)$, $(2, 2)$, ..., $(6, 6)$. So the probability for "either $(1, 1)$ or $(2, 2)$ or ... $(6, 6)$" is given by the addition rule as

$$P(\text{same face}) = P(1, 1) + P(2, 2) + \cdots + P(6, 6) = 6 \times \tfrac{1}{36} = \tfrac{1}{6}.$$

Example 4. What is the probability that with a die thrown twice one gets at least one 6? In this case, we may first use the addition rule to establish the single-die probabilities

$$P(6) = \tfrac{1}{6} \qquad P(\text{not } 6) = \tfrac{5}{6}.$$

Throwing the die twice, we have four possible outcomes

$$\text{``}(6, 6),\text{''} \quad \text{``}(6, \text{not } 6),\text{''} \quad \text{``}(\text{not } 6, 6),\text{''} \quad \text{``}(\text{not } 6, \text{not } 6),\text{''}$$

with probabilities

$$P(6, 6) = \tfrac{1}{6} \times \tfrac{1}{6} = \tfrac{1}{36},$$
$$P(6, \text{not } 6) = P(\text{not } 6, 6) = \tfrac{1}{6} \times \tfrac{5}{6} = \tfrac{5}{36},$$
$$P(\text{not } 6, \text{not } 6) = \tfrac{5}{6} \times \tfrac{5}{6} = \tfrac{25}{36}.$$

All except the last of the four outcomes represents an outcome with at least one 6. So the probability for obtaining at least one 6 is

$$P(\text{at least one } 6) = P(6, 6) + P(6, \text{not } 6) + P(\text{not } 6, 6)$$
$$= \tfrac{1}{36} + \tfrac{5}{36} + \tfrac{5}{36} = \tfrac{11}{36}.$$

Another faster way of getting this result is based on the sum rule, (1-3), that the sum of the probabilities for all possible, mutually exclusive outcomes must be 1. Therefore,

$$P(\text{at least one } 6) = 1 - P(\text{not } 6, \text{not } 6) = 1 - \tfrac{25}{36} = \tfrac{11}{36}.$$

Example 5. The previously mentioned Chevalier de Méré, who started it all, used to spend a lot of time (and money) betting on the outcome of games of dice. He had found a winning combination by betting, for even stakes, that he would get at least one 6 in four throws of a die. And he should win, according to theory.

In one throw of a single (unloaded!) die, we have the probabilities

$$P(6) = \tfrac{1}{6} \qquad P(\text{not } 6) = \tfrac{5}{6}.$$

Throwing the die four times, there are $2 \times 2 \times 2 \times 2 = 16$ possible different outcomes. By the multiplication rule, the particular outcome "not 6, not 6, not 6, not 6" has a probability

$$P(\text{not } 6, \text{not } 6, \text{not } 6, \text{not } 6) = (P(\text{not } 6))^4 = \left(\tfrac{5}{6}\right)^4.$$

Every one of the 15 other possible outcomes has at least one 6. Since these probabilities add up to 1, we infer that

$$P(\text{at least one } 6) = 1 - P(\text{no } 6 \text{ at all})$$
$$= 1 - \left(\tfrac{5}{6}\right)^4 = \frac{671}{1296} = 0.517.$$

Thus, in fact, he has a better than 50–50 chance of winning his bet, and hence after many trials, at even stakes, he should make money. For example, if he bets another player $10 that he will get the winning combination, then after 1000 trials, *on average*, he will win 517 times and lose 483 times. He will have bet a total of $10,000, and pocketed $20 \times 517 = \$10,340$, on average. The loser will have bet the same amount, but will only pocket $9660, on average. The chevalier then thought of varying the game a little. If he could win in betting to get at least one 6 in four throws of a die, he concluded that he could also win by betting to get at least one double 6 in 24 throws of a pair of dice. Here is his reasoning: four trials produced a win when the desired outcome is one of six possible; thus one should also win with 24 trials when the desired outcome is one out of a possible 36, since 24 is to 36 as 4 is to 6. It is manifest that this is a rather hazy kind of reasoning. Anyway, he tried out that new game, and found that he lost money, to his great astonishment. Here is how Pascal comments on it, in his second letter to Fermat, dated July 29, 1654 [2], [3].

> I have no time to send you proof of a difficulty when greatly puzzled M. de Méré, for he is very able, but he is not a geometrician (this, as you know, is a great defect), and he does not even understand that a mathematical line can be divided ad infinitum and believes that it is made up of a finite number of points, and I have never been able to rid him of this idea. If you could do that, you would make him perfect.
>
> He told me that he found a fallacy in the theory of numbers, for this reason:
>
> If one undertakes to get a six with one die, the advantage in getting it in four throws is 671 to 625.*
>
> If one undertakes to throw two sixes with two dice, there is a disadvantage in undertaking it in 24 throws.
>
> And nevertheless, 24 is to 36 (which is the number of pairings of the faces of two dice) as 4 is to 6 (which is the number of faces on one die).
>
> This is what made him so indignant and which made him say to one and all that the propositions were not consistent and that Arithmetic was self-contradictory: but you will very easily see that what I say is correct, understanding the principles as you do.

Indeed, it is very easy for us to see that the attempt to get at least one double 6 in 24 consecutive throws of a pair of dice is not a winning proposition.

The probability for a double 6 in a single throw of a pair of dice is, according to the multiplication rule,

$$P(6, 6) = P(6) \cdot P(6) = \tfrac{1}{6}\tfrac{1}{6} = \tfrac{1}{36}.$$

So, for each throw of the pair, we can establish the probabilities for the two outcomes "6, 6" and "not (6, 6)":

---

*By "advantage" Pascal means the ratio of the probability that the chevalier wins to the probability that his opponent wins. This ratio is $(671/1296)/(625/1296) = (671/625)$.

$$P(6, 6) = \tfrac{1}{36},$$
$$P(\text{not } (6, 6)) = 1 - P(6, 6) = \tfrac{35}{36}.$$

Now throw the pair 24 times. The probability that *each* time the outcome is "not (6, 6)" is, according to the multiplication rule,

$$(P(\text{not } 6, 6))^{24} = \left(\tfrac{35}{36}\right)^{24} = 0.5086.$$

All other of the 2^{24} outcomes of the 24-throw trial contain at least one double 6; hence

$$P(\text{at least one double } 6) = 1 - 0.5086 = 0.4914.$$

Since this is less than one-half, it is not a winning bet when the stakes are even. The reader will also notice that the disadvantage of betting on that outcome is very slight indeed, and the chevalier must be a very assiduous gambler to have found out *empirically* that something was wrong with this assumption. (See Problem 25.)

1.2   Bernoulli Trials. The Binomial Distribution

This simplest case of a trial is one with only *two* possible outcomes. Its prototype is that of flipping a coin, the two outcomes being either "head," (h), or "tail," (t). In such a case it is customary to call the two corresponding probabilities p and q:

$$P(\text{head}) = p,$$
$$P(\text{tail}) = q = 1 - p. \tag{1-14}$$

The importance and interest of this kind of trial lies in the fact that in its N-fold repetition, it serves as a model for many physical processes.

The process of repeating a trial with just two possible outcomes N times is called a Bernoulli process, if the assumption can be made that the outcome of the nth repetition is in no way affected by the outcome of any of the $(n - 1)$ previous trials. Thus repeated coin-flipping is a Bernoulli process.

Not every repetition of a trial with two outcomes is a Bernoulli process. Consider, for example, a random walk of N steps (forward or backward being the two options), governed by the prescription:

$$p = \text{probability that the } n\text{th step is in the } \textit{same}$$
$$\text{direction as the previous step,}$$

$$q = \text{probability that the } n\text{th step is in the opposite}$$
$$\text{direction to the previous step.}$$

This prescription defines a process quite different from a Bernoulli process, and in which the probability for the outcome of a step is controlled by a "remembrance" of what happened in the previous step. Such processes are called *Markov processes*. Despite this extraordinary interest in statistical physics, we must leave them aside since they cannot be described by elementary means.

Bernoulli investigated this process of the N-fold repetition of a trial with just two outcomes to answer a question which any mathematician would have to raise in connection with the definition [(1-2)] of probability: How large must $\mathcal{N}$ be, that is, how often must we repeat a trial in order to establish the value of the probability p with a certain accuracy? And does the ratio $(\mathcal{N}_i/\mathcal{N})$ actually converge to a limiting value as $\mathcal{N}$ gets larger and larger?

Jacob Bernoulli (born 1654) was the oldest of the eight mathematicians who issued from the amazing Bernoulli family in just three generations. The Bernoullis, like so many Protestant families, had fled the Spanish Netherlands in 1583, and had found refuge in the Swiss city of Basel, where the family became wealthy merchants. Jacob was destined for the ministry, got his degree in theology, but at the age of 22 he bolted, went traveling, and became a mathematician. Huygens' work on games of chance aroused his interest in the mathematical theory of probability. He was the first to answer the above-mentioned question, connected with the very definition of probability: In coin flipping, for example, we define the probability p for heads as

$$p = \lim_{\mathcal{N} \to \infty} \left(\frac{\text{number of heads}}{\text{number of trials}} \right) = \left(\frac{\mathcal{N}_h}{\mathcal{N}} \right)_{\mathcal{N} \to \infty}. \tag{1-15}$$

Does p exist, and granted it exists, how large should $\mathcal{N}$ be to enable us to establish the value of p with a certain accuracy? This question led Jacob Bernoulli to investigate finite (N-step) sequences of trials.

Out of these considerations emerged the famous Bernoulli probability distribution that will be the central concern of this chapter. Before investigating this distribution, however, we shall digress briefly to see what answer can be given to the above-stated question posed by Bernoulli. We restate the question in as specific terms as possible.

Assume that heads occur with a certain probability p. If we now flip coins a finite number N of times, and head shows N_h times, then the ratio N_h/N will not be *exactly* p; we know that. But how close will it be to p? Let us express the closeness of these two numbers quantitatively by means of a discrepancy d:

$$d \equiv \left| \left(\frac{N_h}{N} \right) - p \right| . \tag{1-16}$$

Because of the randomness inherent in the coin-tossing process, the answer cannot be given in the form of a deterministic statement, as one might hope. For example, one cannot make a statement like

$$d < 10^{-3} \quad \text{if} \quad N > 10^4.$$

Instead, the answer itself can only be stated in probabilistic terms: The *probability* $P(d < \varepsilon)$ that d will have a value smaller than some number ε (say $\varepsilon = 0.001$) is given by the following inequality

$$P(d < \varepsilon) > 1 - \frac{p(1-p)}{N\varepsilon^2}. \tag{1-17}$$

We simply quote this result and leave its proof as an exercise at the end of the chapter. We wish, however, to explain the meaning and application of this result. Suppose that we require a "confidence level" of say 75%; this means we want $P(d < \varepsilon)$ to be at least 0.75. Having fixed P, (1-17) gives us a relation between the number of trials N required and the desired accuracy ε. This is given by the formula

$$N\varepsilon^2 = \frac{pq}{(1-P)} = \text{constant.}$$

The constant is independent of both the number of trials and the accuracy desired. It depends on the confidence level, and the a priori probabilities p and $q = 1 - p$. This result leads to the important conclusion that the accuracy with which (N_h/N) approximates p is given by

$$\varepsilon = \frac{\text{constant}}{\sqrt{N}}.$$

This is a most important result that occurs again and again in the analysis of experimental data. Namely, that the precision of a measurement increases only in proportion to the *square root* of the number of repeated observations. Thus, for example, to increase the accuracy of a measurement of p by a factor 10, we must increase the number of trials by a factor of 100.

We now leave this digression and return to our principal concern, which is the form and uses of the Bernoulli probability distribution. The analysis of Bernoulli trials will give us the means for a quantitative description of many physical processes. The most important of these is the *random walk*, already mentioned in the introduction. This will be taken up in Chapter 2 where we will show how the study of this phenomenon leads to a quantitative description of the process known as diffusion (which is the spreading of a swarm of molecules in a medium by means of their random motion). Other interesting applications occur in the study of the spatial conformation of linear chain molecules (polymers), consisting of a large number of segments (monomers), and in the study of multiply charged macromolecules in solution (so-called polyelectrolytes, among which are the soluble proteins, such as hemoglobin, the blood plasma proteins, etc.).

Our first example of a Bernoulli trial will be the flipping of a coin with two outcomes "head" (h) or "tail" (t). The results obtained may then, of course, by applied to any process with two possible outcomes; instead of "head or tail" the alternatives may be "step forward or step backward" in the random walk, or "charged or uncharged" for a ionizable molecule in solution.

Our first observation is that since the primitive (one-step) trial has the two outcomes head (h) or tail (t), a two-step trial (consisting of the same coin flipped twice) has the four possible, distinct outcomes

$$hh, ht, th, tt \tag{1-18}$$

and the three-step trial has eight outcomes

$$hhh, hht, hth, thh, htt, tht, tth, ttt. \tag{1-19}$$

In general, the N-step trial has 2^N distinct outcomes. What then is the probability for any specific outcome? Under the assumptions stated, namely, that the outcome of a given coin toss is not affected by the outcome of the preceding tosses, the multiplication rule [(1-12)] applies. Recalling that we called the probabilities $P(\text{head}) = p$, $P(\text{tail}) = q$, we obtain for $N = 2$ the four probabilities

$$\begin{aligned} P(h, h) &= P(h)P(h) = p^2, \\ P(h, t) &= P(h)P(t) = pq, \end{aligned} \tag{1-20}$$

$$P(t, h) = P(t)P(h) = pq,$$
$$P(t, t) = P(t)P(t) = q^2.$$

We see that $P(h, t)$ and $P(t, h)$ are the same. For $N = 3$, the outcomes listed in the tabulation [(1-19)] have the probabilities

$$p^3, p^2q, p^2q, p^2q, pq^2, pq^2, pq^2, q^3. \tag{1-21}$$

The generalization to a N-step trial is now straightforward: Each outcome of such a trial can be represented by a sequence of h's and t's, such as

$$\text{outcome} = (thtththh \ldots ht) \tag{1-22}$$

and the probability of this outcome is

$$P(\text{outcome}) = p^H q^T, \tag{1-23}$$

where H is the number of heads (h) in the sequence [(1-22)], and T the number of tails (t).

Now there are many distinct outcomes in an N-step trial that nevertheless have the same values of H and T, that is, the same total number of heads and tails. So, for example, the tabulations [(1-18) and (1-19)] show that

$$
\begin{array}{lll}
& H = 2,\ T = 0 & \text{once,} \\
\text{for } N = 2,\ \text{one has} & H = 1,\ T = 1 & \text{twice,} \\
& H = 0,\ T = 2 & \text{once,} \\
\\
& H = 3,\ T = 0 & \text{once,} \\
\text{and for } N = 3,\ \text{one has} & H = 2,\ T = 1 & \text{3 times,} \\
& H = 1,\ T = 2 & \text{3 times,} \\
& H = 0,\ T = 3 & \text{once.}
\end{array}
$$

If we are only interested to know the probability that in N tosses, there will be H heads and T tails, irrespective of the order in which h and t show up, we must, according to the addition rule, *add* the probabilities of all distinct outcomes with *common* values of H and T. Let us call this probability $P_N(H, T)$ (the subscript N is in fact redundant, since $H + T = N$). Then, for instance, for $N = 2$:

$$P_2(2, 0) = p^2, \qquad P_2(1, 1) = 2pq, \qquad P_2(0, 2) = q^2,$$

and for $N = 3$:

$$P_3(3, 0) = p^3, \qquad P_3(2, 1) = 3p^2 q, \qquad P_3(1, 2) = 3pq^2, \qquad P_3(0, 3) = q^3.$$

In the general case, $P_N(H, T)$ is given by

$$P_N(H, T) = p^H q^T \times \text{(number of distinct outcomes with } H \text{ heads}$$
$$\text{and } T \text{ tails)}$$
$$\equiv p^H q^T W_N(H, T). \tag{1-24}$$

So our next task is to find this number $W_N(H, T)$. The easiest way to do this is to investigate how this number changes, as the number of tosses N increases. A total of H heads and T tails in N tosses can come about by:

(i) having $(H - 1)$ heads and T tails in $(N - 1)$ tosses, and tossing an "h" in the Nth trial; and

(ii) having H heads and $(T - 1)$ tails in $(N - 1)$ tosses, and tossing a "t" in the Nth trial.

In case (i) we have $W_{N-1}(H - 1, T)$ distinct arrangements; in case (ii) there are $W_{N-1}(H, T - 1)$. Their sum is the number of arrangements of H heads and T tails in N tosses

$$W_N(H, T) = W_{N-1}(H - 1, T) + W_{N-1}(H, T - 1). \tag{1-25}$$

If one starts with the known values

$$W_0(0, 0) = 1, \qquad W_1(1, 0) = W_1(0, 1) = 1.$$

Equation (1-25) can be used to construct all the $W_N(H, T)$. A simple graphical device that does it is the so-called Pascal triangle, shown in Figure 1.2. In this figure the numbers $W_N(H, T)$ are laid out in a pyramid starting with $W_0(0, 0) = 1$:

$$W_0(0, 0);$$
$$W_1(1, 0); \quad W_1(0, 1);$$
$$W_2(2, 0); \quad W_2(1, 1); \quad W_2(0, 2);$$
$$W_3(3, 0); \quad W_3(2, 1); \quad W_3(1, 2); \quad W_3(0, 3).$$

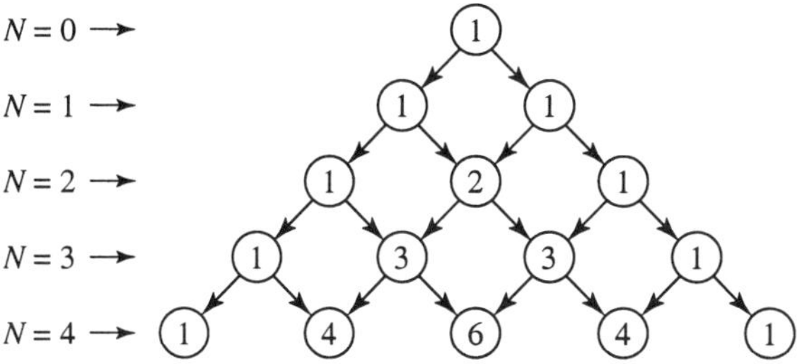

Figure 1.2. Pascal triangle for the coefficients $W_N(H, T)$.

Equation (1-25) for $W_N(H, T)$ then shows that each number in this pyramid is the sum of the two numbers *diagonally above* it. Although the Pascal triangle offers a simple means of finding the numerical values of the coefficients $W_N(H, T)$, it is not sufficient for our purpose: We will need a *formula* for W, which will enable us to calculate these numbers for large values of N. Such a formula can be extracted, rather laboriously, from (1-25); we present the result, and verify, however, that it is a solution of that equation. The formula is

$$W_N(H, T) = \frac{N!}{H!\,T!}, \tag{1-26}$$

where $N!$ (read "n-factorial") is the product

$$N! = 1 \times 2 \times 3 \times \cdots \times (N - 1) \times N. \tag{1-27}$$

By convention $0! \equiv 1$, and the above expression for $W_N(H, T)$ also holds for $H = 0$ and $T = 0$. The proof is very simple; indeed, we have, according to (1-26):

$$W_{N-1}(H - 1, T) = \frac{(N - 1)!}{(H - 1)!\,T!} = (N - 1)!\,\frac{H}{H!\,T!},$$

since $H! = H \times (H - 1)!$. Similarly,

$$W_{N-1}(H, T - 1) = (N - 1)!\,\frac{T}{H!\,T!}.$$

Upon adding these two expressions, we find

$$W_{N-1}(H - 1, T) + W_{N-1}(H, T - 1) = \frac{(N - 1)!}{H!\,T!}(H + T).$$

But $H + T = N$ and $(N - 1)! \times N = N!$, so that this sum is indeed equal to $W_N(H, T)$.

With this we can write the expression for the probability $P_N(H, T)$ as

$$P_N(H, T) = \frac{N!}{H!\,T!} p^H q^T \tag{1-28a}$$

or, observing that $T = N - H$,

$$P_N(H, N - H) = \frac{N!}{H!\,(N - H)!} p^H q^{N-H}. \tag{1-28b}$$

Before going into a discussion of the properties of $P_N(H, T)$, we must establish the identity of the number $W_N(H, T)$, given by (1-26) with the so-called binomial coefficients. These latter arise when we multiply out the Nth power or the "binome" $(a + b)$. Let us recall the familiar formulas

$$(a + b)^2 = a^2 + 2ab + b^2,$$
$$(a + b)^3 = a^3 + 3a^2b + 3ab^2 + b^3.$$

In general, we have a result of the form

$$(a + b)^N = \sum_{H=0}^{N} \binom{N}{H} a^H b^{N-H}. \tag{1-29}$$

The numbers represented by the symbol $\binom{N}{H}$ are the binomial coefficients. So, for instance, for $N = 3$, we have

$$\binom{3}{3} = 1, \qquad \binom{3}{2} = 3, \qquad \binom{3}{1} = 3, \qquad \binom{3}{0} = 1.$$

We demonstrate now that the numbers represented by $W_N(H, N - H)$ and $\binom{N}{H}$ are the *same*. The reason for this identity is simple and can best be seen by looking at an example. If we multiply out $(a + b)^3$, we obtain

$$\begin{aligned}
(a + b) &= (a + b)(a + b)(a + b) \\
&= (aa + ab + ba + bb)(a + b) \\
&= aaa + (aab + aba + baa) + (abb + bab + bba) + bbb \\
&= a^3 + 3a^2b + 3ab^2 + b^3.
\end{aligned}$$

The binomial coefficient $3 = \binom{3}{2}$ in front of a^2b is exactly the number of ways in which two a's and one b can be arranged, namely, three ways:

$$aab \quad aba \quad baa.$$

In general, the binomial coefficient $\binom{N}{H}$ in front of $a^H b^{N-H}$ in (1-28b) is the number of ways that H factors a and $(N-H)$ factors b can be arranged. So let us record this result as

$$\binom{N}{H} = W_N(H, N-H) = \frac{N!}{H!\,(N-H)!}. \tag{1-30}$$

The particular importance of recognizing this equality of the $W_N(H, T)$ with the binomial coefficients lies in the fact that it permits us to see that the probabilities $P_N(H, T)$ add up to one if summed over all possible values of (H, T). Indeed, in each process of tossing a coin N times, the outcome is either $H = 0$ (no head), $H = 1$ (one head), etc., up to $H = N$ (all heads). They represent a complete set of distinct, mutually exclusive outcomes; therefore, the sum of the probabilities

$$P_N(0, N) + P_N(1, N-1) + \cdots + P_N(N, 0) = \sum_H P_N(H, N-H)$$

must be equal to *one*. That this is indeed so is now easily seen by writing

$$P_N(H) \equiv P_N(H, N-H) = \binom{N}{H} p^H q^{N-H}.$$

Then

$$\sum_H P_N(H) = \sum_H \binom{N}{H} p^H q^{N-H}$$
$$= (p+q)^N = 1 \qquad \text{since} \quad (p+q) = 1. \tag{1-31}$$

$P_N(H)$ is now a very well-defined mathematical entity. We are able to calculate the values of $P_N(H)$ for any value of N or H. The function $P_N(H)$ is in fact called a *probability distribution function*. The values $P_N(H)$ of this *distribution function* tell us the probability with which H heads appear after N tosses. To represent this *distribution function*, it is very useful to construct a graph plotting the values of $P_N(H)$ for each of the possible values of H ($H = 0, 1, \ldots, N$). In the following Section 1.3 we will examine mathematically the detailed properties

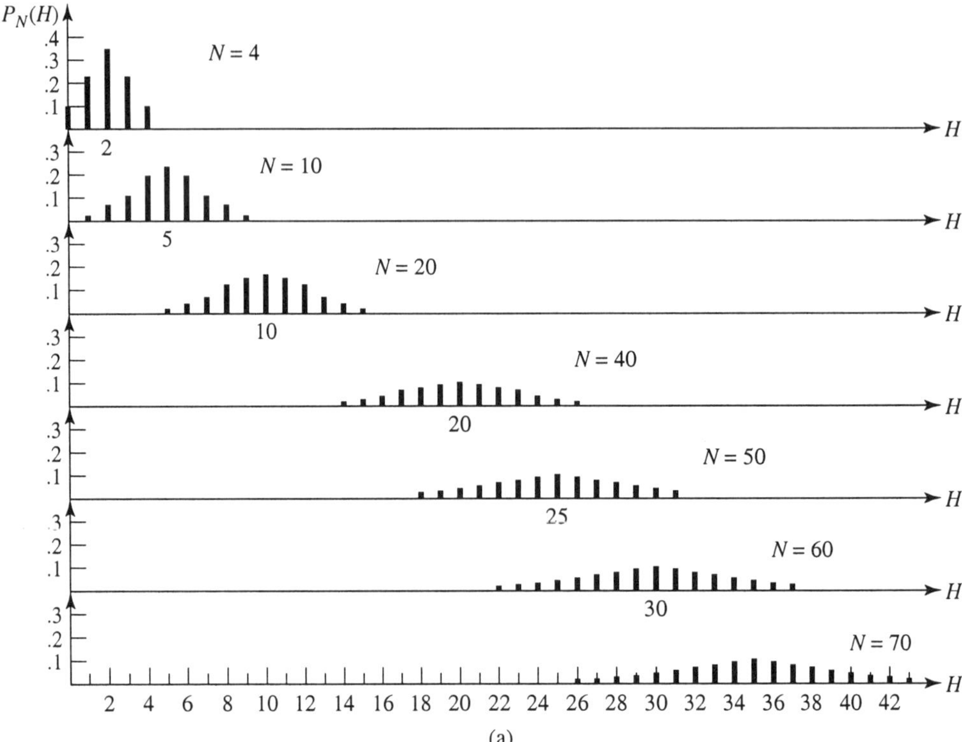

Figure 1.3(a). The Bernoulli distribution $P_N(H)$ for $p = q = \frac{1}{2}$ as a function of H for $N = 4, 10, 20, 40, 50, 60, 70$.

of this Bernoulli probability distribution. At this point it is instructive to examine Figure 1.3(a). In this figure we plot $P_N(H)$ as a function of H for values of $N = 4$, 10, 20, 40, 50, 60, 70, when $p = q = \frac{1}{2}$. All these graphs are plotted using the same horizontal scale for H, and the same vertical scale for $P_N(H)$.

Figure 1.3(a) shows very clearly a number of the principal features of the dependence of $P_N(H)$ on H and on N. First, we observe that when $p = q$ the distribution is symmetrical around the maximum value. Second, we note that the maximum value of each distribution occurs at $H = N/2$ in each of the graphs shown. (In the following section we shall see that this results because the average value of H in each distribution occurs at the point $\overline{H} = pN = N/2$.) As one increases N, then the center of the distribution shifts linearly to the right. A third observation is that the peak value of the distribution falls as N increases. In the following section we will prove that the peak of the Bernoulli distribution

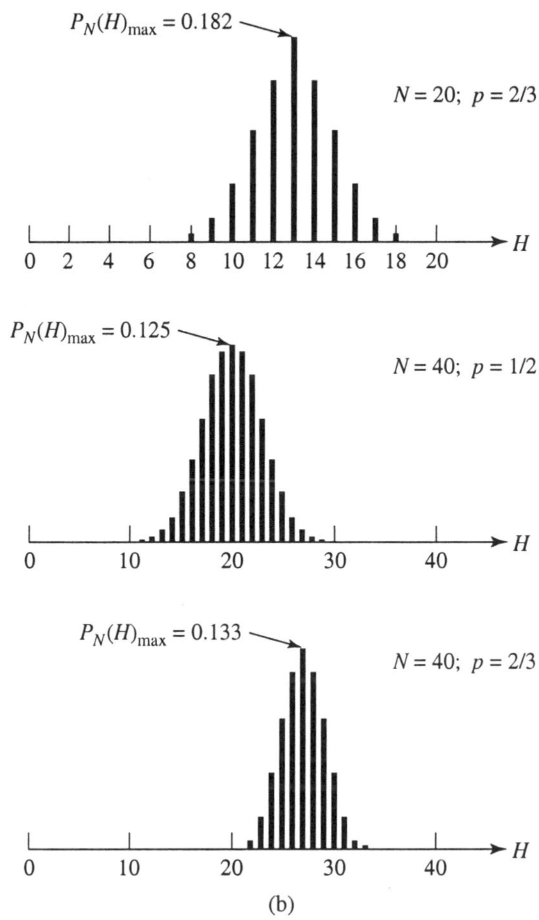

Figure 1.3(b). The Bernoulli distribution $P_N(H)$ for various N.

decreases as $1/\sqrt{N}$. A fourth observation that emerges from an examination of Figure 1.3(a) is that the "width" of the distribution grows larger as N increases. Again, in the section to follow, we shall provide a quantitative characterization of the "width" and show that it grows in proportion to $\sqrt{N}$. Since the center of the distribution occurs at $\overline{H} = N/2$, and the width of the distribution varies as $\sqrt{N}$, we can conclude that, *relative* to its center, the width of the distribution shrinks as $\sqrt{N}/(N/2) \sim 1/\sqrt{N}$. That is, if we changed the horizontal scale for each graph of $P_N(H)$, so that the point $\overline{H} = N/2$ appeared at the same horizontal location for each N, the graphs of the distribution would appear to narrow as N grows, in proportion to $1/\sqrt{N}$. Finally, we should note that as N increases, the probability for extreme values of H, namely, $H = 0$ or $N = N$, gets extremely small. In the table below we list the ratios $P_N(H = 0)$ to $(P_N(H))_{\max}$ for the cases $p = q = \frac{1}{2}$ and $N = 5, 10, 20, 40$:

N	$P_N(H = 0)/(P_N)_{\max}$
5	10^{-1}
10	4×10^{-3}
20	5.4×10^{-6}
40	8×10^{-12}

In Figure 1.3(b) we show the form of $P_N(H)$ for $N = 20$ and $p = \frac{2}{3}$. This graph shows that the maximum of the distribution occurs near $H = pN = \frac{2}{3}N = 13.3$ when $p \neq q$. The lower two graphs show how the symmetric distribution that obtains when $p = q = \frac{1}{2}$, shifts and becomes asymmetrical as p changes. The middle graph is for $N = 40$, $p = \frac{1}{2}$, and the lowest graph is for $N = 40$, $p = \frac{2}{3}$.

1.3   Mean Values and Variance

The probability distribution function $P_N(H, T)$ in (1-28a) expresses the probability that in a trial that consists of flipping a coin N times, head occurs H times. If our attention is focused on how this probability depends on the number H, we shall from now on just write $P_N(H)$ instead of the more elaborate, but redundant, $P_N(H, T)$. (Remember that T is determined by N and H since $T = N - H$.) This $P_N(H)$ represents the probability distribution function for the number H of heads, indicating for each value of H what the probability is that this particular value occurs.

We may then want to know how many heads appear on the average, or equivalently, what the average value of H is. To define this average, we must consider a large number $\mathcal{N}$ of repetitions of the trial, and in each case record the value of H, the number of heads. The average value of H is then *defined* by the relation

$$\text{average value of } H = \left(\frac{\text{sum of all values of } H \text{ recorded in } \mathcal{N} \text{ trials}}{\text{total number of } \mathcal{N} \text{ trials}} \right)_{\mathcal{N} \to \infty}.$$

$$(1\text{-}32)$$

Again, as in the definition of probability, we assume that the ratio in (1-32) will converge to a well-defined value as the number $\mathcal{N}$ of repetitions gets larger and larger.

Now, in the total number of $\mathcal{N}$ trials, a specific value of H occurs $\mathcal{N}_H$ times, and the sum of all H's in $\mathcal{N}$ trials can therefore be written as

$$\text{sum of all } H \text{ in } \mathcal{N} \text{ trials} = \sum_H H\mathcal{N}_H. \tag{1-33}$$

We shall use the notation $\overline{H}$ to indicate the average value of H ($=$ the average number of heads). By combining (1-32) and (1-33), we find

$$\overline{H} = \left(\frac{\sum_H H\mathcal{N}_H}{\mathcal{N}}\right)_{\mathcal{N}\to\infty} = \sum_H H\left(\frac{\mathcal{N}_H}{\mathcal{N}}\right)_{\mathcal{N}\to\infty}. \tag{1-34}$$

Now $(\mathcal{N}_H/\mathcal{N})_{\mathcal{N}\to\infty}$ is, by definition, the probability $P_N(H)$ that H heads occur as the outcome of the trial. We may therefore write for the mean value of H:

$$\overline{H} = \sum_H H P_N(H). \tag{1-35}$$

We can state this result in words in a somewhat generalized form as follows.

If the outcome of a trial can be described by the numerical value of a quantity (for instance, "number of heads"), then the *average* or *mean* value of that quantity is given by the sum of all possible values of that quantity, each multiplied with the *probability* that this value occurs as an outcome. To illustrate this, consider the case of a coin flipped three times; H has then the possible values 0, 1, 2, 3. The associated probabilities are

$$H = 0, \qquad P_3(0, 3) = q^3,$$
$$H = 1, \qquad P_3(1, 2) = 3pq^2,$$
$$H = 2, \qquad P_3(2, 1) = 3p^2q,$$
$$H = 3, \qquad P_3(3, 0) = p^3.$$

The mean value of H is then, according to (1-35):

$$\overline{H} = 0 \times (q^3) + 1 \times (3pq^2) + 2 \times (3p^2q) + 3 \times (p^3)$$
$$= 3p(q^2 + 2qp + p^2) = 3p(q + p)^2 = 3p,$$

since $p + q = 1$.

If $p = \frac{1}{2}$, the average value of H is thus $\overline{H} = \frac{3}{2}$. So we see first that the average need not be an integer. Assume you repeat this triple coin toss 15 times, and record the number H of heads each time. You may get a sequence like

$$H = 1, 2, 0, 2, 2, 1, 3, 1, 2, 2, 1, 1, 0, 2, 2.$$

Then, the sum of all these H values divided by 15, the number of repetition, gives you an approximate mean value

$$(H)_{\text{approx}} = \frac{\sum H}{15} = \frac{22}{15} = 1.49.$$

Again we observe that the "experimental" mean values depend on the number of repetitions used to establish it. If we know that $p = \frac{1}{2}$, we expect that $\overline{H}$ will get closer and closer to the value $\frac{3}{2}$, as we increase the number of repetitions.

The fact that the mean value $\overline{H}$ is not an integer should not be disturbing. It simply expresses the fact that the most likely values of H to occur are the integers just above and below $\overline{H}$.

How do we find the mean value of H for the general case of the Bernoulli distribution $P_N(H, T)$? We can make a guess at what this mean value should be: If the probability for heads in a single toss of a coin is p, we expect to get heads about Np times in N tosses. We shall now prove that Np is indeed the mean value of H. The above example with $N = 3$ has borne this out. To prove the general case, we write $P_N(H)$ in the form

$$P_N(H) = \binom{N}{H} p^H q^{(N-H)}, \tag{1-36}$$

where we have again written $(N - H)$ for T, and where $\binom{N}{H}$ is the binomial coefficient

$$\binom{N}{H} = \frac{N!}{H!\,(N-H)!}.$$

We must now form the sum $\sum_H H P_N(H)$, which we can write as

$$\sum_H \binom{N}{H} (Hp^H) q^{(N-H)}.$$

This sum can easily be done by using a device worth remembering, as it will occur again and again: Take the derivative of p^H:

$$\frac{d}{dp} p^H = H p^{H-1}$$

and multiply this equation again with p:

$$p \frac{d}{dp}(p^H) = pH p^{H-1} = H p^H.$$

Hence Hp^H is obtained from p^H by carrying out the *operation* "$p(d/dp)$" on p^H; the operation consists in first taking the derivative with respect to p, and subsequently multiplying with p again. Now what is the usefulness of all this? It lies in the fact that the operation "$p(d/dp)$" is the same operation for every value of H, and can therefore be carried out *after* the sum over H has been calculated:

$$\overline{H} = \sum_H \binom{N}{H} (Hp^H) q^{(N-H)}$$

$$= \sum_H \binom{N}{H} \left(p \frac{d}{dp} p^H \right) q^{(N-H)}$$

$$= p \frac{d}{dp} \left(\sum_H \binom{N}{H} p^H q^{N-H} \right).$$

Now, this sum in parentheses by the binomial formula [(1-28b)] is just $(p+q)^N$. We have then

$$\overline{H} = p \frac{d}{dp}(p+q)^N$$

$$= pN(p+q)^{N-1}.$$

At this point, we can now use the fact that $p + q = 1$, hence also $(p + q)^{N-1} = 1$, and obtain the result

$$\overline{H} = Np. \tag{1-37}$$

So our expectation that the mean value $\overline{H}$ should be equal to Np is confirmed. We may mention that if $p = \frac{1}{2}$, then $\overline{H} = \frac{1}{2}N$. In this case, the probability distribution is symmetric about the midpoint, $H = N/2$, and for even N, $H = N/2$ also represents the most probable value of H. (For odd N, $H = (N + 1)/2$ and $H = (N - 1)/2$ are the two most probable values of H.) Figure 1.3 also shows the way in which for $p \neq \frac{1}{2}$, the distribution is displaced so that the peak now occurs at $H = Np$.

We have already pointed out the fact, plainly visible in Figure 1.3, that with increasing N the distribution "shrinks." We now develop a quantitative measure of what is loosely called the width of the distribution. This width is the range of H-values for which the probabilities $P_N(H, N - H)$ are substantially different from zero. The half-width is one-half that value. This is illustrated in Figure 1.4. In defining the width quantitatively, we must be guided by the manner in which this width of the probability distribution manifests itself in physical *observations*. Or, to put it another way, the width should be defined in such a way that it represents an observable consequence of the probability distribution. It will be made apparent, in the examples given in the next sections, that a sensible definition is obtained by multiplying the probability $P_N(H, N - H)$ for the occurrence of the value of H

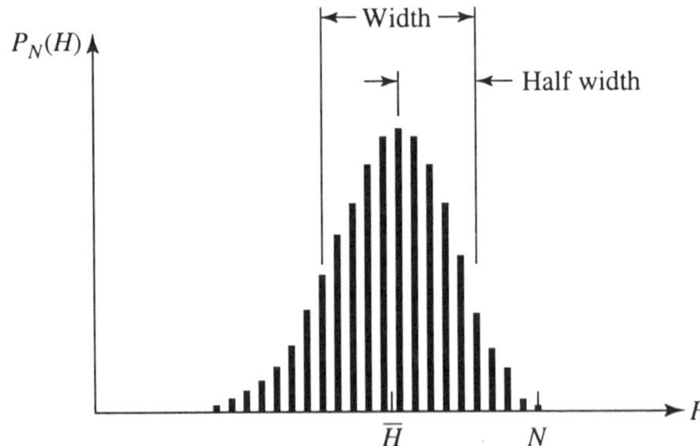

Figure 1.4. Qualitative definition of width and half-width.

with the *square of the deviation*, $(H - \overline{H})^2$, of H from the mean value $\overline{H}$:

$$(H - \overline{H})^2 P_N(H)$$

and summing this over all values of H. The use of the square of $(H - \overline{H})$ ensures that the contribution of H-values both smaller than $\overline{H}$ and larger than $\overline{H}$ enter with positive signs. Summing over all H-values gives the mean square deviation

$$\sum_H (H - \overline{H})^2 P_N(H) \equiv \overline{(H - \overline{H})^2}. \tag{1-38}$$

This quantity is called the *mean square deviation* or the *variance* of H; it represents the mean value of the quantity $(H - \overline{H})^2$, in the sense of the definition given after (1-35). Since the variance is the mean value of the square, $(H - \overline{H})^2$, it represents the *square* of a width, and to get a measure of the width itself, more precisely the half-width, we must take the square root

$$\text{half-width of distribution} \equiv \sqrt{\overline{(H - \overline{H})^2}}. \tag{1-39}$$

This quantity is also referred to as the *root mean square* value of $(H - \overline{H})$, or, in abbreviation, the rms value of $(H - \overline{H})$.

To illustrate that the mean square of $H - \overline{H}$ really measures the width, consider an artificial case of a probability distribution as illustrated in Figure 1.5, where the probabilities $P(H)$ are all the *same* in the range $H_1 < H < H_2$, and zero outside

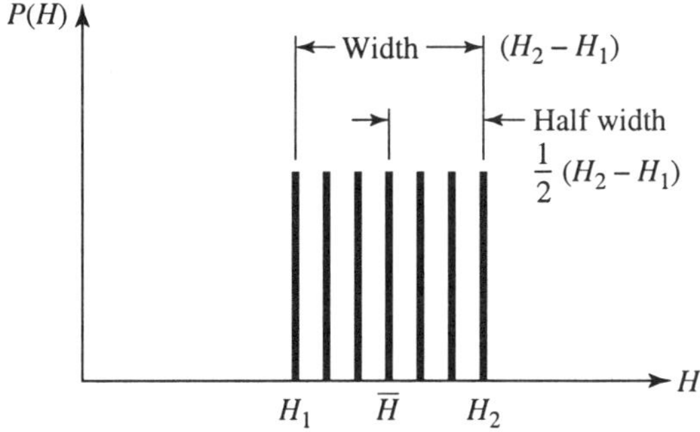

Figure 1.5. Mean value $\overline{H}$ and width for a hypothetical probability distribution $P(H)$.

this range. The average value $\overline{H}$ of H is then

$$\overline{H} = \tfrac{1}{2}(H_1 + H_2)$$

and the variance $\overline{(H - \overline{H})^2}$, as calculated from (1-38) turns out to be

$$\overline{(H - \overline{H})^2} \cong \tfrac{1}{12}(H_2 - H_1)^2$$

(the calculation of these results is tedious, and the above answer is valid if H_1 and H_2 are large numbers). Thus, the rms value of $(H - \overline{H})$ is

$$\sqrt{\overline{(H - \overline{H})^2}} \cong 0.26(H_2 - H_1)$$

and thus indeed a measure of the width $(H_2 - H_1)$, although not identical with $(H_2 - H_1)$.

We proceed now to calculate the width of the Bernoulli distribution $P_N(H, N - H)$. As a preliminary result, we establish the fact that for *any* probability distribution $P(H)$ the mean value of $(H - \overline{H})^2$ is equal to the mean value of H^2 minus the square of the mean value of H:

$$\overline{(H - \overline{H})^2} = \overline{H^2} - (\overline{H})^2. \tag{1-40}$$

The proof of this follows directly from the definition of the mean

$$\overline{(H - \overline{H})^2} \equiv \sum_H (H - \overline{H})^2 P(H),$$

which, upon multiplying out the square, we can write as

$$\sum_H (H - \overline{H})^2 P(H) = \sum_H (H^2 - 2H\overline{H} + \overline{H}^2) P(H).$$

Since $\overline{H}$ is a fixed number that can be taken outside the summation over values of H, this gives

$$\sum H^2 P(H) - 2\overline{H} \left(\sum_H H P(H) \right) + \overline{H}^2 \left(\sum P(H) \right).$$

Now, whatever the distribution $P(H)$ may be, we have $\sum P(H) = 1$; furthermore, $\sum H P(H)$ defines the mean value $\overline{H}$, and $\sum H^2 P(H)$ defines the mean value $\overline{H^2}$ of H^2. Thus

$$\overline{(H - \overline{H})^2} = \overline{H^2} - (\overline{H})^2.$$

This is a most useful result if one has to calculate the variance.

For the Bernoulli distribution, $P_N(H)$, we know already that $\overline{H} = Np$, so all there is to do is to find $\overline{H^2}$. Using again the expression [(1-36)] for $P_N(H)$:

$$P_N(H) = \binom{N}{H} p^H q^{N-H}$$

we have to calculate the sum

$$\overline{H^2} = \sum_H H^2 P_N(H)$$

$$= \sum_H \binom{N}{H} (H^2 p^H) q^{N-H}.$$

The way to achieve this summation is again to use the device already employed in calculating H. There we had found that the identity

$$H p^H = p \frac{d}{dp} p^H$$

made it possible to reduce the sum

$$\sum H P_N(H) \quad \text{to} \quad p \frac{d}{dp}(p+q)^N.$$

Here, we have to deal with $H^2 p^H$ which we write as

$$H(H p^H) = H \left(p \frac{d}{dp} p^H \right) = p \frac{d}{dp}(H p^H)$$

$$= p \frac{d}{dp} \left(p \frac{d}{dp} \right) p^H.$$

In brief, to calculate $\overline{H^2}$, we have to apply the "operation" $p(d/dp)$ twice in succession:

$$\overline{H^2} \equiv \sum_H H^2 P_N(H) = p\frac{d}{dp}\left(p\frac{d}{dp}\right)\sum_H P_N(H) = p\frac{d}{dp}\left(p\frac{d}{dp}\right)(p+q)^N.$$

Let us now carry out these derivatives

$$p\frac{d}{dp}(p+q)^N = p \cdot N(p+q)^{N-1}.$$

The second application of $p(d/dp)$ gives

$$Np\frac{d}{dp}\{p(p+q)^{N-1}\} = Np\{(p+q)^{N-1} + p(N-1)(p+q)^{N-2}\}.$$

After all the derivatives are calculated, we are allowed to remember that $p + q$ is actually equal to 1. Hence

$$\begin{aligned}\overline{H^2} &= Np\{1 + p(N-1)\}\\ &= (Np)^2 + Np(1-p).\end{aligned}$$

Subtracting $(\overline{H})^2 = (Np)^2$ from $\overline{H^2}$, we find that the variance is given by

$$\overline{(H - H)^2} = \overline{H^2} - (\overline{H})^2 = Np(1-p) = Npq. \tag{1-41}$$

So if we use the rms value of $(H - \overline{H})$ as a measure of the *half-width* of the Bernoulli distribution, we find the astonishing result that this width is proportional to the *square root* of N:

$$\text{half-width} = \sqrt{\overline{(H - \overline{H})^2}} = \sqrt{Npq}. \tag{1-42}$$

This is a most important result and deserves some careful discussion.

We will consider the most important case where N is a large number

$$N \gg 1.$$

In this case, it is easily seen that $\sqrt{N}$ is small compared to N itself

$$\frac{\sqrt{N}}{N} = \frac{1}{\sqrt{N}} \ll 1.$$

For example, if $N = 10^4$, then $\sqrt{N} = 10^2$, and

$$\sqrt{N} = \frac{1}{100} N \ll N.$$

Thus, the width of the Bernoulli distribution, $\sqrt{Npq}$, is for large N only a small fraction of the total range of values of H, which extends from zero to N. This is represented in Figure 1.6 which we shall take as the basis for our discussion. We consider first the case where $p = q = \frac{1}{2}$, so that $\overline{H} = N/2$; we shall use the

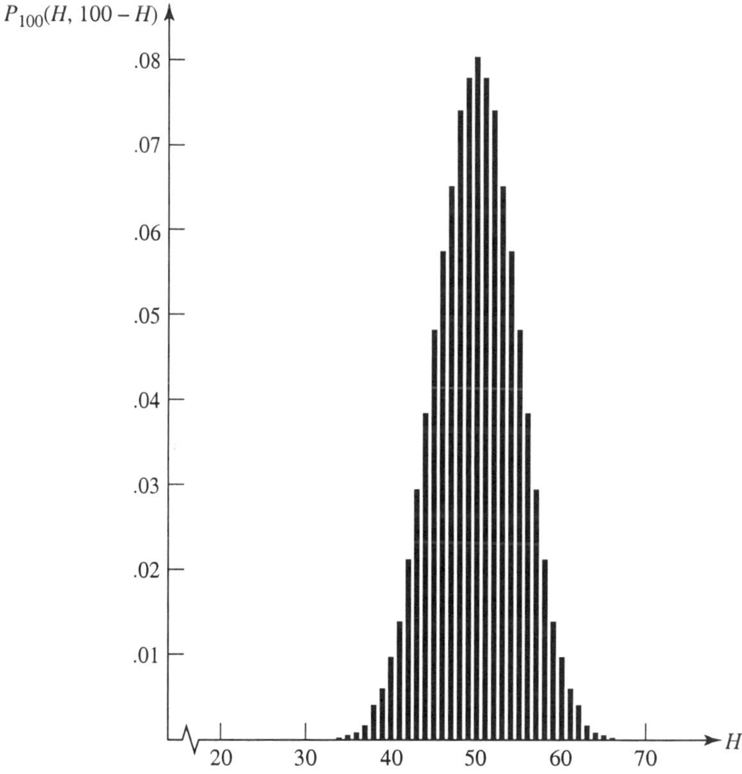

Figure 1.6. The Bernoulli distribution for $N = 100$, $p = q = \frac{1}{2}$, $H = N/2 = 50$, $\Delta = \frac{1}{2}\sqrt{N} = 5$.

symbol Δ for the half-width, (1-42); for $p = q = \frac{1}{2}$, this is

$$\Delta \equiv \sqrt{\overline{\left(H - \frac{N}{2}\right)^2}} = \frac{1}{2}\sqrt{N}.$$

We shall now consider and answer the following questions:

(i) What is the maximum probability for any value of H? Because of $p = q = \frac{1}{2}$, this maximum probability occurs at $H = \overline{H} = N/2$ (if N is even, otherwise at $(N \pm 1)/2$). Since P_N is given by

$$P_N(H) = \left(\tfrac{1}{2}\right)^N \binom{N}{H} = \left(\tfrac{1}{2}\right)^N \frac{N!}{H!\,(N - H)!},$$

the maximum probability is (assume even N)

$$(P)_{\max} = P_N\left(\frac{N}{2}, \frac{N}{2}\right) = \frac{1}{2^N} \frac{N!}{\left(\dfrac{N}{2}\right)! \left(\dfrac{N}{2}\right)!}.$$

To evaluate this for large values of N, we must make use of the so-called *Stirling approximation* to $N!$, a most important expression in statistical physics, which we shall use again and again. It is

$$N! \cong \sqrt{2\pi N}\, e^{N \ln N - N}. \tag{1-43}$$

This formula approximates $N!$ with an error that is of order $\frac{1}{12N}$, so that already for $N = 10$ the error is less than 1%, and for really large N it becomes a fantastically good approximation. We shall make use of it here without going into all the derivations of our results, which would be disruptive, and will be presented more appropriately in Chapter 2.

With the help of (1-43) we can establish the result that the maximum probability is

$$(P)_{\max} \cong \sqrt{\frac{2/\pi}{N}}. \tag{1-44}$$

Notice that the product of peak probability $(P)_{\max}$ and the width 2Δ is independent of N, and of order 1:

$$(P)_{\text{max}} \times 2\Delta = \sqrt{\frac{2}{\pi}} \cong 0.8. \tag{1-45}$$

(ii) If we move away from $H = N/2$, where P is maximum, to $H = (N/2) + \Delta$ or $H = (N/2) - \Delta$, to how much has the probability dropped from its maximum value? This again can be answered using Stirling's formula with the result that

$$P_N \left(\frac{N}{2} + \Delta, \; \frac{N}{2} - \Delta \right) = e^{-1/2}(P)_{\text{max}} = 0.607(P)_{\text{max}}. \tag{1-46}$$

So P has dropped to about 60% of its peak value.

(iii) We also want to know what the total probability is that a value of H falls *within* the boundaries defined by the width of the distribution

$$\frac{N}{2} - \Delta < H < \frac{N}{2} + \Delta.$$

This too can be estimated by using (1-43) and replacing the sum over H-values by an integral. One finds that the probability that H lies within the above limits is

$$P \left(\frac{N}{2} - \Delta < H < \frac{N}{2} + \Delta \right) \cong 0.68. \tag{1-47}$$

So in about two-thirds of the cases H will lie in the range between

$$\left(\frac{N}{2} - \frac{\sqrt{N}}{2} \right) \quad \text{and} \quad \left(\frac{N}{2} + \frac{\sqrt{N}}{2} \right)$$

and only in about one-third of the cases will it be found outside that range. If that interval is *doubled*, that is, if we consider the probability that H lies in the interval

$$\frac{N}{2} - \sqrt{N} < H < \frac{N}{2} + \sqrt{N},$$

then we find that this probability is now 95%. This is the quantitative formulation of the "shrinking" of the distribution with increasing values of N: The odds are 95% in favor of an H-value lying within a distance $\sqrt{N}$ from its mean value $\overline{H} = N/2$.

(iv) As a final observation, we observe that the width of the distribution is largest for $p = q = \frac{1}{2}$. This is also visible in Figure 1.3, where for $N = 20$ and $N = 40$, the width for $p = \frac{2}{3}, q = \frac{1}{3}$ is visibly narrower than for $p = q = \frac{1}{2}$. Indeed, according to (1-42) the width is proportional to $\sqrt{pq}$, and this is equal to

$$\sqrt{pq} = \tfrac{1}{2}\sqrt{1 - (p - q)^2}.$$

As it should be for $p = 0$ or $p = 1$, we have *certainty*, and the width of the distribution is zero.

1.4   Illustrative Applications

In this section we discuss three examples of statistical phenomena that can be understood by means of the Bernoulli distribution. They are:

1.4.A. The distribution of boys and girls in families with N children.

1.4.B. The size distribution of randomly coiled chain-polymer molecules.

1.4.C. The distribution of electric charge on a protein molecule in solution (in this case hemoglobin).

1.4.A.   The Sex Distribution of Children

The sex of a child is determined ultimately by chance. Half of the sperm cells that fertilize the ovum carry the X-chromosome, producing a girl, half of them carry the Y-chromosome, leading to a boy. An enormous number of sperm cells compete for the fertilization of an ovum, and chance decides which type of sperm cell will merge with the ovum.

We have thus a trial with two possible outcomes, "boy" and "girl." Let us call the associated probabilities p and q:

$$p = \text{probability for a boy},$$
$$q = \text{probability for a girl}.$$

It is well known that p and q are not the same in this case; p is slightly larger than $\frac{1}{2}$. A possible reason for this is a slight difference in the mobility of the two types of sperm cells.

We shall base our discussion on data compiled by Geissler in Saxony (Germany) during the years 1876–1885 [4]. Large families were much more numerous at that time, and Geissler's data contain sex distribution for families with up to 12 children.

Consider a family with N children, and let

$$B = \text{number of boys,}$$
$$G = \text{number of girls.}$$

If the sex of children is determined by chance, then the probability that there are B boys and G girls in a family with $N = B + G$ children should be

$$P_N(B, G) = \frac{N!}{B!\,G!} p^B q^G. \tag{1-48}$$

This probability is given by an expression exactly analogous to the expression for the probability for tossing H heads and T tails, when a coin is tossed $N = H + T$ times (see (1-27)).

As before, we shall use the shorter notation $P_N(B)$ for the distribution [(1-48)], if attention is focused on the number B of boys. Using $B + G = N$, we get

$$P_N(B, G) \equiv P_N(B) = \frac{N!}{B!\,(N - B)!} p^B q^{N-B}. \tag{1-48a}$$

Greissler's data for the sex of the first-born child are

$$\begin{aligned}
\text{boys:} \quad & \mathcal{N}_B = 114{,}609 \\
\text{girls:} \quad & \mathcal{N}_G = 108{,}719 \\
\hline
\text{total:} \quad & \mathcal{N} = 223{,}328
\end{aligned}$$

From this we get an estimate of p and q:

$$p \cong \frac{\mathcal{N}_B}{\mathcal{N}} = 0.513,$$
$$q \cong \frac{\mathcal{N}_G}{\mathcal{N}} = 0.487. \tag{1-49}$$

By means of (1-17) we can find out what confidence to attach to these numbers. A simple calculation shows that the difference between p and $\mathcal{N}_B/\mathcal{N}$:

$$\left|\frac{\mathcal{N}_B}{\mathcal{N}} - p\right|$$

is less than 0.0035 with 90% probability, less than 0.0015 with 50% probability. We can therefore not quite trust the third decimal in (1-49). Nevertheless, we shall use these data, together with (1-48), to predict the distribution of boys and girls in a family of 12, and then compare our prediction with the data that contain a total of 6125 families with 12 children. We expect the number $\mathcal{N}_B$ of families with B boys out of 12 children to be equal to the total number of families with 12 children times the probability $P_{12}(B)$ for B boys in a family of 12 children

$$\text{expected } \mathcal{N}_B = \mathcal{N} \times P_{12}(B) = 6125\,P_{12}(B). \tag{1-50}$$

In Table 1.1 we show in column (2) the value of $P_{12}(B)$ calculated from (1-48a) with $N = 12$. In column (3) we show the expected number $\mathcal{N}_B$ of families with B

Table 1.1. Distribution in the Number of Boys in Family of 12 Children.

(1) Number of boys B	(2) Probability $P_{12}(B)$	(3) Expected number $\mathcal{N}_B$ [(1-50)]	(4) Observed number $\mathcal{N}_B$(obs)	(5) Difference $\mathcal{N}_B$(obs) $-$ $\mathcal{N}_B$(calc)
12	0.0003	2	7	+ 5
11	0.0038	23	45	+ 22
10	0.0198	121	181	+ 60
9	0.0627	384	478	+ 94
8	0.1337	819	829	+ 10
7	0.2030	1243	1122	−121
6	0.2247	1376	1343	− 33
5	0.1827	1119	1033	− 86
4	0.1083	663	670	+ 7
3	0.0456	279	286	+ 7
2	0.0130	80	104	+ 24
1	0.0022	14	24	+ 10
0	0.0002	1	3	+ 2

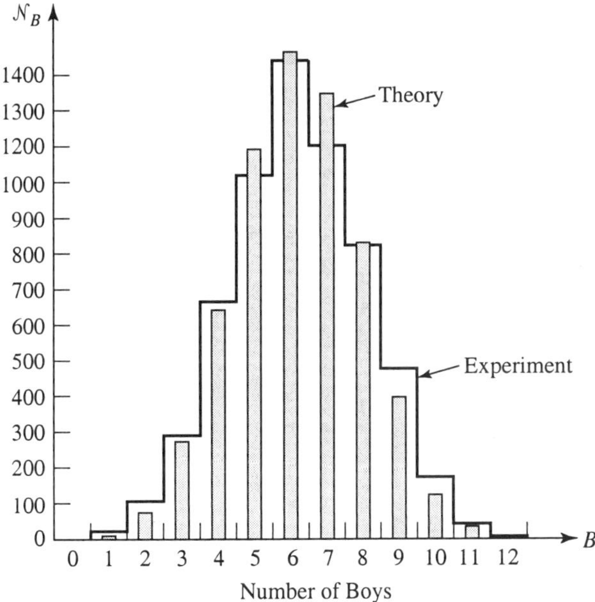

Figure 1.7. Theoretical and experimental values of the number of boys in 6115 families having 12 children.

boys (out of 12 children). In column (4) we present the actual number of families recorded by Geissler. Finally, in column (5) we show the difference between the expected and the observational value.

In Figure 1.7 these results are shown in graphical form. The solid bars are the theoretical results, and the step graph represents the data of Geissler. We see that the prediction based on (1-50) is a reasonable fit to the data. Since the sample— 6125 families—is not very large, the discrepancies between prediction and observation fluctuate somewhat.

The average number of boys per family computed from these data is found to be

$$\overline{B} = \sum B \left(\frac{\mathcal{N}_B(\text{obs})}{\mathcal{N}} \right) = 6.190.$$

This number should be equal to $12p$, according to (1-37). If we take p from (1-49), we find

$$12p = 12 \times 0.513 = 6.156.$$

One might think of using the observed value 6.190 of B together with (1-37) to define a probability p for that sample of families with 12 children. This gives

$$p = \frac{6.190}{12} = 0.516.$$

But this value is based on a much smaller sample than the value of p based on first child data, and thus inherently less reliable.

One may speculate whether the results in Table 1.1 do indicate a systematic deviation from Bernoulli in favor of the occurrence of extreme values of B. This suspicion appears to be borne out by looking at the more numerous data for smaller families. Table 1.2 gives the data for families with six children. The total number of such families in Greissler's report is 72,069. We see from the two last columns of this table that indeed the extremes, "all boys" or "all girls," are about 17% more numerous than expected. We may apply (1-17) to the case of "all boys" for which we find

$$\frac{\mathcal{N}_6}{\mathcal{N}} = \frac{1,579}{72,069} = 0.0219.$$

This is to be compared with $P_6(6)$ which is only 0.0186. We then have a discrepancy

$$d = \frac{\mathcal{N}_6}{\mathcal{N}} - P_6(6) = 0.0219 - 0.0186 = 0.0033.$$

Table 1.2. Distribution of Boys in Families with Six Children.

Number of boys B	Probability $P_6(B)$	Expected $\mathcal{N}_B$	Observed $\mathcal{N}_B$	Difference $\mathcal{N}_B(\text{obs}) - \mathcal{N}_B(\text{calc})$	% Diff.
6	0.0186	1,343	1,579	+236	+17%
5	0.1053	7,589	1,908	+319	+ 4.2%
4	0.2481	17,879	17,332	−547	− 3.1%
3	0.3117	22,462	22,221	−241	− 1.1%
2	0.2203	15,874	15,700	−174	− 1.1%
1	0.0830	5,983	6,233	+250	+ 4.0%
0	0.0130	939	1,096	+157	+16.7%

From (1-17), on the other hand, we can conclude with 75% confidence that this discrepancy should be less than 0.001.

This is a clear indication that the a priori probability p for "boy" is not a universal constant; there are more "all boy" and "all girl" families than predicted by a universal value of p. But to press this point any further, we would need a better estimate of confidence than is provided by (1-16) which would bring us into the topic of statistical analysis of finite data. This is a field of study in its own right with important applications in all fields of science, physics, biology, physiology, economics, etc., but outside the main topic of this book.

1.4.B.   Random Coils: The Conformation of Chain Polymers

Chain polymers are macromolecules arising from the repeated linkage of small units—the so-called monomers—so as to form a long chain. Such polymers are of extraordinary importance. Examples range from the synthetic polymers like nylon or polyethylene to the natural biopolymers like the proteins. Indeed, all proteins, as well as DNA and RNA, are chain polymers. The biochemical machinery within cells appears to be singularly well equipped to synthesize long chain molecules. It is well known that many proteins assume characteristic conformations that lead to their so-called secondary and tertiary* spatial structure. These conformations are stabilized by the cross linking of monomers through van der Waals forces, hydrogen bonds, etc. If the ensuing structure has a rather tightly packed, well-defined architecture, we speak of globular proteins. All enzymes, and the oxygen carrier molecules, myoglobin and hemoglobin, are globular proteins. Under suitable conditions, however, such proteins can be *denatured*, that is, reduced to their primary structure of a long, flexible chain of monomers. Many synthetic polymers in solution also have this characteristic conformation of a *random chain*.

In this Section 1.4.B we shall analyze some properties of polymers, both synthetic and biopolymers, in the random chain conformation. These random conformations can be described statistically by means of a probability distribution for parameters like the end-to-end distance of the chain, or other measures of size. We shall see that a suitably defined *mean square radius* is a useful characterization of macromolecular size. This mean square radius can be determined experimentally. By studying how this mean square radius varies with the number of monomers in the polymer, we can obtain insight into the conformation of the molecule. The Bernoulli distribution in particular will be shown to play a central role in relating, theoretically, the number of monomers to the mean square radius.

---

*The primary structure being the order of monomers in the chain.

In the study of macromolecules one relies on a large variety of physical methods to determine properties like molecular weight, size, and shape. Such physical methods are: X-ray diffraction, ultracentrifugation, osmotic pressure measurements, and the measurements of electric polarizability and the conductivity of macromolecules in solution.

A particularly effective method for determining molecular size is the method of light scattering. It is based on the observation that a solution of macromolecules is not quite transparent, but slightly turbid ("hazy"). What this haziness of the solution expresses is the fact that a beam of light passing through the solution gets partly deflected ("scattered") by the macromolecules. From the intensity of this scattered light one can determine the number of molecules per cm^3 in solution; from the dependence of this intensity on the angle θ of observation, one can find their size. In Figure 1.8 we show the experimental setup. The reason that the scattered intensity should depend on the size of the molecules lies in the wave nature of light. The incident light beam is a traveling wave of wavelength λ ($\lambda = 5890$ Å for the light of a sodium lamp); if this wave encounters an obstacle like a macromolecule, a secondary wave is produced spreading in all directions and representing the scattered light. This secondary wave has a pattern that depends on the size of the obstacle in relation to the wavelength λ. This is illustrated in Figure 1.9. Wave theory gives a simple quantitative expression for the manner in which the intensity of the scattered light, $I(\theta)$, depends on the angle θ at which it is observed

$$I(\theta) = \text{constant}\left(1 - \frac{16\pi^2}{3}\frac{R_g^2}{\lambda^2}\sin^2\frac{\theta}{2}\right). \qquad (1\text{-}51)^*$$

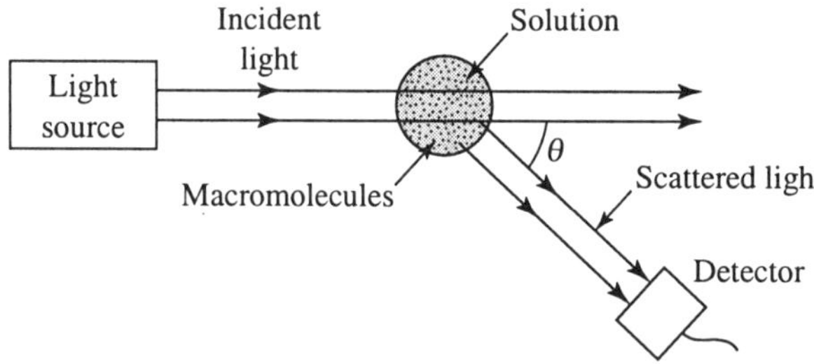

Figure 1.8. Schematic diagram of a light scattering experiment.

---

*This is strictly true only for polarized light with the plane of polarization perpendicular to the plane defined by the incident and scattered beams.

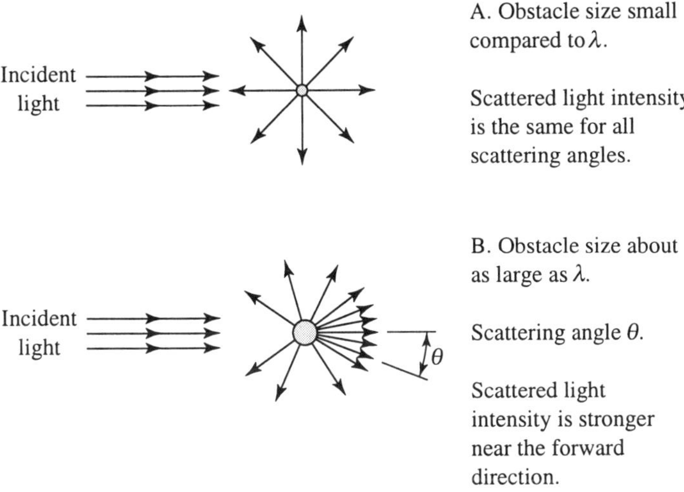

Figure 1.9. Secondary waves produced by an obstacle.

In this formula λ is the wavelength of the light used, and typically of the order of 5000 to 6000 Å. R_g is the so-called radius of gyration of the molecule; it is a measure of the size of the molecules responsible for scattering, and will be defined presently. Equation (1-51) is valid in the case that $R_g^2 \ll \lambda^2$. We shall see that even very large macromolecules have values of $R_g \lesssim 500$ Å, so that this condition is satisfied. The importance of (1-51) is then in the fact that, by measuring $I(\theta)$ for several angles θ, we can experimentally determine the value of R_g^2.

To give a precise definition of R_g, we show in Figure 1.10 a folded chain molecule. We draw as small circles the monomers of the polymer chain that is

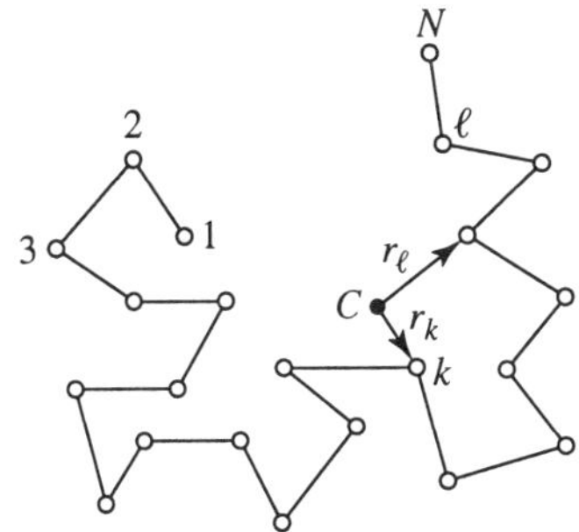

Figure 1.10. A folded polymer chain. The point C is the center of the molecule.

folded up in some manner. Let r_k be the distance of the kth monomer from the geometric center C of the molecule.* Then R_g is defined by

$$R_g^2 = \frac{1}{N} \sum r_k^2. \tag{1-52}$$

(If the N monomer units were regularly distributed over the inside of a sphere of radius R, we could have $R_g^2 = \frac{3}{5} R^2$.)

Formula (1-51) shows that the scattering experiments provide a direct measurement of R_g^2. What now, if the individual molecules in the solution are not all identical in size, because the chains have a random arrangement? In this case, we must use the fact that the observed intensity is the *sum* of the contributions of all $\mathcal{N}$ molecules exposed to the incident light beam. There are millions of them, all existing under identical conditions (chemical composition of solvent, temperature, etc.). But they will have differing values of R_g^2 if they are random chains. We must, therefore, replace the number R_g^2 appearing in (1-51) by a *mean* value $\overline{R_g^2}$, defined by

$$\overline{R_g^2} = \sum_{\text{all molecules}} \frac{R_g^2}{\mathcal{N}}, \tag{1-53}$$

that is, we add the values of R_g^2 for each molecule, and divide by the number of molecules. Now assume that the diverse values of R_g^2 which occur are ordered according to size and put in groups i with an approximately common value $R_g \cong R_i$. If $\mathcal{N}_i$ of the $\mathcal{N}$ molecules have a value of $R_g \cong R_i$, then we can write the average $\overline{R_g^2}$ also in the form

$$\overline{R_g^2} = \sum_{\text{all groups } i} R_i^2 \frac{\mathcal{N}_i}{\mathcal{N}},$$

$(\mathcal{N}_i/\mathcal{N})$ is the fraction of the molecules for which the value of R_g is $\cong R_i$; since $\mathcal{N}$ is enormously large, the number $(\mathcal{N}_i/\mathcal{N})$ accurately represents the *probability* $P(R_i)$ that R_g has the value R_i. Thus we can write the mean value $\overline{R_g^2}$ in the form

$$\overline{R_g^2} = \sum_{\text{all groups } i} R_i^2 P(R_i) \tag{1-53a}$$

---

*The center is defined by the condition $\sum \vec{r}_k = 0$.

Thus, we see that the experimentally observed mean value of R_g^2 is identical with the mean value adopted in probability theory where we stated that the mean value of any quantity x, having a probability distribution $P(x)$, is given by

$$\bar{x} = \sum_{\text{all values } x} x P(x).$$

The probability distribution for values R_g cannot be determined experimentally. It results, in general, from a theoretical *model* of the behavior of the molecules as we will show. But having constructed such a distribution, we know now that in testing it against experiment, we must use (1-53a) which describes accurately the *kind* of average which is contained in the experimental information.

It should also be clear from this argument that in tabulations of the value R_g of the radius in gyration, it is understood that "R_g" represents the root mean square value

$$R_g \equiv \sqrt{\overline{R_g^2}}.$$

Before attempting to construct a theoretical model for P, let us start with presenting some data, and then see whether we can interpret these data by means of a statistical model for the behavior of random chains. In Table 1.3 we present examples that demonstrate the striking difference in the values of R_g for a tightly packed structure, as it occurs in globular proteins, and the R_g values for random coils. The latter conformation can be easily *identified* by R_g—values which are *much larger* than what one would expect for a tight structure: In this table the first two polymers, serum albumin and catalase, are tightly packed. Their molecular weight is in the range 5×10^4 to 2×10^5, and their radii of gyration are small, $\sim$ 30 to 40 Å.

Table 1.3. Radius of Gyration R_g for Various Polymers (after Tanford [5]).

Polymer	Molecular weight	R_g (Å) Expt.	R_g (Å) Packed ball	Comments
Serum albumin	6.6×10^4	29.8	21	tightly
Catalase	2.25×10^5	39.8	31	packed
Myosin	4.93×10^5	468	41	random
Polystyrene[†]	3.2×10^6	494	84	coils
DNA	4×10^6	1170	74	

[†]Dissolved in cyclohexane.

The latter three polymers, myosin, polystyrene, and DNA, have molecular weights M in the range $5 \times 10^5 < M < 5 < 10^6$ which is somewhat larger than the first two. However, their radii of gyration are enormous compared to those of the tightly packed polymers. These latter molecules are clearly random coil polymers.

Having satisfied ourselves as to the fact that just from the size of the radius of gyration one can classify molecules crudely into tightly packed ("globular") polymers and random coil polymers, let us now examine more analytically how we can expect the radius of gyration to depend on the molecular weight for these two different classes. We shall then also present experimental data on R_g as a function of N, the number of monomer units.

If M_0 is the molecular weight of a monomer, N the number of monomers in the chain, then the molecular weight M of the polymer is

$$M = N M_0.$$

For a *tightly packed* or *globular* structure we expect the volume of the molecule to be proportional to N, and hence to M. Since the volume is also proportional to R_g^3, we expect M to be proportional to R_g^3, $M \propto R_g^3$. This leads us to expect that R_g will be proportional to the cube root of M:

$$R_g \propto M^{1/3}. \tag{1-54}$$

In Table 1.4 we present some data on polystyrene chains obtained from light scattering experiments that display the variation of R_g with the number of segments in

Table 1.4. Radius of Gyration of Polystyrene Chains (in Butanone at 22 °C; after Tanford [5]).

Molecular weight M	No. of monomers N	R_g (Å) observed	$R_g/\sqrt[3]{M}$	$R_g/\sqrt{M}$
3.2×10^6	30,800	494*	3.35*	0.276*
1.77×10^6	17,000	437	3.61	0.33
1.63×10^6	15,700	414	3.51	0.32
1.32×10^6	12,700	367	3.34	0.32
0.94×10^6	9,000	306	3.12	0.32
0.524×10^6	5,040	222	2.75	0.31
0.23×10^6	2,200	163	2.66	0.34

*In cyclohexane.

the chain. The monomer of polystyrene is

$$-CH_2-\underset{\underset{\textstyle C_6H_5}{|}}{CH}-$$

and has a molecular weight $M_0 = 104$. These data show, not surprisingly, that (1-54) is not satisfied. The size of the polystyrene polymer does not increase in direct proportion to the cube root of the molecular weight as one expects for tightly packed structures. Polystyrene is a random coil polymer, and the data shown in the last column of Table 1.4 shows that its radius of gyration actually increases in direct proportion to the *square root* of the molecular weight or number of monomers. This fact is demonstrated dramatically in the graphical data of Figure 1.11. In this figure we plot measurements of the radius of gyration for polystyrene in butanone,

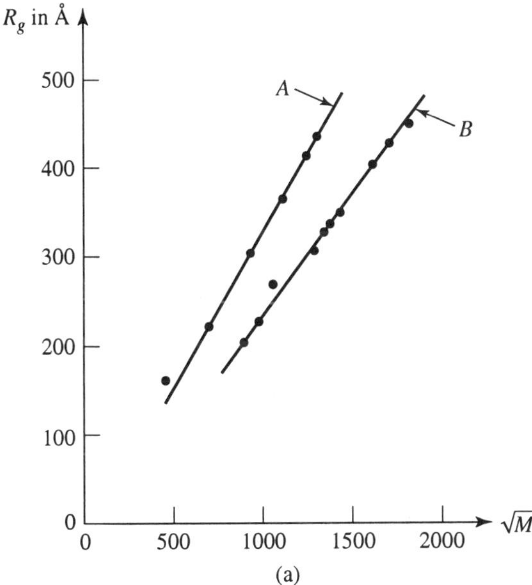

Figure 1.11(a). Radius of gyration for polystyrene and polyvinyl acetate polymers as a function of the square root of the molecular weight. (A) is the data on polystyrene in butanone (Outer, Carr, and Zimm, *J. Chem. Phys.* **18**, 830 (1959)); (B) is the data on polyvinyl acetate in methyl ethyl ketone (Schultz, *J. Am. Chem. Soc.* **76**, 3422 (1954)).

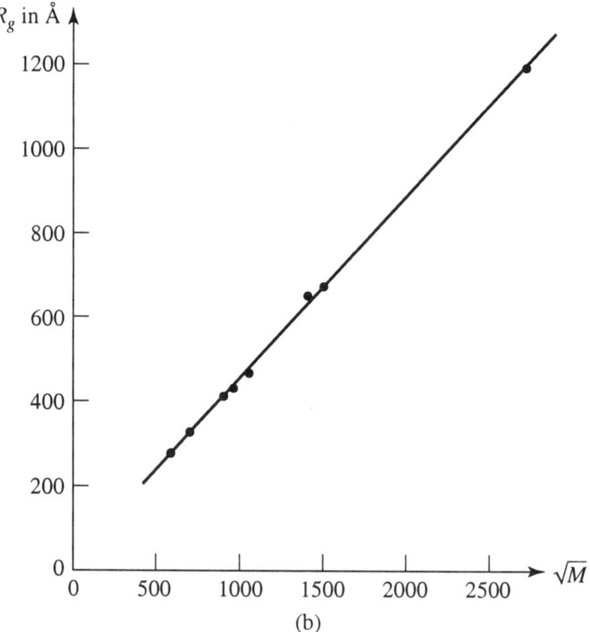

Figure 1.11b. Radius of gyration of polystyrene polymers as a function of the square root of the molecular weight. Data for polystyrene in benzene (Kunst, *Rec. Trav. Chim.* **69**, 125 (1950)).

polystyrene in benzene, and polyvinyl acetate in methyl ethyl ketone, as a function of the *square root* of the molecular weight. In all three cases, we have a straight line within the accuracy of the experiment. We conclude from this data that R_g varies experimentally in direct proportion to $\sqrt{M}$, i.e.,

$$R_g \equiv \sqrt{\overline{R_g^2}} \propto \sqrt{M}.$$

Alternatively, since M is proportional to N, the total number of monomers, we can conclude that the mean square radius of gyration is proportional to N, i.e.,

$$\overline{R_g^2} = \text{constant } N. \tag{1-55}$$

In the remainder of this section we shall use the Bernoulli distribution to show why the mean square radius of gyration of *random coil* polymers is proportional to the number of monomers. We will, in fact, derive theoretically the experimental result summarized in (1-55) by showing first of all that the mean square end-to-end distance $\overline{D_N^2}$ of a chain of N monomers is directly proportional to N. We will then show that the mean square end-to-end distance is in turn directly proportional to the mean square radius of gyration of the polymers.

To establish these results, we will first construct some simplified models of chain molecules. We shall proceed from the simplest, somewhat unrealistic model, to a gradually more precise understanding of the cases closest to reality.

First Model: Suppose that a large number of chains, each having N segments of length l, are laid out in a plane "at random," subject to the following rule: Each segment may be oriented in any of four perpendicular directions with equal a priori probability. Figure 1.12 illustrates two cases: We start the chain at point $(0, 0)$ of an x–y-coordinate system. For each possible configuration the position of the end point of the chain can be specified by two integers m, n, $x = ml$ being the x-coordinate of the end point, and $y = nl$ its y-coordinate.

Therefore, the end-to-end distance of the chain will be given by

$$D_N^2 = l^2(m^2 + n^2). \qquad (1\text{-}56)$$

Assuming that each chain link points with equal probability in the $+x$, $-x$, $+y$, $-y$-directions, we now attempt to find the *probability* that the end point of the chain has the coordinates $x = ml$, $y = nl$. Call this probability

$$P_N(m, n),$$

N referring to the number of segments of the chain.

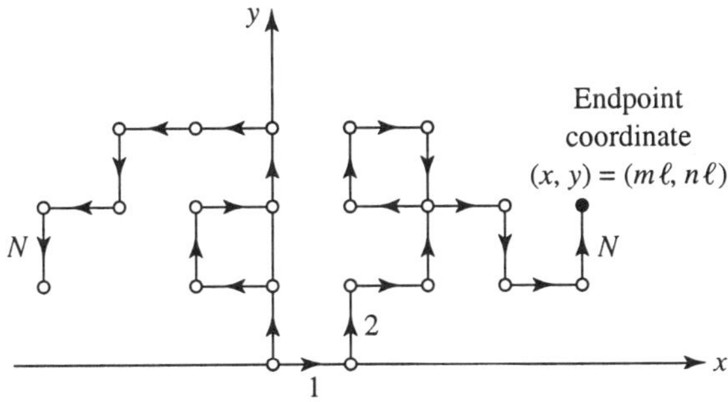

Figure 1.12. A random chain in the x–y plane.

We can best find an expression for P_N by considering a random chain being built segment by segment. Then, upon each addition of a segment, a pair of binary decisions have to be made:

(i) Shall the segment be parallel to the x- or y-axis?

(ii) Once (i) is decided, will the segment point in the $+$ or $-$ direction, along the chosen axis?

If the number of segments is large, about one-half of the decisions (i) will produce a segment pointing in the $\pm x$-directions, and one-half of them pointing in the $\pm y$-directions. Since the subsequent decision (ii) is independent of the previous decision (i), we have approximately (for large N):

$$P_N(m, n) \cong P_{N/2}(m) \times P_{N/2}(n), \qquad (1\text{-}57)$$

where $P_{N/2}(m)$ is the probability that in $N/2$ steps along the x-axis, a net x-displacement $x = ml$ results. $P_{N/2}(n)$ is the corresponding probability for a displacement $y = nl$.

Our problem is then to determine the form of the probability distribution $P_{N/2}(m)$. For this purpose, let us consider the $N/2$ segments pointing along the $\pm x$-axes. If there are R_x of them pointing in the $+x$-direction (to the right), L_x of them pointing to the left, the end point of the chain will be located at

$$x = l(R_x - L_x) = lm. \qquad (1\text{-}58)$$

We see that

$$m = (R_x - L_x). \qquad (1\text{-}59)$$

It should be clear that the probability for having R_x of the $N/2$ segments pointing to the right, L_x of them pointing to the left, is the same as tossing R_x heads and L_x tails in $N/2$ coin tosses with $p = q = \frac{1}{2}$. In the (H) head, (T) tail cases, this probability would be

$$P_{N/2} = \frac{(N/2)!}{H!\,T!} p^H q^T \qquad (\text{see } (1\text{-}27)).$$

In the present case then the desired probability is

$$P_{N/2}(R_x, L_x) = \frac{(N/2)!}{R_x!\,L_x!} \left(\tfrac{1}{2}\right)^{N/2}. \qquad (1\text{-}60)$$

If we just write this in terms of $m = R_x - L_x$, we have an expression for the desired probability $P_{N/2}(m)$. Using $R_x + L_x = N/2$:

$$R_x = \frac{N}{4} + \frac{m}{2}, \qquad L_x = \frac{N}{4} - \frac{m}{2},$$

and inserting this into (1-60), we get the expression for

$$P_{N/2}\left(\frac{N}{4} + \frac{m}{2}, \frac{N}{4} - \frac{m}{2}\right) \equiv P_{N/2}(m)$$

as

$$P_{N/2}(m) = \frac{\left(\frac{N}{2}\right)!}{\left(\frac{N}{4} + \frac{m}{2}\right)! \left(\frac{N}{4} - \frac{m}{2}\right)!} \left(\tfrac{1}{2}\right)^{N/2}, \tag{1-61}$$

which, we repeat, is the probability that the end point of the chain is located at $x = ml$.

In complete analogy we get a corresponding expression for the probability $P_{N/2}(n)$ that the end point of the chain is located at $y = nl$. These probabilities are normalized so that

$$\sum_{m \equiv -N/2}^{+N/2} P_{N/2}(m) = 1. \tag{1-62}$$

We are now ready to calculate the mean square end-to-end distance of the chain. For given values m, n, the end point is located at $x = ml$, $y = nl$, and the square of the end-to-end distance is

$$D^2 = x^2 + y^2 = l^2(m^2 + n^2). \tag{1-63}$$

The mean of D^2 is thus

$$\overline{D^2} = l^2(\overline{m^2} + \overline{n^2}). \tag{1-64}$$

Now we calculate $\overline{m^2}$. This is defined as

$$\overline{m^2} = \sum_m m^2 P_{N/2}(m), \tag{1-65}$$

with the probability distribution $P_{N/2}(m)$ given by ((1-61)). This distribution is really the same as ((1-60)), except that it is written as a function of $m = R_x - L_x = 2R_x - N/2$ rather than as a function of R_x. But to evaluate the mean values $\overline{m}$ and $\overline{m^2}$, we can use the familiar results (see Eqns. (1-37) and (1-42), with $p = q = \frac{1}{2}$) that

$$\overline{R_x} = \left(\frac{N}{2}\right) p = \frac{N}{4},$$

$$\overline{R_x^2} = \left(\frac{N}{2}\right) pq + \overline{(R_x)^2} = \frac{N}{8} + \left(\frac{N}{4}\right)^2,$$

to find that

$$\overline{m} = 2\overline{R_x} - \frac{N}{2} = 0$$

and

$$\overline{m^2} = 4\overline{R_x^2} - 2N\overline{R_x} + \left(\frac{N}{2}\right)^2 = \left(\frac{N}{2}\right).$$

We thus see that the Bernoulli distribution $P_{N/2}(m)$ has the two characteristic properties:

(i) The mean value $\overline{m}$ of m is zero. $\overline{m} = 0$ (1-66)

(ii) The mean square value of m, which because of (i) is equal to the variance, is proportional to N:

$$\overline{m^2} = \frac{N}{2}. \tag{1-67}$$

Since $\overline{m^2}$ is the variance, its square root $\sqrt{\overline{m^2}}$ is a measure of the width of the distribution. This width, characteristically, is $\propto \sqrt{N}$.

The same property also holds, of course, for $P_{N/2}(n)$ where we find that $\overline{n^2} = N/2$. The mean square end-to-end distance $\overline{D^2}$, as given by (1-64), is therefore in this model

$$\overline{D^2} = Nl^2. \tag{1-68}$$

It is hard to overstate the importance of this result. It shows that the *mean square end-to-end distance* of a random chain, as described in our model, is proportional to the number N of segments. This is a direct consequence of the fact that the probability distribution of values m and n is a Bernoulli distribution with a mean value $\overline{m} = 0, \overline{n} = 0$. The mean square values $\overline{m^2}$ and $\overline{n^2}$ are thus the variances of the distribution and, characteristically, proportional to N rather than to N^2. We are shortly going to show that the mean square radius of gyration $\overline{R_g^2}$ is proportional to $\overline{D^2}$, and hence also proportional to N. In this way, we have already established an explanation of the fundamental experimental result [(1-55)]. Refinements of our model will do little more than firm up the value of the proportionality constant between $\overline{D^2}$ and N or $\overline{R_g^2}$ and N.

Refinements of the Model. The next obvious step is to make the previous model three-dimensional by adding an option for orientation of the segments along the z-axis. The end point coordinates of the chain will then be

$$x = ml,$$
$$y = nl,$$
$$z = sl,$$

and the mean square end-to-end distance D^2 is

$$\overline{D^2} = l^2(\overline{m^2} + \overline{n^2} + \overline{s^2}).$$

But now only one-third of the total number N of segments will point (on the average) in the x-, y-, or z-direction, so that instead of ((1-77)) we will have

$$\overline{m^2} = \overline{n^2} = \overline{s^2} = \left(\frac{N}{3}\right)$$

and hence again

$$\overline{D^2} = Nl^2. \tag{1-68}$$

Before we go any further, let us state the relation between the mean square end-to-end distance $\overline{D^2}$ and the mean square radius of gyration $\overline{R_g^2}$, which is measured.

With a bit of algebra one can show that R_g^2, as defined in (1-52), can also be written as

$$R_g^2 = \frac{1}{2N^2} \sum_{ij} r_{ij}^2,\tag{1-69}$$

where r_{ij} is the distance between the ith and jth monomers on the chain, assuming these monomers numbered from 1 to N. The sum extends over all pairs (i, j). (The proof of this is elementary, but tedious, and will be left as a problem assignment.) Now the piece of chain from monomer i to monomer j consists of $N_{ij} = |i - j|$ segments of length l each, and according to (1-68), the mean square distance r_{ij}^2 is given by

$$\overline{R_{ij}^2} = N_{ij}l^2 = |i - j|l^2.$$

Therefore

$$\overline{R_g^2} = \frac{1}{2N^2} \sum_{ij} \overline{r_{ij}^2}$$

or

$$\overline{R_g^2} = \frac{1}{2N^2} l^2 \sum_{i,j} |i - j|.$$

The double sum is found to have the value $N^3/3$ for large N. (Again we leave the demonstration of this as a problem assignment.) It follows that

$$\overline{R_g^2} = \tfrac{1}{6}Nl^2 = \tfrac{1}{6}\overline{D_N^2}.\tag{1-70}$$

We see therefore that in a three-dimensional model of the chain molecule, in which N segments of length l are each oriented at random in either the $\pm x$-, $\pm y$- or $\pm z$-directions, the mean square radius of gyration is indeed proportional to N:

$$\overline{R_g^2} = \tfrac{1}{6}l^2 N,\tag{1-71}$$

in agreement with the experimental finding [(1-55)].

The model on which (1-71) is based also predicts the proportionality constant between $\overline{R_g^2}$ and N. In dealing with a specific chain molecule, $\overline{R_g^2}$ is measured, N is known from a molecular weight determination, and l should be known from the chemical structure. (Bond lengths in organic molecules are well known.) So, in addition to establishing the proportionality of R_g^2 with N, one can extract from the *data* a length L_0 defined as

$$L_0 \equiv \sqrt{\frac{6\overline{R_g^2}}{N}}. \tag{1-72}$$

If our simplified model of the random coil polymers was correct, the quantity L_0 would be identical to the chain length l.

To examine the applicability of the theory, let us take the example of polystyrene, data for which are given in Table 1.4. Now, the polystyrene monomer is itself made of two linked subunits, and the polystyrene chain looks as shown in Fig. 1.13, where the dot stands for CH_2 and the square for $CH-C_6H_5$. So, in a molecule of N monomers, there are actually $2N$ links of length l. We must therefore—for polystyrene—insert $2N$ rather than N in (1-72). Doing this, and using the data of Table 1.4, we find

$$L_0 \cong 5.6 \text{ Å},$$

whereas l, the carbon–carbon bond length, is well known to have the value

$$l = 1.54 \text{ Å} \qquad (\text{C}-\text{C bond length}).$$

There is a discrepancy of almost a factor of 4 in the two numbers. Additional data on polyethylene (see Chiang [20]):

$$-CH_2-CH_2-(CH_2)-\cdots$$

also lead to values of L_0 of the order of 5.4 to 6 Å.

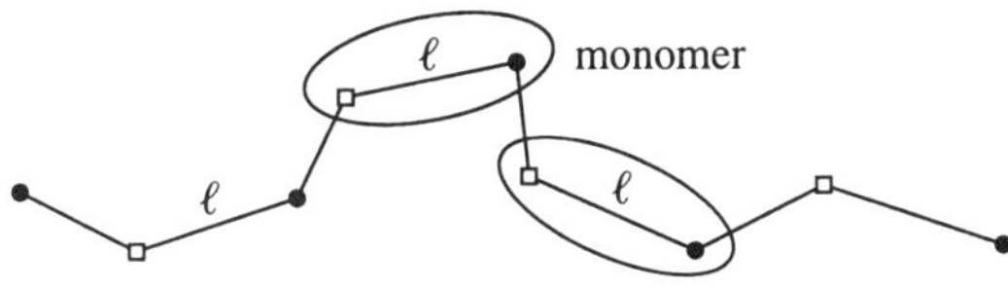

Figure 1.13. Model of a polystyrene chain.

How do we explain this discrepancy? It must be due, in part, to the lack of flexibility of the chain links. In the model which led to (1-71), we assumed that in essence each orientation of a segment relative to the previous segment is equally probable. This is patently not the case. In fact, we *know* that in a carbon–carbon–carbon backbone successive segments must be at a fixed angle $\alpha = 70°32'$ relative to each other (see Figure 1.14). One can calculate what the effect of this restriction is. (See Tanford [5], or Volkenstein [7], or Birshtein and Ptitsyn [6], for how this is done.) One finds that (1-71) is modified into

$$\overline{R_g^2} = \frac{1}{6}\left(\frac{1+\cos\alpha}{1-\cos\alpha}\right) l^2 N. \tag{1-73}$$

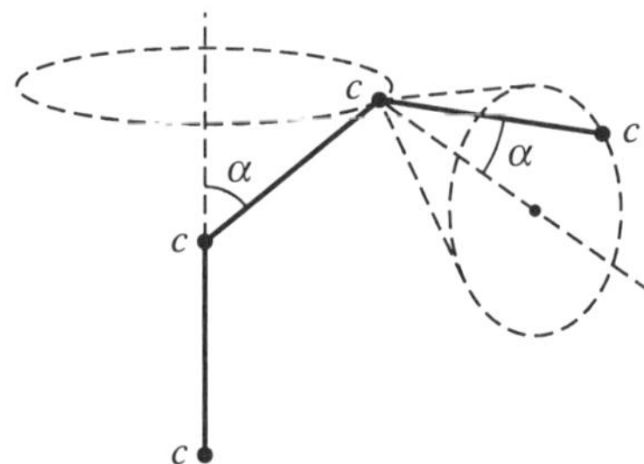

Figure 1.14. Bond angle α in a C−C backbone of a chain polymer.

In the polyethylene or polystyrene chain with $\alpha = 70°32'$, $(1+\cos\alpha)/(1-\cos\alpha) = 2$. The discrepancy between L_0 [(1-72)] and $\sqrt{(1+\cos\alpha)/(1-\cos\alpha)}\,l$ is now only a factor of 2.6. This residual factor is mainly due to the *bulkiness* of the monomers that exclude certain configurations of the random coil in which a piece of the chain would fold back upon itself. Our theoretical model allowed that, but nature does not. The chain is therefore less "flexible" than the theoretical model. It is important to realize that this reduced flexibility does *not* invalidate the basic result that $\overline{R_g^2}$ is proportional to the number of segments N provided N is large enough. Assume we determine experimentally the value of L_0 [(1-72)], and that L_0 is found to be larger than the segment length (= bond distance) l. Let us express this discrepancy by means of a coefficient k:

$$L_0^2 = kl^2 \qquad \text{(with } k > 1). \tag{1-74}$$

We can then write the experimental result

$$(\overline{R_g^2})_{\exp} = \tfrac{1}{6}L_0^2 N$$

in the form

$$(\overline{R_g^2})_{\exp} = \tfrac{1}{6}kl^2 N = \tfrac{1}{6}(kl)^2 \left(\frac{N}{k}\right). \qquad (1\text{-}75)$$

In this last form the result has again a simple statistical interpretation (due to W. Kuhn; see [5, p. 159]): Divide the N-monomer chain into N/k segments of k monomers, each segment now having length kl. These multimonomer segments behave as if their orientation relative to each other were at random. It is easily seen from a case, as illustrated in Figure 1.12, that although the relative orientation of two *consecutive* bonds may be restricted, the relative orientation of pieces consisting of k bonds may indeed be quite unrestricted. For most polymers in solution, k appears to be of the order of 3 to 6 (see Birshtein and Ptitsyn [6, p. 14]). Thus, even for polymers, in which the relative orientation of *consecutive monomers* is restricted, we may view the polymer as divided into pieces of three to six consecutive monomers, and regard such consecutive pieces as if they were able to assume a random orientation relative to each other.

1.4.C.   The Distribution of Electric Charges on the Hemoglobin Molecule

Hemoglobin, the oxygen carrier in red blood cells, is one of the proteins whose structure has been fully established. It is one of the "globular" proteins, and consists of two pairs of closely packed, intricately folded polymer chains, commonly called α- and β-chains. In human hemoglobin, the α-chain has 141, the β-chain 146 segments, or monomers. These monomers are *amino acids* which have the general structure

$$\mathrm{NH_3^+ - \overset{\displaystyle H}{\underset{\displaystyle R}{\overset{|}{\underset{|}{C}}} - COO^-}$$

R being the so-called *residue* whose structure characterizes the individual amino acid. (For a description of the biologically important amino acids, see Lehninger et al. [8].)

Amino acids polymerize by forming the peptide bond according to the scheme

$$NH_3^+ - CH - COO^- + NH_3^+ - CH - COO^-$$
$$\quad\quad\quad | \quad\quad\quad\quad\quad\quad\quad\quad\quad |$$
$$\quad\quad\quad R_1 \quad\quad\quad\quad\quad\quad\quad\quad R_2$$

$$\longrightarrow NH_3^+ - CH - CO - NH - CH - COO^- + H_2O$$
$$\quad\quad\quad\quad\quad | \quad\quad\quad\quad\quad\quad\quad\quad |$$
$$\quad\quad\quad\quad R_1 \quad\quad\quad\quad\quad\quad\quad R_2$$

A popular account of protein structure may be found in the book by Dickerson and Geis [12]. In hemoglobin, the four chains do not form random coils, but have a precise architecture giving the molecule the overall shape of a box-like structure with dimensions 64 Å × 55 Å × 50 Å, see Figure 1.15. Each of the four chains binds a so-called *heme* group, capable of attaching one oxygen molecule.

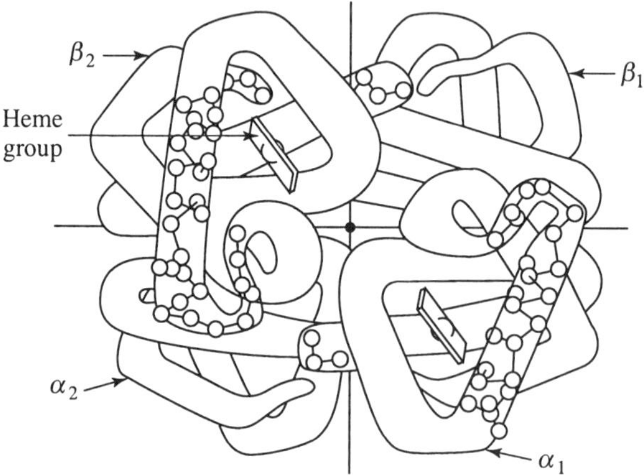

Figure 1.15. Architecture of the hemoglobin molecule with the two α- and β-chains displayed.

Hemoglobin is readily soluble in water, and in solution behaves as a polyelectrolyte, that is, the molecule carries in general a net electric charge. (A separation method for proteins, known as electrophoresis, to be discussed in Volume 3, is based on the fact that soluble proteins are polyelectrolytes.) The electric charge of a protein is carried in part by the COO^- and NH_3^+ groups of the terminal monomers of the polymer chains, and in part by amino acid residues R. Charge carrying residues are called *polar*. Polar residues may be either *acidic* or *basic*, and the distinction between them and the *neutral* (nonpolar) residues is illustrated by a few examples given below:

(i) Nonpolar (neutral) amino acid residues R:

$$\text{alanine:} \quad -R = -CH_3,$$

$$\text{valine:} \quad -R = -CH\Big\langle \begin{matrix} CH_3 \\ CH_3 \end{matrix},$$

the residue contains only covalent bonds.

(ii) Polar acidic (A) amino acid residues R:

$$\text{aspartic acid:} \quad -R = -CH_2 - COO^-,$$

$$\text{glutamic acid:} \quad -R = -CH_2 - CH_2 - COO^-,$$

(iii) Polar basic (B) amino acid residues R:

$$\text{lysine:} \quad -R = -(CH_2)_4 - NH_3{}^+,$$

$$\text{arginine:} \quad -R = -(CH_2)_3 - NH - C(NH_2)_2{}^+.$$

In solution a fraction of the polar residues will be uncharged. The passage from the charged to the uncharged state is described by the reactions

$$\text{acidic R:} \quad COOH \rightleftarrows COO^- + H^+, \tag{1-76}$$

$$\text{basic R:} \quad NH_3{}^+ \rightleftarrows NH_2 + H^+. \tag{1-77}$$

To the left of the equation we have listed the residues in their "protonated" form (in this form the residue has a hydrogen ion H^+, or proton, attached). To the right the residue appears "deprotonated"; it has lost its attached hydrogen ion. Equations (1-76) and (1-77) are written deliberately in this form so that the reaction arrow $\rightarrow$ describes the loss of a unit of charge, and the inverse reaction $\leftarrow$ the acquisition of a unit of charge.

In solution a dynamic equilibrium is established between the two charge states of each polar residue. This equilibrium can be described by stating for each type of residue the probability that it will be in the protonated state (COOH for acidic, $NH_3{}^+$ for basic residues) or in the deprotonated state (COO$^-$ for acidic, NH_2 for basic residues). These probabilities depend on the *hydrogen ion concentration* in the solution, and on an equilibrium constant characteristic of the residue.

Let us then define

$$p = \text{probability for protonated state} \left.\vphantom{\begin{array}{c}a\\b\end{array}}\right\} \text{of residue.}$$
$$q = 1 - p = \text{probability for deprotonated state}$$

In appendix to Section 1.4C we develop a simple kinetic argument to show that

$$p = \frac{[H^+]}{[H^+] + K}, \qquad q = 1 - p = \frac{K}{[H^+] + K}, \tag{1-78}$$

where $[H^+]$ is the concentration of hydrogen ions in the solution, and K is the effective equilibrium constant. Each polar residue can be characterized by a value of K. Equation (1-78) represents only an approximation sufficient for our purpose. For a more detailed discussion we refer the reader to texts in physical chemistry, for instance, Daniels and Alberty [10] or Williams and Williams [11].)

Notice that equations (1-78) hold for both acidic and basic residues. The values of the equilibrium constant K (whose dimension is mol/l) range from about 10^{-2} for acidic residues to 10^{-10} for basic residues, and a corresponding range of values of $[H^+]$ occurs in going from acid to basic (alkaline) solutions. For this reason a logarithmic concentration scale has been introduced by chemists. They define

$$pH = -\log_{10}[H^+],$$
$$pK = -\log_{10}K. \tag{1-79}$$

Since K and $[H^+]$ are not dimensionless, it is important to remain aware that in equations (1-79) $\log[H^+]$ and $\log K$ refer to the logarithm of the numerical value of $[H^+]$ and K, if measured in units of mol/l. So, for example, $pK = 2$ means, in fact, $\log_{10} K = -2$, $K = 10^{-2}$ mol/l. The inversion of these relations is then

$$[H^+] = 10^{-pH}\,\text{mol/l}, \qquad K = 10^{-pK}\,\text{mol/l}. \tag{1-80}$$

In terms of pH and pK we can now express the probabilities p and q as

$$p = \frac{10^{-pH}}{10^{-pH} + 10^{-pK}} = \frac{1}{1 + 10^{pH-pK}}, \tag{1-81a}$$

$$q = \frac{10^{-pK}}{10^{-pH} + 10^{-pK}} = \frac{1}{1 + 10^{pK-pH}}. \tag{1-81b}$$

Considered as functions of the variable (pH–pK), p and q are therefore *universal* functions which can be plotted once and for all. This is done in Figure 1.16. The hydrogen ion concentration of pure water at 25 °C is $[H^+] = 10^{-7}$ mol/l, and its pH is therefore 7.0. This value of pH $= 7$ defines a reference point on the pH scale: solutions with $[H^+] \gg 10^{-7}$, or pH < 7, are called acid; those with $[H^+] \ll 10^{-7}$, or pH > 7, are called alkaline or basic. We may note in passing that the hydrogen ion concentration in body fluids (blood plasma and interstitial fluid) are closely controlled so as to stay in the range

$$7.35 < \mathrm{pH_{body}} < 7.45$$

(see Pitts [13]).

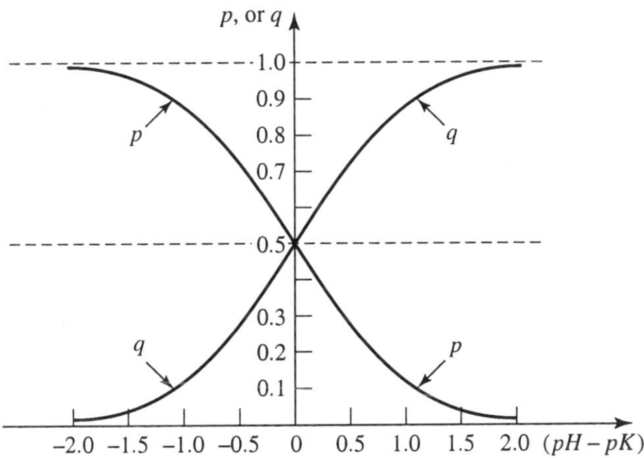

Figure 1.16. The probabilities p for protonation and $q = 1 - p$ for deprotonation of a polar residue in a function of pH–pK.

We are now in a position to make some statements about the charge state of the polar amino acid residues of hemoglobin in a solution having a certain value of pH. Consulting Figure 1.16, we see that if the pK of a residue is near the pH of the solution, then the probabilities p and q will be near $\frac{1}{2}$, and the residue will be charged in about half the cases. On the other hand, if its pK is 2 units *below* pH, the residue will have *lost* its proton H^+ 99 out of 100 times. Conversely, if pK is 2 units above pH, the residue will be protonated with 99% probability. To see what this argument implies for the case of hemoglobin in solution, we now present a table of the polar residues in Hb, along with their pK values, the number of such

Table 1.5. Numbers and pK Values of Acidic and Basic Polar Groups in Hemoglobin.

Residue A = acidic B = basic	pK value (approximate)	Number of residues in Hb	Electric charge at pH = 7	Comments
α-COOH (A)*	~ 2	4	− 4	sum of
Aspartic acid (A)	~ 4	30	−30	negative
Glutamic acid (A)	~ 4	24	−24	charges
Heme COOH (A)	~ 4	8	− 8	−66
Cysteine (A)	~11	6	0	
Tyrosine (A)	~10	12	0	
α-NH$_2$ (B)*	~ 7	4	variable	range of charges:
Histidine (B)**	6–7	38**	variable	0–24
Lysinc (B)	~10	44	+44	sum
Arginine (B)	~12	12	+12	+56

*These are the *terminal* groups at the end of the polymer chain.
**It is known that eight histidine residues partake in the binding of the heme groups; these have a fixed charge zero. Of the remaining 30, 10 appear to be "masked," that is, buried in the folded chains, therefore not exposed to the solvent, and permanently uncharged.

residues in the Hb molecule, and their electric charge at pH = 7 (see Table 1.5). It must be mentioned that although the pK values of the polar residues of *free* amino acids are precisely known, these values are modified by building the amino acid monomer into a chain, and modified again by the folding of the chain. For this reason, only approximate values are listed in Table 1.5.

The four acid residues listed first in Table 1.5 have a pK at least 3 units below 7, and are therefore deprotonated, and negatively charged at a pH near 7. Their total charge is −66. Two acid residues have a pK at least 3 units higher than 7, and are therefore uncharged at pH = 7. The two last listed basic residues have a pK which is 3 units higher than 7, will be protonated at pH = 7, and positively charged; their total charge is +56.

There remain 24 residues with pK near 7, and the probability of their being charged will vary as the pH value deviates from the neutral value of pH = 7 according to (1-81). The experimental procedure of determining the relation between charge and pH is called titration: The concentration of H$^+$ ions is changed in a controlled way by either adding a strong acid (like hydrochloric acid HCl which in solution is completely dissociated into H$^+$ and Cl$^-$, and thus adds a known amount of H^+ ions), or by adding a strong base (like NaOH which in solution ionizes into $Na^+ + OH^-$ and removes H^+ ions according to the reaction $Na^+OH^- + H^+ \rightarrow Na^+ + H_2O$).

After these preliminaries let us now see what predictions we can make about the probability distribution of charges on the hemoglobin molecule for values of the pH in the range from about $pH = 6$ to $pH = 8$. These predictions will be based on the following model:

(i)   There are N residues characterized by a common constant K. All these residues are basic.

(ii)   Each residue carries a charge $+1$ with probability p, and is uncharged with probability $q = 1 - p$, with p and q given by Eqns. 1-78 and 1-81.

(iii)   The charge state ($+$ or 0) of one residue has no influence on the charge state of a neighboring residue.

(iv)   The molecule also carries a fixed charge $-z_0$.

With these premises we see at once that the probability for z' of the N residues being charged is the same as the probability for z' heads to show up in a toss of N identical coins with p the probability for showing head, namely, the binomial distribution

$$P_N(z') = \frac{N!}{z'!\,(N - z')!}\, p^{z'} q^{N-z'}. \tag{1-82}$$

Since $p + q = 1$, it is useful to express this distribution in terms of the single parameter

$$x \equiv \frac{p}{q} = \frac{[H^+]}{K} \tag{1-83}$$

(using (1-78)). Also from (1-78) we find that $p = x/(1 + x)$, $q = 1/(1 + x)$, and (1-82) may be written as

$$P_N(z') = \frac{N!}{z'!\,(N - z')!}\, \frac{x^{z'}}{(1 + x)^N}. \tag{1-84}$$

We know already that for such a distribution, the mean value $\overline{z'}$ for z' is given by Np:

$$\bar{z'} = Np = N \frac{x}{1+x} \tag{1-85}$$

and the mean total charge $\bar{z}$ on the molecule is therefore

$$\bar{z} = \bar{z'} - z_0 = N \frac{x}{1+x} - z_0. \tag{1-85a}$$

Since $x = 10^{\mathrm{pK-pH}}$, (1-85a) represents the relation between the mean charge of the molecule and the pH value of the solution. This relation is called the *titration curve* of the dissolved molecule.

For hemoglobin we expect (1-85) to be valid for the range

$$6 \gtrsim \mathrm{pH} \gtrsim 8$$

and z_0 to have the value $z_0 = 10$, and $N = 24$. This latter value, it must be said, is of course the result of a comparison of a complete titration of Hb covering a pH range from 2 to 12, and an interpretation of this result in terms of a generalization of our (1-85). Let us now turn to the data. The most recent results appear to be those of Jannsen and coworkers [15]. Their results are presented in the form of a *differential* titration curve in which

$$\frac{-1}{(d\bar{z}/d(\mathrm{pH}))} \text{ is plotted as a function of } \bar{z}.$$

We shall see why this is a useful procedure. In order to calculate $d\bar{z}/d(\mathrm{pH})$, which is the rate of change of $\bar{z}$ with respect to a change of pH, we use the rule

$$\frac{d\bar{z}}{d(\mathrm{pH})} = \frac{d\bar{z}}{dx} \cdot \frac{dx}{d(\mathrm{pH})}.$$

The first factor follows at once from (1-85)

$$\frac{d\bar{z}}{dx} = N \left(\frac{1}{1+x} - \frac{x}{(1+x)^2} \right) = N \frac{1}{(1+x)^2}. \tag{1-86}$$

To obtain the second factor, we write

$$x = 10^{-\mathrm{pH}+\mathrm{pK}} = 10^{-(\mathrm{pH}-\mathrm{pK})} = e^{-2.303(\mathrm{pH}-\mathrm{pK})},$$

which gives

$$\frac{dx}{d(\text{pH})} = -2.303x. \tag{1-87}$$

Equations (1-86) and (1-87) combine to give

$$\frac{d\bar{z}}{d(\text{pH})} = -2.303N\frac{x}{(1+x)^2}. \tag{1-88}$$

We can express the right-hand side of (1-88) in terms of $\bar{z}'$ by using (1-85) which gives

$$x = \frac{\bar{z}'}{(N-\bar{z}')}, \qquad (1+x) = \frac{N}{(N-\bar{z}')}, \tag{1-89}$$

and, hence,

$$\frac{d\bar{z}}{d(\text{pH})} = -2.303\frac{\bar{z}'(N-\bar{z}')}{N} \tag{1-90}$$

or, by taking the reciprocal on both sides,

$$-\frac{d(\text{pH})}{d\bar{z}} = \frac{1}{2.303}\frac{N}{\bar{z}'(N-\bar{z}')}. \tag{1-91}$$

Experimentally the relation between $\bar{z}$ and pH is obtained by titration, as previously mentioned. For example, the addition of a known small amount of hydrochloric acid will add a known amount of hydrogen ions. Part of them will be picked up by the hemoglobin molecules, the remainder will stay in solution and increase the concentration $[H^+]$, that is, lower the value of pH. Using the measured change of pH (see [10] and [11]), the known amount of acid added, and the known number of Hb molecules in solution, one can determine the amount of charge picked up by each Hb molecule. This gives $\Delta\bar{z}/\Delta(\text{pH})$.

In Figure 1.17, the data of Jannsen et al. [15] are plotted, and at the same time the result of (1-91) with $z_0 = 10$ and $N = 24$. [In plotting the curve [(1-91)], a constant value of 0.023 has been added to the right-hand side; this represents an electrostatic correction that improves the agreement of the two curves.]

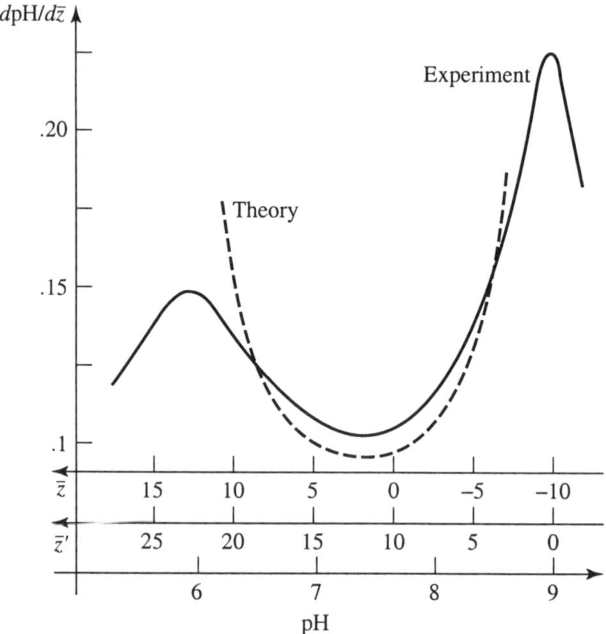

Figure 1.17. The differential titration curve of hemoglobin versus the actual charge z. The solid curve is the experimental data. The dotted curve is the theory with $N = 24$ residues of common pK. $z' = z + 10$ gives the best fit.

Notice that according to (1-91), $-d(\mathrm{pH})/d\bar{z}$ goes to infinity at $\bar{z}' = 0$ and $\bar{z}' = N = 24$. In Jannsen's curve this does not happen; only two peaks show up. These infinite values of $-d(\mathrm{pH})/d\bar{z}$ occur in our model because at $\bar{z}' = 0$ all residues are uncharged, and no increase of pH can reduce the charge any further. In reality, however, a further increase of pH leads to the gradual removal of the 56 positive charges mentioned in Table 1.5, and instead of $-d(\mathrm{pH})/d\bar{z}$ going to infinity, all we see is a sharp peak.

Let us consider what our analysis has yielded up to this point:

(i) Equation (1-90), based on the assumption of N equivalent polar groups, gives a reasonable account of the differential titration curve in the range $6.5 < \mathrm{pH} < 8.5$.

(ii) Calculations based on (1-91) with various N-values give $N = 24$ as a best overall fit. It is also the conclusion of the more detailed analysis by Jannsen and coworkers [15] that 24 units of charge are removed between the two peaks of their differential titration curve.

In this analysis no property of the probability distribution $P_N(z')$ of charges on the molecule has yet been used except the mean value $\overline{z'}$. It is clearly of interest to look for data that do depend on the *form* of the distribution $P_N(z')$ as given by (1-82) or (1-84). There are indeed two pieces of such data that we will now discuss.

The first emerges from an alternative interpretation that may be given to (1-88) for $d\overline{z}/d(\mathrm{pH})$. In (1-83) the quantity x was defined as $x = p/q$ so that

$$p = x/(1+x) \qquad \text{and} \qquad q = 1/(1+x), \quad p+q = 1.$$

Hence, (1-88) may be written as

$$\frac{d\overline{z}}{d(\mathrm{pH})} = -2.303 N pq.$$

Now, in Section 1.3 of this chapter, we showed that Npq is just the *variance* of z' for the binomial distribution $P_N(z')$:

$$\overline{(z' - \overline{z'})^2} \equiv \sum_{z'} (z' - \overline{z'})^2 P_N(z') = Npq.$$

The variance of the partial charge z' is also the variance of the total charge $z = z' - z_0$, since $z - \overline{z} = z' - \overline{z'}$. We therefore reach the conclusion that

$$-\frac{d\overline{z}}{d(\mathrm{pH})} = 2.303\,\overline{(z - \overline{z})^2}. \tag{1-92}$$

This relation, which is remarkable by both its simplicity and by its content, was discovered by K. Linderstom-Lang, and first reported in a seminar at the Harvard Medical School in 1939 (see [16]). Equation (1-92) states that the mean square fluctuation of the charge, whose distribution obeys the Bernoulli formula [(1-82) and (1-84)], can be inferred from the measured titration curve. It is important to make the observation that the validity of ((1-92)) is more general than its derivation suggests; in particular, it remains valid if the molecule carries several distinct unequivalent groups of polar residues, even if electrostatic distortions of the probability distributions of charge are included.

We may therefore view (1-92) as giving a semi-experimental value of the variance $\overline{(z - \overline{z})^2}$ of the charge distribution, based on the *measured* values of

$d\bar{z}/d(\text{pH})$. We may compare this value of $\overline{(z - \bar{z})^2}$ with the predictions of our model of the probability distribution of charges. This model, according to (1-90), predicts

$$\overline{(z - \bar{z})^2} = \frac{\bar{z}'(N - \bar{z}')}{N}$$

or, with $N = 24$, $\bar{z}' = \bar{z} + 10$:

$$\overline{(z - \bar{z}')^2} = \frac{(\bar{z} + 10)(14 - \bar{z})}{24}. \tag{1-93}$$

This comparison is made in Table 1.6. We see that in general our model predicts a variance that is too large, a result that is mainly due to our neglect of electrostatic forces. These forces distort the Bernoulli distribution into a somewhat narrower distribution of charges.

Table 1.6. Comparison of Variance $\overline{(z - \bar{z})^2}$ of
Electric Charge on Hemoglobin. A: Exp. Titration
Curve and (1-92). B: From Bernoulli Fit to Titration
Curve: (1-90) and (1-93).

Mean charge z	$\overline{(z - \bar{z})^2}$ from A	$\overline{(z - \bar{z})^2}$ from B
-5	3.2	3.96
0	4.13	5.83
5	4.05	5.62
10	3.21	3.34

In essence, the data in Table 1.6 represent in another way the degree of agreement between our simple model for the probability distribution of charge and the titration data.

The large values of the variance shown in Table 1.6 are a reflection of the fact that most of the molecules are actually carrying a charge different from the mean charge. To demonstrate this, let us calculate as an example the probability that the Hb molecule will be charged when the mean charge $\bar{z}$ is zero. This can be derived from (1-84) for $P_N(z')$. We must then put $z' = z_0$ since $z = z' - z_0$, and adjust the value of x so as to give $z' = 0$, or $\bar{z}' = z_0$. From (1-89) this value of x is

$z_0/(N - z_0)$, and the probability that the molecule is uncharged at $z = 0$ turns out to be

$$P_N(z' = z_0)|_{\bar{z}=0} = \frac{N!}{z_0! \, (N - z_0)!} \, \frac{z_0^{z_0}(N - z_0)^{N-z_0}}{N^N}.$$

This horrible expression in Stirling's approximation reduces to

$$\sqrt{\frac{N}{2\pi z_0(N - z_0)}}$$

and for $N = 24$, $z_0 = 10$, we find the probability for zero charge at $\bar{z} = 0$ to be 0.165, and therefore the probability that the molecule will be charged at $\bar{z} = 0$:

$$P(\text{charged})|_{\bar{z}=0} = 0.835.$$

It would clearly be of interest to have an actual *direct* measurement of the variance $\overline{(z - \bar{z})^2}$, which could be compared with the semitheoretical result [(1-92)] based on titration data. Such measurements do exist, and are again based on the method of light scattering.

Imagine a large number of protein molecules in solution. If they were all uncharged, they would be randomly distributed in the solution. Now, if some are positively charged, others negatively charged, those of equal charge would stay away from each other due to electrostatic repulsion, and those of opposite charge would tend to cluster due to mutual attraction. It is also observed that any solution of macromolecules, such as proteins, is slightly turbid ("hazy"); that is, a beam of light passing through the solution is partly scattered out of its path by the macromolecules. We discussed this in Section 1.4.B. The effect of molecular *clustering* on the amount of scattered light can be calculated, and is found to depend on $\overline{z^2}$ which is the mean square charge, but is equal to the variance $\overline{(z - \bar{z})^2}$ at the *isoelectric point* where $\bar{z} = 0$. Such measurements of light scattering at the isoelectric point and an appendant determination of $\overline{z^2}$ have been made on solutions of *bovine syrum albumin* by Timasheff and coworkers [19], who also compared their results with predictions of $\overline{z^2}$ based on titration data and (1-92). The results are as follows:

titration and (1-92) gives $\sqrt{\overline{z^2}} = 3.4$,

light scattering gives $\sqrt{\overline{z^2}} = 3.5$.

This is a dramatic confirmation of the view that the electric charge on polyelectrolytes in solution have a distribution of a measurable finite width.

As a final demonstration of the usefulness of the preceding analysis, we will use it to help understand the *solubility* of proteins. The results obtained above for the probability distribution for the charge of a macromolecule can be used to show why the solubility of a protein varies very markedly with the pH, and why the solubility is a *minimum* at the isoelectric point, i.e., the pH at which the mean charge on the molecule is zero.

We begin by recognizing that the solubility S of a protein is defined as the maximum total concentration (regardless of charge) of all the protein molecules. This maximum is established when we have solute in contact with the solid, crystallized phase of the molecular species in which case the "equilibrium" between solution and solid phase is established. To understand the nature of this equilibrium, it is essential to realize that the molecules in the solid phase are *uncharged*, so that all molecules which go from the solid phase into solution, or vice versa, are uncharged molecules. A solution is "in equilibrium" with its solid phase when the concentration (S_0) of uncharged molecules has a value such that during each unit of time the same number of uncharged molecules leave the solid phase (and enter the liquid) as solute molecules leave the liquid and enter the solid phase. The concentration S_0 of *uncharged* molecules will be regarded as a constant independent of the pH of solution, and dependent only on the temperature.

As the pH of the solution is changed, though the concentration of uncharged molecules (S_0) will remain constant, the total concentration S of all proteins charged *and* uncharged must change. In fact the variation of S with pH must be just that for which the concentration (S_0) of uncharged molecules remains the same. This requirement is the basis of our present discussion of protein solubility. We can compute how S varies with pH keeping S_0 constant in the following way: The total concentration of the molecules of all charges z is given by

$$S = \sum_z S_z. \tag{1-94}$$

Here S_z is the concentration of molecules having charge z. The charge z is made up of two contributions

$$z = z' - z_0, \tag{1-95}$$

$-z_0$ is the total number of fixed charges, and z' is the pH dependent variable charge. The concentration S_z is related to the probability distribution $P_N(z')$ (see

(1-84)) by

$$S_z = SP_N(z') = SP_N(z + z_0). \tag{1-96}$$

Here N is the total number of charge variable residues. P_N is given as a function of z' by the (1-84), viz:

$$P_N(z') = \frac{N!}{z'!(N - z')!}\frac{x^{z'}}{(1 + x)^N}. \tag{1-97}$$

Here, of course, x depends directly on the H^+ ion concentration through its definition

$$x = \frac{[H^+]}{K}. \tag{1-98}$$

From (1-96) we find that the concentration S_0 of uncharged molecules is

$$S_0 = SP_N(z = 0) = SP_N(z' = z_0). \tag{1-99}$$

But

$$P_N(z' = z_0) = \frac{N!}{z_0!(N - z_0)!}\frac{x^{z_0}}{(1 + x)^N}. \tag{1-100}$$

Equations (1-98), (1-99), and (1-100) provide the desired relationship between S, S_0, and pH or $[H^+]$. Writing out the relationship explicitly, we find

$$S \equiv \text{"solubility"} = \frac{S_0}{P_N(z' = z_0)} \tag{1-101}$$

or

$$S = S_0 z_0!\frac{(N - z_0)!(1 + x)^N}{N!\,x^{z_0}}. \tag{1-102}$$

As the pH of the solution changes, x changes, S_0, z_0, and N remain constant, and S, the solubility, should change with pH in accordance with (1-102).

Let us now examine the behavior of this result both in the limits $x \to 0$ and $x \to \infty$:

$$\text{for} \quad x \to \infty, \quad S \to x^{N-z_0},$$
$$\text{for} \quad x \to 0, \quad S \to x^{-z_0}.$$

Since, in our model of hemoglobin, $N = 24$, $z_0 = 10$, we see that in both these limits the solubility becomes very large. In between, S has a minimum value. We can locate the value of $[H^+]$ concentration at this minimum by differentiating Eqns. (1-102) and setting the result equal to zero

$$\left(\frac{dS}{dx}\right) = \frac{S_0 z_0! \, (N - z_0)!}{N!} \left\{ \frac{N(1 + x)^{N-1}}{x^{z_0}} - \frac{z_0(1 + x)^N}{x^{z_0+1}} \right\} \tag{1-103}$$

or

$$\left(\frac{dS}{dx}\right) = S \left\{ \frac{N}{1 + x} - \frac{z_0}{x} \right\}. \tag{1-104}$$

Setting this equal to zero, we find that the value of x at which the solubility is minimum is

$$\left(\frac{x}{1 + x}\right)_{\min} = \frac{z_0}{N}. \tag{1-105}$$

We now need only note that the mean charge $\bar{z}$ of the molecule is given by

$$\bar{z} = -z_0 + N \left(\frac{x}{1 + x}\right). \tag{1-106}$$

On using the value of x from (1-105) in (1-106), we see at once that $\bar{z} = 0$ when the solubility has its minimum value. Thus, we have demonstrated a fact well known experimentally, namely, that the solubility of proteins is minimum at the isoelectric point.

We can now examine the dependence of S on pH around the isoelectric point. In Figure 1.18 the *ratio* $R = S/(S)_{\text{iso}}$ of the solubility relative to that at the isoelectric point is plotted as a function of the pH (recall $\text{pH} - \text{pK} = -\log_{10} x$). Unfortunately, good solubility data are very few. The best data are on β-lactoglobulin, and reported in Lehninger et al. [8, p. 133]. It is seen there that solubility depends not only on pH, but on a *second* parameter, the so-called *ionic strength* of the

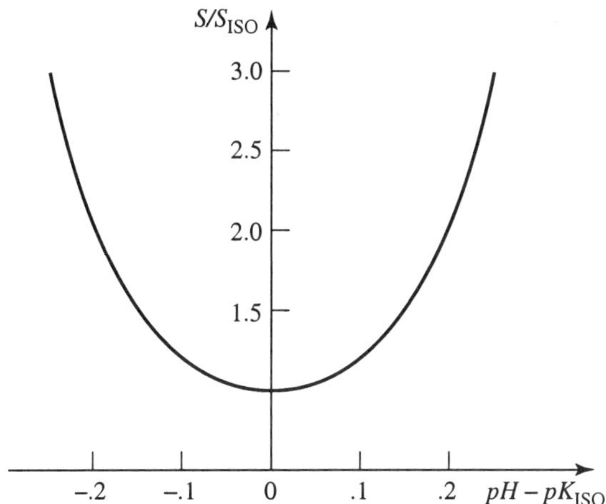

Figure 1.18. Solubility of hemoglobin according to (1-102) as a function of pH–pH$_{\text{iso}}$.

solution. As was to be expected, there are electrostatic energies involved in the equilibrium whose effect is to modify the distribution $P_N(z')$. This subject will be taken up in Volume 3 of this text (*Electricity and Magnetism*). However, even with this shortcoming of the present analysis, the characteristic feature of large changes of solubility accompanying a small change in pH is well reproduced by our very simplified model.

Appendix to Section 1.4.C: Probabilities for the State of Ionization of a Polar Residue

In this Appendix we give a derivation of equations (1-78) for the probabilities p and q that polar residue is protonated or deprotonated. These values of p and q are determined by the equilibrium of the reactions [(1-76) or (1-77)]. To deal with a specific case, consider the first reaction. In solution the processes indicated by the two arrows in (1-76) are constantly operating: Protonated residues (COOH) are breaking up at a certain rate, and deprotonated residues (COO$^-$) are combining with available hydrogen ions to reform COOH. In equilibrium the turnover in the two opposite processes is the same, and the numbers of COOH and COO$^-$ remain constant on the average.

We now formulate this idea in a quantitative way, and show how epxressions (1-78) follow. Consider a solution containing $\mathcal{N}$ acidic residues, $\mathcal{N}_0$ of which are in the protonated form COOH, $\mathcal{N}_-$ of them in the deprontonated from COO^-. The solution also contains hydrogen ions H^+; we use the symbol $[H^+]$ to designate the concentration (in mol/l) of hydrogen ions. Consider now the rate of change of the number $\mathcal{N}_0$ of protonated residues. This rate $d\mathcal{N}_0/dt$ is equal to

$$\frac{d\mathcal{N}_0}{dt} = \text{rate of recombination } (COO^- + H^+ \to COOH)$$
$$- \text{rate of dissociation } (COOH \to COO^- + H^+). \qquad (1\text{-}107)$$

The rate of dissociation must be proportional to the number of undissociated residues, namely, $\mathcal{N}_0$:

$$\text{rate of dissociation} = +k_2\mathcal{N}_0.$$

The rate of recombination must be proportional to the number $\mathcal{N}_-$ of dissociated residues, and the concentration $[H^+]$ of hydrogen ions needed as partners in the recombination process:

$$\text{rate of recombination} = k_1[H^+]\mathcal{N}_-.$$

Inserting these terms into the rate equation (1-107) gives

$$\frac{d\mathcal{N}_0}{dt} = k_1[H^+]\mathcal{N}_- - k_2\mathcal{N}_0. \qquad (1\text{-}108)$$

The value of $d\mathcal{N}_-/dt$ is equal and opposite to that, so that $d\mathcal{N}/dt = d\mathcal{N}_0/dt + d\mathcal{N}_-/dt = 0$.

Now, if equilibrium has been reached, $\mathcal{N}_0$ is constant, and its time rate of change zero. In this case, we have

$$k_1[H^+]\mathcal{N}_- - k_2\mathcal{N}_0 = 0$$

or

$$\left(\frac{\mathcal{N}_0}{\mathcal{N}_-}\right) = \left(\frac{k_1}{k_2}\right)[H^+] \equiv \frac{[H^+]}{K}, \qquad (1\text{-}109)$$

where $K = (k_1/k_2)$, the equilibrium constant of the reaction (1-76). Since $\mathcal{N} = \mathcal{N}_0 + \mathcal{N}_-$, we have from (1-109):

$$\left(\frac{\mathcal{N}_0}{\mathcal{N}}\right) = \frac{[H^+]}{K + [H^+]}, \qquad \left(\frac{\mathcal{N}_-}{\mathcal{N}}\right) = \frac{K}{K + [H^+]}. \qquad (1\text{-}110)$$

The number $\mathcal{N}$ of acidic residues in solution is very, very large; all residues exist under manifestly identical conditions. The numbers $\mathcal{N}_0$ and $\mathcal{N}_-$ can therefore be viewed as the number of outcomes "protonated" or "deprotonated" in $\mathcal{N}$ independent trials, and the ratios $(\mathcal{N}_0/\mathcal{N})$ and $(\mathcal{N}_-/\mathcal{N})$ therefore represent the probabilities that a residue appears in the protonated or unprotonated form.

1.5   References and Supplementary Reading

1. As a general introduction into probability theory, we recommend W. Feller's *An Introduction to Probability Theory and its Applications*, Vol. 1. Wiley, New York (1957). Chapters 1 to 7 are quite elementary and contain many interesting examples.

2. *Games, Gods and Gambling*, by N. F. David. Griffin, London (1962). This work contains the complete Pascal–Fermat correspondence. Pascal's letters are also reproduced in [3].

3. *Oeuvres Complètes* by B. Pascal; text annotated by J. Chevalier. Bibliothèque de la Pléiade, Gallimard, Paris (1954).

4. An Analysis of Geissler's Data on the Human Sex Ratio, by A. W. F. Edwards. *Annals of Human Genetics* **23**, pp. 6–15 (1958).

5. *Physical Chemistry of Macromolecules*, by C. Tanford. Wiley, New York (1967).

6. Conformations of Macromolecules, by T. M. Birshtein and O. B. Ptitsyn. In *High Polymers*, Vol. XXII. Interscience, New York (1966).

7. Configurational Statistics of Polymer Chains, by M. V. Volkenstein. In *High Polymers*, Vol. XVII. Interscience, New York (1963).

8. *Principles of Biochemistry*, by A. L. Lehninger, D. L. Nelson, and M. M. Cox. Worth Publications, New York (1993).

9. *Principles of Biochemistry*, 6th ed., by A. White, Ph. Handler, E. L. Smith, and I. R. Lehman. McGraw-Hill, New York (1978).

10. *Physical Chemistry*, 6th ed., by R. A. Alberty. Wiley, New York (1983).

11. *Basic Physical Chemistry for the Life Sciences*, by V. R. and H. B. Williams. W. H. Freeman, San Francisco (1967).

12. *The Structure and Action of Proteins*, by R. E. Dickerson and I. Geis. Harper & Row, New York (1969).

13. *The Physiology of the Kidney and Body Fluids*, 2nd ed., by R. F. Pitts. Year Book Medical Publishers, Chicago (1968).

14. *Renal Function*, by H. Valtin. Little, Brown, Boston (1973).

15. Titration Studies of Human Hemoglobin, by L. H. M. Jannsen, S. H. de Bruin and G. A. J. van Os. *Biochim.-Biophys. Acta* **221**, 214 (1970); and Titration Behaviour of Histitdines in Human, Horse and Bovine Hemoglobin, by the same authors, *Journal of Biol. Chem.* **247**, 1743 (1970).

16. *Acid-Base Equilibria of Proteins*, by K. Linderstrom-Lang and S. O. Nielsen. In *Electrophoresis*, edited by Milan Bier. Academic Press, New York (1959).

17. The Principles of Acid-Base Chemistry and Physiology, Chap. 12. In *A Primer of Water, Electrolyte and Acid-Base Syndromes*, 6th ed., by E. Goldberger. Lea and Febiger, Philadelphia (1980).

18. *pH Homeostasis: Mechanisms and Control*, edited by D. Häussinger. Academic Press, New York (1988).

19. Studies of Molecular Interactions in Isoionic Protein Solutions by Light Scattering, by S. N. Timasheff, H. M. Dintzis, J. G. Kirkwood, and B. D. Coleman. *Proc. Nat. Acad. Sci. USA.* **41**, 710, (1955).

20. Intrinsic Viscosity-Molecular Weight Relationships for Fractions of Linear Polyethylene, by R. Chiang. *Journal of Physical Chemistry.* **69**, 1645 (1965).

1.6 Problems

(These problems are grouped topically, and within each group they are arranged roughly in increasing order of difficulty.)

Probability of Composite Events. Addition and Multiplication Rules. The Meaning of Probability

1. From a shuffled deck of 52 cards, a single card is drawn. Find the probability:

    (a) that the card pulled is the king of diamonds;

    (b) that the card pulled is a red jack;

(c) that the card pulled is an ace; and

(d) that the card pulled is a club.

2. Show that the probability of pulling the queen of hearts from a shuffled deck of cards is the product of the probability of pulling any queen, and the probability of pulling any heart.

3. Two unloaded dice are thrown. Find the probability:

(a) that the sum thrown is seven;

(b) that the sum thrown is an odd number; and

(c) that two unequal numbers are thrown.

Hint: For each question tabulate the *set* of outcomes that satisfy the specification given. Add up the probabilities of the outcomes in each such set.

4. A single die is thrown three times. Find the probability:

(a) for throwing a 6 three times;

(b) that in the three throws, the same face shows three times;

(c) for throwing a 6 in the first and third throws, but not in the second;

(d) for throwing two 6's and a face different from 6, irrespective of order.

5. A single die is thrown six times. Find the probability that "1" is up at *least* once.

Hint: This question looks difficult. The key to the answer to all "at least" type questions is to find first the probability for the outcomes *excluded* by the question. In the present case, this is the probability that "1" never shows up. Then use

$$P(\text{"1" occurs at least once}) + P(\text{never "1"}) = 1.$$

6. Find the probability that in six throws of a die, six different faces appear. The order in which they appear does not matter.

7. Proof of the general multiplication theorem, (1-13). This theorem states that if a trial T, in which outcome "i" has probability $P(i)$, is repeated N times, then the probability for the outcome "$i_1, i_2, \ldots, i_N$" is

$$P(i_1, i_2, \ldots, i_N) = P(i_1)P(i_2)\ldots P(i_N).$$

Recall that the outcome called "$i_1, i_2, \ldots, i_N$" represents the occurrence of "i_1" in the first trial, "i_2" in the first repetition, "i_N" in the Nth trial. Such proofs are given by the method of induction: Consider the N-fold repetition as the combination of just *two* trials, T_{N-1} and T, where T_{N-1} is the $(N-1)$-fold repetition of T with outcomes

$$\text{``}j\text{''} \equiv \text{``}i_1, i_2, \ldots, i_{N-1}\text{''},$$

which have probability $P(j)$.

(a) You can now show that the probability for the outcome "$i_1, i_2, \ldots, i_{N-1}$, i_N" $\equiv$ "j, i_N" is

$$P(j, i_N) = P(j)P(i_N).$$

(b) Now decompose the trial T_{N-1} into a combination of T_{N-2} and T. Repeat the argument, and show that it proves the theorem (M).

8. Proofreading a manuscript. If a trial has two outcomes, "success" or "failure," then the probability p for "success" is established by the limiting value of the ratio $\mathcal{N}_s/\mathcal{N}$ of the number of successes $\mathcal{N}_s$ to the number of trials $\mathcal{N}$:

$$p = \left(\frac{\mathcal{N}_s}{\mathcal{N}}\right)_{\mathcal{N}\to\infty}.$$

If $\mathcal{N}$ is finite, then the ratio $(\mathcal{N}_s/\mathcal{N})$ gives at least an estimate of p. In this problem, $\mathcal{N}$ is the (unknown, but large) number of misprints in a manuscript. Two proofreaders, 1 and 2, each read the manuscript. Reader 1 discovers $\mathcal{N}_1$ misprints, reader 2 $\mathcal{N}_2$.

(a) Taking this as an indication that they discover a misprint with a probability p_1 and p_2, respectively, obtain the probability p_{12} that they *both* discover a given misprint. Keep in mind the fact that the discovery of a misprint by reader 1 is independent of its discovery by reader 2.

(b) After reading and comparing notes, the proofreaders find a total of $\mathcal{N}_{12}$ misprints found by both of them. Obtain an estimate of the *actual* number $\mathcal{N}$ of misprints in the manuscript in terms of $\mathcal{N}_1$, $\mathcal{N}_2$, and $\mathcal{N}_{12}$.

(c) Numerical example: $\mathcal{N}_1 = 120$, $\mathcal{N}_2 = 80$, $\mathcal{N}_{12} = 48$. How many misprints went *undetected*?

9. The incidence of color blindness. Color blindness is a sex-linked defect; the defective gene is located in the X-chromosome. Recall that chromosomes occur in pairs, and that females carry the chromosome pair XX, whereas males have a single X-chromosome, paired with a Y-chromosome.

Let p stand for the fraction of defective X-chromosomes, which we call X_c. The following chromosome pairings are possible:

$$\text{male} \begin{pmatrix} X & Y \\ X_c & Y \end{pmatrix}, \qquad \text{female} \begin{pmatrix} X & X \\ X_c & X \\ X_c & X_c \end{pmatrix}.$$

The trait for color blindness is recessive, that is, the presence of *one* good X-chromosome allows color vision. Only the males with the $(X_c Y)$ pair and females with the $X_c X_c$ pair are color-blind.

(a) In a large population, about 8% of the males are found to be color-blind. The number of color-blind females is much smaller. What percentage do you expect? Data, not very accurate, indiate the incidence of female color blindness to be of the order of $\frac{1}{2}$%. Is this consistent with your calculated percentage?

(b) What is the percentage of females carrying the chromosome combination $X_c X_c$, $X X_c$, and $X X$, respectively? Remember that the sum of all the probabilities must add up to unity.

(c) There are many other fascinating questions one may ask. The chromosome set of offspring is made up of pairs, with one member of the pair donated by the father, the other by the mother. Transmission of sex (and hence, of sex-linked traits) then takes place according to the scheme shown below. This scheme gives the opportunity to establish a large number of probabilistic statements; they are rather easily found, if you keep the basic definition of probability in mind.

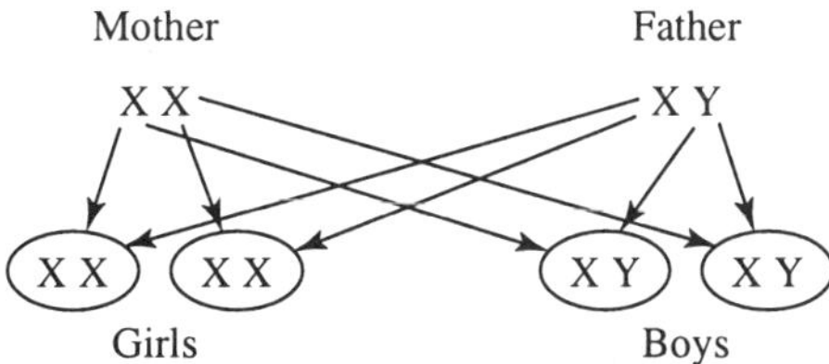

Establish the following results:

(i) If both parents are color-blind, *all* children will be.

(ii) If the mother is color-blind, the father not, then *all* male children will be color-blind, but *none* of the female children.

(iii) If the father is color-blind, and the mother *not*, the probability that a child will be color-blind is $2/27 = 7.4\%$, irrespective of sex.

(iv) If none of the parents is color-blind, then the probability of a male child being color-blind is $2/27 = 7.4\%$, that for a female child is zero.

10. Birthdays: At a certain day a number N of people are assembled. Find the probability that this particular day is the birthday of at least one of the people assembled.

   Hint: The word "at least" should be a tip-off. Find the probability that this day is *not* the birthday of any of the N persons. Then, by subtraction, find the probability asked for. You get a more compact answer by using

$$(1 - \varepsilon)^k \cong e^{-k\varepsilon} \quad \text{if} \quad |\varepsilon| \ll 1.$$

   How many people should be assembled so that there is a 50% chance that at least one has a birthday on that day?

Probability Distributions and Mean Values

11. The probability distribution $P(n)$ for the occurrence of "n" (ranging from 1 to 6) in a single throw of a die has a constant value $P(n) = \frac{1}{6}$ for all $1 \leq n \leq 6$. Find the mean value $\bar{n}$ thrown in a single throw of a die.

12. In throwing two dice, you may throw a sum n which ranges from $n = 2$ to $n = 12$.

   (a) *Find* the probability distribution function $P(n)$ for throwing a sum n with two dice.

   (b) Calculate the mean value $\bar{n}$ of the sum thrown.

Probability Distributions for the First Occurrence
of a Certain Outcome in Repeated Trials

This is a familiar type of question of which we will give some examples. One wants to know the probability $P(n)$ that a particular outcome of a trial, call it "i," occurs for the first time in the nth trial. In order to find $P(n)$, consider the two outcomes of the trial, namely "i," with probability p, and "not i," with probability $q = 1 - p$. The desired probability $P(n)$ is then

$$P(n) = q^{n-1} p \qquad (n = 1 \text{ represents the first trial}).$$

In Problem 13, some general properties of $P(n)$ are to be investigated; the subsequent problems are applications. In the problems that follow, you will find it useful to recall the expression for the sum of a geometric series

$$\sum_{n=0}^{\infty} x^n = \frac{1}{1-x}, \qquad \text{provided} \quad |x| < 1.$$

Remember also the device introduced in the text to calculate sums of type $\sum nx^n$:

$$\sum_{n=0}^{\infty} nx^n = x\frac{d}{dx}\sum x^n = x\frac{d}{dx}\frac{1}{1-x} = \frac{x}{(1-x)^2}.$$

13. Using the form of $P(n)$ presented above:
    (a) Show that $P(n)$ is normalized, i.e.,

$$\sum_{n=1}^{\infty} P(n) = 1.$$

    State clearly the operational significance of this normalization.
    (b) Show that the mean value $\bar{n}$ of n has the value

$$\bar{n} = \sum_{n=1}^{\infty} nP(n) = \frac{1}{p}.$$

14. How often must a die be thrown, on the average, until a "6" shows up for the first time? Read the *introduction* to Problem 13, and read Problem 13. You may use the results stated in Problem 13, and the task is then just to find p and q.

15. Russian roulette. In this deadly game, the player is given a revolver with six chambers holding one cartridge. Before each trial, the player spins the cylinder, and then fires against his head:
    (a) Find the probability that the player will kill himself in the nth trial?
    (b) Find the probability that the player will survive at least $(n-1)$ trials. Tabulate these probabilities for $n = 1, 2, 3, \ldots$, and draw a graph of $P(n)$ versus n.

16. The St. Petersburg game. In the St. Petersburg Casino, after paying an admission fee, you were offered the following bargains: You were given a coin to flip, and if heads showed up *for the first time* after n tosses, you gained 2^n rubles:

(a) Let the probability for "heads" be p. Find the probability $P(n)$ that "heads" shows up for the first time in the nth toss of the coin.

(b) Find the mean number of tosses needed to get the first "heads."

(c) Find the average gain of the player engaging in this game. Show that this gain would be infinite, unless p, the probability for "heads," is larger than 50%.

(d) If $p = 0.6$, what admission fee must be charged for the casino to break even?

The Bernoulli Distribution. Mean and Variance

17. A numerical exercise. Bernoulli distribution with $p = q = \frac{1}{2}$. This distribution is

$$P_N(H) = \frac{N!}{H!\,(N - H)!} \cdot \frac{1}{2^N}.$$

(a) Use $N = 6$, and *tabulate* the number $P(H)$ from $H = 0$ to $H = 6$.

(b) The rms width $\Delta = \sqrt{Npq}$ is 1.22 in this case. Calculate the total probability P for all H values that lie within the range $\overline{H} - \Delta < H < \overline{H} + \Delta$.

(c) Compare this result with the approximation given in the text as (1-45):

$$(P)_{\text{max}} \times 2\Delta \simeq 0.8.$$

18. Bernoulli distribution with $p \neq q$:

(a) The Bernoulli distribution $P_N(H)$ is given by

$$P_N(H) = \binom{N}{H} p^H q^{N-H}.$$

Show that if $x = (p/q)$, that

$$P_N(H) = \binom{N}{H} \frac{x^H}{(1 + x)^N}.$$

(b) Using the result found in part (a), calculate and tabulate the values of $P_N(H)$ for $H = 1, \ldots, 6$ when $N = 6$, $p = \frac{3}{4}, q = \frac{1}{4}$.

(c) Make a graph of the distribution function $P_6(H)$ as a function of H. Locate on your graph the position of $\overline{H}$, as well as the position of $\overline{H} + \Delta$ and $\overline{H} - \Delta$, where Δ is the rms width of the distribution.

(d) Find the total probability that H is within the range $\overline{H} - \Delta < H < \overline{H} + \Delta$.

19. **Traffic lights.** A commuter has to pass 10 unsynchronized traffic lights on his way from home to work. He finds that, on the average, he has to stop at four of the 10 lights.

(a) Show that the probability distribution $P_{10}(s)$ of being stopped s times by a light is a Bernoulli distribution.

(b) Find the values of $p =$ probability of being stopped at a single light, and of $q = 1 - p$.

(c) Find the probabilities that the commuter:

  (i) arrives at work without being stopped even once; and

  (ii) is stopped by *every* light.

(d) Assuming the driver commutes 250 days of the year, how often does it happen that he arrives at work without being stopped by a light?

20. **Properties of binomial coefficients.** The binomial coefficients $\binom{n}{k}$ are defined as

$$\binom{n}{k} = \frac{n!}{k!\,(n-k)!} \qquad (0 \le k \le n).$$

(a) Show that $\binom{n}{k} = \binom{n}{n-k}$ (symmetry property of the binomial coefficients).

(b) The binomial coefficients occur in the expansion of $(1 + x)^n$:

$$(1 + x)^n = \sum_{k=0}^{n} \binom{n}{k} x^k.$$

Writing $(1 + x)^n = (1 + x)^{n_1} \cdot (1 + x)^{n_2}$, $n_1 + n_2 = n$, and expanding all three factors in powers of x, show that

$$\sum_{k'} \binom{n_1}{k'}\binom{n_2}{k - k'} = \binom{n_1 + n_2}{k}.$$

(c) As a special case of (b), show that

$$\sum_k \binom{n}{k}^2 = \binom{2n}{n}.$$

(d) An application of (c) is given by following problem. Two players independently toss a coin N times. What is the probability that they will come up with the *same* number of heads H? (regardless of the value of H). Find the numerical answer for $N = 3$.

21. A rifleman with a lot of target practice hits a bullseye in 20% of his attempts, on the average.

(a) Find the probability that he hits a bullseye:

(i) once in five shots; and

(ii) twice in 10 shots.

To answer these questions, you must establish the fact that the probability of H hits (bullseyes) in N attempts is a Bernoulli distribution. What are the values of p and q?

(b) Analyze in detail the reason why the probability for (ii) is smaller than that for (i).

22. Density fluctuations in a gas. A container of volume V_0 contains N molecules of a gas. Consider a *small* volume element ΔV inside V_0: $\Delta V \ll V_0$. The probability p that any particular molecule is found inside ΔV is

$$p = \frac{\Delta V}{V_0},$$

$q = 1 - p$ is then the probability for the molecule being outside ΔV.

(a) Show that the probability $P_N(N_1)$, that N_1 of the N molecules are inside ΔV, is given by

$$P_N(N_1) = \frac{N!}{N_1!\,(N - N_1)!} p^{N_1} q^{N - N_1}.$$

(b) Find the average number $\overline{N}_1$ of molecules in ΔV.

(c) Find the mean square fluctuation

$$\Delta^2 = \overline{(N_1 - \overline{N}_1)^2}$$

of the number of particles in ΔV. Find an expression for Δ^2 in terms of the *mean* density $C = (N/V_0)$, the volume ΔV, and the mean number $\overline{N}_1$ of particles in ΔV.

(d) Find the ratio between the rms fluctuation Δ and the mean number $\overline{N}_1$ of particles in ΔV.

(e) The ratio $(\Delta/\overline{N}_1)$ determines the accuracy with which the local density of the gas, in volume element ΔV, can be defined. Let ΔV be a cube with sides of length $l = 10^{-5}$ cm, which is the order of the mean free path of gas molecules at atmospheric pressure and room temperature. In this case, the mean density C is 2.7×10^{19} particles/cm^3. Find the value of $\Delta/\overline{N}_1$ for the above value of ΔV.

23. For families with four children, the following data on sex distribution are given by Geissler (see [4]):

4 boys	8,628
3 boys	31,611
2 boys	44,793
1 boy	28,101
all girls	7,004
total	120,137

(a) Determine the mean number $\mathcal{N}_B$ of boys, and deduce from it a value for p, the a priori probability for "boy," by means of

$$\overline{\mathcal{N}}_B = p\mathcal{N}.$$

(b) Using this value of p, construct the probability distribution $P_4(B)$, where B is the number of boys among the four children. Compare $P_4(B)$ with the experimental values of $(\mathcal{N}_B)/\mathcal{N}$.

24. Experimental determination of the probability $P(1)$ for throwing a "1" with a die: This exercise is an application of (1-17). In the present problem

$p = P(1)$ is the probability for throwing a "1."

$q = P(\text{not "1"})$ is $1 - p$, the probability for throwing "2 or 3 or 4 or 5 or 6."

The question is: How often must the die be thrown to establish the value of p (which we "expect" to be $\frac{1}{6}$), with an accuracy of 5%, with 75% confidence? Let $\mathcal{N}$ be the required number of trials, and $\mathcal{N}_1$ the number of outcomes "1." We want

$$d = \left| \frac{\mathcal{N}_1}{\mathcal{N}} - p \right| \quad \text{to be less than or equal to} \quad \varepsilon = \left(\frac{5}{100} \right) p,$$

and we want the probability $P(d \le \varepsilon)$ that d is smaller than or equal to ε to be at least $\frac{3}{4}$. The main point in this problem is not the algebra, but the understanding of the question asked. To get an estimated value of the needed number $\mathcal{N}$ of throws, you may insert the approximate value $p \sim \frac{1}{6}$ on the right-hand side of the equation. Find a numerical estimate for the required number $\mathcal{N}$ of trials.

25. In the game played by the Chevalier de Méré, and in which he was losing money, the probability p for success was

$$p = 0.4914 = 0.5 - 0.0086.$$

How often would he have to play to find out, with 75% confidence, that the probability for winning was less than 50%?

   Hint: Let $\mathcal{N}$ be the number of games played, and $\mathcal{N}_s$ the number of successes (for the Chevalier). There will be a "discrepancy" d:

$$d = \left| \left(\frac{\mathcal{N}_s}{\mathcal{N}} \right) - 0.4914 \right|.$$

To be 75% confident that p is less than $\frac{1}{2}$, you want to be 75% certain that d is less than $\varepsilon = 0.0086$. On that basis, estimate $\mathcal{N}$. Also estimate the time the Chevalier had to spend playing before he realized that p was less than $\frac{1}{2}$.

   Note: Equation (1-17) is the simplest, but by no means the *best* estimate for the number $\mathcal{N}$ needed. A more accurate estimate, based on the Gaussian form of the probability distribution of the number $\mathcal{N}_s$ of success leads, for the 75% confidence level, to a value of $\mathcal{N}$ about three times smaller than (1-17).

26. The mean square length of a polymer chain. Mean square length of randomly coiled polymer chain. Consider a chain of n segments of length l each. We

may consider each segment i as a vector $\vec{l}_i$. The vector $\vec{L}$, giving the relative position of the end points, is then the *sum* of all segment vectors

$$\vec{L} = \vec{l}_1 + \vec{l}_2 + \vec{l}_3 + \cdots + \vec{l}_N.$$

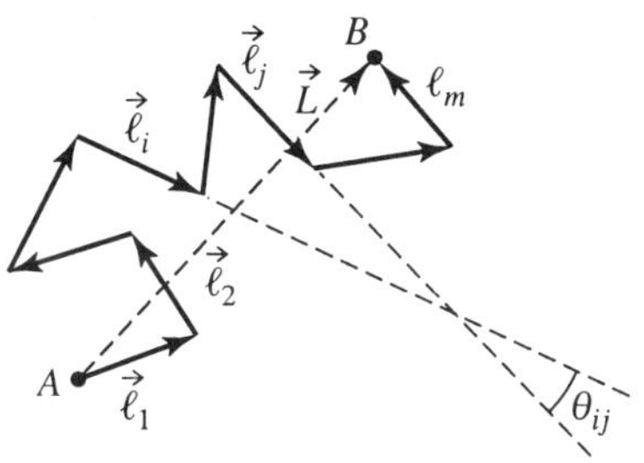

The square of the end-to-end distance, D^2_{AB}, is then just the square length of $\vec{L}$, and expressed as the scalar product of $\vec{L}$ with itself

$$D^2_{AB} = (\vec{L} \cdot \vec{L}).$$

The scalar product of a sum of vectors can be "multiplied out"

$$\vec{L} \cdot \vec{L} = (\vec{l}_1 + \vec{l}_2 + \cdots + \vec{l}_N) \cdot (\vec{l}_1 + \vec{l}_2 + \cdots + \vec{l}_N)$$

$$= \left(\sum_i \vec{l}_i\right) \cdot \left(\sum_j \vec{l}_j\right) = \sum_{ij} (\vec{l}_i \cdot \vec{l}_j).$$

(a) Carrying this further, show that

$$D^2_{AB} = Nl^2 + 2\sum_i \left\{ \vec{l}_i \cdot \vec{l}_{i+1} + \vec{l}_i \cdot \vec{l}_{i+2} + \cdots \right\}.$$

(b) Now consider the mean square distance $\overline{D^2_{AB}}$. Using $(\vec{l}_i \cdot \vec{l}_j) = l^2 \cos(\theta_{ij})$, θ_{ij} being the angle between segment i and segment j (see figure), show that:

(i) $\overline{D^2_{AB}} = Nl^2$ if any pair of segments i, j have a random orientation relative to each other; and

(ii) $\overline{D^2_{AB}} = N^2 l^2$ if all segments are parallel.

(c) In most chain polymers, the angle between *consecutive segments* is fixed: $\theta_{i,i+1} = \alpha = $ constant. Try to show (by a geometric construction) that in this case the mean value of $(\vec{l}_i \cdot \vec{l}_{i+2})$ for the *next to nearest* segments is

$$\overline{(\vec{l}_i \cdot \vec{l}_{i+2})} = l^2 \cos^2 \alpha.$$

For this, consult Figure 1.14 in the text.

(d) Show that there are $(N-1)$ pairs of nearest neighbors, $(N-2)$ pairs of next to nearest neighbors, etc., and that

$$\overline{D_{AB}^2} = Nl^2 + 2(N-1)l^2 \cos \alpha + 2(N-2)l^2 \cos^2 \alpha + \cdots$$
$$+ \text{ contribution of more distant neighbors.}$$

NB: For large N, this series can be approximately summed and gives

$$Nl^2 \frac{(1 + \cos \alpha)}{(1 - \cos \alpha)},$$

provided $\alpha \gg 1/\sqrt{N}$.

27. The Chebyshev inequality. Let $P(m)$ be a probability distribution with a mean $\overline{m}$ and a variance Δ^2. Furthermore, let us denote as $Q(h)$ the total probability that m will be found anywhere in the range $\overline{m} - h\Delta \leq m \leq \overline{m} + h\Delta$. (Here h is any number greater than unity, and

$$Q(h) \equiv \sum_{\overline{m}-h\Delta}^{\overline{m}+h\Delta} P(m)).$$

Chebyshev's inequality simply states that

$$Q(h) \geq \left(1 - \frac{1}{h^2}\right). \tag{1}$$

Thus $(1 - 1/h^2)$ is a lower limit to $Q(h)$. This lower limit approaches Q more accurately as $h \gg 1$.

(a) Prove the correctness of (1).
   Hint: Start from the definition of the variance as

$$\Delta^2 = \sum_{m=0}^{M} (m - \overline{m})^2 P(m).$$

Then show that the sum on the right-hand side is certainly greater than

$$(h\Delta)^2 \left(\sum_{m=0}^{\overline{m}-h\Delta} P(m) + \sum_{m=\overline{m}+h\Delta}^{M} P(m) \right).$$

By relating the quantity in parentheses to $Q(h)$, you can obtain (1).

(b) Suppose $P(m)$ is the Bernoulli distribution $(P_N(m))$ with $\overline{m} = pN$ and $p \neq q$. Show that the probability $(Q(N\varepsilon))$, that m will have a value anywhere in the region $\overline{m} - N\varepsilon < \overline{m} < \overline{m} + N\varepsilon$, satisfies the inequality

$$Q(N\varepsilon) \geq \left(1 - \frac{pq}{N\varepsilon^2} \right).$$

28. In this problem you will prove the correctness of (1-17). By definition the a priori probability p that a Bernoulli trial will register a success is

$$p = \lim_{\mathcal{N} \to \infty} \frac{\mathcal{N}_s}{\mathcal{N}}.$$

Here

$\mathcal{N}$ is the number of trials,

$\mathcal{N}_s$ is the number of successes.

For a finite number of trials the values of $\mathcal{N}_s$ form a Bernoulli distribution $P_{\mathcal{N}}(\mathcal{N}_s)$. Denote $Q(N\varepsilon)$ as the probability that after $\mathcal{N}$ trials $(\mathcal{N}_s/\mathcal{N})$ will differ from p by a number between $p - \varepsilon$ and $p + \varepsilon$.

(a) Show, using the Chebyshev inequality (Problem 27) that the number $\mathcal{N}$ of trials required to determine p within $p \pm \varepsilon$ with a confidence level Q is given by

$$N\varepsilon^2 = \frac{pq}{(1 - Q(N\varepsilon))}.$$

This is (1-17).

(b) Show that the fractional error ε/p with which p can be determined after $\mathcal{N}$ trials with a confidence level Q is just

$$\left(\frac{\varepsilon}{p}\right) = \left(\sqrt{\frac{q}{p}}\right)\sqrt{\frac{1}{\mathcal{N}(1-Q)}}$$

provided that the Chebyshev inequality is used as an approximation to the correct value for $Q(\mathcal{N}\varepsilon)$.

Diffusion and Transport Processes

2.1 Molecular Movement and the Physical Properties of Gases: A Short Survey

In this introductory section we review briefly how the physical properties of gases are interpreted from a molecular kinetic point of view. This view is basic to the understanding of the process of diffusion which will be the main topic of this chapter.

2.1.A. Ideal Gas Law. Kelvin Temperature. Avogadro's Number

It is well known that gases such as hydrogen, oxygen, nitrogen, and helium all obey (except under extreme conditions) an equation known as the "ideal gas law," which states that the product of the gas pressure p and molar volume V_M are only a function of the temperature t:

$$pV_M = f(t). \tag{2-1}$$

The function $f(t)$ is the *same* for all gases; this fact has been used to define a temperature scale, T, in which $f(t)$ is *proportional* to this temperature

$$f(t) = RT. \tag{2-2}$$

The temperature T so defined is the ideal gas or Kelvin temperature. The choice of the proportionality constant R is a matter of convention, and was determined by the condition that a temperature *increment* ΔT on the Kelvin scale should be equal to the increment Δt on the Centigrade (or Celsius) scale

$$\Delta T = \Delta t$$

At temperature $t = 0\,°C$ (freezing point of water), a small increment Δt of temperature produces a fractional increment (pV_M) proportional to Δt:

$$\frac{\Delta(pV_M)}{(pV_M)_{t=0}} = \alpha\,\Delta t \tag{2-3}$$

(Gay-Lussac's law). The "expansion coefficient" has the value

$$\alpha = \left(\frac{1}{273}\right) \Big/ \text{degree Centigrade.}$$

According to (2-1) and (2-2), we can write (2-3) in the form

$$\frac{\Delta(pV_M)}{pV_M} = \frac{R\Delta T}{RT_0} = \frac{\Delta T}{T_0}, \tag{2-4}$$

where T_0 is the Kelvin temperature corresponding to $t = 0$. Comparison with (2-3) shows that

$$T_0 = 273\,°C$$

and since $\Delta t = \Delta T$, the Kelvin scale (T) and the Centigrade scale (t) are related by

$$T = t + 273\,°C.$$

Thus, room temperature ($t = 20\,°C$ or $68\,°F$) corresponds to $293\,°K$.

The constant R in (2-2) is known as the *universal gas constant*. It has the dimension of an energy per unit temperature, and its value is

$$R = 82.06 \text{ atm cm}^3/°K$$
$$= 8.314 \times 10^7 \text{ erg}/°K$$
$$= 8.314 \text{ J}/°K$$
$$= 1.987 \text{ cal}/°K.$$

If the amount of gas considered is n moles rather than 1 mole, then at a given temperature T the value of (pV) is n times the value of pV_M. We can thus state the ideal gas law in the general form

$$pV = nRT, \tag{2-5}$$

n being the number of moles.

Let us recall that the concept of the mole arose from the stoichiometric relations observed in chemical reactions; in reactions of gases these relations manifested themselves as simple volume relations. This is the experimental basis for the hypothesis, already stated in 1811 by Avogadro, that equal volumes of gases contained, under conditions of equal pressure and temperature, have the same number of *molecules*. This statement is now long past the stage of a hypothesis. But in its time it stimulated attempts to determine what that number of molecules per mole actually is. In fact, mutually consistent determinations of that number by various means at the beginning of this century were crucial for establishing the general acceptance of Avogadro's view. We shall come back to the question of how that number can be determined, and give here only its value

$$N_0 = 6.023 \times 10^{23} \text{ molecules/mol},$$

N_0 is called Avogadro's number.

2.1.B.   Mean Kinetic Energy of a Molecule. The Boltzmann Constant

In the kinetic picture of the ideal gas developed by Boltzmann and Maxwell in the second part of the nineteenth century, the gas pressure is viewed as being due to the bombardment of the container walls by the gas molecules. It was shown by Clausius that if one takes this view, the product pV of pressure p and gas volume V had to be equal to two-thirds of the total kinetic energy of the gas molecules

$$pV = \frac{2}{3} \sum \frac{m}{2} v^2, \tag{2-6}$$

where the sum extends over all molecules in the gas; m and v are mass and speed of an individual molecule, respectively. An elementary derivation of this equation is given in Appendix 2.A1.

By introducing the number $N = nN_0$ of molecules, and using the gas law [(2-5)], we can derive from (2-6) an expression for the *average* kinetic energy of a molecule in a gas at temperature T:

$$\overline{\tfrac{1}{2}mv^2} = \frac{1}{N}\sum \tfrac{1}{2}mv^2 = \frac{1}{N_0 n}\tfrac{3}{2}pV = \frac{1}{N_0 n}\tfrac{3}{2}nRT$$

or

$$\overline{\tfrac{1}{2}mv^2} = \tfrac{3}{2}\left(\frac{R}{N_0}\right)T = \tfrac{3}{2}kT. \tag{2-7a}$$

This is a most important result that deserves a few comments. It states that the mean kinetic energy of a gas molecule is proportional to the Kelvin temperature T of the gas. The proportionality constant k introduced in (2-7a) is known as the *Boltzmann constant*, and is the ratio of R and N_0. Using the numerical values previously given for these two constants, we find that

$$k = 1.381 \times 10^{-16} \text{ erg/}^{\circ}\text{K}.$$

Just as R and N_0, k is a *universal* constant, independent of the particular ideal gas. This means specifically that at a given temperature T the mean kinetic energy of a molecule of hydrogen is the same as that of oxygen or any other ideal gas. Since the two molecules mentioned have different masses, $m(O_2) = 16m(H_2)$, it follows that the mean square velocity of an oxygen molecule must be 16 times smaller than that of a hydrogen molecule for the same temperature.

    This result also gives us already some insight into the nature of thermal equilibrium. It is known that any two bodies in contact, and therefore capable of exchanging heat, will do so until their temperatures have become equal. Applied to two gases, separated by a heat conducting barrier, this fact of experience can now be described as follows: Two gases in thermal contact will exchange energy until the mean kinetic energies per molecule of the two gases have become equal.

    Put in this form, this is a highly suggestive statement. It suggests that the state of thermal equilibrium of a system is characterized by the fact that on average the total available thermal energy is *equally shared* by the constituent molecules or atoms.

    A general discussion of the topic of thermal equilibrium (in the general sense, that which includes chemical equilibrium) will be a given in Chapter 4 of this volume. At present we will confine ourselves to just a few additional aspects of thermal equilibrium which will be needed in this chapter.

2.1.C. The Equipartition Law. Specific Heats

We shall, in this section, extend the fundamental result [(2-7)], and then describe in which way this result and its generalizations may be verified by observation.

In order to introduce one first extension, let us consider a mixture of two gases; let it be a mixture of N_1 molecules of mass m_1, and N_2 molecules of mass m_2, and let $N_1 + N_2 = N_0$ so that we have one mole of the mixture. It is well known experimentally that this mixture satisfies the ideal gas law

$$pV = RT.$$

But if we now apply Clausius' formula (2-6) to this mixture, we obtain instead of (2-7):

$$\left(\frac{N_1}{N_0}\right) \tfrac{1}{2}\overline{m_1 v_1^2} + \left(\frac{N_2}{N_0}\right) \tfrac{1}{2}\overline{m_2 v_2^2} = \tfrac{3}{2}kT. \tag{2-8}$$

This result alone does *not* allow us to conclude that the mean kinetic energies of the molecules of gas 1 and gas 2 are *separately* equal to $\tfrac{3}{2}kT$. We would expect this to be so, but it does *not follow* from (2-8). To settle the issue we shall, in Appendix 2.A2, demonstrate a result, a weak form of the so-called *Equipartition Law*. This law states that *any* object of mass M suspended in a gas temperature T will acquire (as a result of collisions of gas molecules) a kinetic energy whose *mean value* is

$$\frac{M}{2}\overline{v^2} = \tfrac{3}{2}kT. \tag{2-9}$$

This object may be a gas molecule of a different species, and this resolves the question left unanswered by (2-8). But it may also be a grain of dust, or a pollen grain, or a virus particle, anything that manages to stay suspended in the gas.

This is a result of far-reaching implications. Dust grains, etc., can be *observed* in the microscope, and make it possible to verify directly some of the consequences of the result [(2-9)], such as the random motion (Brownian motion) of individual particles, or the sedimentation equilibrium of a suspension of such particles.

Without any attempt to give a proof, we state some further extension of the equipartition law. Consider a diatomic molecule, like oxygen, O_2. The mean energy $\tfrac{3}{2}kT$ in this case refers to the motion of the molecule as a whole (the so-called translational motion). But the molecule is also capable of oscillation and rotation about its center of mass. (See Figure 2.1) In Figure 2.1 we show the three perpendicular velocity components $\vec{v}_1$, $\vec{v}_2$, and $\vec{v}_3$ of internal motion for one atom. The

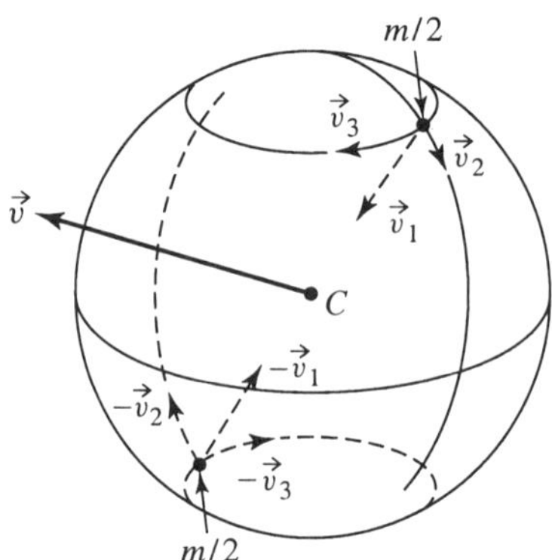

Figure 2.1. Diatomic molecules showing the three velocity components $\vec{v}_1$ (oscillation), $\vec{v}_2$ and $\vec{v}_3$ (rotation), and the translational (center of mass) velocity $\vec{v}$.

velocities of the second atom are $-\vec{v}_1$, $-\vec{v}_2$, and $-\vec{v}_3$. The pair $(\vec{v}_1, -\vec{v}_1)$ describes an *oscillation*, and pairs $(\vec{v}_2, -\vec{v}_2)$, $(\vec{v}_3, -\vec{v}_3)$ describe a *rotation* of the molecule.

The generalized equipartition law now states that the mean kinetic energy associated with each *velocity component* is equal to $\frac{1}{2}kT$, whether it be translational, oscillatory, or rotational. Since the translational velocity $\vec{v}$ has three components v_x, v_y, v_z, and the total translational kinetic energy is

$$\frac{m}{2}v^2 = \frac{m}{2}(v_x^2 + v_y^2 + v_z^2)$$

and its mean value, therefore, $3 \times \frac{1}{2}kT = \frac{3}{2}kT$. In the case of translational motion we can get the result $(m/2)\overline{v_x^2} = \frac{1}{2}kT$ from (2-6) by reasoning backward since it is intuitively clear that the mean values of v_x^2, v_y^2, and v_z^2 will all be the same, and equal to one-third of $\overline{v^2}$ so that indeed $(m/2)\overline{v_x^2} = \frac{1}{2}(m/2)\overline{v^2} = \frac{1}{2}kT$.

As applied to the oscillation and rotation of a diatomic molecule, it is considerably less intuitive that there should be a mean energy of $\frac{1}{2}kT$ associated with every velocity component, and there is no elementary proof of it. Since there are two velocity components for rotation (see Figure 2.1), the mean rotational kinetic energy of the molecule is

$$\overline{E}_{\text{rot}} = 2\left(\tfrac{1}{2}kT\right) = kT.$$

The vibrational or oscillatory motion takes place along the line joining the two atoms, and a mean kinetic energy of $\frac{1}{2}kT$ is invested in this oscillation. But we must not forget that there is an oscillatory potential energy. This latter, as we know from *Mechanics* (Volume I, Section 2.7.B) is equal, on average, to the kinetic energy of oscillation, and hence also has an average value of $\frac{1}{2}kT$. We can therefore establish the following tabulation of mean energies for a diatomic molecule in a gas at temperature T:

$$\text{mean translational energy } \frac{m}{2}\overline{v^2} = \tfrac{3}{2}kT, \tag{2-7b}$$

$$\text{mean rotational energy} = kT, \tag{2-10}$$

$$\text{mean energy of oscillation} = kT. \tag{2-11}$$

How do we verify these assertions? Simply by measuring the specific heat C_V of the gas. C_V, the specific heat at constant volume V, is defined as the ratio between heat input ΔQ per mole and associated temperature change ΔT:

$$\Delta Q = C_V \Delta T. \tag{2-12}$$

By the first law of thermodynamics we know that a heat input ΔQ is equal to the energy change ΔU of the system if no work is done by the system. To fulfil this latter condition, the gas is heated at constant volume. Then, we have indeed

$$\Delta U = \Delta Q = C_V \Delta T. \tag{2-13}$$

In a gas of single atoms, like helium or argon, the only energy item in U is the translational kinetic energy of the atom. In this case, the energy per mole is

$$U = N_0 \cdot \tfrac{3}{2}kT = \tfrac{3}{2}RT$$

and therefore

$$\Delta U = \tfrac{3}{2}R\Delta T. \tag{2-14}$$

Comparison with (2-13) shows that the specific heat at constant volume for helium and argon should be

$$C_V = \tfrac{3}{2}R = 3 \text{ cal/}°\text{K}.$$

This is in excellent agreement with the data for these gases.

For a gas of diatomic molecules, like oxygen, the internal energy, at temperature T, is expected to be [(2-7b), (2-10),(2-11)]:

$$U = N_0 \left(\tfrac{3}{2}kT + kT + kT\right) = \tfrac{7}{2}RT, \tag{2-15}$$

giving a specific heat

$$C_V = \tfrac{7}{2}R = 7 \text{ cal/}^\circ\text{K}.$$

Data for oxygen give only $C_V \simeq 5$ cal/°K, or $\tfrac{5}{2}R$, at room temperature. This discrepancy is now explained as a *quantum* effect; it has its roots in the fact that the energy of oscillation of the molecule is *quantized*: instead of a continuous range of energies of oscillation, only discrete energy states (energy levels) do in fact exist. This fact of energy quantization does not altogether invalidate the equipartition law, but restricts its validity to cases where the separation ΔE between energy levels is small compared to kT:

$$\Delta E \ll kT.$$

This condition is *not* satisfied for most diatomic molecules at room temperature. What happens in this case is that the mean energy of oscillation remains far below the value kT; the oscillatory motion is "frozen." By frozen we mean that the vibrational contribution to the internal energy of the molecule is a *constant* independent of temperature (in the vicinity of room temperature). As a result $U = N_0(3kT/2 + kT + \text{constant})$. Thus $C_V = (\Delta U/\Delta T) = N_0\tfrac{5}{2}k = \tfrac{5}{2}R$ cal/°K.

As a final item, we must raise and answer the question of the validity of the equipartition law for liquids and solids. We shall, in later parts of this chapter, deal extensively with solutions of molecules or suspensions of small particles, in water or other fluids. We shall attempt to give credible arguments for the view that a molecule in solution, or a particle suspended in a fluid at temperature T will again have a mean translational kinetic energy $\tfrac{3}{2}kT$. In other words, (2-9) holds also for particles in a fluid!

In support of this view we can analyze data on the specific heats of solids and liquids which shows that the individual atoms of the solid or liquid have energies consistent with the law of equipartition of energy. Under these circumstances it becomes rather plausible that any foreign particle or molecule suspended in a fluid will also get its due share of $\tfrac{3}{2}kT$ for its translational kinetic energy.

For solids it has long been known that, at high temperatures, the specific heat per mole (C_V) of solids has rather uniformly the value

$$C_V \cong 6 \text{ cal/}^\circ\text{K} = 3R.$$

This is the so-called Law of Dulong and Petit. This is consistent with the view that each atom is oscillating in three dimensions with a mean translational kinetic energy of $(\frac{1}{2}kT)$ for each direction of its motion. This amounts to $\frac{3}{2}kT$ for kinetic energy. Since these atoms are oscillating around their equilibrium positions, there is also an equal contribution of $3(kT/2)$ from the potential energy. Thus, the total internal energy is $3kT$ per atom, and the specific heat per mole is $3R$: the Dulong–Petit value. (For temperatures T considerably below room temperature, the value of C_V for solids drops, and reaches $C_V = 0$ at $T = 0$. This again is a quantum effect and corresponds to the "freezing out" or the settling of the atomic oscillations into their lowest energy state.)

In the case of fluids it is not so clear whether one may regard each atom as an *oscillator* with $\frac{3}{2}kT$ for its kinetic energy and $\frac{3}{2}kT$ for its potential energy. Nevertheless, this view does apply to a number of liquids. Water, for instance, has a heat capacity of 1 cal/g (this defines the caloric!), and hence a specific heat C_V of 18 cal/mol of water = $9R$. But since the water molecule contains three atoms, this amounts to $3R$ per mole of atomic constituents, which is the Dulong–Petit value. A similar situation prevails for some organic solvents like liquid carbon tetrachloride, CCl_4, which has a specific heat of 30 cal/mol = $15R$. Since there are five atoms per molecule, this reduces to $3R$ per mole of atoms.

We can *summarize* these results by stating that there is wide evidence in support of the statement that in any material at temperature T, each atomic constituent has a mean translational kinetic energy of $\frac{3}{2}kT$ ($\frac{1}{2}kT$ for each of the three velocity components) and an additional mean potential energy of $\frac{1}{2}kT$ for each direction of motion, along which there is a restoring force resulting in an oscillatory motion.

Any isolated particle, whatever its mass, suspended in a gas or liquid has a mean translational kinetic energy of motion equal to $\frac{3}{2}kT$.

2.1.D.   Random Motion of a Gas Molecule, Root Mean Square Velocity, Mean Free Path, and Collision Frequency

If we ask for the motion of an *individual* gas molecule, we see at once that Newton's second law is not of much help. Any given molecule will in a short time collide with millions of others, and as a result perform a zigzag motion through the gas. And since every gas molecule has a similar fate, and there is a very large number of them, a *statistical* description of this motion is appropriate. In this description, we consider the path of any molecule as a *random walk*, consisting of straight line segments terminated by a collision which leads to an abrupt and unpredictable change in direction. (See Figure 2.2.) The straight segments between two

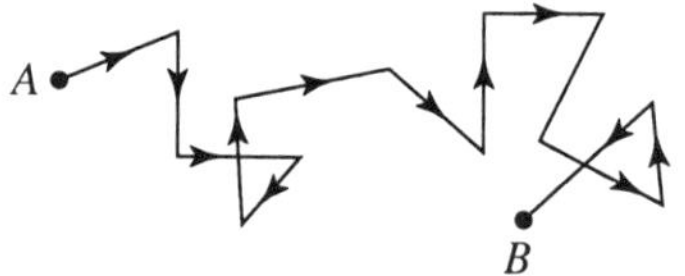

Figure 2.2. Path of a molecule in a gas.

consecutive collisions will vary in length, but have a well-defined average value l_0, the so-called *mean free path*. The speed with which the molecule goes through its zigzag path is also variable; in every collision the molecule may either pick up or lose some energy, and hence speed, but again, a well-defined mean value exists. This mean speed is close to the root mean square (rms) speed (but not exactly equal to it); the latter has the advantage that it can be inferred directly from the temperature through (2-7):

$$v_{\text{rms}} = \sqrt{\overline{v^2}} = \sqrt{\frac{3kT}{m}}. \tag{2-16}$$

In (2-16), m is the mass of one molecule, and is related to the so-called "molecular weight" M, or mass of one mole, by $m = M/N_0$. Hence, v_{rms} may also be written as

$$v_{\text{rms}} = \sqrt{\frac{3RT}{M}}. \tag{2-17}$$

Table 2.1 gives some values of v_{rms} for various molecules.

The first five molecules listed in Table 2.1 are well-known gases with molecular weights ranging from 2 g/mol for hydrogen to 200 g/mol for mercury (its vapor is a monoatomic gas). It is good to remember that the average root mean square velocity of an air molecule is of the order of 500 m/s or 1125 mph at room temperature.

The bottom part of Table 2.1 lists macromolecules with molecular weights from 10^4 to 10^6 g/mol, and viruses. Their rms velocities are calculated by using the equipartition law which attributes to them a mean kinetic energy of $\frac{3}{2}kT$ if they are suspended in a gas of temperature T. This holds irrespective of their mass. It is astonishing to see how large the rms velocities of such "large" objects still are. For the Tobacco Mosaic Virus (TMV), whose molecular weight is 5×10^7 g/mol, one

Table 2.1. Root Mean Square Velocities for Various Molecules at $t = 15\,°\mathrm{C}$ ($T = 288\,°\mathrm{K}$).

Molecule		m (g)	M (g/mol)	v_{rms} (m/s)
Hydrogen	H_2	3.35×10^{-24}	2.016	1880
Helium	He	6.65×10^{-24}	4.002	1340
Nitrogen	N_2	4.65×10^{-23}	28.02	506
Oxygen	O_2	5.32×10^{-23}	32.0	474
Mercury	Hg	3.33×10^{-22}	200.6	186
Macromolecules		1.67×10^{-20}	10^4	26 m/s
		1.67×10^{-18}	10^6	2.6 m/s
Viruses		1.67×10^{-16}	10^8	26 cm/s
		1.67×10^{-14}	10^{10}	2.6 cm/s

finds $v_{\mathrm{rms}} \sim 35$ cm/s. Like any gas molecule, they too, therefore, perform an irregular, random motion through the gas (or liquid) in which they are suspended. This is the famous "Brownian Motion" first observed in 1827 by the botanist Robert Brown with pollen grains. Its correct interpretation was first given by Einstein in 1905 when he showed that this motion was a consequence of the equipartition law applied to these particles [1].

Our next task is to form an idea of the value of the mean free path length of this random motion. We can easily *estimate* it as follows: Consider a spherical particle of radius a. In moving a distance l, it "sweeps up" a cylindrical volume of size (see Figure 2.3):

$$V_l = \pi a^2 l. \tag{2-18}$$

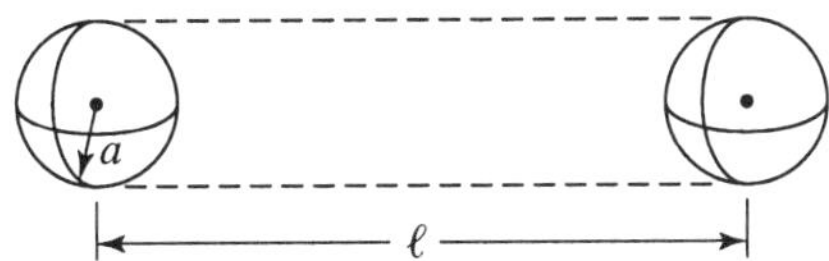

Figure 2.3. Volume swept up by a spherical object of radius a, moving a distance l

If l is the mean free path $\bar{l}$, then the volume V_l will contain, on average, just one gas molecule. So $\pi a^2 \bar{l}$ must be equal to the gas volume per particle, namely, V_M/N_0 where V_M is the molar volume. At $0\,°C$ and $p = 1$ atm, V_M is 22.4 l; this gives

$$\pi a^2 \bar{l} = \frac{22.4 \times 10^3 \text{ cm}^3}{6 \times 10^{23}} = 3.7 \times 10^{-20} \text{ cm}^3. \tag{2-19}$$

On further thought we realize that this argument would be correct if the object of radius a were to collide with negligibly small ("point-like") gas molecules. If the latter have a radius a' themselves, then (2-19) must be amended by augmenting a to $(a + a')$:

$$\pi (a + a')^2 \bar{l} = \frac{V_M}{N_0} = 3.7 \times 10^{-20} \text{ cm}^3. \tag{2-20}$$

Gas molecules have radii of order $r \sim 10^{-8}$ cm, which gives as an estimate for the mean free path of a gas molecule

$$\bar{l} \simeq 10^{-5} \text{ cm} \quad \text{for } T = 273°, \qquad p = 1 \text{ atm.} \tag{2-21}$$

These preliminary arguments have given us two important parameters for our future analysis of the random walk: *the mean free path* [(2-21)] and the velocity of motion, as given in Table 2.1. Combining the two, we obtain an estimate of the *mean time interval* t_c between two consecutive collisions

$$t_c \sim \frac{\bar{l}}{v_{\text{rms}}}. \tag{2-22}$$

For oxygen and nitrogen and their mixture, air, we have $v_{\text{rms}} \sim 500$ m/s which gives, with (2-21):

$$t_c \simeq \frac{10^{-5} \text{ cm}}{0.5 \times 10^5 \text{ cm/s}} = 2 \times 10^{-10} \text{ s.} \tag{2-23}$$

This means that a gas molecule in air goes through about five billion collisions every second.

Considering now a typical zigzag path of an individual molecule, we ask the question: What distance does this molecule actually travel away from its point of

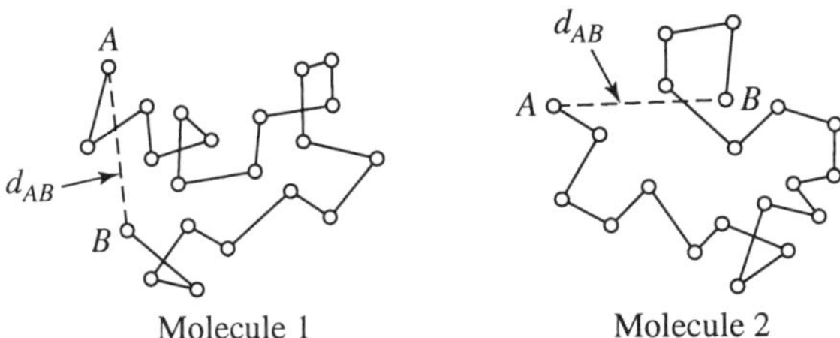

Figure 2.4. Distance d_{AB} traveled by a molecule in a fixed time interval.

departure in a given time interval, or equivalently, in a given number of collisions? This will, of course, be different for every molecule (see Figure 2.4), but if we sample a very large number of molecules, and *count* the fraction of them that travel a certain distance in a fixed time interval, we expect a result as in Figure 2.5: As we increase the number of molecules sampled, we expect the distribution in Figure 2.5 to become more and more smooth. This smooth distribution is then by definition the *probability distribution* of distances: It gives the probability that any given molecule will be found after an elapsed time t a distance between l and $l+\Delta l$ away from its point of departure. In the next section of this chapter, we proceed to establish the main features of this distribution, starting with a simplified model of the random walk process.

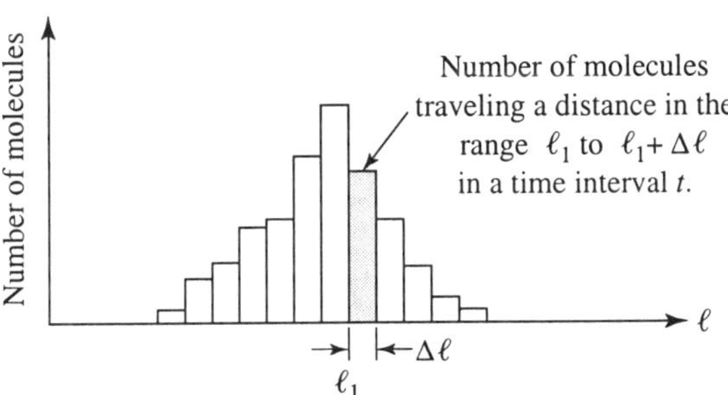

Figure 2.5. Result of an "experiment" tabulating the number of molecules traveling a certain distance in a fixed time interval.

2.2   Random Walk in One and Three Dimensions

2.2.A.   The Bernoulli Distribution for the Probability $P_N(x)$ of a Displacement x in N Steps

In this section we construct a one-dimensional, simplified *model* of the random walk. Despite this apparently unrealistic restriction to a motion along a straight line (one dimension), this model contains all the characteristic features of the full, three-dimensional case. In particular, it will help us to understand how it comes that the *mean square* displacement of a random walking particle increases only linearly with the number N of steps.

Let us now describe the model to be used for the random walk of a particle:

(i)   The particle moves on a straight line.

(ii) The motion consists of steps of length l. Each step is either to the right, or to the left, and right or left steps follow each other in a random fashion. It is like the walk of a drunkard along the curbstone of a sidewalk who after each step ponders whether to go forward or backward.

(iii) The probability for a step to the right is p, that for a step to the left, q. We shall here assume that $p = q = \frac{1}{2}$; that is, we consider an unbiased random walk.

It is apparent that with these rules a random walk of N steps is a Bernoulli process, like the tossing of a coin N times. Each specific outcome

$$htththhhtth \ldots t$$

of N coin tosses has its parallel here in a specific sequence of right or left steps, r or l:

$$rllrlrrrllr \ldots l.$$

Let us designate by R the total number of steps to the right in a sequence of N steps, and by L the total number of steps to the left. R and L are the analogues of H and T in the coin-tossing process.

The two processes, tossing a coin N times, and a random walk of N steps, thus have exactly the same structure or anatomy; they are *isomorphic* (of "identical form"). The recognition of isomorphisms is of great importance in science: Two isomorphic problems have the same solution. It is rather evident, for instance, that the models of a random folding of a polymer chain are isomorphic to suitable random walk models. The reader is well advised to keep this in mind, and to compare statements on coin tossing, chain folding, and random walk.

This being said, it should be understood that no further derivations are needed to state the result that $P_N(R, L)$, defined as the probability that in a random walk of N steps, R of them are to the right and $L = N - R$ of them to the left, is given by

$$P_N(R, L) = \frac{1}{2^N} \frac{N!}{R!\, L!} = \frac{1}{2^N} \binom{N}{R} \tag{2-24}$$

in full analogy to the probability $P_N(H, T)$ for H heads, T tail in N tosses of a coin.

What distance does the particle travel in N steps? If we lay out its path along the x-axis, and start at $x = 0$, then after N steps, R of which are to the right, L to the left, it will end up at position

$$x = (R - L)l \equiv ml. \tag{2-25}$$

Since $R + L = N$, and $(R - L) = m$, we can write

$$R = \left(\frac{N + m}{2}\right), \qquad L = \left(\frac{N - m}{2}\right), \tag{2-26}$$

and the probability $P_N(m)$, that the particle will be displaced by ml away from the origin, is obtained by inserting (2-26) into (2-24):

$$P_N(m) \equiv P_N\left(R = \frac{N + m}{2}, \; L = \frac{N - m}{2}\right) = \frac{1}{2^N} \frac{N!}{(N + m/2)!\,(N - m/2)!} \tag{2-27}$$

Observe that we had encountered this distribution already in Chapter 1 in the problem of the folded polymer chain.

In Figure 2.6 we plot this distribution again for a large value of N, and reiterate some of its characteristic features: This distribution is centered about $m = 0$, and therefore the mean value of m,

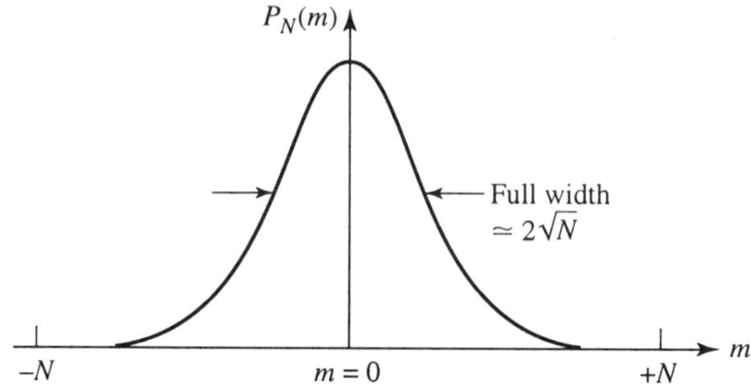

Figure 2.6. Bernoulli distribution $P_N(m)$ for large N.

$$\overline{m} = \sum m\, P_N(m), \tag{2-28}$$

is zero. The variance $\overline{(m - \overline{m})^2} = \overline{m^2}$ is equal to N, as we saw in Section 1.4:

$$\overline{m^2} = \sum_m m^2 P_N(m) = N. \tag{2-29}$$

Hence the root mean square of m, $\sqrt{\overline{m^2}}$, which is a good measure of the *half width* of the Bernoulli distribution, is equal to $\sqrt{N}$. This means that values of m significantly larger than $\sqrt{N}$ occur only with very small probability. Since for large values of N, $\sqrt{N}$ is a number much smaller than N (e.g., for $N = 10^6$, $\sqrt{N} = \frac{1}{1000}N$) the distribution $P_N(m)$ is very *narrowly* peaked about $m = 0$ for $N \gg 1$.

This fact is of profound importance. A molecule, although capable *in principle* of a displacement (Nl) in N steps, is *more likely* to move only a distance $\sqrt{N}\,l$, that is, a very much lesser distance. We shall show in Sections 2.2.D and 2.3 how this characteristic feature of the random walk can be verified experimentally.

Before we go into this, however, we must first rewrite the results obtained in this section by relating the number of steps N to the elapsed time t, so as to obtain a probability distribution for the displacement $x = ml$ in a given time t.

To do this, we also need a more convenient representation of $P_N(m)$ than the one offered by (2-27). We, therefore, develop in this next Section 2.2.B the so-called *Gaussian* form of $P_N(m)$. This will pave the way for obtaining a simple form for the probability $P(x_1 t)$ of a given displacement x in an elapsed time t.

2.2.B.   The Gaussian Form of the Bernoulli Distribution

We have already mentioned that the number of steps N that occur in molecular diffusion is very, very large, of the order of 5×10^9 steps per second. Also, the length of the individual steps is microscopically short. We, therefore, cannot count the number of steps N, nor can we determine the exact displacement $x_m = ml$ along the x-axis.

What we really would like to know is the probability that a particle, starting its random walk at $x = 0$ and time $t = 0$, will end up at time t somewhere in the interval of size Δx at a distance x away from the point of departure. This probability can be derived from the knowledge of $P_N(m)$.

As a first step toward this objective, we are answering a question that rather naturally arises upon contemplation of the distribution $P_N(m)$ as shown in Figure 2.6: Is it possible to find a smooth, *continuous* function of m that represents the *envelope* of the Bernoulli distribution? (See Figure 2.7.) This envelope should have the property that it is defined for *all* values of m, and for integer values closely approximates the value of $P_N(m)$. The advantage of using the envelope is that it allows us to describe the broadening of the distribution with increasing N as a continuous process in time and space. To construct this envelope, we need a formula which represents with sufficient accuracy the manner in which the value of the factorial $n!$ depends on the value of n. This is provided by the already mentioned *Stirling formula* for $n!$:

$$n! \cong \sqrt{2\pi n}\, n^n e^{-n} = \sqrt{2\pi n}\, e^{n \ln n - n}. \tag{2-30}$$

The relative error with which this formula approximates $n!$ is of the order $\frac{1}{12}n$ for large values of n. Its accuracy therefore improves with increasing values of n. Already at $n = 10$ it represents $n!$ with a better than 1% accuracy; and even

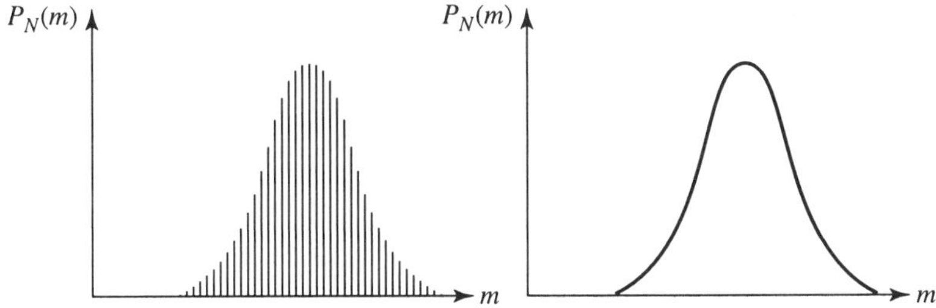

Figure 2.7. Bernoulli (a) distribution and (b) its envelope.

at $n = 1$ it predicts $1! = 0.92$. It is a truly amazing formula without which the description of random processes would be very much more difficult.

Before we apply (2-30) to the expression [(2-27)] for the probability distribution $P_N(m)$, we may do well to recall some properties of natural logarithms which will be needed in order to cast the expression for $P_N(m)$ into a simple form. The formulas that we shall need are

$$\ln a + \ln b = \ln(ab),$$
$$n \ln a = \ln(a^n), \tag{2-31}$$
$$\ln\left(\frac{1}{a}\right) = -\ln a.$$

In addition, we shall need the approximation, valid if $|\alpha|$ is a number small compared to 1:

$$\ln(1 + \alpha) \cong \alpha. \tag{2-32}$$

Let us then start with applying Stirling's formula to $P_N(R, L)$ as given by (2-24)

$$P_N(R, L) = \frac{1}{2^N} \frac{N!}{R!\,L!}.$$

Remember that $P_N(m)$ derives from this expression by the simple substitution: $R = (N + m)/2$, $L = (N - m)/2$. Inserting (2-30) into (2-24), we find

$$\frac{N!}{R!\,L!}\frac{1}{2^N} \cong \sqrt{\frac{N}{2\pi RL}}\, e^{N \ln N - R \ln R - L \ln L}\, e^{-N \ln 2},$$

where we used $\left(\frac{1}{2}\right)^N = 2^{-N} = e^{N \ln 2}$. Upon writing $N \ln N = (R + L) \ln N$, $N \ln 2 = (R + L) \ln 2$, and using (2-31), this can be rewritten as

$$P_N(R, L) \cong \sqrt{\frac{N}{2\pi RL}}\, e^{-R \ln(2R/N) - L \ln(2L/N)}. \tag{2-33}$$

At this point we introduce the variable $m = R - L$, (2-25), giving $2R = N + m$, $2L = N - m$, and hence

$$\frac{2R}{N} = 1 + \frac{m}{N}, \qquad \frac{2L}{N} = 1 - \frac{m}{N}.$$

This gives for the exponent in (2-33):

$$
\begin{aligned}
R \ln\left(\frac{2R}{N}\right) + L \ln\left(\frac{2L}{N}\right) &= \frac{N+m}{2} \ln\left(1 + \frac{m}{N}\right) + \frac{N-m}{2} \ln\left(1 - \frac{m}{N}\right) \\
&= \frac{N}{2} \ln\left(1 - \frac{m^2}{N^2}\right) + \frac{m}{2} \ln\left(1 + \frac{m}{N}\right) \\
&\quad - \frac{m}{2} \ln\left(1 - \frac{m}{N}\right).
\end{aligned}
\tag{2-34}
$$

We can simplify this further by observing that for very large values of N, we need an exact expression for $P_N(m)$ only for values of $|m|$ that are *small* compared to N: Indeed, we know that $P_N(m)$ gets very small if $|m|$ is only a few times $\sqrt{N}$; but "a few times $\sqrt{N}$" still represents a number $\ll N$ if N is really large. (See Figure 2.6.)

With this in mind we now apply the approximation formula [(2-32)] with $-(m/N)^2$ and $(\pm m/N)$ playing the role of the small number α:

$$\ln\left(1 - \frac{m^2}{N^2}\right) \cong -\frac{m^2}{N^2},$$

$$\ln\left(1 + \frac{m}{N}\right) \cong \frac{m}{N},$$

$$\ln\left(1 - \frac{m}{N}\right) \cong -\frac{m}{N}.$$

Inserted into (2-34), this gives

$$R \ln\left(\frac{2R}{N}\right) + L \ln\left(\frac{2L}{N}\right) \cong -\frac{m^2}{2N} + \frac{m^2}{2N} + \frac{m^2}{2N} = \frac{m^2}{2N} \tag{2-35}$$

In the same spirit of approximation we write

$$\sqrt{\frac{N}{2\pi RL}} = \sqrt{\frac{4N}{2\pi(N^2 - m^2)}} = \sqrt{\frac{2}{\pi N(1 - m^2/N^2)}} \cong \sqrt{\frac{2}{\pi N}}, \tag{2-36}$$

since $m^2/N^2 \ll 1$.

We now have all the pieces. Remember that we started with (2-33) for $P_N(R, L)$ which, if written in terms of $m = (R - L)$, becomes $P_N(m)$:

$$P_N(m) \equiv P_N\left(R = \frac{N+m}{2},\; L = \frac{N-m}{2}\right).$$

So, inserting (2-35) and (2-36) into (2-33), we find

$$P_N(m) \cong \sqrt{\frac{2}{\pi N}}\, e^{-m^2/2N}. \tag{2-37}$$

This is the surprisingly simple Gaussian approximation to the probability distribution function $P_N(m)$. It is valid for large values of N, and for values of (m) small compared to N. Yet, even for modest values of N, it is surprisingly accurate for most values of (m). This is shown in Table 2.2 [2].

Table 2.2. Exact Values of $P_N(m)$ and Gaussian Approximation for $N = 10$.

m	P_{10} (m)	Gaussian approx.
0	0.246	0.252
2	0.205	0.207
4	0.117	0.113
6	0.044	0.042
8	0.010	0.010
10	0.001	0.002

2.2.C. Space–Time Evolution of the Probability Distribution. The Diffusion Constant. The Mean Square Displacement as a Function of Time

We now proceed to carry out the second part of the program outlined in the beginning of Section 2.2.B. We will find the probability that after N steps the particle is found in an interval of width Δx, a distance x away from the point of departure. For this purpose let us draw a portion of the probability distribution, and use the fact that each step is of length l. This is done in Figure 2.8. Since each step is of

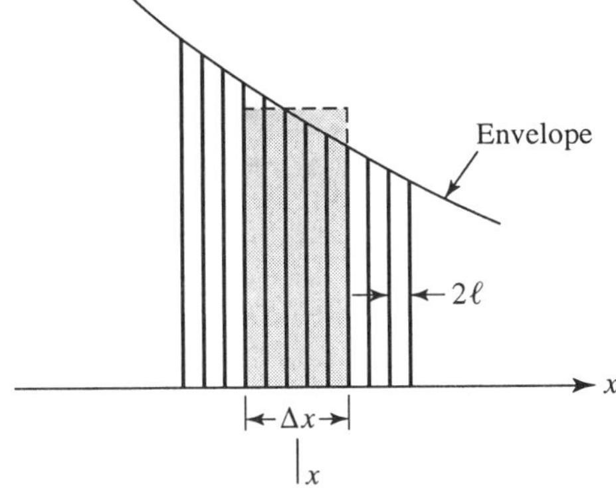

Figure 2.8. Probability for displacement $x < ml < x + \Delta x$.

length l, a net displacement of $(R - L) = m$ steps leads to the position

$$x = ml$$

giving

$$m = \frac{x}{l}. \tag{2-38}$$

We can therefore write the envelope function [(2-37)] as a function of x by simply replacing m by x/l:

$$P_N(m) = P_N\left(\frac{x}{l}\right) \cong \sqrt{\frac{2}{\pi N}}\, e^{-x^2/2Nl^2}. \tag{2-39}$$

Now, if we ask for the probability that the particle be found in the interval Δx of Figure 2.8, the addition rule tells us to *add* all those $P_N(m)$ for which the position $x = ml$ lies *in* the interval Δx:

$$P(\text{particle in } \Delta x) = \sum_{\text{all } m\text{'s in } \Delta x} P_N(m).$$

For this group of neighboring m-values $P_N(m)$ has *almost the same value*; a representative value, somewhere in the middle, is $P_N(m = x/l)$. We may then write approximately

$$\sum_{\substack{\text{all } m \text{ in} \\ \text{interval } \Delta x}} P_N(m) = \text{a representative value of } P_N \text{ in interval } \Delta x \quad \text{times the number of } m\text{-values in } \Delta x. \tag{2-40}$$

To find the number of m-values in Δx, we must recall the definition of m as $m = R - L$, the difference between the number R of steps to the right and L, the number of steps to the left. $N = R + L$ is the total number of steps. So, for $N = 2$, the possible values of (R, L) are

$$(R, L) = (2, 0), (1, 1), (0, 2),$$

and the possible values of m therefore

$$m = 2, 0, -2.$$

For $N = 3$ there are the possibilities

$$(R, L) = (3, 0), (2, 1), (1, 2), (0, 3),$$

and hence the possible m-values

$$m = 3, 1, -1, -3.$$

We see that the values of m are always all even or all odd, and therefore, two consecutive m-values are two units apart. Therefore, the distance between consecutive displacements

$$x = (m - 2)l, \qquad x = ml, \qquad x = (m + 2)l, \ldots,$$

is $2l$. An interval of length Δx on the x-axis therefore includes the number

$$\left(\frac{\Delta x}{2l} \right)$$

of m-values. It follows that we can write the "verbal formula" [(2-40)] in the form

$$\sum_{(\text{all } m \text{ in } \Delta x)} P_N(m) = P_N\left(m = \frac{x}{l} \right) \times \left(\frac{\Delta x}{2l} \right). \tag{2-40a}$$

We see that the probability in question is proportional to Δx, the width of the interval. We must therefore introduce a notation that will express this fact, and write

$$\sum_{(m\text{'s in }\Delta x)} P_N(m) = \left(\frac{1}{2l} P_N\left(m = \frac{x}{l}\right) \right) \cdot \Delta x$$

$$\equiv P_N(x)\Delta x. \tag{2-41}$$

this last equation defines $P_N(x)$. Inserting (2-39), we find

$$P_N(x) = \sqrt{\frac{1}{2\pi\, Nl^2}}\, e^{-x^2/2Nl^2}. \tag{2-42}$$

A quantity like $P_N(x)$ is called a *probability density*. Multiplied with the length Δx of an *interval* located at position x, it becomes a probability that the particle is located inside that interval.

What we have achieved with (2-42) is the elimination of any reference to the number m, which cannot be determined in an experiment anyhow.

In a similar fashion we can eliminate the reference to the number of steps N ont he assumption that these steps follow each other regularly in time intervals t_c. Again, in all realistic situations these intervals are so short that we cannot see the individual steps, but we can measure the *total elapsed time.*

$$t = Nt_c. \tag{2-43}$$

We therefore express N in terms of t_c and t by

$$N = \frac{t}{t_c}.$$

For the expression $P_N(x) = P_{(t/t_c)}(x)$ we use the neater symbol $P(x, t)$; this gives

$$P(x, t) = \sqrt{\frac{t_c}{2\pi l^2 t}}\, e^{-(t_c/2l^2 t)x^2}. \tag{2-44}$$

It is customary to condense the combination of symbols l^2 and t_c occurring in (2-44) into a single symbol D, *defined* as

$$D = \frac{l^2}{2t_c}. \tag{2-45}$$

In terms of this we may write

$$P(x, t) = \sqrt{\frac{1}{4\pi\, Dt}}\, e^{-x^2/4Dt}. \tag{2-46}$$

Let us at this point state again what the *meaning* of $P(x, t)$ is:

> $P(x, t)\,\Delta x$ represents the probability that a particle, starting at $x = 0$, will by random walk be found a time t later in the interval Δx located a distance x from the coordinate origin.

The most important feature of (2-46) is that all reference to the individual step length l, and to the time interval t_c, has been eliminated by combining them into a single parameter D, the so-called *diffusion constant*. All the observable information on the statistical properties of the random walk process must therefore be contained in the value of D. As an illustration of this, let us find the mean square distance traveled in random walk in time t. Our point of departure is then (2-29) which states that

$$\overline{m^2} = N.$$

But $m = x/l$, and $N = t/t_c$, and the above equation may be written as

$$\overline{m^2} = \frac{\overline{x^2}}{l^2} = \frac{t}{t_c}.$$

It follows that the mean square displacement $\overline{x^2}$ may be written as

$$\overline{x^2} = l^2 \frac{t}{t_c} = \left(\frac{l^2}{2t_c}\right) 2t.$$

Again the parameters l and t_c occur only in the combination $D = l^2/2t_c$, and we can write the mean square displacement as

$$\overline{x^2} = 2Dt. \tag{2-47}$$

Equation (2-46) for $P(x, t)$ and (2-47) for the mean square displacement are the most important results of this section. These equations show that the quantitative features of the random walk depend on a *single* parameter D, the diffusion constant. According to (2-47), D relates the mean square displacement to the elapsed time.

Both (2-46) and (2-47) are subject to experimental verification, and we shall present some results shortly. The distribution $P(x, t)$ can be *tested* by using a suspension of particles large enough to be visible, so that individual displacements in a fixed time interval can be sampled and plotted. In this experimental test of $P(x, t)$ we thus use again the operational definition of probability: Since $P(x, t)\Delta x$ is the probability that the displacement of a particle is in the range x to $x + \Delta x$, it must also represent the *fraction* $\Delta \mathcal{N}/\mathcal{N}$ of all $\mathcal{N}$ particles sampled whose displacement falls into that interval. This point is again explained in Figure 2.9, where $P(x, t)$ is plotted for three consecutive values $t_1 < t_2 < t_3$ of the time t. As a final comment in this section, we draw attention to the analogy between the result that the root mean square distance traveled in random walk increases only with the square root of elapsed time

$$\sqrt{\overline{x^2}} = \sqrt{2Dt} \tag{2-48}$$

and the result obtained in Chapter 1 on polymer chains that the root mean square end-to-end distance of a random polymer coil of N segments is only proportional to the square root $\sqrt{N}$ of N.

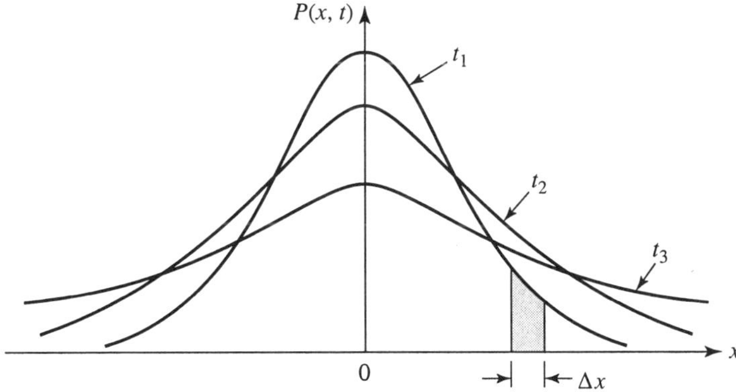

Figure 2.9. Gaussian distribution $P(x, t)$ for three consecutive times $t_1 < t_2 < t_3$. The shaded area represents the *fraction* of particles which in time t_1 have undergone a displacement in the range x to $x + \Delta x$.

In the next section, we extend the random walk model to three dimensions, and then present some experimental evidence for the results obtained.

2.2.D. Probability of Displacements for the Three-Dimensional Random Walk. Numerical Values for Diffusion Constants. Some Elementary Applications

In this section we take the step from the one-dimensional model of random walk to the random walk in real, three-dimensional space.

In order to establish the three-dimensional analog of our main result [(2-46)], we consider a still somewhat idealized model that can be described as follows (see Figure 2.10):

(i) Each step may take place either along the x-axis, the y-axis, or the z-axis; these choices occur with equal probability.

(ii) Along the chosen axis, each step may be either forward or backward with equal probability.

(iii) Each step is of length l, and takes a time t_c.

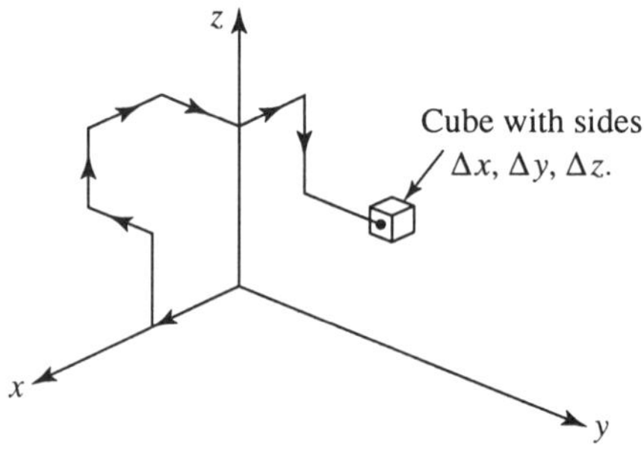

Figure 2.10. Model of a random walk in three dimensions.

Now, if we consider a very large number N of steps, then roughly one-third of them will be along the x-axis, one-third along the y-axis, and one-third along the z-axis. What is then the probability for a net displacement $x = m_x l$ along the x-axis, after N steps? It will be to a good approximation given by

$$P_{N/3}(m_x),$$

where $N/3$ is the assumed total number of steps in the $\pm x$-directions. For the probability of a net displacement $y = m_y l$ in the y-direction, and $z = m_z l$ in the z-direction, we have the corresponding expressions

$$P_{N/3}(m_y) \quad \text{and} \quad P_{N/3}(m_z).$$

Now the displacements m_x, m_y, m_z in the x-, y-, and z-directions occur independently of each other. The probability for the occurrence of a displacement m_x in the x-direction, *and* a displacement m_y in the y-direction, and a displacement m_z in the z-direction is then by the multiplication rule given by the product $P_{N/3}(m_x) \cdot P_{N/3}(m_y) \cdot P_{N/3}(m_z)$. Let us call this probability $P_N(m_x, m_y, m_z)$ and write it in the Gaussian limit [(2-37)]

$$P_N(m_x, m_y, m_z) \equiv P_{N/3}(m_x) \cdot P_{N/3}(m_y) \cdot P_{N/3}(m_z)$$

$$\cong \left(\sqrt{\frac{2}{\pi N/3}} \right)^3 e^{m_x^2/(2N/3)} e^{-m_y^2/(2N/3)} e^{-m_z^2/(2N/3)}$$

$$= \left(\frac{6}{\pi N} \right)^{3/2} e^{-(3/2N)(m_x^2 + m_y^2 + m_z^2)}. \tag{2-49}$$

It is now a straightforward matter to carry out the two steps which led in the one-dimensional case to the definition of the probability density $P(x, t)$. What we have to do is:

(i)   write x/l for m_x, y/l for m_y, z/l for m_z in (2-49);

(ii)   write t/t_c for the number of steps N in (2-49); and

(iii)   take the *sum* of all $P_N(m_x, m_y, m_z)$ that represents a displacement of the particle in the *interval* Δx along the x-axis, Δy along the y-axis, and Δz along the z-axis.

This sum is to a good approximation given by

$$P_{(t/t_c)}\left(m_x = \frac{x}{l}, m_y = \frac{y}{l}, m_z = \frac{z}{l} \right) \left(\frac{\Delta x}{2l} \right) \left(\frac{\Delta y}{2l} \right) \left(\frac{\Delta z}{2l} \right).$$

The result of this gives the probability that after an elapsed time t a particle starting from the origin $x = y = z = 0$ will be found in the interval $\Delta x, \Delta y, \Delta z$; that is, inside the small cube drawn in Figure 2.10. This probability will be written as

$$P(xyz, t)\,\Delta x\,\Delta y\,\Delta z$$

and is given by

$$P(xyz, t)\,\Delta x\,\Delta y\,\Delta z = \left(\frac{3t_c}{2\pi t l^2}\right)^{3/2} e^{-(3t_c/2tl^2)(x^2+y^2+z^2)}\,\Delta x\,\Delta y\,\Delta z. \qquad (2\text{-}50)$$

The expression for $P(xyz, t)$ can be simplified again by introducing the diffusion constant D. In this case, we define it as

$$D = \frac{l^2}{6t_c}, \qquad (2\text{-}51)$$

which will give the expression for $P(xyz, t)$ a form very similar to (2-46):

$$P(xyz, t) = \left(\frac{1}{4\pi Dt}\right)^{3/2} e^{-(x^2+y^2+z^2)/4Dt}. \qquad (2\text{-}52)$$

This is our final result. Let us also calculate the mean square displacements in x, y, z. The probability distribution of the displacements in the x-direction is $P_{N/3}(m_x)$ which gives as we know

$$\overline{m_x^2} = N/3,$$

an expression which we rewrite as

$$\frac{\overline{x^2}}{l^2} = \frac{1}{3}\frac{t}{t_c}.$$

So we see that by using our definition of D, (2-51), we have again

$$\overline{x^2} = 2Dt, \qquad (2\text{-}53)$$

and similarly

$$\overline{y^2} = 2Dt, \qquad \overline{z^2} = 2Dt.$$

The mean square distance away from the origin is then

$$\overline{r^2} = \overline{x^2 + y^2 + z^2} = \overline{x^2} + \overline{y^2} + \overline{z^2} \qquad \text{or} \qquad \overline{r^2} = 6Dt. \qquad (2\text{-}54)$$

We must now again devote some attention to the proper interpretation of these results, and to a comparison with experimental data.

First, we notice that the distribution function $P(xyz, t)\Delta x \Delta y \Delta z$ is the product $P(xt)\Delta x$, $P(yt)\Delta y$, $P(zt)\Delta z$:

$$P(xyz, t)\Delta x \Delta y \Delta t = \left(\sqrt{\frac{1}{4\pi Dt}}\, e^{-x^2/4Dt}\Delta x\right) \cdot \left(\sqrt{\frac{1}{4\pi Dt}}\, e^{-y^2/4Dt}\Delta y\right)$$

$$\cdot \left(\sqrt{\frac{1}{4\pi Dt}}\, e^{-z^2/4Dt}\Delta z\right). \qquad (2\text{-}55)$$

This expresses the *statistical independence* of the displacements in the x-, y-, and z-directions. We also observe that each of these factors in (2-26) is *normalized* to 1:

$$\sum_{\text{all }\Delta x} P(xt)\Delta x = \sum_{\text{all }\Delta x} \sqrt{\frac{1}{4\pi Dt}}\, e^{-x^2/4Dt}\Delta x = \int \frac{1}{\sqrt{4\pi Dt}}\, e^{-x^2/4Dt} dx = 1.$$

$$(2\text{-}56)$$

[See Appendix 2.A3 for Gaussian integrals.]

What is then the physical significance of each factor in (2-55)? We see this by answering the question, for instance: What is the probability that the particle after time t has a displacement in the x-direction between x and $x + \Delta x$, *irrespective* of what its displacement in the y- and z-directions are. According to the addition rule, we obtain this probability by taking the probabilities

$$P(xyz, t)\Delta x \Delta y \Delta z$$

and *summing* over all intervals Δy and Δz: This gives

$$\sum_{\substack{\text{all }\Delta x \\ \text{all }\Delta z}} P(xyz, t)\Delta x \Delta y \Delta z = P(x, t)\Delta x \sum_{\text{all }\Delta y} P(y, t)\Delta y \sum_{\text{all }\Delta z} P(z, t)\Delta z$$

$$= P(x, t)\Delta x \int_{-\infty}^{+\infty} P(y, t)\, dy \int_{-\infty}^{+\infty} P(z, t)\, dz$$

$$= P(x, t)\Delta x = \sqrt{\frac{1}{4\pi Dt}}\, e^{-x^2/4Dt}\Delta x. \qquad (2\text{-}57)$$

Thus, $P(x, t)\Delta x$ emerges as the probability that the x-displacement lies in the interval x to $x + \Delta x$, *irrespective* of what the y- or z-displacement is. Somewhat more generally, we can say (see Figure 2.11) that

$$\int_{x_1}^{x_2} P(x, t)\, dx \tag{2-58}$$

is the probability that the x-displacement lies between x_1 and x_2, whatever the values of the displacement in the y- and z-directions are.

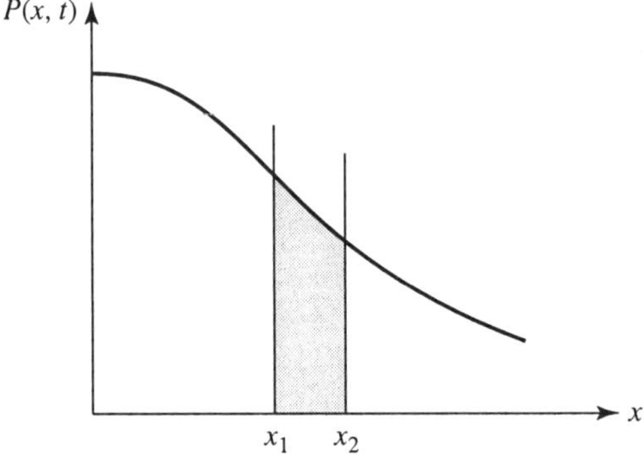

Figure 2.11. Shaded area defines the probability that a particle is found in the interval $x_1 \le x \le x_2$.

The prediction [(2-57)] can be verified experimentally. Such tests have been carried out by J. Perrin and his collaborators, and are reported in his book *Atoms* [2]. In one of the test runs carried out with latex droplets of radius $a = 2.12 \times 10^{-5}$ cm, the interval from $x = 0$ to $x = 17\ \mu$ $(1\ \mu = 10^{-4}$ cm) is divided into 10 "bins" of width $\Delta x = 1.7\ \mu$ each, and the numbers N_k of particles that migrated into each bin from $x = 0$, recorded. The numbers are compared with the prediction based on (2-57), according to which one expects that

$$N_k = \text{number of particles in the } k\text{th bin}$$

$$= \text{total number of observed particles} \times \int dx\, P(x, t) \qquad (k\text{th bin}).$$

Table 2.3. Verification of the Form of the Distribution $P(x, t)$, (2-57).

Bin number (see text)	Observed number of particles	Calculated number of particles
1	48	44
2	38	40
3	36	35
4	29	28
5	16	21
6	15	15
7	8	10
8	7	5
9	4	4
10	4	2

To compare the two, the theoretical expression is written as

$$P(xt) = \sqrt{\frac{1}{2\pi \overline{x^2}}}\, e^{-x^2/2\overline{x^2}}$$

by making use of relation (2-53), and adjusting $\overline{x^2}$ to the experimental value. The results are listed in Table 2.3. The observed and calculated data fit as well as can be expected with such a small sample of only 205 particles.

Let us turn now to the *numerical* values of the diffusion constant D, defined in (2-51). In Section 2.1.D of this chapter we gave an estimate both of the path length l, and of the time interval t_c between two collisions. This leads to a correct *order of magnitude* estimate of the diffusion constant D for small molecules in a dilute gas. In Table 2.4 we present some numerical values. Notice that the dimension

Table 2.4(a). Diffusion Constants of Small Molecules in Air at Atmospheric Pressure.

Molecule	Temp. (°C)	D (cm^2/s)
Hydrogen	0°	0.634
Water vapor	8°	0.239
Oxygen	0°	0.178
Carbon dioxide	0°	0.139
Alcohol vapor	40°	0.137

Table 2.4(b). Diffusion Constants of Molecules in Water (20 °C).

Molecule	M (g/mol)	Radius (Å)	$D(\mathrm{cm}^2/\mathrm{s})$
Water (H_2O)	18	~ 1.5	2.0×10^{-5}
Oxygen (O_2)	32	~ 2	1.0×10^{-5}
Urea $CO(NH_2)_2$	60	~ 4	1.12×10^{-5}
Glucose ($C_6H_{12}O_6$)	180	~ 5	6.7×10^{-6}
Ribonuclease	13,683	~ 18.0	1.2×10^{-6}
β-lactoglobulin	35,000	~ 27.4	7.82×10^{-7}
Hemoglobin	68,000	~ 31.0	6.9×10^{-7}
Catalase	250,000	~ 52.2	4.1×10^{-7}
DNA	6,000,000		0.13×10^{-7}
Bushy Stunt Virus	10,700,000		1.15×10^{-7}
Tobacco Mosaic Virus	50,000,000		0.39×10^{-7}

of D is that of (length)^2/time, and values of D are generally given in cm^2/s. In Section 2.1.D we had estimated l to be of order 10^{-5} cm, and for oxygen, $t_c \sim 2 \times 10^{-10}$ s. This would give a value of D of order $D \sim 0.1$ cm^2/s in order of magnitude agreement with the data.

In Table 2.4 we present some data for the diffusion constant D of particles in water: This table contains a wealth of interesting and important information.

In the *first group* of molecules tabulated we have water, oxygen, urea, glucose, all light molecules. Typically, their diffusion constant is of the order

$$D \simeq 10^{-5} \text{ cm}^2/\text{s}.$$

This is about a factor 10^4 smaller than the diffusion constant of small molecules in air at atmospheric pressure, as a comparison with the data in Table 2.4(a) shows. This smallness of D in water, as compared to air, reflects the difference in density of the two media that are 1.2 kg/m^3 for air, 10^3 kg/m^3 for water, giving a ratio of 10^3 in densities. Since the mean free path is inversely proportional to the density, we expect a typical value of l for a small molecule in water to be 10^{-8} cm, a reduction of 10^{-3} from its value in a gas at atmospheric pressure. Also, since

$$D \propto \frac{l^2}{t_c} = l\left(\frac{l}{t_c}\right) = lv \cong l\sqrt{3mkT},$$

we expect the value of D to scale with l. On that basis, we would expect for small molecules

$$\frac{(D) \text{ in water}}{(D) \text{ in air}} \sim 10^{-3},$$

whereas the actual ratio is $\sim 10^{-4}$. This discrepancy signals the breakdown of the picture of sequential, "binary" collisions of molecules, valid for dilute gases. If the supporting medium is as dense as water, or if the random walking particle gets big enough, the dominant type of collision is the "multiple" collision. This is sketched in Figure 2.12.

As the number of simultaneous collisions of the particle with gas or water molecules increases, it becomes possible to describe the net effect of these multiple collisions in terms of a hydrodynamic drag described by Stokes' law (see Volume I, Section 2.6.B)

$$\vec{F}_{\text{drag}} = -6\pi a \eta \vec{v},$$

a being the radius of the molecule, η the viscosity of the fluid (gas or liquid). It will be shown in Section 2.5.C of this chapter that under these circumstances the diffusion constant D has the value

$$D = \frac{kT}{6\pi \eta a}, \tag{2-59}$$

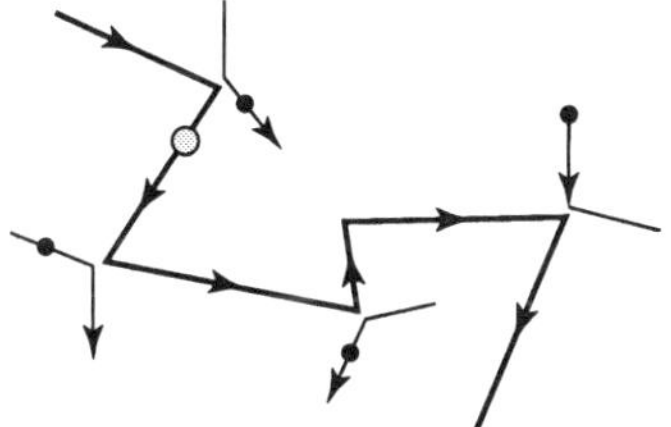

(a) Binary collisions of random walking particles with molecules in dilute gas.

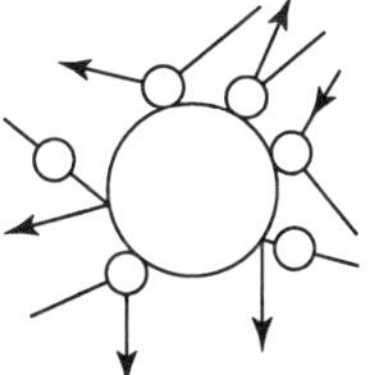

(b) Multiple collisions of random walking large molecule in dense gas or fluid.

Figure 2.12. Random walk of small molecules in a dilute gas, and of a large molecule in dense mediums. (a) Binary collisions of random walking particles with molecules in a dilute gas. (b) Multiple collisions of a random walking large molecule in a dense gas or fluid.

a result known as the Einstein–Stokes relation. The second group of molecules listed in Table 2.4(b) gives four examples for which the Einstein–Stokes expression for D applies. They are all globular proteins of roughly spherical shape, and a diameter large compared to that of the water molecule. It can indeed be verified from the data in the table that the product Da of diffusion constant D and radius a in this group has the rather constant value

$$Da \sim 21 \times 10^{-14} \text{ cm}^3/\text{s}.$$

It is left as a problem assignment to verify that the Einstein–Stokes law actually holds for these four molecules.

In the last group listed in Table 2.4(b), finally, we give the values of D for DNA in the random coil form, and for two viruses. The Tobacco Mosaic Virus is a rod, and (2-59) does not apply, but we leave it as a problem assignment to determine the size of the Bushy Stunt Virus, assuming it is a sphere.

Notice that the diffusion constant of DNA is much smaller than that of the Bushy Stunt Virus despite its smaller molecular weight. The latter is a compact object, DNA a random coil filling a larger volume of space.

We close this discussion on diffusion constants with a few remarks about the means by which they are actually determined. Since information on the value of D is most important for macromolecules (proteins), where its knowledge contributes to the determination of molecular weight and size, we mention methods particularly suitable for such molecules.

The most direct method is based on the observation of the density fluctuations of a solution of macromolecules. Because of their random motion the number of particles in a small volume element of fluid fluctuates in time. If monochromatic (single frequency) light from a laser beam is sent through such a solution, one observes that in the *scattered light* the frequency of the light wave is spread out into a *band*. The width of this frequency distribution is proportional to the diffusion constant D which can therefore be measured in such a scattering experiment [3]. A second method is based on the determination of the time required for a discontinuity in the concentration of macromolecules in solution to smoothen out. This process, and its utilization to measure D, will be described in Section 2.3.A.

2.2.E.   Elementary Application: The Transfer of Oxygen and Carbon Dioxide in the Human Lung

In the next Sections 2.3 and 2.4 of this chapter, we will develop in detail the precise manner in which the random walk of particles leads to a bulk flow of such particles in solution or suspension. This bulk flow is called *diffusive flow*, and the characteristic smoothing out of an uneven concentration of suspended particles constitutes the phenomenon commonly called *diffusion*.

In this present section we will consider a case of diffusive transport in a non-rigorous, semiquantitative way only, based on (2-53) and (2-54), mainly to dramatize the importance of a good quantitative understanding of transport phenomena.

A large number of molecular exchanges between compartments in plants and animals occurs by simple diffusion, and the dimension and surface areas of these compartments must be designed so as to allow the required amounts of materials to be exchanged. Equation (2-54) plays a crucial role for our understanding of the design of systems in which such exchanges take place.

As an example, let us consider the human lung in which oxygen is transferred, by diffusion, from the lung into the blood capillaries and carbon dioxide moved from the blood into the lung.

The exchange of these gases takes places in the so-called *alveoli* of the lung (of which there are about 3×10^8), little sacs surrounded by a network of blood capillaries. Figure 2.13 shows an alveolar sac and adjoining blood capillary. We now estimate the time required for oxygen or carbon dioxide molecules to travel a distance of the order of the alveolar radius, $R_A \sim 100 \, \mu = 10^{-2}$ cm. Using the data from Table 2.4(a), we find that for oxygen we can estimate this time as

$$t = \frac{r^2}{6D} \simeq \frac{R_A^2}{6D}$$

$$= \frac{10^{-4} \text{ cm}^2}{6 \times 0.178 \text{ cm}^2/\text{s}} \sim 10^{-4} \text{ s.}$$

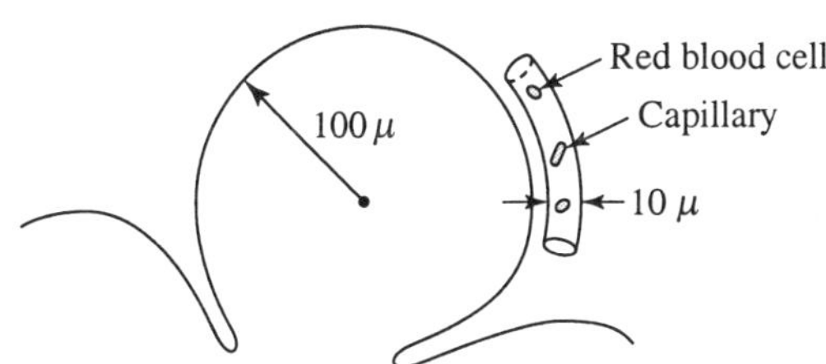

Figure 2.13. Alveolar sac and blood capillary.

The alveolar and capillary walls are only a fraction of a micron in thickness. Assuming that gases diffuse through them as through water, we see that the oxygen has to diffuse through another $5 \, \mu$ of water (corresponding to the capillary radius

$R_c \sim 5\ \mu$) for which we estimate a time, according to Table 2.4(b):

$$t \sim \frac{(5 \times 10^{-4}\ \text{cm})^2}{6 \times 10^{-5}\ \text{cm}^2/\text{s}} \sim 4 \times 10^{-3}\ \text{s}.$$

We have thus an estimate of $t \gtrsim 10^{-3}$ s as the required transit time for an oxygen molecule from the alveolar center into the blood capillary. (Inside the capillary oxygen diffuses into the red blood cells, and then is bound by the hemoglobin molecules.)

This transit time estimated above must be *matched* to the transit time of the red blood cells as they move through the capillaries, if these cells are to pick up the available oxygen. A capillary hugs an alveolus over a length of about $100\ \mu = 10^{-2}$ cm, and the speed of blood flow in a capillary is about 0.1 cm/s. So the transit time for blood is of the order of 10^{-1} s. This then is the time during which a red blood cell is in proximity to the alveolar surface. It is clear that there is indeed adequate time for the oxygen to be picked up.

2.3   The Diffusion Equation

2.3.A.   The Space–Time Evolution of Particle Distribution. Integral Representations for Concentration $C(x, t)$

So far, we have discussed the motion of a *single* particle undertaking a random walk, and found that we could assign a probability $P(x, t)\Delta x$ that such a particle, starting at $x = 0$, would end up in the interval Δx on the x-axis, a time t later. We also found the three-dimensional generalization of this in (2-50) or (2-55).

In this section we propose to determine what happens in the course of time to the distribution in space of a *large number* of independent particles engaged in random walks. Such a situation occurs, for instance, when we consider a dilute solution of a substance in a solvent, say sugar in water, or serum albumin in blood plasma, etc. On a macroscopic scale, such a distribution would be characterized by a *concentration* $C(xyz, t)$ of particles. The general definition of concentration C is

$$C = \text{number of particles/unit volume}.$$

Very often we have to deal with a situation where the concentration is not *uniform*, that is, it is different in different regions of space; it may also be *nonstationary*, that is, it changes with time. To define C properly in these cases, we must then consider a small volume element ΔV of solution containing a number ΔN of dissolved molecules. The *local concentration* $C(x, y, z, t)$ at time t is then defined by the ratio of ΔN to ΔV as ΔV gets very small:

$$C(x, y, z, t) = \lim_{\Delta V \to 0} \left(\frac{\Delta N}{\Delta V} \right). \qquad (2\text{-}60)$$

If this concentration is uniform on a macroscopic scale, then experience shows that it remains so. If, on the contrary, the concentration varies, say, as a consequence of carefully placing pure water on top of a solution of sugar in a test tube, then experience again shows that in due time the original difference of concentration smoothes out and eventually disappears. This *diffusion* of the sugar molecules from the regions of higher concentration into those of lower concentration is a consequence of their random walk. We are therefore now in a position to answer the question: How does an originally *non*uniform concentration of particles $C(xyz)$ evolve in time? How and how fast does it smoothen out?

For simplicity of notation let us consider a case where nonuniformity is only in the x-direction. Consider then the case where at some initial time $t = 0$ we have a concentration $C(x, 0)$ of particles in solution, or in suspension in a gas. Our aim is then to *predict* the concentration $C(x, t)$ of particles after a time t has elapsed.

The answer to this question is surprisingly simple if we make use of our description of the random walk in terms of the probability distribution function $P(x, t)\Delta x$ of (2-46) or (2-57). Let us reiterate the physical meaning of $P(x, t)\Delta x$:

$P(x, t)\Delta x$ represents the probability that a particle in the time interval t will have undergone a displacement in the x-direction, whose value is between x and $x + \Delta x$. To construct the value of the concentration $C(x, t)$, we make use of a space–time diagram of the random walk, as shown in Figure 2.14. In this diagram, we plot elapsed time in the vertical direction, and the position of the particle in the horizontal direction. The irregular line thus represents the trajectory of a particular random walking particle. This particle, at $t = 0$, is located at a position x' in the interval $\Delta x'$, and ends up at time t in the interval Δx, undergoing a net displacement $(x - x')$ in the elapsed time t.

We now first state in words how the concentration $C(x, t)$ can be found, and then subsequently express these statements in quantitative terms. Since all particles are random walking through the solvent medium, the number $\Delta N(x)$ of particles that will be located in interval Δx at time t is the *sum* of all particles that walk into that interval over a period of time t. The ones that were in Δx at time $t = 0$ will all have walked away in time t if the interval is small enough (small compared

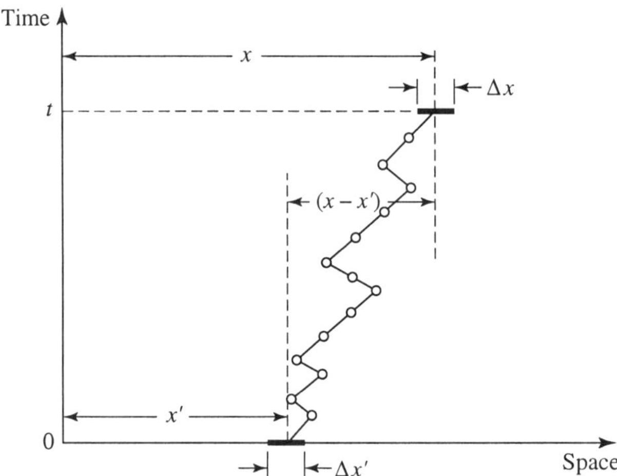

Figure 2.14. Space–time diagram of a random walk.

to $\sqrt{6\Delta t}$). Let $N_0(x')$ stand for the number of particles located in interval $\Delta x'$ at time 0. The probability that any one of these ends up in Δx at time t is given by

$$P(x - x', t)\Delta x,$$

according to the interpretation of P just given. Notice that $(x - x')$ represents the displacement a particle has to achieve to get from $\Delta x'$ into Δx. Now, since $\Delta N_0(x')$ particles are starting from $\Delta x'$, the number of those that will arrive in Δx is

$$\Delta N_0(x')P(x - x', t)\Delta x.$$

To get the *total* number of particles, $\Delta N(x)$, that will arrive in Δx, we must simply add the contribution from every starting interval $\Delta x'$:

$$\Delta N(x) = \sum_{\text{all intervals } \Delta x'} \Delta N_0(x')P(x - x', t)\Delta x. \tag{2-61}$$

If we now express this last result in terms of concentration, we will have the desired answer. This requires two steps:

(i) The number of particles $\Delta N_0(x')$ in interval $\Delta x'$ at time 0 is equal to $\Delta x'$ times the concentration $C_0(x')$ of particles at time 0:

$$\Delta N_0(x') = C_0(x')\Delta x'. \tag{2-62}$$

(ii) Similarly, the number $\Delta N(x)$ of particles which at time t are in interval Δx is

$$\Delta N(x) = C(x, t)\Delta x, \tag{2-63}$$

where $C(x, t)$ is the concentration at x at time t. Inserting this into (2-61), this equation reads

$$C(x, t)\Delta x = \sum_{\text{all } \Delta x'} \Delta x' C_0(x') P(x - x', t)\Delta x. \tag{2-64}$$

Lopping off the common factor Δx, and writing the sum

$$\sum_{\Delta x'} \Delta x' \quad \text{as the integral} \quad \int dx',$$

we have

$$C(x, t) = \int dx' C_0(x') P(x - x', t). \tag{2-65}$$

Let us pause here to see what we have achieved with this equation (2-65). It expresses the concentration $C(x, t)$ of particles at time t in terms of the *original* concentration at $t = 0$ at all points x' in space. The knowledge of $C_0(x')$ of all points is needed because in the finite interval t of time, particles can random walk their way into position x from everywhere (although with diminishing probability from far-away points). Equation (2-65) states in exact mathematical terms the very simple proposition that the number of particles in interval dx at position x and time t is simply the number of particles that have random walked their way into Δx from wherever they were at time $t = 0$.

It is important to recognize that we *do* know the expression for $P(x - x', t)$ which is

$$P(x - x', t) = \sqrt{\frac{1}{4\pi Dt}}\, e^{-(x-x')^2/4Dt}.$$

So, if we actually know what the spatial dependence $C_0(x)$ of the concentration is at time zero, (2-65) gives us a means to find $C(x, t)$ at any later time by simple integration. What we have in (2-65) is called an *integral representation* of the process of concentration change with time. It gives us $C(x, t)$ for *any* elapsed time, provided we know the initial value $C_0(x)$ *everywhere*.

Useful as it is to have a description of space–time evolution of concentration in this form, one often wishes for a *differential* representation of the process. This latter would answer the question: What *local* information about the concentration is it relevant to know how $C(x, t)$ is going to change, locally, with time?

We shall answer this question shortly by constructing from (2-65) a differential equation for $C(x, t)$ which answers the above question. But first let us illustrate the use and usefulness of the integral form [(2-65)].

2.3.B.　Application of the Integral Representation for $C(x, t)$. The Experiment of Lam and Polson. Determination of Diffusion Constant D

As an illustration of the use of (2-65), consider a long tube whose lower half is filled with a solution of concentration C_0, and its upper half with pure water, producing (at time $t = 0$) a sharp boundary. (See Figure 2.15.) The interest of this experiment lies in the fact that a definite experimental verification of the predicted evolution of the concentration is possible. Let us see first what theory predicts. The concentration $C_0(x)$ is zero for $x > 0$, and has a constant value C_0 from $x = 0$ to

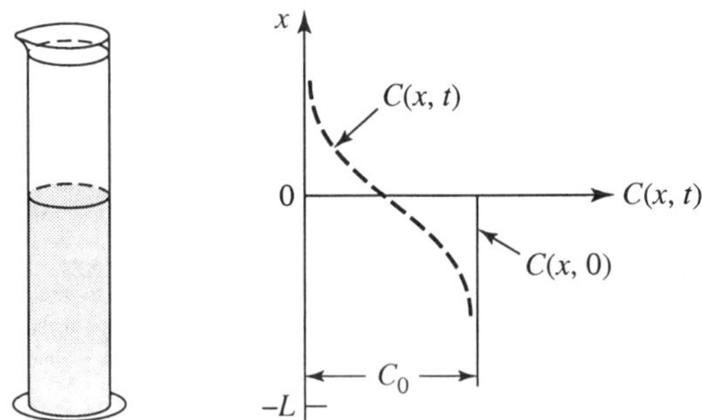

Figure 2.15. Evolution of nonuniform concentration formed by placing pure solvent on solution.

$x = -L$. According to (2-65), this gives

$$C(x, t) = C_0 \int_{-L}^{0} dx' P(x - x', t)$$

and upon inserting the expression [(2-46)] for $P(x, t)$:

$$C(x, t) = \frac{C_0}{\sqrt{4\pi Dt}} \int_{-L}^{0} dx' e^{-(x-x')^2/4Dt}. \tag{2-66}$$

Let us rewrite this in a more useful way by changing the integration variable for x' to $s' = (x - x')/\sqrt{4Dt}$. We must then write dx' as

$$dx' = -\sqrt{4Dt}\, ds'.$$

Thus we can write

$$C(x, t) = -\frac{C_0}{\sqrt{\pi}} \int_{x'=-L}^{x'=0} e^{-s'^2} ds' = \frac{C_0}{\sqrt{\pi}} \int_{x'=0}^{x'=-L} e^{-s'^2} ds'.$$

At the upper limit, $x' = -L$, of the last integral, s' is

$$s' \equiv s_2 = \frac{(x + L)}{\sqrt{4Dt}}.$$

At the lower limit ($x' = 0$), of the last integral, s' is

$$s' \equiv s_1 = \frac{x}{\sqrt{4Dt}}.$$

Thus s' takes on the range of values

$$x/\sqrt{4Dt} \leq s' \leq \frac{(x + L)}{\sqrt{4Dt}}. \tag{2-67}$$

The concentration $C(x, t)$ can now be written as

$$C(x, t) = \frac{C_0}{\sqrt{\pi}} \int_{s_1}^{s_2} ds'\, e^{-s'^2} \tag{2-68}$$

or

$$C(x, t) = \frac{C_0}{\sqrt{\pi}} \left(\int_0^{s_2} ds'\, e^{-s'^2} - \int_0^{s_1} ds'\, e^{-s'^2} \right).$$

The integrals over a Gaussian e^{-s^2} cannot be expressed in terms of familiar elementary functions. However, it occurs so often in many problems that the integrals of e^{-s^2} are defined as the Gauss-error function and are tabulated in most books of integrals (see ref. [29] at end of chapter). The Gauss-error integral is defined as

$$\phi(s) = \frac{2}{\sqrt{\pi}} \int_0^s ds'\, e^{-s'^2}. \tag{2-69}$$

In Figure 2.16 we plot $\phi(s)$ as a function of s. Notice that for $s \gtrsim 2$, $\phi(s) \to 1$. In terms of this error integral, the concentration $C(x, t)$ can be written as

$$C(x, t) = \frac{C_0}{2}\big(\phi(s_2) - \phi(s_1)\big).$$

If the observation time t is restricted to values

$$t \ll \frac{L^2}{4D}$$

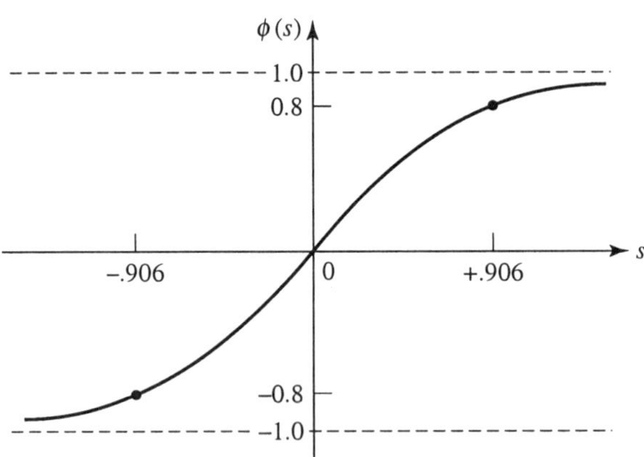

Figure 2.16. Plot of the Gauss error integral $\phi(s) = (2/\sqrt{\pi}) \int_0^s ds\, e^{-s^2}$.

or if L is made very long, the value $s_2 = (x + L)/\sqrt{4Dt}$ will be large compared to 1, and $\phi(s_2)$ will be ≈ 1. Under these conditions, we find

$$C(x, t) \cong \frac{C_0}{2}\left(1 - \phi\left(\frac{x}{\sqrt{4Dt}}\right)\right). \tag{2-70}$$

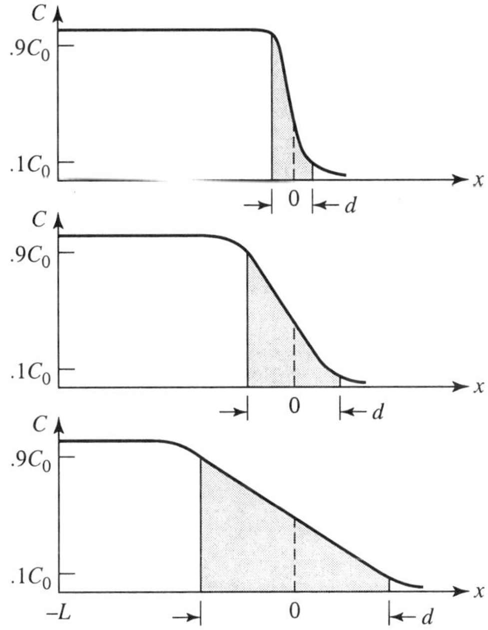

Figure 2.17. Concentrations profile at three different times showing the growth of the domain for which $10\%C_0 < C < 90\%C_0$.

We must now appraise the contents of this result. In Figure 2.17 the form of $C(x, t)$ is plotted for three consecutive values of the time t. In this plot we also marked the distance d over which the concentration changes from 90% of C_0 to 10% of C_0. According to (2-70), $C(x, t) = 0.9C_0$ for $\phi = -0.8$, and $0.1C_0$ for $\phi = +0.8$. Math tables (see ref. [29] at end of chapter 2), and Figure 2.16 show that

$$\phi(s) = 0.8 \quad \text{for} \quad s = 0.906$$

and

$$\phi(s) = -0.8 \quad \text{for} \quad s = -0.906.$$

The 90% points and 10% points occur therefore at

$$s = \frac{x}{\sqrt{4Dt}} = \pm 0.906,$$

that is,

$$x_{10\%} = +0.906\sqrt{4Dt}, \qquad x_{90\%} = -0.906\sqrt{4Dt}.$$

We thus see that these two points move away from $x = 0$ in proportion with the *square root* of the elapsed time. Let us point out that the growth of the width of the 90%–10% transition region with the square root of t is a direct expression of the fact that in random walk the root mean square distance $\sqrt{x^2}$ traveled by a particle is proportional to $\sqrt{t}$.

We now turn to the experimental verification of some aspects of this concentration change with time. In one experiment the slope $\partial C/\partial x$ of the profile at $x = 0$ has been measured as a function of time. This slope is, according to (2-70):

$$\frac{\partial C}{\partial x} = -\frac{C_0}{2}\frac{\partial \phi(x/\sqrt{4Dt})}{\partial x} = -\frac{C_0}{2}\left(\frac{d\phi}{ds}\right)\left(\frac{\partial s}{\partial x}\right),$$

where $s \equiv x/\sqrt{4Dt}$. From the definition of $\phi(s)$ in (2-69), it follows that

$$\left(\frac{d\phi}{ds}\right) = \frac{2}{\sqrt{\pi}}e^{-s^2} = \frac{2}{\sqrt{\pi}}e^{-x^2/4Dt}$$

and since $\partial s/\partial x = 1/\sqrt{4Dt}$, we have

$$\left(\frac{\partial C}{\partial x}\right) = \frac{-C_0}{\sqrt{4\pi Dt}}e^{-x^2/4Dt}. \tag{2-71}$$

At the point $x = 0$, the slope of the profile has its maximum value

$$\left(\frac{\partial C}{\partial x}\right)_{x=0} = \frac{-C_0}{\sqrt{4\pi Dt}}. \tag{2-72}$$

At the onset, $t = 0$, this slope is ∞. As time increases, the slope flattens out in inverse proportion to the square root of the elapsed time.

This relation has been verified experimentally by Lam and Polson [4]. They used solutions of the proteins ovalbumin and hemoglobin, and measured the concentration gradient $(\partial C/\partial x)_{x=0}$ optically by means of the deflection of a light beam; in their experiment this deflection, H, is proportional to the concentration gradient. According to (2-72), the product

$$\left(\frac{\partial C}{\partial x}\right)_{x=0} \times \sqrt{t} \tag{2-73}$$

should have the same value at all times. Lam and Polson's data, presented in Table 2.5, confirm this expectation.

Table 2.5. Values of the Product $(\partial C/\partial x)_0 \times \sqrt{t}$ as a Function of Time for Ovalbumin and Hemoglobin.

Ovalbumin (in buffer at pH = 4.6)				Hemoglobin (in buffer at pH = 6.5)			
t (h)	$\sqrt{t}$	$(\partial C/\partial x)_0$*	$\sqrt{t}\left(\frac{\partial C}{\partial x}\right)_0$	t (h)	$\sqrt{t}$	$(\partial C/\partial x)_0$*	$\sqrt{t}\left(\frac{\partial C}{\partial x}\right)_0$
8	2.828	25.2	71.26	15 h 50 m	3.958	42.2	167.0
12	3.464	20.8	72.05	24	4.796	34.4	165.0
24	4.796	14.95	71.70	28	5.292	32.7	173.0
32	5.657	12.9	72.97	40	6.325	27.4	173.3
48	6.928	10.4	72.05	48	6.928	25.0	173.2

*$(\partial C/\partial x)_0$ as measured by a beam deflection H in unspecified units. We see that indeed the product [(2-73)] is constant within the accuracy of the experiment.

It is also apparent that a *quantitative* determination of the relative concentration profile

$$\frac{1}{C_0}C(x,t) = \tfrac{1}{2}\left(1 - \phi(x/\sqrt{4Dt})\right)$$

or of the derivative

$$\frac{1}{C_0}\left(\frac{\partial C}{\partial x}\right) = \frac{-1}{\sqrt{4\pi Dt}}e^{-x^2/4Dt}$$

allow a determination of the diffusion constant D.

Such quantitative measurements of the concentration change are easily carried out with optical, mainly interferometric methods, using the fact that the optical index of refraction of a dilute solution changes linearly with the concentration of solute molecules. Lam and Polson found a value of D for ovalbumin ($M = $ 44,000 g/mol)

$$D = 7.55 \times 10^{-7}\ \mathrm{cm}^2/\mathrm{s},$$

that is consistent with the other data reported for globular proteins in Table 2.4(b).

2.3.C.   The Diffusion Equation for $C(x, t)$

In this section we derive the *rate* equation which describes the manner in which the concentration $C(x, t)$ changes with time. This equation, as we shall see, relates the local rate of change of C, as expressed by the time derivative $\partial C(x, t)/\partial t$, to the *local* spatial variation of C. This equation reveals aspects of the behavior of $C(x, t)$ that are not at all transparent in the integral expression, (2-65), for C.

This equation can easily be obtained from (2-65). Let us make clear again what this equation says: It expresses the concentration $C(x, t)$ at position x, at time t in terms of the concentration everywhere in space, at an initial time $t = 0$. We chose this time to be zero for convenience, but this is really immaterial; the equation also relates the concentration $C(x, t_1)$ at any time t_1 to the concentration $C(x', t_0)$ at some previous time t_0 provided we insert into (2-65) the elapsed time $(t_1 - t_0)$:

$$C(x, t_1) = \int dx' C(x', t_0) P(x - x', t_1 - t_0). \tag{2-72a}$$

In the same vein we can use this equation to relate the concentration $C(x, t + \Delta t)$ at time $t + \Delta t$ to the concentration $C(x', t)$ at time t, Δt being a small increment in time; this gives

$$C(x, t + \Delta t) = \int dx' C(x', t) P(x - x', \Delta t). \tag{2-73a}$$

We have the option of using, instead of x', the quantity

$$s = x' - x$$

as an integration variable in (2-73a). We must then write $C(x', t)$ as $C(x + s, t)$, and $P(x - x', \Delta t)$ as $P(-s, \Delta t) = P(+s, \Delta t)$, and finally dx' as ds. This gives an alternative, but equivalent expression

$$C(x, t + \Delta t) = \int ds\, C(x + s, t) P(s, \Delta t). \qquad (2\text{-}73b)$$

This expression is *exact*. We can, however, now exploit the fact that we intend the interval Δt to be very small. In this case, the function $P(x, \Delta t)$ is a very narrowly peaked Gaussian, whose width in s is of order $\sqrt{D\Delta t}$. This is shown in Figure 2.18 where $P(s, \Delta t)$ is plotted along with the function $C(x + s, t)$. The product $P(s, \Delta t)$ and $C(x + s, t)$ in (2-73a) will be vanishingly small as soon as $|x|$ is larger than the very small value $\sim \sqrt{D\Delta t}$. The value of the integral depends therefore only on the values of $C(x+s, t)$ in the *immediate* neighborhood of $s = 0$. In this small neighborhood $C(x + s, t)$ can be represented in terms of $C(x, t)$ and its first and second derivative at x:

$$C(x + s, t) = C(x, t) + s\frac{\partial C(x, t)}{\partial x} + \tfrac{1}{2}s^2\frac{\partial^2 C(x, t)}{\partial x^2} + \cdots . \qquad (2\text{-}74)$$

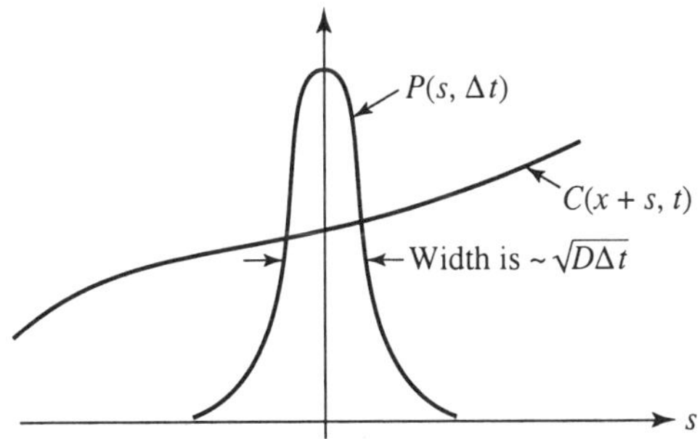

Figure 2.18. Plot of $P(s, \Delta t)$ and $C(x + s, t)$.

The fact that (2-74) is a good enough representation of $C(x + s, t)$ in (2-73b), provided Δt is very small, is what makes it possible to get local characterization of the space–time behavior of $C(x, t)$.

Inserting (2-74) into (2-73b), we find

$$C(x, t + \Delta t) = C(x, t) \int ds\, P(s, \Delta t)$$

$$+ \frac{\partial C(x, t)}{\partial x} \int ds\, s\, P(s, \Delta t)$$

$$+ \frac{1}{2} \frac{\partial^2 C(x, t)}{\partial x^2} \int ds\, s^2 P(s, \Delta t)$$

$$+ \text{negligible terms.} \tag{2-75}$$

Now let us look at the various s-integrals in (2-75): The first s-integral is the *normalization* of the probability distribution, and has the value 1:

$$\int ds\, P(s, \Delta t) = 1. \tag{2-75a}$$

The second defines the mean value $\bar{s}$ of s, and this is zero

$$\bar{s} = \int ds\, s\, P(s, \Delta t) = 0. \tag{2-75b}$$

The third defines the variance $\overline{s^2}$ of s, and this has the value $2D\Delta t$ (see Appendix 2.A.3 of this chapter):

$$\overline{s^2} = \int ds\, s^2 P(s, \Delta t) = 2D\Delta t. \tag{2-75c}$$

We can insert these results into (2-75) and find

$$C(x, t + \Delta t) \cong C(x, t) + D\frac{\partial^2 C(x, t)}{\partial x^2}\Delta t. \tag{2-76}$$

Moving $C(x, t)$ to the left and dividing by Δt, we have on the left the ratio

$$\frac{C(x, t + \Delta t) - C(x, t)}{\Delta t}.$$

which, as we let Δt shrink to zero, defines the partial time derivative $\partial C(x, t)/\partial t$. With this we have obtained the (one-dimensional) *diffusion equation* for the concentration $C(x, t)$:

$$\frac{\partial C(x, t)}{\partial t} = D\frac{\partial^2 C(x, t)}{\partial x^2}. \tag{2-77}$$

This is then the alternative *differential* description of the diffusion process. In this differential form, the equation states that the *local rate of change* of the concentration $C(x, t)$ is proportional to the second derivative of the spatial variation of C. This in itself is an important piece of information, as we shall see, and reveals a property of the behavior of $C(x, t)$ in space–time which is hidden in (2-65).

It is important to realize, however, that (2-65) and (2-77) are equivalent descriptions of the diffusion process. The first equation gives a *global* (integral) description, the second a *local* (differential) one.

Let us finally mention, without derivation, what the three-dimensional form of the diffusion equation is. It may be obtained from the three-dimensional generalization of (2-65):

$$C(x, y, z, t) = \int dx' \int dy' \int dz' C_0(x', y', z') P(x-x', t) P(y-y', t) P(z-z', t).$$

By repeating all the steps that lead to (2-77), one finds the equation

$$\frac{\partial C(xyz, t)}{\partial t} = D\left(\frac{\partial^2 C}{\partial x^2} + \frac{\partial^2 C}{\partial y^2} + \frac{\partial^2 C}{\partial z^2}\right) \tag{2-78}$$

Most of the applications will make use of the one-dimensional form only, that is, they will describe situations where the concentration C varies only in one direction, and therefore depends on a single space variable only.

2.3.D.   An Application: Smoothing Out of Sinusoidal Variations in Concentration

One of the advantages of having a differential description of the diffusion process, as given by the diffusion equation (2-77), is the possibility of identifying some general properties of the behavior of the concentration $C(x, t)$.

For instance, we see right away that the only *time-independent* concentration profile must have the form

$$C(x) = C_0 + C_1 x,$$

that is, the concentration must either be the same everywhere, or change linearly with position x. We shall deal in Section 2.4 with profiles of this type.

Another property of the diffusion equation that can be exploited is its *linearity*: Only the first power of C and its derivatives occur in (2-77). This linearity makes it possible to superpose solutions. Let us describe this precisely.

If a function $C_1(x, t)$ is a solution of the diffusion equation, and another function $C_2(x, t)$ is a solution too, then the superposition

$$C(x, t) = aC_1(x, t) + bC_2(x, t)$$

is a solution too. a and b are arbitrary constants. The solution $C(x, t)$ is called a superposition of the solutions $C_1(x, t)$ and $C_2(x, t)$. This superposition principle makes it possible to construct solutions $C(x, t)$ of (2-77) by making use of a "catalogue" of known solutions. We shall show presently that the evolution of all concentration profiles that vary *sinusoidally* in space can be simply determined. The superposition principle then yields at once the evolution of *any* concentration profile that can be written as a sum of sinusoidal functions. Representation of a function of x as a sum of sinusoidally varying functions is called in mathematics a *Fourier series* representation. Such representations are of great importance in the description of physical processes governed by *linear* equations, like diffusion, heat conduction, wave propagation, etc.

We now proceed to construct the solution $C(x, t)$ of the diffusion equation for a concentration profile which initially at $t = 0$ is sinusoidal:

$$C(x, t = 0) = C_0 + \Delta C_0 \sin\left(\frac{2\pi}{L}x\right). \tag{2-79}$$

What will happen as time goes on? We expect that the amplitude ΔC_0 of the wiggles of the concentration will gradually decrease and ultimately disappear. We therefore make the educated guess that the concentration $C(x, t)$ will have the general representation valid for all times

$$C(x, t) = C_0 + \Delta C(t) \sin\left(\frac{2\pi}{L}x\right) \tag{2-80}$$

and use the diffusion equation to determine the manner in which $\Delta C(t)$ depends on t. So we calculate the space and time derivatives

$$\frac{\partial C(x,t)}{\partial t} = \frac{d\,\Delta C(t)}{dt}\sin\left(\frac{2\pi}{L}x\right),$$

$$\frac{\partial^2 C(x,t)}{\partial x^2} = -\Delta C(t)\left(\frac{2\pi}{L}\right)^2\sin\left(\frac{2\pi}{L}x\right).$$

We see these derivatives have a common factor $\sin((2\pi/L)x)$ which we can lop off after inserting them into the diffusion equation, and obtain

$$\frac{d\,\Delta C(t)}{dt} = -D\left(\frac{2\pi}{L}\right)^2\Delta C(t) \equiv -\frac{1}{t_0}\Delta C(t). \tag{2-81}$$

The combination of factors $D(2\pi/L)^2$ has the dimension of a reciprocal time, and we call this $1/t_0$. Equation (2-81) is the differential equation for the decreasing exponential function with the solution

$$\Delta C(t) = \Delta C_0 e^{-t/t_0}. \tag{2-82}$$

This shows that the wiggles in the concentration *do* die out, exponentially, in fact; their amplitude is reduced by a factor e in the time

$$t_0 = \frac{L^2}{4\pi^2 D}. \tag{2-83}$$

The longer the wavelength (L) of the sinusoidal fluctuation, the longer it takes for the concentration fluctuation to die away.

If, in fact, one were able to measure experimentally both the time constant t_0 and the wavelength L of the sinusoidal fluctuation, one could determine the diffusion constant D using (2-83). This is actually the basis of the light scattering spectroscopy experiments mentioned in Section 2.2.D. These experiments exploit the fact that in an apparently uniform solution there exist spontaneous fluctuations in the concentration. These spontaneous concentration fluctuations can be expressed as a Fourier series, or the sum of "Fourier components" with many different wavelengths. Light can be scattered by these components, but light scattered into a direction θ is produced solely by a Fourier component of wavelength L where

$$L = \lambda/(\sin\theta/2).$$

Here λ is the wavelength of *light* in the medium. L is the wavelength of the scattering fluctuation. Thus, from the angle of scattering one obtains L. The damping time t_0 is obtained from the *spectrum* of the light scattered in the direction θ. In fact, if the spectral width (Δv) is measured in frequency units (cycles/s), the time t_0 is equal to $(2\pi \Delta v)^{-1}$. Thus from the *spectrum* of the scattered light, one finds the time t_0 for the decay of the sinusoidal fluctuation, and from the scattering angle θ one obtains L. The diffusion coefficient is then found using (2-83).

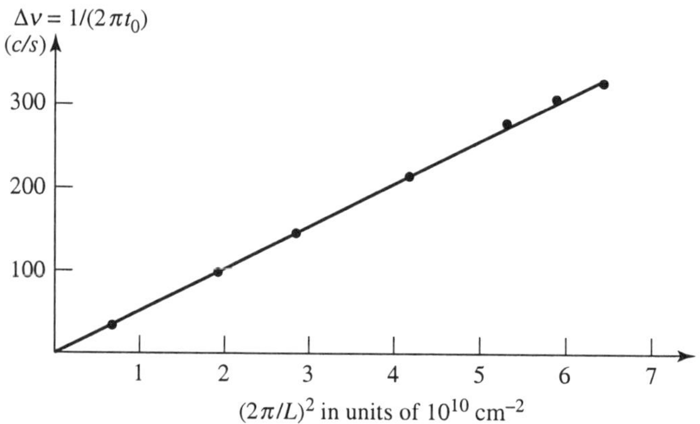

Figure 2.19. Graph of $1/(2\pi t_0)$ versus $(2\pi/L)^2$ for the light scattered from a solution of bacteriophage T_4 particles. This graph demonstrates the experimental verification of (2-83) and permits the determination of the diffusion coefficient D. (S. B. Dubin et al., *J. Mol. Biol.* **54**, 547 (1970).)

In Figure 2.19 we see a graph of $(2\pi t_0)^{-1}$ versus $(2\pi/L)^2$ for the light scattered from a nominally uniform solution of bacteriophage particles. The graph shows clearly that t_0 is proportional to L^2 as is predicted in (2-83). The slope of the graph yields a value of $D = (0.295 \pm 0.003) \times 10^{-7}$ cm^2/s.

2.4   Particle Conservation, Particle Current, and Fick's Law

2.4.A.   Particle Conservation, Current, and the Continuity Equation

Particle conservation is the simple statement of the fact that in the absence of a chemical reaction the total number of molecules of a given kind remains constant.

Molecules, like people, cannot just disappear. In this section, we shall formulate some implications of this fact.

To begin with, let us consider two rooms connected by a door. At a given time t, room 1 contains N_1 people, room 2, N_2 people. At some later time, t', it is found that there are now N_1' people in room 1, N_2' people in room 2. What can we conclude from this? We conclude, obviously, that there has been traffic through the door, and that the difference $N_1' - N_1$ can be accounted for as follows:

$$N_1' - N_1 = \text{number of people moving } in \text{ from room 2}$$
$$- \text{ number of people moving } out \text{ of room 1.}$$

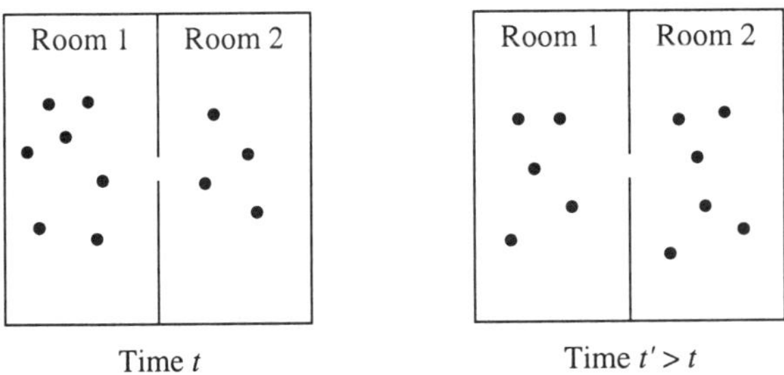

Figure 2.20. Change of population in two rooms in the time interval $(t' - t)$.

A similar statement can be made for $N_2' - N_2$. The point made here is that the change of population in either room is accompanied by a net *flow* of people across the doorway. The fact that people are conserved implies that the change of population in one room and the net flow across the doorway are not independent—from the knowledge of one, one can infer what the other is.

The same argument applies to molecules in solution. There, the number N of molecules in a volume V is given by the integral

$$N = \text{number of particles in } V$$
$$= \int_V dV\, C(xyz, t). \tag{2-84}$$

Let us again consider the example discussed at the end of Section 2.3, and illustrated in Figures 2.15 and 2.17. From Figure 2.17 we see that the number of molecules located in the upper part of the container ($x \geq 0$) increases, those in the lower part decreases. We conclude that there must be a *flow* of particles from the bottom towards the top.

Let us then *define* the concept of flow quantitatively. In Figure 2.21 we introduce the plane $x = 0$ that divides the tube of length $2L$ into two parts (1) and (2). Let N_1 be the number of particles in (1); N_2 that in (2). These numbers are

$$N_1(t) = A \int_{-L}^{0} dx\, C(x, t), \tag{2-85a}$$

$$N_2(t) = A \int_{0}^{L} dx\, C(x, t). \tag{2-85b}$$

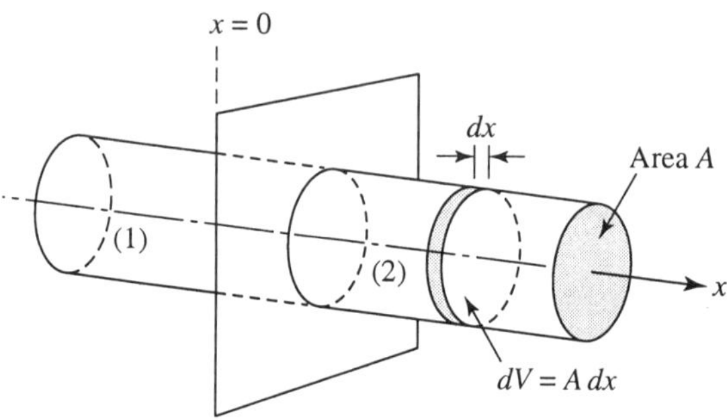

Figure 2.21. Tube with a solution of concentration $C(x, t)$, divided into two parts by plane $x = 0$.

(We wrote $dV = A\, dx$, see Figure 2.21.) Now let us *define* the *total current J* at $x = 0$ as

$$J(x = 0) = \begin{cases} \text{number of particles } per\ second \text{ crossing plane} \\ x = 0 \text{ from (1) into (2),} \\[6pt] -\text{ number of particles per second crossing that plane} \\ \text{from (2) into (1).} \end{cases} \tag{2-86}$$

So $J(x = 0)$ is a quantity which a "gatekeeper" stationed at $x = 0$ could determine. Since particles are conserved, J also represents the *rate of loss* particles in compartment (1):

$$J(x = 0) = -\left(\frac{dN_1}{dt}\right) \tag{2-87}$$

and the rate of gain of particles in compartment (2). A straight generalization of this arises when we consider two planes, located at x_1 and x_2. Let us call $N_{12}(t)$ the number of molecules in the space *between* the two planes. Let $J(x_1)$ stand for the total particle current at position $x = x_1$, and defined as in (2-86). Let $J(x_2)$ be the total current at position $x = x_2$. A positive $J(x_1)$ then represents a net *inflow* into the space between the planes; similarly, a positive $J(x_2)$ represents a net *outflow*. This gives us the relation

$$\frac{dN_{12}}{dt} = J(x_1) - J(x_2). \tag{2-88}$$

This equation is the integral form of the law of conservation of particles. If we can assume that the flow is the same at every point inside the circle C_1 and C_2, we can also introduce a *current density* or *flow per unit area*

$$j = \frac{J}{A} = \text{current density.} \tag{2-89}$$

Along with this, we write N_{12} in terms of the concentration as

$$N_{12}(t) = \int_{x_1}^{x_2} dV\, C(x, t) = A \int_{x_1}^{x_2} dx\, C(x, t).$$

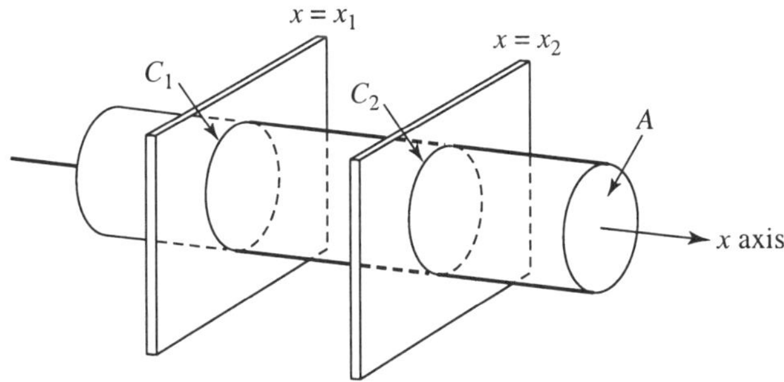

Figure 2.22. Tube intersected by two planes $x = x_1$ and $x = x_2$.

Thus we can write $(1/A)dN_{12}/dt$, using (2-88) and (2-89), as

$$\frac{d}{dt}\int_{x_1}^{x_2} dx\, C(x,t) = \int_{x_1}^{x_2} dx\, \frac{\partial C(x,t)}{\partial t} = j(x_1) - j(x_2). \tag{2-90}$$

With this result we are well on our way to establish the so-called differential form of the conservation law [(2-88)].

Consider the case where x_1 and x_2 are very close together

$$x_2 = x_1 + \Delta x,$$

and that $C(x,t)$ changes slowly enough with increasing x so that it has approximately the same value in the interval $x_1 \leq x \leq x_1 + \Delta x$. The left-hand side of (2-90) can then be written as

$$\Delta x \frac{\partial C(x_1,t)}{\partial t}.$$

On the right-hand side, we have

$$j(x_1) - j(x_1 + \Delta x) = j(x_1) - \left(j(x_1) + \Delta x \left(\frac{\partial j}{\partial x}\right)_{x_1} + \cdots \right) \cong -\Delta x \left(\frac{\partial j}{\partial x}\right)_{x_1}.$$

So, we get

$$\Delta x \frac{\partial C(x_1,t)}{\partial t} = -\Delta x \left(\frac{\partial j}{\partial x}\right)_{x_1}.$$

This must hold wherever x_1 is located. We therefore drop the subscript 1 on x_1 and write the result in the form

$$\frac{\partial C(x,t)}{\partial t} = -\frac{\partial j(x)}{\partial x}. \tag{2-91}$$

This is the so-called *continuity equation*, or differential form, of the conservation law of particles. Let us repeat that this equation states that the rate of change of the number of particles inside a volume element is equal to the excess of inflow over outflow of particles through the *boundaries* of that volume element. In

(2-91) we have expressed this statement in terms of the rate of change of the local concentration $C(x, t)$, and the derivative of the current density $j(x)$.

It is important to realize that the two forms of the statement of particle conservation, the integral form [(2-88)], and the differential form [(2-91)], have the same contents. Depending on the problem at hand, the one or the other formulation will be more useful. Whichever form is used, it states the fact that particles cannot disappear or be created, but only move about in space.

There is also a three-dimensional form of the particle conservation law that we shall state without proof. In formulating it, we must recognize that the particle current density is a vector in three dimensions, and must be written as $\vec{j}(x, y, z, t)$; its Cartesian components along the x-, y-, and z-axes are j_x, j_y, j_z:

$$\vec{j} = j_x \hat{\imath} + j_y \hat{\jmath} + j_z \hat{k}.$$

The flow of particles through a small surface area dA_x perpendicular to the x-axis is then given by

$$j_x \, dA_x$$

and corresponding expressions $j_y \, dA_y$, $j_z \, dA_z$ hold for the flow of particles through small areas dA_y, dA_z perpendicular to the y- and z-axes, respectively.

By repeating the "bookkeeping" argument which leads to (2-91), one establishes the general differential form of the particle conservation law

$$\frac{\partial C}{\partial t} = -\left(\frac{\partial j_x}{\partial x} + \frac{\partial j_y}{\partial y} + \frac{\partial j_z}{\partial z} \right).$$

Equation (2-91) is recovered as a special case in which j_y and j_z are zero, and $j = j_x$ is independent of the coordinates y and z of the particles.

2.4.B.   The Relation Between Current and a Concentration Gradient. Fick's Law

In this section we show how the diffusion equation (2-77), if used in conjunction with the conservation law of particles, gives us an expression for the current density in terms of the local concentration gradient.

Let us first consider the example illustrated in Figure 2.21, and use (2-85a) and (2-87). The rate of change of the number of particles in compartment (1) of the tube is

$$\frac{dN_1}{dt} = \frac{d}{dt} A \int_{-L}^{0} dx\, C(x, t) = A \int_{-L}^{0} dx \frac{\partial C(x, t)}{\partial t}.$$

Since the concentration $C(x, t)$ obeys the diffusion equation (2-77), the integrand $\partial C/\partial t$ can be written as $D\, \partial^2 C/\partial x^2$:

$$\frac{dN_1}{dt} = DA \int_{-L}^{0} dx \frac{\partial^2 C}{\partial x^2} = DA \int_{-L}^{0} dx \frac{\partial}{\partial x}\left(\frac{\partial C}{\partial x}\right). \qquad (2\text{-}92)$$

The integral of the derivative of a function of x is equal to the difference of the value of that function at the end points of the integration interval

$$\int_{a}^{b} dx \frac{df}{dx} = f(b) - f(a).$$

Applied to (2-92), this gives

$$\frac{dN_1}{dt} = DA\left(\left(\frac{\partial C}{\partial x}\right)_{x=0} - \left(\frac{\partial C}{\partial x}\right)_{x=-L}\right) = DA\left(\frac{\partial C}{\partial x}\right)_{x=0}, \qquad (2\text{-}93)$$

since at $x = -L$, the concentration profile is flat. According to (2-87), the right-hand side must represent the negative of the particle current $(J(x))_{x=0}$. Thus

$$\left(J(x)\right)_{x=0} = -DA\left(\frac{\partial C}{\partial x}\right)_{x=0}. \qquad (2\text{-}94)$$

This result shows that the current is proportional to the negative of the slope of the concentration profile, $(\partial C/\partial x)$. Reference to the cross section A of the tube can be eliminated by again introducing the current *density* (of dimension: number of particles/cm^2 s):

$$j(x) = \frac{J(x)}{A}$$

in terms of which (2-94) takes the form

$$j(x, t) = -D \frac{\partial C(x, t)}{\partial x}. \tag{2-95}$$

This is *Fick's law*, sometimes also called *Fick's first law*, Fick's second law being the diffusion equation (2-77). The reader may be puzzled by the apparent reversal of what comes first and what comes second in this terminology. Its roots are in history. Adolph Fick (1829–1901) was a physiologist. He made numerous contributions to muscle physiology. Circulatory physiologists know him as the originator of the "Fick principle" for the determination of cardiac output. To ophthalmologists, he is known for the development of tonometry, i.e., the measurement of the intraocular pressure from the force required to produce a given deformation of the cornea.

Fick's first law was stated in 1855 [5], that is, before the time when Maxwell and Boltzmann had formulated the kinetic theory of gases, and therefore, long before the random walk picture of the diffusion process was established. Fick's approach was therefore purely phenomenological, and in fact, based on an analogy with the phenomenon of *heat flow*. In his own words:

> The spreading of a dissolved material in a solvent obeys the same law that was claimed by Fourier for the spreading of heat in a conductor of heat. ...
> In the law of Fourier one has only to exchange the expression for the quantity of heat by the expression for the dissolved material, and the word temperature with the word solvent concentration.

The law of Fourier referred to above is the statement that the heat current density j_H is proportional to the gradient of temperature $T(x)$:

$$j_H = -\lambda \frac{\partial T}{\partial x}. \tag{2-96}$$

We may mention that another such relation was well known at the time; this is *Ohm's law*, that relates the electric current density j_e to the gradient of the electrostatic potential $\phi(x)$:

$$j_e = -\sigma \frac{\partial \phi}{\partial x}. \tag{2-97}$$

Let us briefly return to Fick's *second* law. Once the first law, (2-95), is accepted, it need only be combined with the continuity equation (2-91) to yield the diffusion equation as a consequence! Indeed, from

$$j(x) = -D\frac{\partial C}{\partial x} \quad \text{and} \quad \frac{\partial C}{\partial t} = -\frac{\partial j}{\partial x}$$

it follows at once that

$$\frac{\partial C}{\partial t} = D\frac{\partial^2 C}{\partial x^2}, \tag{2-77a}$$

so that in fact this becomes a second law.

Let us assess what we have gained by having Fick's second law. Starting with the random walk of a single particle, which has no preferred direction, we have arrived at a demonstration that in a spatial distribution of such particles, expressed by means of the concentration $C(x, t)$, there is a *net* flow in the direction of decreasing density. This flow is quantitatively expressed by Fick's first law, (2-95). It may appear strange at first sight that the *undirected* random walk of individual particles should result in a *directed* flow. But consider a case where the concentration $C(x)$ decreases from left to right, so that there is a higher concentration at the left, and the flow is toward the right. It is now important to realize that this flow represents the *excess* of particles random walking from left to right over those which go the other way. This excess exists simply because there are more particles at the left than at the right. The net flow represents an *average result* of a statistical process which acquires the nature of a "law," by virtue of the fact that an extremely large number of particles is generally involved, so that deviations from the average behavior are small.

The importance of Fick's law in physiology can hardly be overstated. Every living system is compartmentalized into subsystems. At the lowest level of organization we have cells, separated from each other, and from the outside, by cell walls and cell membranes. On a higher level of organization we have the circulatory, respiratory, digestive, etc., systems. Across the boundaries of all these systems material must constantly be exchanged: Nutrients move from the digestive tract into the blood stream, and later across the walls of capillary blood vessels into the tissues; oxygen in the alveoli of the lung moves into the blood capillaries, and carbon dioxide out of them into the lung, and so forth. Much of this traffic is governed by Fick's law. Even in the cases when so-called pumps operate to move ions from regions of low concentration into regions of high concentration, the diffusive counterflow described by Fick's law is there, and must be taken into account to determine the net efficiency of the pump.

The next section will be devoted to a description of the diffusive flow of molecules across membranes. Biological membranes are astonishing and complex structures. Their selectivity in allowing diffusive transport across them, and their ability to increase the concentration difference of certain ions in apparent violation of Fick's law are well known.

We shall therefore start our discussion of transport across membranes first with a description of artificial ("industrial") membranes that exhibit much simpler properties. This will be done in Section 2.4.C.

2.4.C.   Flow and Diffusion Across Porous Membranes in the Presence of Either a Concentration Difference ΔC or a Pressure Difference ΔP

At the end of the previous section we gave a sketch of the role of membranes in physiology. In this section we will establish a quantitative description of the passage of fluids and solutes across porous membranes.

There are two agents that can drive the transmembrane flow. The first is a difference of *hydrostatic pressure* between the two fluid compartments on either side of the membrane. Such pressures are physiologically important. They occur in the human body, for instance, across the walls of blood capillaries, between blood plasma, and interstitial fluid. (The familiar "swelling" of feet after prolonged upright posture is a manifestation of fluid transfer induced by a hydrostatic pressure difference.)

If the membrane pores are large compared to the size of the dissolved solute molecules, the fluid driven through the pores will have the same composition as that in the compartment with the higher pressure. However, if the pore size is small, say of the order of 50–100 Å diameter, then the passage of dissolved macro-molecules will be blocked, whereas smaller dissolved molecules will still pass through along with water. This process is known as ultrafiltration. It occurs, for instance, in the glomerulus of the kidney (see Section 2.4.D). We will show, however, that even molecules quite a bit smaller than the pore size are partly withheld by membranes. This property of a membrane is quantitatively expressed in terms of the so-called reflexion coefficient σ.

The second agent generating transmembrane flow is a gradient of solute concentration between the two compartments on either side of the membrane. This concentration gradient includes a *diffusive* flow of solute, as described by Fick's law. We shall show in Section 2.6 that a concentration gradient also generates a bulk flow opposite in direction to the diffusive solute flow.

It is now clear that prior to a quantitative description of these phenomena, we must give a careful definition of what we mean by "flow" across a membrane, since the phenomena deal both with flow of fluid ("bulk" flow), and flow of *solute* particles. Let us define these two flows:

(a) The *bulk* flow (or *volume* flow) J_v is defined simply as the total *volume* of fluid across the membrane per second and unit membrane area

$$J_v = \frac{\text{volume of fluid moving across membrane}}{\text{seconds} \times \text{cm}^2 \text{ of membrane surface}}. \qquad (2\text{-}98)$$

(b) In addition to this bulk flow, we must introduce a quantity describing the movement of solute particles. We therefore define the solute flow J_s as the *number of solute molecules* crossing a unit membrane area per second

$$J_s = \frac{\text{number of solute molecules moving across membrane}}{\text{seconds} \times \text{cm}^2 \text{ of membrane area}}. \qquad (2\text{-}99)$$

These two quantities are the ones most easily measured, the first, J_v, by volume change, the second by using radioactively labeled solute molecules. (This latter method can also be used to determine J_v; see part 2.4.C(iii) of this section.) It is important to realize that according to the definitions above, J_v and J_s have *different* units.

In Section 2.6 we will describe, in quantitative fashion, how these two flows J_v and J_s are related to *both* a pressure gradient and a concentration gradient across the membrane. The discussion in this present section will deal only with the simplest cases, considering *either* a pressure *or* a concentrated gradient, along with structurally well-defined membranes suitable for experimental verification of the relation we shall derive.

(i)   Volume Flow Across a Porous Membrane Under the Influence of a Pressure Gradient. The Hydraulic Permeability L_p

A porous membrane is a thin sheet of material of thickness Δx, pierced by small openings ("pores"), which in the simplest case are of cylindrical shape and perpendicular to the surface of the membrane. This is illustrated in Figure 2.23. Let us first assume that all solute molecules are small compared to the pore size. All components of the fluid will then pass through the pores at the same rate. The problem of calculating the volume flow J_v is then one of hydrodynamics, the flow of fluid through a cylindrical pipe. This flow is described by the so-called *Poiseuille* formula. This formula, derivable from elementary fluid mechanics, states that the volume of flow J in cm^3/s of a fluid of viscosity η through a cylindrical pipe of

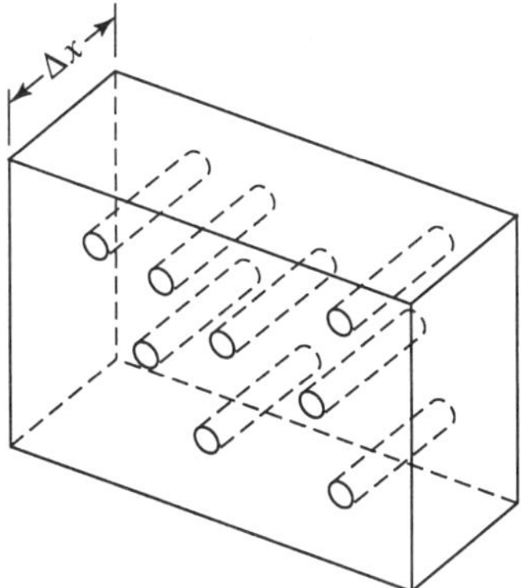

Figure 2.23. Enlarged view of a porous membrane with cylindrical pores.

radius a and length l is given by

$$J = \frac{\pi a^4}{8\eta l}\Delta p, \qquad (2\text{-}100)$$

Δp being the difference in fluid pressure between the ends of the pipe.

Consider now a membrane with n pores per cm^2 of membrane surfaces. Let a be the pore radius and Δx the membrane thickness. If a pressure difference Δp is set up between the two sides of a membrane, the flow J_v of fluid *per unit membrane area* will be

$$J_v = \text{flow per pore} \times \frac{\text{number of pores}}{\text{membrane area}}$$

$$= \frac{\pi a^4}{8\eta \Delta x}\Delta p \times n$$

or

$$J_v = L_p \Delta p, \qquad (2\text{-}101)$$

where

$$L_p = \left(n\frac{\pi a^4}{8\eta \Delta x}\right). \qquad (2\text{-}102)$$

L_p is called the *filtration coefficient* or *hydraulic permeability* of the membrane. It is expressed in units of (cm/s)/atm. The term "filtration coefficient" recalls the fact that a porous membrane may always be considered as a filter for coarse, suspended particles, large compared to the pore size. Membranes with pore sizes small enough to screen out even macromolecules (or virus particles), act as *ultra*filters.

Experimental verification of the validity of (2-102) will be given in part 2.4.C(iii) of this section.

(ii)   Solute Flow Across a Porous Membrane Due to a Concentration Gradient. The Membrane Permeability p

We now consider a different situation altogether, where the membrane separates two compartments containing solutions of *different* concentrations, but the same pressure. The membrane then serves as a barrier preventing rapid mixing of the fluids in the two compartments. Inside the membrane pores, however, a *concentration gradient* is set up, and solute particles will diffuse across with a net flow given by Fick's law.

Let us calculate this solute flow for the membrane introduced in 2.4.C(i). In Figure 2.24, the concentration profile for compartments (1) and (2), and the membrane is drawn. Two assumptions are made: First, that in compartments (1) and (2) the concentration remains *uniform*. This will be the case if the fluid is either stirred or circulates, as would be the case for blood plasma in a capillary. The second assumption is that the concentration profile in the pore is *linear*, that is, $\partial C/\partial x$ is constant. This is always the case if the two compartments are large enough so that the solute concentrations C_1 and C_2 change very slowly only in time. In this case, one has a quasi-stationary situation, where $\partial C/\partial t \cong 0$, and hence by (2-77), $\partial^2 C/\partial x^2 \cong 0$, which in turn implies that $\partial C/\partial x$ is constant. The concentration

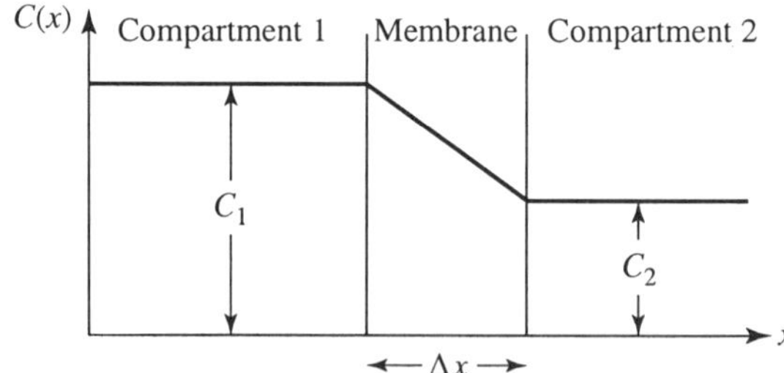

Figure 2.24. Concentration profile for a porous membrane separating two compartments.

gradient in the pore can therefore be written as

$$\frac{\partial C}{\partial x} = -\frac{C_1 - C_2}{\Delta x} = -\frac{\Delta C}{\Delta x}. \tag{2-103}$$

The flow of solute through a single pore, according to Fick's law, (2-95), is equal to

$$J/\text{pore} = -\text{pore area} \times \text{diffusion constant} \times \text{concentration gradient}$$
$$= -\pi a^2 D \left(\frac{C_2 - C_1}{\Delta x} \right)$$

As the number of pores per unit membrane area is n, the *solute* flow *per unit membrane area*, J_s, is given by

$$J_s = n(J/\text{pore}) = n\pi a^2 \frac{D}{\Delta x}(C_1 - C_2).$$

Using the symbol ΔC for the concentration difference $C_1 - C_2$, we can write this result as

$$J_s = p \Delta C. \tag{2-104}$$

In this relation, p is a constant whose value is given by the expression

$$p = n\pi a^2 \frac{D}{\Delta x}, \tag{2-105}$$

p is called the *membrane permeability*. Since the concentration C is measured in molecules/cm^3, and the solute flow J_s in molecules/cm^2, p has the dimension of cm/s, a velocity! The value of p remains unchanged, if C is measured in mol/cm^3 *and J_s in units of mol/cm^2 s.*

(iii)   Numerical Values for the Filtration Coefficient L_p and Permeability p. Theory and Experiment Compared. The Hindrance Factor

Most synthetic membranes used in the laboratory or in medicine, like cellophane, have a structure more complex than the membrane model used in 2.4.C(i) and 2.4.C(ii). In cellophane, for instance, the "pores" are actually irregular openings of nonuniform size, and providing a tortuous channel through the membrane that is longer than the membrane thickness. The volume flow J_v produced by a pressure difference Δp is still given by (2-101):

$$J_v = L_p \Delta p, \qquad (2\text{-}101a)$$

but the interpretation of L_p is now not a simple matter any more. Similarly, the *solute* flow J_s generated by a concentration difference ΔC is still given by (2-104)

$$J_s = p \Delta C, \qquad (2\text{-}104a)$$

but again it no longer is possible to express the permeability coefficient p by a formula like (2-105). It is therefore of importance to be able to verify expressions [(2-102)] for L_p and (2-105) for p for membranes, that actually do have cylindrical pores of uniform size, and for which the number of pores n per unit area, the pore radius a, and the membrane thickness Δx, can be measured. This has only recently become possible (see Bean [8], and Beck and Schultz [7]). Membranes with the required characteristics have been produced by irradiating very thin sheets of mica with the fission fragments of uranium or californium. Along the fission track, a cylindrical hole can be etched out with hydrofluoric acid. The number of pores/cm can be measured, and pore sizes estimated with an electron microscope.

Bean [8] reports an experiment carried out with pore radii ranging from $a = 150$ Å to 1500 Å. Both L_p and p are determined in the same experiment; the "solute" is a small amount of radioactivity labeled water (HTO), admixed in water in compartment (1). Compartment (2) has pure water initially. The transfer rate of radioactivity is measured as a function of pressure difference Δp. For large Δp, this transfer is mainly due to the volume flow of water; for $\Delta p = 0$, it is due to diffusion alone. It should be clear that the permeability p so measured is *that of water*, since the diffusion process involved is that of labeled water in water. We shall report also on experiments in which the permeability p for solutes other than labeled water are measured. These experiments reveal unambiguously that the equation (2-105) for the permeability p is adequate for water, but definitely predicts too large values of p for other, larger solute molecules.

Let us begin with a report on one set of Bean's data [8]:

membrane data: number of pores: $(nA) = 7.54 \times 10^4$,
$(A =$ membrane area$)$,
membrane thickness: $\Delta x = 4.9 \times 10^{-4}$ cm;

physical data: water temperature: $T = 293\,°\mathrm{K} \sim 86\,°\mathrm{F}$,
viscosity of water: $\eta = 1.0 \times 10^{-2}$ g/cm/s,
diffusion constant D_0 for water in water:
$$D_0 = 2.0 \times 10^{-5}\ \mathrm{cm^2/s}.$$

In Figure 2.25, we present Bean's results for the transmembrane flow as a function of pressure difference Δp. It is possible to analyze the data shown in this figure to obtain the filtration coefficient L_p and the permeability p that characterize the membrane. In fact, since both L_p and p, according to (2-102) and (2-105), can be related to the pore radius (a), we can examine the validity of these equations by separately computing (a) from the values of L_p and p deduced from Figure 2.25. If the values of (a) so obtained are in agreement with one another, we can have some confidence in the applicability of (2-102) and (2-105) to the flow of water across these thin mica membranes.

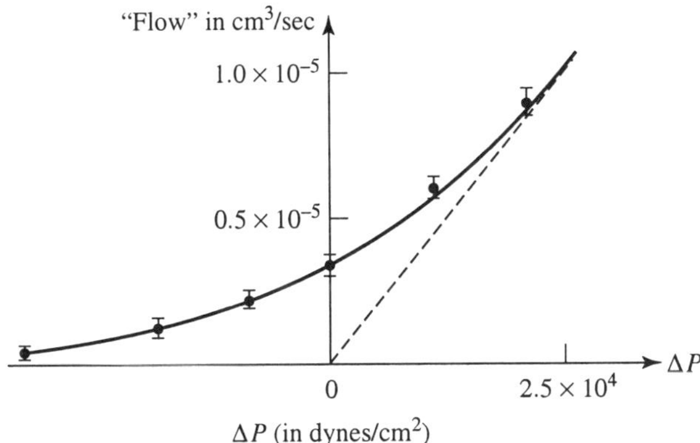

Figure 2.25. Observed transmembrane flow of radioactively labeled water as a function of pressure difference Δp between two compartments separated by porous membrane.

To extract L_p and p from Figure 2.25, we must first understand clearly what is plotted there. The quantity plotted vertically called "flow" is proportional to the solute flow J_s and is obtained from measurements of the transfer of radioactivity from compartment (1) into compartment (2). Assuming a concentration C_1 of radioactive water HTO in compartment (1) and an initial concentration $C_2 = 0$ in compartment (2), the rate at which radioactive (solute) molecules are transferred is

$$\text{transfer rate} = \text{membrane area} \times \text{solute flow/unit area}$$
$$= A J_s \ (\text{molecules/s}).$$

This rate determines the amount of radioactivity that is measured in compartment (2) as a function of time. By measuring the transfer rate radioactively and by knowing the concentration C_1 (in molecules/cm^3) of HTO in compartment (1), one can obtain the quantity $A J_s / C_1$ (cm^3/s). This is called "flow" in Figure 2.25, and is plotted vertically. Horizontally one plots the hydrostatic pressure Δp across the membrane. We can obtain the permeability p from the value of the "flow" when $\Delta p = 0$. In this case, only a concentration difference ΔC is present, and according to the definition of J_s [(2-104)], we have

$$\text{"flow"} = \frac{J_s A}{C_1} = \frac{p \Delta C A}{C_1} = p A, \qquad (2\text{-}106)$$

since $\Delta C = C_1 - C_2$, and we assume $C_2 = 0$: Thus the value of the flow at $\Delta p = 0$ is the product of the membrane permeability p and the area A. $p = (\text{flow }(\Delta p = 0)/A)$. Instead of numerically computing p, let us compute instead the radius of the pore using (2-105):

$$p = \frac{n \pi a^2 D}{\Delta x} = \frac{\text{"flow" } (\Delta p = 0)}{A}.$$

This gives

$$a^2 = \frac{\text{flow } (\Delta p = 0)}{\pi (n A)} \cdot \left(\frac{\Delta x}{D} \right).$$

or

$$a^2 = \frac{0.33 \times 10^{-5}}{(3.14)(7.54 \times 10^4)} \left(\frac{4.9 \times 10^{-4}}{2.0 \times 10^{-5}} \right) = 3.42 \times 10^{-10} \ \text{cm}^2$$

or

$$a = 1.85 \times 10^{-5} \ \text{cm}$$

from the membrane permeability.

We can now use Figure 2.25 to determine L_p. As you will recall, in the presence of a pressure difference Δp alone, the volume flow J_v is given by $J_v = L_p \Delta p$. In Bean's experiment *both* a concentration gradient *and* a pressure gradient are present. If *only* a pressure difference existed across the membrane, the "flow" would be linearly proportional to Δp as shown by the dotted line in Figure 2.25. The difference between the dotted line and the experimental line shows the effect of diffusion on the flow. For pressures greater than about 2.5×10^4 dyn/cm^2, the flow is determined almost entirely by the pressure difference across the membrane. Thus, for $\Delta p > 2.5 \times 10^4$ dyn, if we can relate the "flow" to J_v, we could deduce L_p. In fact, we can make this relation quite simply as follows. The solute flow J_s is equal to the bulk flow J_v times the concentration of solute C_1:

$$J_s \cong C_1 J_v, \tag{2-107}$$

because the volume flow of water also carries with it the solute flow of HTO. (Remember in checking the units of (2-107) that J_s is measured in molecules/cm^2, J_v in cm^3/cm^2 s, and C_1 has the units of molecules/cm^3.) In the regime that the "flow" is linearly proportional to Δp in the figure we have, on using (2-107)

$$\text{"flow"} \equiv \frac{A J_s}{C_1} = A J_v = A L_p \Delta p, \tag{2-108a}$$

$$L_p = \left(\frac{\text{"flow"}}{A \Delta p} \right)_{\Delta p > 2.5 \times 10^4 \ \text{dyn/cm}^2}. \tag{2-108b}$$

This permits a determination of L_p provided that the surface area A of the membrane is specified. Let us use (2-108b) along with (2-102) to compute the pore radius now from L_p. That is,

$$L_p = \frac{\text{"flow"}}{A \Delta p} = \frac{n \pi a^4}{8 \eta \Delta x} \tag{2-109}$$

or

$$a^4 = \frac{8 \eta \Delta x}{\pi (n A)} \frac{\text{"flow"}}{(\Delta p)}.$$

Using now the parameters given on pages 156 and 157, we find

$$a^4 = \frac{8(1.0 \times 10^{-2})(4.9 \times 10^{-4})(1 \times 10^{-5})}{(3.14)(7.54 \times 10^4)(2.5 \times 10^4)},$$

$$a^4 = 6.62 \times 10^{-20} \text{ cm}^4,$$

$$a = 1.6 \times 10^{-5} \text{ cm} \quad \text{(from hydraulic permeability)}.$$

Thus we have the two values of a, namely $a = 1.85 \times 10^{-5}$ cm and $a = 1.6 \times 10^{-5}$ cm, as determined from the diffusive membrane permeability p, or the hydraulic permeability L_p. These two values are in quite satisfactory agreement with one another. In fact, the two deductions of the pore size can be brought into even better agreement if one takes into account the fact that the pore is actually diamond-shaped rather than circular. a represents the radius of an "equivalent" circular pore profile. The Poiseuille flow through a pore with a diamond-shaped profile is somewhat less than that through a pore with circular profile of the same area, as given by (2-100).

The importance of these data, as already mentioned, lies in the fact that membranes can be built for which all parameters can be reliably measured, and for which the validity of the formula (2-105) for the membrane permeability p can be established, at least when the solute is radioactively labeled water.

For solute molecules *other* than water (sugars, urea, etc.), the value of the permeability p can then be *predicted* by means of (2-105), and using the known diffusion constant D of these molecules in water. How does this prediction stand up against a *measurement* of p for such molecules?

In Table 2.6 we present data recently obtained by Beck and Schultz [7], carried out with fission track etched mica membranes. Pore numbers and membrane thick-

Table 2.6. Membrane Permeability of a Porous Membrane (Pore Radius $a = 66$ Å) for Various Solutes.

Solute	Diffusion coeff. $D \times 10^6$ (in cm²/s)	Molecular radius r (in Å)	Permeability $\times 10^4$ (in cm/s)	$D_{\text{eff}}/D =$ Hindrance factor
Urea	13.8	2.04	7.09	0.86
Glucose	6.73	4.44	3.42	0.82
Sucrose	5.21	5.55	2.45	0.78
Raffinose	4.34	6.45	1.92	0.72
2-Dextrin	3.44	8.00	1.33	0.61
β-Dextrin	3.22	8.98	1.17	0.55
Ribonuclease	1.18	21.6	0.093	0.13

ness were measured, and pore radii calculated from (2-109). (Even the electron microscope does not give a reliable measurement of pore size.) The first column of this table lists molecules in order of increasing molecular weight. The second column lists the "free" diffusion constant, D, of the molecule in water at 25 °C. The third column gives the radii of the tabulated molecules that are seen to range from 2.64 Å for urea to 21.6 Å for ribonuclease. The fourth column lists the *experimentally* determined permeability p of the membrane. Since all the parameters of the membrane are known in this experiment, p can also be calculated from (2-105). This calculation gave values for p larger than the measured values, if the "free" diffusion constant D is used. The authors then fitted the calculated values of p to the data by using an effective diffusion constant

$$D_{\text{eff}} = D \times \text{hindrance factor},$$

which, if introduced into (2-105), reproduces the observed value of p:

$$p(\text{observed}) = D_{\text{eff}} \frac{n\pi a^2}{\Delta x} = D \frac{n\pi a^2}{\Delta x} \times \text{hindrance factor.} \qquad (2\text{-}110)$$

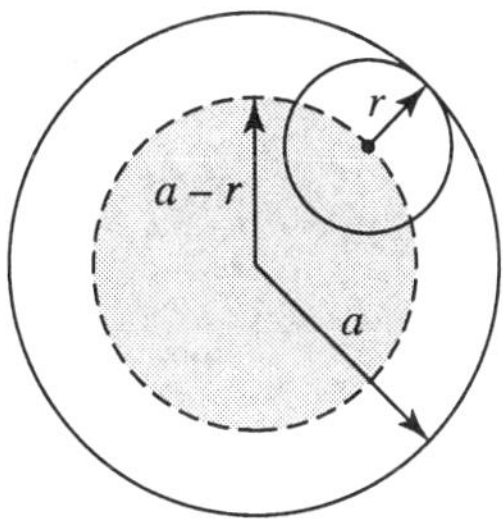

Figure 2.26. Effective pore area for molecules of radius r.

In the last column of the table this hindrance factor, the ratio of D_{eff} and D, is tabulated. We see that it systematically increases with the molecular radius. There are two effects which contribute to the hindrance factor. The first is easily understood by a simple geometric argument. (See Figure 2.26.) The molecule, of radius r, will not enter the pore if in its random motion it hits the rim of the pore. The effective pore area is therefore not πa^2, but only $\pi (a - r)^2$:

$$\text{effective pore area} = \pi a^2 \left(1 - \frac{r}{a}\right)^2 . \tag{2-111}$$

This argument, then, gives a hindrance factor of $(1 - r/a)^2$. A look at Figure 2.27, in which we plot D_{eff}/D as a function of (r/a), shows that the data points lie below the curve $(1 - r/a)^2$. An additional contribution was indeed calculated by Renkin [6] based on a calculation of an additional drag on the diffusing molecule due to the proximity of the pore walls: Renkin's complete expression for the hindrance factor is

$$\frac{D_{\text{eff}}}{D} = \left(1 - \frac{r}{a}\right)^2 \times \left(1 - 2.1\left(\frac{r}{a}\right) + 2.09\left(\frac{r}{a}\right)^3 - 0.95\left(\frac{r}{a}\right)^5\right) . \tag{2-112}$$

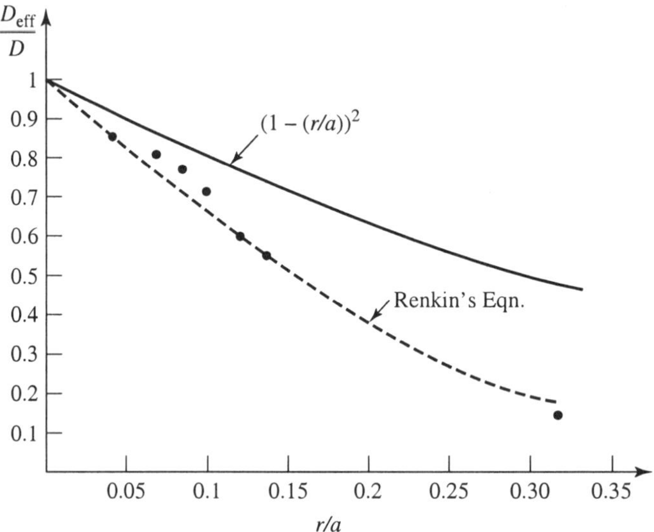

Figure 2.27. Hindrance factors (D_{eff}/D) in diffusion through a porous membrane. The circles represent the data points given in Table 2.6. The solid line is $(1 - (r/a)^2)$. The dashed line is the theory for (D_{eff}/D) given by (2-112).

This hindrance factor is plotted in Figure 2.26 as a dashed line, and seems to give a better, though not perfect, account of the observations. Ribonuclease, which has a molecular radius about one-third of the pore radius, shows the hindrance effect very dramatically. The permeability of the membrane to the passage of this molecule is reduced by nearly a factor of 10 because of the hindrance of the molecular motion by the proximity of the walls. For such large molecules we see that the volume flow *cannot* be so simply connected to the solute flow as we assumed for the case of HTO in water. There we had $J_s = C_1 J_v$. The hindrance effect described above shows, however, that the actual flow of solute molecules through a membrane can be much smaller than $J_s = C_1 J_v$ because of the hindrance effect. In the next subsection we will consider the effect of this hindrance on the flow of solute.

**(iv) Molecular Sieving by Membranes. The Reflection Coefficient σ.
Introduction to the Relation Between Solute Flow J_s,
Volume Flow J_v, and the Concentration and Pressure Differences
ΔC and Δp Across the Membrane**

In this subsection we will summarize the results on solute and volume flow across membranes and describe their extension to the more general case, where *both* a pressure difference Δp *and* a solute concentration difference ΔC exists across a membrane. In providing this generalization, we will distinguish three cases:

(a) The membrane pore cannot distinguish a solute from a solvent. This is the case, for instance, for radioactive water (HTO) as a solute.

(b) The membrane is impermeable to the solute. The solvent will pass through the pores, but the solute is completely "sieved" out. Such a membrane is called a semipermeable membrane. An example of this is the blood capillaries which are impermeable to the blood plasma proteins, but permit passage of water.

(c) The membrane is somewhat permeable to the solute. This is the "general" case. Solute permeability depends on the size of solute particles, as expressed by Renkin's "hindrance factor."

In case (a), it is easy to establish the general relations for volume flow and solute flow. Assume fluid flows in the pore with velocity v. The solute *flow* density $j_s(x)$ (solute particles/cm^2 s) at point x in the pore is then given by the sum of two terms:

(1) $C_s(x)v = $ solute carried by bulk flow; and

(2) $-D\dfrac{\partial C_s}{\partial x} = $ diffusional solute flow according to Fick's law.

So the solute flow density j_s is

$$j_s = C_s(x)v - D\frac{\partial C_s}{\partial x}. \tag{2-113}$$

If this flow is stationary ($C_s(x)$ depends on x only, but does not change with time), then j_s and v are independent of the location x along the pore. We can then average (2-113) over the pore, replacing

$$C_s(x) \quad \text{by} \quad \overline{C}_s$$

and

$$\frac{\partial C_s(x)}{\partial x} \quad \text{by} \quad \overline{\left(\frac{\partial C_s}{\partial x}\right)} = -\frac{\Delta C_s}{\Delta x},$$

Δx being the membrane thickness. This gives us the equation

$$j_s = \overline{C}_s v + \frac{D}{\Delta x}\Delta C_s. \tag{2-114}$$

Recall now that the volume flow J_v is defined as the volume of fluid passing the unit membrane area per second. This is related to v by

$$J_v = \text{number of pores/unit membrane area} \times \text{pore area}$$
$$\times \text{ flow velocity } v,$$
$$J_v = n(\pi a^2)v.$$

Similarly, the solute flow J_s is defined as the number of solute particles passing the unit membrane area per second

$$J_s = \text{number of pores/unit membrane area} \times \text{pore area}$$
$$\times \text{ number of solute particles per unit area flowing through pore}$$

or

$$J_s = n(\pi a^2)j_s$$

We can therefore use (2-114) and express J_s in the form

$$J_s = \overline{C}_s J_v + p\Delta C, \tag{2-115a}$$

where

$$p = n(\pi a^2)\frac{D}{\Delta x}$$

is the previously defined (unhindered) membrane permeability (see (2-105)). Equation (2-115a) and the hydrodynamic equation for volume flow

$$J_v = L_p \Delta p \tag{2-115b}$$

are the generalized flow equations for case (a) of a membrane that does *not* discriminate between solvent and solute.

Case (b) represents the other limit, where the membrane does not allow passage of the solute (the so-called semipermeable membrane). In this case we have, instead of (2-115a), the equation

$$J_s = 0. \tag{2-116a}$$

The equation for the volume flow is strongly modified in this case by the presence of a concentration difference ΔC across the membrane. We will show in Section 2.6 that instead of (2-115b), we have now an equation of the form

$$J_v = L_p \Delta p_{\text{eff}},$$

where the effective pressure difference Δp_{eff} is given by the expression

$$\Delta p_{\text{eff}} = \Delta p - kT\,\Delta C.$$

The term $kT\,\Delta C$ is referred to as the *osmotic pressure difference*, and we will establish its origin in Section 2.6. So we have the relation

$$J_v = L_p(\Delta p - kT\,\Delta C) \tag{2-116b}$$

and

$$J_s = 0. \tag{2-116a}$$

Equations (2-116a, b) represent the general flow equations for the case of the semipermeable membrane, i.e., case (b) mentioned above.

The *general* case (c) is still more complex, and we will just *state* the results, for completeness, deferring a detailed derivation to Section 2.6. Then, we will show that the solute dragged along by the fluid flowing through a pore at speed v is no longer given by $C_s v$, but by a reduced quantity $(1 - \sigma)C_s v$, σ being a pure number between 0 and 1. σ is called the reflection coefficient. ($\sigma = 1$ represents case (b), the semipermeable membrane.) At the same time, the effective pressure difference responsible for volume flow J_v is in this case given by

$$\Delta p_{\text{eff}} = \Delta p - \sigma kT \Delta C.$$

The equations relating solute flow J_s and volume flow J_v to the differences Δp and ΔC of pressure and solute concentration across the membrane are thus

$$J_s = (1 - \sigma)C_s J_v + p\Delta C \tag{2-117a}$$

$$J_v = L_p(\Delta p - \sigma kT \Delta C) \tag{2-117b}$$

We see that the two limiting cases (a) and (b) are covered by these equations, provided we observe that for $\sigma \to 1$, the permeability p also goes to zero.

These relations are of great importance in the physiology of plants.

(v) Equalization Time for the Concentration Difference Across a Membrane, a Two-Compartment Problem

The physiological role of membranes is twofold. On the one hand, they serve as barriers; on the other hand, they must permit the *transfer* of selected solutes from one compartment to another. For instance, blood capillaries must deliver glucose and oxygen to tissues, pick up carbon dioxide and metabolic waste. It is clear that by simple diffusion no *complete* transfer of material from one compartment to another is possible. The best that can be done is that an initial concentration unbalance can be equalized.

In this subsection, we address the question: Given a fluid compartment (1) of volume V_1, containing a solute at concentration C_1, and, separated by a membrane of area A, a compartment of volume V_2, filled with solute at concentration C_2, what is the time required for the two concentrations to become substantially equal? In Figure 2.28, we sketch the situation schematically. We assume that initially, $C_1 > C_2$, so that the flow of solute will be from compartment (1) to (2). Let $N_1(t)$ and $N_2(t)$ stand for the number of molecules in compartments (1) and (2), respectively. In the process of diffusion, the total number $N = N_1 + N_2$ remains constant. But N_1 decreases with time, and N_2 increases with time due to the solute flow from (1) to (2). The solute flow J_s per unit membrane area is given by (2-104). The total solute flow per second for the entire membrane of area A is then

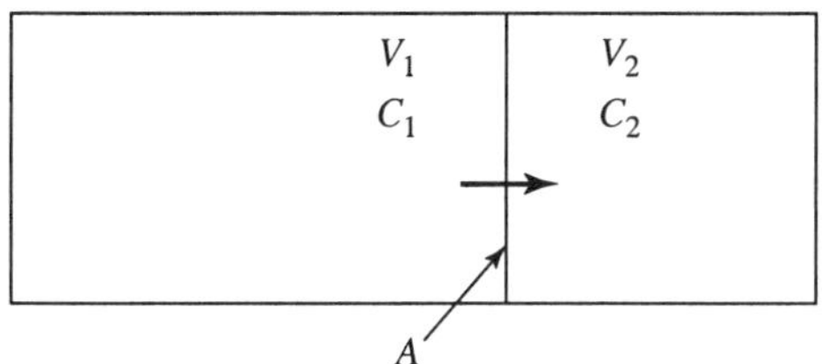

Figure 2.28. Two compartments separated by a permeable membrane.

$J_s A$ particles/s. This is the rate of decrease of N_1 and the rate of increase of N_2:

$$\frac{dN_1}{dt} = -J_s A, \qquad \frac{dN_2}{dt} = +J_s A. \tag{2-118}$$

From these continuity equations, we can get equations for the concentrations $C_1(t)$ and $C_2(t)$ by using (2-104) for J_s in terms of $(C_1 - C_2)$, and by expressing $N_1(t)$ and $N_2(t)$ in terms of C_1 and C_2:

$$N_1(t) = C_1(t)V_1,$$
$$N_2(t) = C_2(t)V_2. \tag{2-119}$$

In writing these last equations, we assume that the concentrations in compartments (1) and (2) are and remain *uniform*, which requires circulation or stirring of the fluid. Inserting (2-104) and (2-119) into (2-118), we find

$$\frac{dN_1}{dt} = V_1\frac{dC_1}{dt} = -Ap(C_1 - C_2),$$
$$\frac{dN_2}{dt} = V_2\frac{dC_2}{dt} = +Ap(C_1 - C_2). \tag{2-120}$$

These two coupled equations for C_1 and C_2 must now be solved. At first, we exploit the already mentioned conservation law for the total number of particles $N = N_1 + N_2$:

$$\frac{dN}{dt} = \frac{dN_1}{dt} + \frac{dN_2}{dt} = 0, \tag{2-121}$$

a property which, as can be seen, is built into the system of (2-120). Written in terms of the concentrations C_1 and C_2, (2-121) takes the form

$$V_1 \frac{dC_1}{dt} + V_2 \frac{dC_2}{dt} = \frac{d}{dt}(V_1 C_1 + V_2 C_2) = 0. \tag{2-122}$$

That is, the quantity $V_1 C_1 + V_2 C_2$ retains the same value at all times. Initially, we start with two different concentrations $C_1(0)$ and $C_2(0)$; ultimately, we will have a single, uniform concentration C_∞ for both compartments. Equation (2-122) then tells us that

$$V_1 C_1(0) + V_2 C_2(0) = V_1 C_1(t) + V_2 C_2(t) = (V_1 + V_2)C_\infty. \tag{2-123}$$

The *final*, common concentration C_∞ is therefore related to the initial concentrations $C_1(0)$ and $C_2(0)$ by

$$C_\infty = \left(\frac{V_1}{V_1 + V_2}\right) C_1(0) + \left(\frac{V_2}{V_1 + V_2}\right) C_2(0). \tag{2-124}$$

As a second step, we construct from the two equations (2-120) a single equation for the concentration *difference* $C_1 - C_2$. This requires dividing the first equation by V_1, the second by V_2, and subtraction of the two equations

$$\frac{dC_1}{dt} - \frac{dC_2}{dt} = -Ap\left(\frac{1}{V_1} + \frac{1}{V_2}\right)(C_1 - C_2),$$

which we can rewrite as

$$\frac{d}{dt}(C_1 - C_2) = -Ap\left(\frac{V_1 + V_2}{V_1 V_2}\right)(C_1 - C_2). \tag{2-125}$$

The combination of constants multiplying $(C_1 - C_2)$ on the right-hand side of this equation has the dimension of a reciprocal time, as can easily be verified. We therefore use for it the symbol $1/\tau_0$:

$$Ap\frac{V_1 + V_2}{V_1 V_2} \equiv \frac{1}{\tau_0}. \tag{2-126}$$

We now recognize in this equation for $\Delta C = C_1 - C_2$ the differential equation for the exponential function

$$\frac{d\Delta C}{dt} = -\frac{1}{\tau_0}\Delta C(t), \tag{2-127}$$

which has the solution

$$\Delta C(t) = \Delta C(0)e^{-t/\tau_0}. \tag{2-128}$$

This illuminates the interpretation of the constant τ_0 introduced in (2-126): τ_0 is the time needed for the original concentration difference to drop to $1/e = 1/2.7$ of its initial value. It represents the characteristic *time constant* of the equalization process. In time $2\tau_0$, the difference ΔC will have dropped to $1/e^2 = 1/7.4$ of its initial value, etc.

The characteristic time τ_0, as given by (2-126) depends on the membrane area, membrane permeability, and the volumes V_1 and V_2 of the compartments. It is always useful, in such a case, to test such an expression against one's intuition. Does τ_0 depend on the various factors as we expect? Writing τ_0 as

$$\tau_0 = \frac{1}{Ap}\,\frac{1}{\left(\dfrac{1}{V_1} + \dfrac{1}{V_2}\right)} \tag{2-126a}$$

we see that τ_0 decreases as the membrane area is increased; this is OK. It also decreases if the permeability is increased. The dependence on volume is more complex, and we must distinguish several cases. Consider first the case where the two volumes V_1 and V_2 are the same: $V_1 = V_2 = V$. In this case

$$\tau_0 = \frac{1}{2}\frac{V}{Ap} \quad (V_1 = V_2 = V),$$

that is, the characteristic time is in direct proportion to the volume of the two compartments. Since this volume determines the amount of material which must be transferred to equalize concentrations, it is intuitively clear that τ_0 should be proportional to V.

Another important special case arises when V_1, the compartment with the higher initial concentration, is much larger than V_2: $V_1 \gg V_2$. Then $1/V_1$ is much smaller than $1/V_2$, and τ_0 is approximately equal to

$$\tau_0 \simeq \frac{V_2}{Ap} \quad (V_1 \gg V_2),$$

that is, independent of V_1. This is so because V_1 now represents an essentially infinite reservoir of particles at concentration C_1. As can be seen from (2-124), the final common concentration will be substantially equal to C_1 in this case, and τ_0 measures the characteristic time to bring the originally lower concentration C_2 (in volume V_2) up to that level. The concentration in V_1 hardly changes in this process. The opposite limit $V_2 \gg V_1$ gives

$$\tau_0 \simeq \frac{V_1}{Ap}$$

and again the independence of τ_0 in V_2 has a similar interpretation.

We will see, in the next section on the artificial kidney, what values the time constant τ_0 may have. But before going into this, let us pursue further how the individual concentrations $C_1(t)$ and $C_2(t)$ evolve in time. To obtain an expression for $C_1(t)$ separately, we must combine the results (2-123) and (2-130): (2-123) gives

$$\frac{V_1}{V_2}C_1(t) + C_2(t) = \frac{V_1 + V_2}{V_2}C_\infty.$$

Equation (2-128) gives

$$C_1(t) - C_2(t) = \left(C_1(0) - C_2(0)\right)e^{-t/\tau_0},$$

from which we can eliminate $C_2(t)$ and obtain

$$\left(\frac{V_1}{V_2} + 1\right)C_1(t) = \frac{V_1 + V_2}{V_2}C_\infty + \left(C_1(0) - C_2(0)\right)e^{-t/\tau_0}.$$

Using (2-123), $C_1(0) - C_2(0)$ can be written as

$$\frac{V_1 + V_2}{V_2}(C_1(0) - C_\infty),$$

which gives the final result

$$C_1(t) = C_\infty + \left(C_1(0) - C_\infty\right)e^{-t/\tau_0}. \tag{2-129}$$

In a similar way, we find

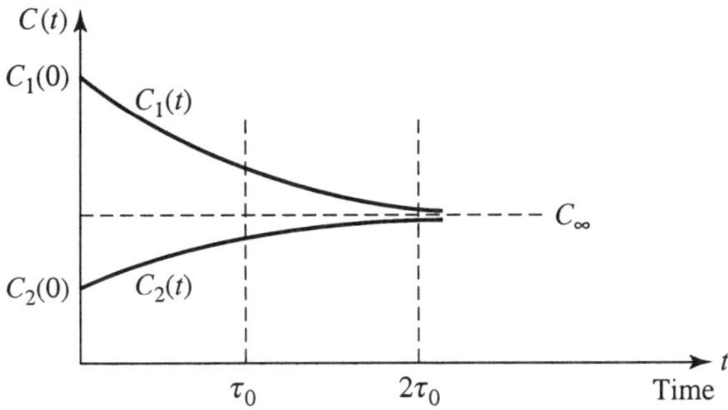

Figure 2.29. Equalization of concentration in two compartments connected by a permeable membrane.

$$C_2(t) = C_\infty + \left(C_2(0) - C_\infty\right)e^{-t/\tau_0}. \tag{2-130}$$

In Figure 2.29, we plot the behavior of $C_1(t)$ and $C_2(t)$ as a function of time.

2.4.D.   Hemodialysis. The Artificial Kidney

In this subsection we discuss an important application of diffusion across a porous membrane: We will describe the design and function of the artificial kidney, a device used by patients afflicted with loss or impairment of the function of their natural kidneys.

(i)   Physiological Role of the Kidney

The kidney is an organ of vital importance in the human body. Its main function is to maintain and regulate the composition of solutes in the blood plasma, and thus indirectly, also that of the other body fluids (interstitial fluid, intracellular fluid).

The principal functional unit in the kidney is the nephron. (For a more detailed description of the anatomy and physiology of the kidney, see Leaf [14], Vander, Sherman, and Luciano [11], or the article by H. W. Smith in *Scientific American* [12].) Figure 2.30 is taken from the latter article.

A human kidney contains about one million nephrons. For our purposes, it is sufficient to consider two functional stages of the nephron:

(a) Bowman's capsule. It contains the *glomerulus*, consisting of a balloon-shaped *basement membrane*, whose inside part is lined by blood vessels circulating

The nephron:

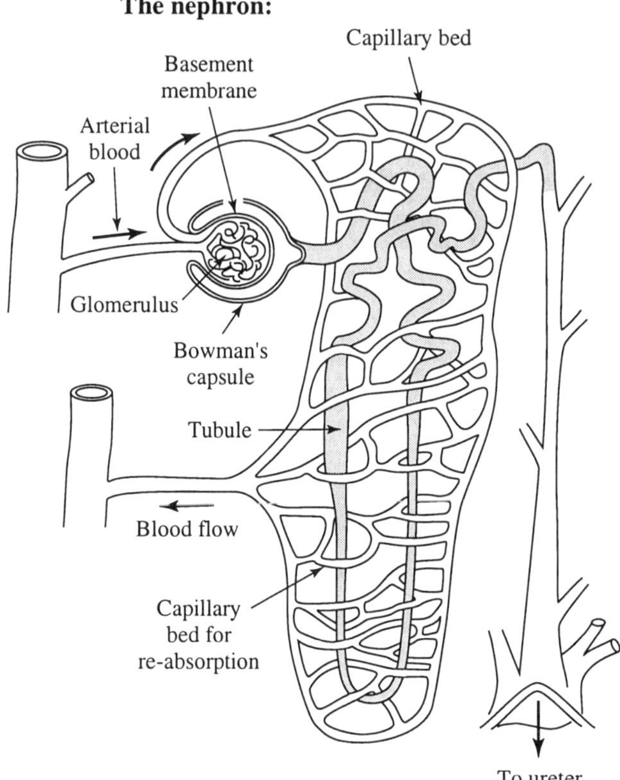

Figure 2.30. Schematic drawing of the nephron, the principal functional unit of the kidney. Much of the arterial blood entering the glomerulus is filtered through the basement membrane into the renal tubules. Reabsorption of the filtrate takes place between the tubules and the capillary bed surrounding it.

arterial blood; its outside is surrounded by the walls of the capsule to which is attached a collecting duct, the tubule.

(b) The *tubule* is a long (2–4 cm), convoluted duct with walls consisting of a single layer of cells. It is surrounded by a dense network of blood vessels, infused by the blood which previously circulated through Bowman's capsule.

The basement membrane in the capsule serves as an ultrafilter. The entering arterial blood is under a hydrostatic pressure of about 70 mmHg. The size of the membrane pores permits passage of the blood plasma and low molecular weight solutes. Red blood cells and the plasma proteins, however, are retained in the blood.

The total amount of blood circulating through the kidney is of the order of 1700 l/day, or about 20% of the *total* daily circulated blood volume. Since the total blood volume is about 6 l, all the blood circulates through the kidneys about 280 times each day. The really remarkable fact, however, is the volume of the filtrate collected by all the Bowman capsules. It amounts to 180 l/day. This vastly exceeds the volume of secreted urine per day ($\sim$ 1.5 l). Indeed, $\sim$ 99% of this filtrate is reabsorbed as it passes through the tubules, and returned to the bloodstream.

There are now enough data on this filtration process to have a fairly accurate characterization of the basement membrane; these data are given below:

total basement membrane area: A $= 1.5$–4.5 m^2.

fractional pore area: 5–10%.

pore radius: 35–50 Å.

pore length: $\Delta x = 400$–600 Å.

Pressure difference for filtration: $\Delta p' \sim 25$–45 mmHg ($\Delta p'$ is the hydrostatic pressure difference Δp minus an "osmotic" pressure difference $\Delta \pi$; for the latter, consult Section 2.6. Δp is generated by the systolic pressure of the heart).

It is left as a problem assignment to verify that these data are consistent with a filtrate volume of 180 l/day.

We must now reflect on the possible reason for this enormous volume of filtrate, and the appendant task of reabsorbing it. This task is vividly illustrated by the fact that the blood capillaries reabsorb from the tubules 2.5 lb of sodium chloride daily, whereas only about 5 to 10 g are excreted.

One principal reason for this setup lies in the stringent requirements of the maintenance of a constant composition ("homeostasis") of the body fluids with regards to the concentration of various ions, pH, and glucose level. The kidney, far from being just an organ of excretion of body wastes, is an essential *control* organ, capable of maintaining the blood plasma solute composition within the bounds required for the functioning of the organism. The enormous amount of filtrate makes it possible to achieve a rapid and precise control of the blood plasma composition.

In Table 2.7 we list the average composition of blood plasma: The normal blood glucose level is 800 mg/l. We see from this table that the total ion concentration is about 300 milli-eq/liter, and that the most abundant ions are Na$^+$ and Cl$^-$. The phosphates and bicarbonates are the crucial "buffers" for maintaining the plasma pH, that is normally kept within the limits $7.35 < \text{pH} < 7.45$. This is

Table 2.7. Composition of Blood Plasma.

Cations (milli-eq/liter)		Anions (milli-eq/liter)	
Na^+	142	Cl^-	103
K^+	4	Bicarbonate	27
Ca^{++}	5	Phosphate	2
Mg^{++}	3	SO_4^{--}	1
		Organic acids	5
		Proteins	16

achieved by means of reactions

$$H_2PO_4^- \rightleftharpoons H^+ + HPO_4^{--} \qquad (pK = 7.25),$$
$$H_2CO_3 \rightleftharpoons H^+ + HCO_3^- \qquad (pK = 6.3).$$

The regulatory function of the kidney is exercised through selectivity in the re-absorption of the plasma electrolytes in the tubules. This reabsorption is not by simple diffusion, but actually aided by the mechanism of *active* transport. This is an energy-consuming process, only beginning to be understood in detail, in which solutes can be moved in the direction *opposite* the one indicated by Fick's law, uphill, so to speak. Such a mechanism is called a *pump.*

Of particular interest is the reabsorption of glucose (blood sugar). It has been established that the capacity for glucose reabsorption (aided by an active transport, a "pump" mechanism) is limited to about 400 mg/min. This is sufficient to re-absorb essentially all the plasma glucose if its concentration is below 1800 mg/l. In a person suffering from diabetes, the normal control mechanism for the blood plasma glucose level has broken down, and ingestion of food may be followed by a raise in glucose concentration above 1800 mg/l. In this case, tubular re-absorption will be incomplete, and sugar appears in the urine.

Let us now turn to a description of the eliminatory function of the kidney, and state the types and amounts of waste products that must be gotten rid of per day in normal circumstances: In Tables 2.7 and 2.8 the unit milli-equivalent/liter is equal to the number of millimoles per liter of the material in question times z, where z is the number of electron charges on the ion. Table 2.8 shows that in addition to metabolic wastes like urea, substantial amounts of electrolytes must be disposed of, the most abundant being again sodium chloride. If these materials cannot be disposed of, they upset the homeostasis of the body

Table 2.8. Composition of Urine in Normal Adult for 24 h Period.

Substance		Quantity	
Water		1000–2500 cm^3	
Electrolytes:	Na$^+$	100–250	milli-eq/liter
	Cl$^-$		
	K$^+$	40–80	milli-eq/liter
	Mg^{++}	8–16	milli-eq/liter
	Ca^{++}	3–8	milli-eq/liter
	NH$_4{}^+$	< 80	milli-eq/liter
	SO$_4{}^{--}$	50–200	milli-eq/liter
Buffers	Bicarbonate	0–50	milli-eq/liter
	Phosphate	20–50	milli-eq/liter
Nitrogenous materials:			
Urea		20–40 gms	
Creatinine		1–2 gms	
Uric acid		0.5–0.8 gms	
Amino acids		0–1.5 gms	

fluids. Toxic quantities of metabolites accumulate, and death follows eventually.

Impairments of this vital function of the kidney can be classified into four groups:

(a) Acute renal shutdown in which the kidneys stop working entirely.

(b) Renal insufficiency. The nephrons are being destroyed little by little, reducing the effectiveness of the kidney.

(c) Kidney nephrosis syndrome. Here the basement membrane becomes too permeable, permitting passage of considerable amounts of plasma proteins.

(d) Specific tubular abnormalities, including incapacity to reabsorb certain solutes.

In all cases, the protection normally offered by the kidney is impaired. The artificial kidney may then be used in its place.

In concluding our description of the anatomy and function of the kidney, it may be worth stressing that we have made no attempt to explain how the kidney actually senses the level of various ions in the blood, compares these with the desired levels, and then controls them by means of selective tubular reabsorption. The understanding of this control system is a central theme of renal physiology.

(ii)   Description and Function of the Artificial Kidney

The artificial kidney is essentially a very simple device. A fraction of the total circulating blood is shunted outside the body by an external arterio-venous shunt. The externally routed blood is pumped by the heart with the assistance of a peristaltic pump through what is essentially a porous bag with a large area to volume ratio. This bag is made of several types of artificial membrane, mostly cellophane.

Circulating outside this bag is a solution called dialyzate. It consists of a solution of glucose, sodium chloride, and other electrolytes at a concentration *equal* to that desired in the blood plasma. This dialyzate is pumped around the outside of the dialysis bag.

With this setup, solutes contained in the blood plasma, but absent in the dialyzate, will diffuse into the latter across the cellophane membrane, and be carried off by the flowing dialyzate. The addition of glucose and the desired electrolytes to the dialyzate will assure, on the other hand, that these solutes are not removed from the blood plasma, since no concentration gradient exists for them across the membrane.

This process is called hemodialysis. We see that it is a way of removing undesirable solutes that uses a principle rather different from that on which the normal

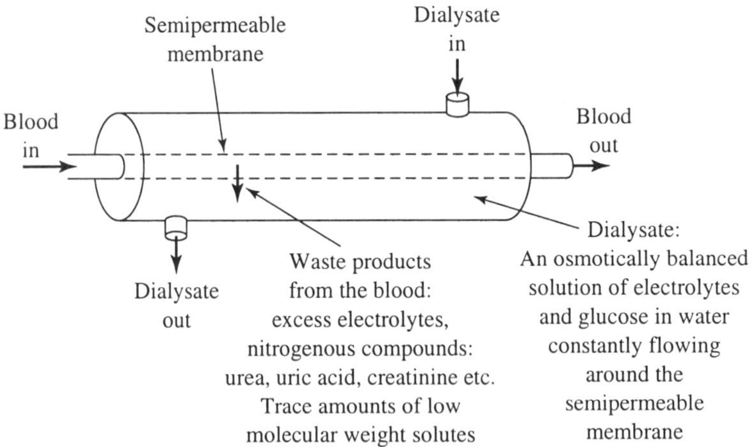

Figure 2.31. Schematic representation of the process of hemodialysis. (From C. Colton, PhD Thesis, MIT [13].)

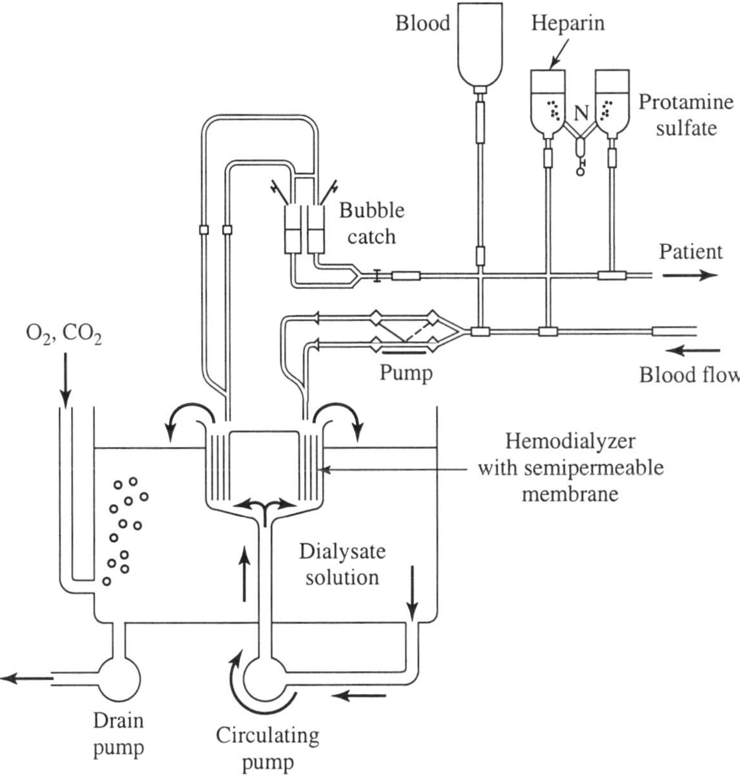

Figure 2.32. Typical artificial kidney hemodialyzer. (From C. Colton, PhD Thesis [13].)

kidney function is based. Instead of the combination of ultrafiltration and subsequent reabsorption, aided by an active transport or pump mechanism, the artificial kidney relies on simple diffusion. Figure 2.31 gives a schematic representation of the process, and Figure 2.32 shows a model of an actual hemodialyzer. Notice that before blood is allowed to enter the apparatus, an anticlotting agent, heparin, is added. This agent is neutralized by protamine sulfate before the blood returns to the body. These two figures are taken from the PhD Thesis of Professor Clark Colton at MIT [13], one of the leading experts on this problem of medical engineering.

Let us now look at some of the design parameters of a hemodialyzer. Basically we deal with a process of diffusion in a two-compartment system, as discussed in Section 2.4.C(iv). In that section, it was shown that the concentration difference

ΔC of solutes in two fluid compartments (1) and (2) equalizes exponentially

$$\Delta C(t) = \Delta C(0)e^{-t/\tau_0}$$

with a time constant τ_0 given by (2-126a):

$$\tau_0 = \frac{1}{Ap}\left(\frac{1}{V_1} + \frac{1}{V_2}\right)^{-1},$$

A being the membrane area, p its permeability, and V_1, V_2 the sizes of the two fluid compartments. A useful design should give a time constant not much longer than 1 hour, so that in a few hours most of the undesirable solutes have passed into the dialyzate.

To start with, consider the volumes V_1 and V_2. At first, one would think of V_1 as being simply the blood volume, 4.5 l. This is not correct. Wastes are produced in the tissues, pass into the interstitial fluid, and hence into the blood-stream. This represents a total volume of fluid more like 30 to 40 l, since body fluid accounts for about 63% of body weight in adult human males, 52% in females. The volume of dialyzate must be much larger than this since the ultimate equalized concentration of undesirable solutes should be made very small. For further discussion, we therefore assume $V_2 \gg V_1$, which allows us to write the time constant τ_0 as

$$\tau_0 \simeq \frac{V_1}{Ap}.$$

As the material for membranes, cellophane of 2.0 mil ($= 5.08 \times 10^{-3}$ cm) thickness is commonly used. It has sufficient mechanical strength, and a sufficiently large permeability. Table 2.9 lists its permeability for various solutes: We see in this table again the effect of hindered diffusion mentioned in Section 2.4.C(iii). Urea diffuses through most easily and plasma proteins (albumin) are almost totally retained.

With these data, it can now be seen that a time constant τ_0 of about 1 hour for urea can be achieved, if the membrane area A is of the order of 2 m^2. Indeed, with

$$V_1 = 40\,\mathrm{l} = 4 \times 10^4\ \mathrm{cm}^3,$$
$$p = 5.63 \times 10^{-4}\ \mathrm{cm/s\ (urea)},$$
$$A = 2\,\mathrm{m}^2 = 2 \times 10^4\ \mathrm{cm}^2,$$

Table 2.9. Permeability p
of 2 mil Cellophane.

Solute	$10^4 \, p$ (cm/s)
Urea	5.63
Creatinine	3.03
Uric acid	2.79
Sucrose	0.953
Vitamin B	0.380
Heparin	0.023
Albumin	0.00022

we find

$$\tau_0 = \frac{4}{2 \times 5.63} \times 10^4 \, \text{s} = 3.5 \times 10^3 \, \text{s} \simeq 1 \, \text{h}.$$

For uric acid τ_0 is about double that value. If the apparatus is operating for 4 h, 98% of the urea will have been removed, and 87% of the uric acid. The blood flow rate through the dialyser is generally set at about 200 cm^3/m, or 50 l in 4 h. This is 10 times the total blood volume, but only slightly larger than the total body fluid volume. It is not possible to even approximately match the blood flow volume through the natural kidneys.

2.5 Flow and Diffusion of Particles Under the Action of External Forces and Collisions with Solvent Molecules

The objective of this section is to generalize Fick's law for the flow of solute particles in a concentration gradient to the case where these particles are also subject to an external force (f). This force can be gravity, an electric force, or a centrifugal force. We will carry out this generalization using a simple one-dimensional model of the diffusion process, that reveals clearly all the essential features of the flow of particles under the influence of concentration gradients, collisions with solvent molecules, and externally applied forces.

The one-dimensional model is very rich in content. We can use it to gain a deeper insight into the equation of continuity and Fick's law. It will also provide us with alternative expressions for the diffusion coefficient D in terms of the average kinetic energy of the particle and its rate of collision with solvent molecules. The model, furthermore, can show in detail how flowing particles move under the action of concentration gradients, external forces, and collisions with the solvent molecules.

2.5.A.   Flow, Collisions, and Momentum Transfer in a Concentration Gradient

We will begin by considering a situation in which only a concentration gradient of solute molecules is present in a solution. We will present the model for the particle motion in this situation and apply it as mentioned above. Furthermore, we will demonstrate, in particular, that the existence of a current density $j(x)$, induced by the concentration gradient, implies that the flowing solute particles exert a force on the *solvent*. The force per unit volume (F/V) on the solvent will be shown to be equal to

$$\left(\frac{F}{V}\right) = \frac{kT}{D} j. \tag{2-131}$$

Our one-dimensional model for a diffusion process is as follows. The solute particles move only in the x-direction, with either velocity $+v$ or the opposite velocity $-v$. The solvent medium is represented by a spatial distribution of "obstacles." Any solute molecule colliding with an obstacle reverses its velocity from v to $-v$ or from $-v$ to v. These collisions generate a one-dimensional random walk for each solute molecule. Though this model may appear oversimplified, it contains all the principal features of the diffusion process, it keeps the mathematics to a minimum, and it permits a clear insight into the meaning of general results, such as (2-131).

Let us begin by using this model to examine the rate at which the number of solute particles changes inside some small volume V, of width Δx and area A (see Fig. 2.33). This consideration will yield the equation of continuity and will provide the basis for further examination of the momentum transfer and forces acting on the solute molecules. Let $N_+(t)$ stand for the number of solute particles in V that move with velocity v (to the right), and $C_+(x, t)$ their concentration in V, i.e.,

$$N_+(t) = C_+(x, t)V. \tag{2-132}$$

Similarly, $N_-(t)$ and $C_-(x, t)$ represent the number and concentration of particles having velocity $-v$ (moving to the left). The numbers $N_+(t)$ and $N_-(t)$ may

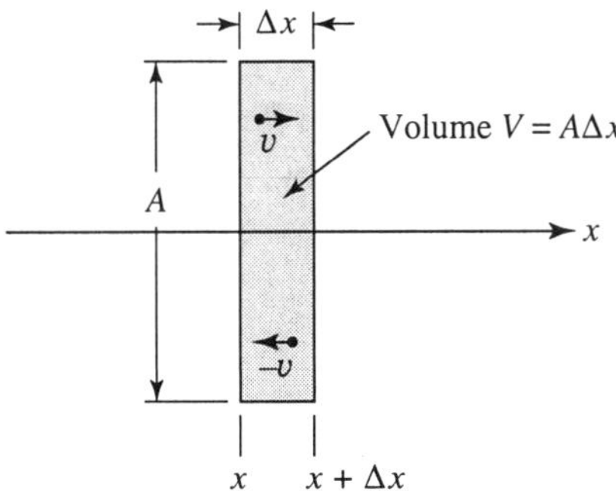

Figure 2.33. Small volume through which diffusing particles move. Note that the symbol V represents *volume*, while v represents *velocity*.

change with the time, and we now itemize the four ways this may happen. The rate of change of N_+ is given by

$$\frac{dN_+(t)}{dt} = V\frac{\partial C_+}{\partial t} = \text{inflow of particles at left boundary } (J_{\text{in}}(x))$$

$$- \text{ outflow of particles at right boundary} \equiv (J_{\text{out}}(x + \Delta x))$$

$$+ \text{ rate at which particles with velocity } -v \text{ collide and are converted to particles with } +v \equiv (dN/dt)_{+\text{coll}}$$

$$- \text{ Rate at which particles with velocity } +v \text{ collide and are converted to particles with } -v \equiv (dN/dt)_{-\text{coll}}.$$

$$(2\text{-}133)$$

These four items can be expressed quantitatively as follows:

(i) The inflow of particles at the left is given by the current

$$J_{\text{in}}^+(x) = vAC_+(x).$$

(ii) The outflow of particles at the right is given by

$$J_{\text{out}}^+(x + \Delta x) = vAC_+(x + \Delta x).$$

Thus the *net* due to these two effects is

$$J_{\text{in}}^+(x) - J_{\text{out}}^+(x + \Delta x) = vA\big(C_+(x) - C_+(x + \Delta x)\big)$$

$$= vA\left(C_+(x) - \left[C_+(x) + \left(\frac{\partial C_+}{\partial x}\right)\Delta x\right]\right)$$

or

$$J_{\text{in}}^+(x) - J_{\text{out}}^+(x + \Delta x) = -vA\,\Delta x\left(\frac{\partial C_+}{\partial x}\right) = -Vv\left(\frac{\partial C_+}{\partial x}\right). \qquad (2\text{-}134)$$

(iii) The rate of conversion of particles from $-v$ to $+v$ is proportional to the number N_-. The proportionality factor (γ) has the dimension (s^{-1}). Since the solute molecule will change velocity from $+v$ to $-v$ as the result of a collision with a solvent "obstacle," we see that γ is the collision rate $1/\tau_c$, where τ_c is the time between collisions. We then write

$$\left(\frac{dN}{dt}\right)_{+\text{coll}} = \gamma N_- = \gamma V C_-. \qquad (2\text{-}135\text{a})$$

(iv) The rate at which particles are converted from $+v$ to $-v$ is

$$\left(\frac{dN}{dt}\right)_{-\text{coll}} = \gamma N_+ = \gamma V C_+. \qquad (2\text{-}135\text{b})$$

Thus the net rate of change of dN_+/dt due to collisions is

$$\left(\frac{dN}{dt}\right)_{+\text{coll}} - \left(\frac{dN}{dt}\right)_{-\text{coll}} = \gamma V(C_- - C_+). \qquad (2\text{-}135\text{c})$$

We now can use this result and (2-134) in (2-133) to obtain the following results for (dC_+/dt):

$$\left(\frac{\partial C_+}{\partial t}\right) = -v\frac{\partial C_+}{\partial x} - \gamma(C_+ - C_-). \qquad (2\text{-}136\text{a})$$

A perfectly similar argument for the rate of change (dN_-/dt) of the number of particles moving to the left gives

$$\left(\frac{\partial C_-}{\partial t}\right) = +v\frac{\partial C_-}{\partial x} + \gamma(C_+ - C_-). \tag{2-136b}$$

These two equations represent quantitatively the verbal statement made in (2-133).

As a first application of this model, let us compute the rate of change of the concentration of both particles. The total particle concentration is

$$C(x) = C_+(x) + C_-(x).$$

We obtain (dC/dt) then simply by adding together (2-136a) and (2-136b). This gives

$$\left(\frac{\partial C}{\partial t}\right) = -v\left(\frac{\partial C_+}{\partial x} - \frac{\partial C_-}{\partial x}\right) = -\frac{\partial}{\partial x}\left(v(C_+ - C_-)\right).$$

On the other hand, the net current density $j = J/A$ at any *point* in the fluid is

$$j = j^+ - j^- = (vC_+ - vC_-). \tag{2-137}$$

Using this in the equation above, we find that

$$\left(\frac{\partial C}{\partial t}\right) = -\left(\frac{\partial j}{\partial x}\right). \tag{2-138}$$

This is just the "equation of continuity" that represents the fact that the rate of change of the number of particles inside V is determined by the net flow of particles into V. Notice that in obtaining the continuity equation, the terms in (2-136a, b), corresponding to collisions with the solvent $[-\gamma(C_+ - C_-)$ and $+\gamma(C_+ - C_-)]$, cancel. The collisions do not affect the net number of particles in V since a particle removed from N^+ on collision appears as N^- so that $N^+ + N^-$ is unaffected by the collisions.

In order to see clearly the effect of collisions, we must compute the rate of change of the momentum P inside the volume V. By interpreting the equation for (dP/dt), we will obtain a host of important results: We shall see how a current builds up *in time* in the presence of a fixed concentration gradient. We shall obtain

Fick's law as the final steady current established in the presence of the gradient. We shall obtain an alternative microscopic interpretation for the diffusion coefficient D. And finally, we shall show that the steady current (j) of solute particles exerts a force per unit volume on the solvent given by $(F/V) = (kT/D)j$.

The total momentum P inside the volume V is the sum of the momenta of each of the individual particles. If m is the mass of each solute molecule, we have

$$P = N_+(mv) + N_-(-mv) \tag{2-139}$$

The total momentum in this model changes because of changes in the numbers N_+ and N_- of particles moving with the constant momentum mv or $-mv$. Thus

$$\left(\frac{dP}{dt}\right) = \left(\frac{dN_+}{dt}\right)mv - \left(\frac{dN_-}{dt}\right)mv. \tag{2-140}$$

Using $N_\pm = VC_\pm$, we can write this as

$$\left(\frac{dP}{dt}\right) = V\left(mv\frac{\partial C_+}{\partial t} - mv\frac{\partial C_-}{\partial t}\right)$$

We may now use our previous results, (2-136a, b), to reexpress (dP/dt) in the form

$$\frac{1}{V}\left(\frac{dP}{dt}\right) = -mv^2\left(\frac{\partial C_+}{\partial x} + \frac{\partial C_-}{\partial x}\right) - 2m\gamma v(C_+ - C_-). \tag{2-141}$$

Using $C_+ + C_- = C$ and $j = v(C_+ - C_-)$, we find

$$\frac{1}{V}\left(\frac{dP}{dt}\right) = -mv^2\left(\frac{\partial C}{\partial x}\right) - 2m\gamma j. \tag{2-142}$$

Let us now pause to interpret this result in terms of Newton's second law of mechanics. The left-hand side of (2-142) is $(1/V)$ times the rate of change of momentum of all the solute particles in the volume V. According to the right-hand side of this equation $(1/V)(dP/dt)$ comes from two effects. The first is expressed by the term $mv^2(\partial C/\partial x)$. We can show quite easily that this term represents the *flow of momentum* into V carried by particles with velocity $\pm v$. The momentum flow across area A at any point x is given by

$$[mv(vC^+(x)) + (-mv)(-vC^-(x))]A = mv^2(C^+(x) + C^-(x))A = mv^2C(x)A.$$

The *net* flow of momentum into volume V across the boundaries at x and $x + \Delta x$ is

$$mv^2 C(x)A - mv^2 C(x + \Delta x)A = mv^2 \left[C(x) - \left(C(x) + \left(\frac{\partial C}{\partial x} \right) \Delta x \right) \right] A$$

$$= -mv^2 \left(\frac{\partial C}{\partial x} \right) \Delta x A = -mv^2 \left(\frac{\partial C}{\partial x} \right) V.$$

Thus we see that the flow of momentum into V gives a contribution to $(1/V)$ (dP/dt) that is equal to $-mv^2(\partial C/\partial x)$.

Let us now examine the physical meaning of the second term $-2m\gamma j(x)$ on the right-hand side of (2-142). This term represents the net force per unit volume that the solvent molecules exert on the solute as a result of the collisions between them. At each collision a solute molecule experiences a strong force which in our model reverses the momentum of the molecule. The term $(-2m\gamma j(x))V$ is the time average force exerted by the solvent "obstacles" on all the solute molecules in V. This is seen as follows: The change in momentum that occurs when a single solute molecule, initially moving with $+v$, collides with a solute "obstacle" is $-2mv$. The total number of such changes that occur per second is γN_+. Thus the total averaged rate of change of momentum produced by collisions of $+v$ molecules is $-2mv\gamma N_+ = -2mv\gamma C_+ V$. Similarly, the collision of solute molecules moving with velocity $-v$ produces a rate of change of momentum equal to $+2mv\gamma N_- = 2mv\gamma C_- V$. The *net* rate of change of momentum produced by the collisions is the time average force exerted inside the volume V by the solvent obstacles is $F = -2mv\gamma(C_+ - C_-)V$. And, since j = particle current $= v(C_+ - C_-)$, we have

$$\left(\frac{F}{V} \right)_{\text{collision}} = -2m\gamma j. \tag{2-143}$$

which is precisely the form of the second term in (2-142). This term does indeed represent the average force exerted on the solute molecules by collision with the solvent molecules.

Equation (2-142) is in fact a differential equation for the particle current density j at the point x, as a function of time. We can see this very clearly, and obtain a very useful form for this momentum flow equation, by observing that (dP/dt) can be related to the rate of change of the particle current density j. In our model the total momentum in V, according to (2-139), is

$$P = N_+(mv) - N_-(mv)$$

or

$$P = mv(C_+ - C_-)V = mjV \tag{2-144}$$

since $j = (vC_+ - vC_-)$. Using (2-144) in (2-142), we find that the equation for the rate of change of momentum in V gives us a differential equation for the time-dependence of j at each point x in space, namely:

$$m\left(\frac{dj}{dt}\right) = -mv^2\left(\frac{\partial C}{\partial x}\right) - 2m\gamma j. \tag{2-145}$$

This can be expressed in a mathematically more useful form by canceling the m, and factoring the coefficients to yield the very simple form

$$\left(\frac{dj}{dt}\right) = (j_F - j)\frac{1}{\tau}, \tag{2-146a}$$

where

$$\frac{1}{\tau} = 2\gamma \tag{2-146b}$$

and

$$j_F = -\left(\frac{v^2}{2\gamma}\right)\left(\frac{\partial C}{\partial x}\right). \tag{2-146c}$$

These equations now enable us to see clearly the temporal buildup of the particle current, and permits us to obtain the various results mentioned previously.

Suppose that a concentration gradient $(\partial C/\partial x)$ of solute particles is established at $t = 0$ at some point (x), and that the current j is initially zero. How does the current j build up? This can be seen from (2-145) or (2-146). According to (2-145) the existence of the concentration gradient produces a flow of momentum. Thus, just following $t = 0$, we have an initial growth rate: $(dj/dt) = -v^2(\partial C/\partial x) = j_F/\tau$. Thus, at first, the current increases linearly with time. As

j grows, however, the term $-j/\tau$ builds up and reduces the growth rate from its initial value j_F/τ, to the value $(j_F - j)/\tau$. Physically this occurs because the net force produced by the collisions is proportional but in opposite direction to the flow of momentum. Thus, the rate of growth of the current decreases until finally (dj/dt) becomes equal to zero. At this stage the flow of momentum produced by the concentration gradient is exactly canceled by the force exerted by the solvent on the solute molecules through the mechanism of collisions. This full range of behavior is seen quite directly by solving (2-146) for $j(t)$, assuming $j = 0$ at $t = 0$. The solution is simply

$$j(t) = j_F \left(1 - e^{-t/\tau}\right)$$

and this variation of j with time is shown in Fig. 2.34. Thus the current relaxes exponentially to a quasi-static value equal to j_F in a time of the order of the collision time $\tau_c = 1/\gamma$:

$$\tau = \frac{1}{2\gamma} = \tfrac{1}{2}\tau_c$$

It is proper to call $j_F = -(v^2/2\gamma)(\partial C/\partial x)$ a quasi-static current because for extremely long times, unless the concentration gradient is externally maintained at its initial value, the current of solute particles will reduce the gradient and establish ultimately a situation in which the concentration gradient becomes zero. In the situation under consideration, however, since the relaxation time τ is very short, i.e., of the order of the collision time ($\sim 10^{-10}$ s), we see that for any concentration gradient ($\partial C/\partial x$) a steady flow with current density given by (2-146c) is very quickly established. If we now compare this current with that expressed in

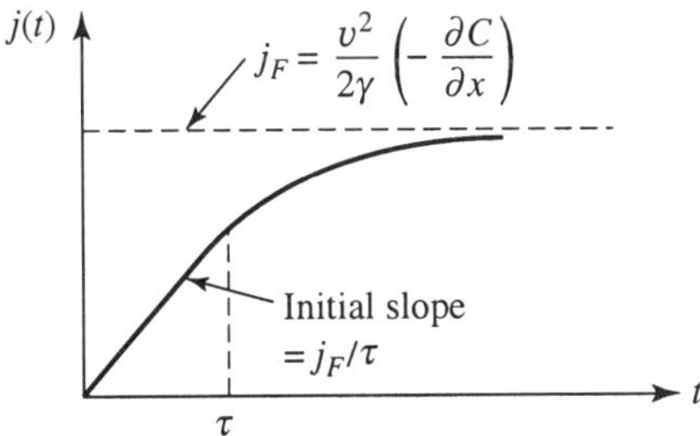

Figure 2.34. Temporal evolution of the current j following the establishment of a concentration gradient ($\partial C/\partial x$).

Fick's law

$$ j = -D \left(\frac{\partial C}{\partial x} \right) $$

we see at once that the diffusion coefficient D can be identified in an alternative way, namely,

$$ D = \left(\frac{v^2}{2\gamma} \right) = \tfrac{1}{2} v^2 \tau_c \tag{2-147} $$

Previously, in our definition of the diffusion coefficient using the random walk model, we had found that $D = L^2/2\tau_c$ for one dimension. Here L^2 is the mean square step length and τ_c is the time between steps. Our present result gives us another insight into the meaning of the diffusion coefficient. Equation (2-147) in effect expresses the diffusion coefficient in our one-dimensional model as the product of the mean square velocity of the random walking particle and the collision time $\tau_c = 1/\gamma$. We can use (2-147) as a means of expressing the collision rate γ in terms of the diffusion coefficient and the temperature by recognizing that the mean square velocity is related to the temperature by the equipartition theorem. In our model, since the velocity is either $\pm v$, the mean square velocity is v^2, thus

$$ \tfrac{1}{2} m v^2 = \tfrac{1}{2} kT. $$

according to the equipartition theorem (Section 2.1.C). Using this to eliminate v^2 from (2-147), we find

$$ D = \left(\frac{m v^2}{2} \right) \left(\frac{1}{m\gamma} \right) = \frac{1}{2} \frac{kT}{m\gamma}. \tag{2-148} $$

Thus we can express the collision time τ_c or collision rate γ in terms of D as

$$ \tau_c = \left(\frac{1}{\gamma} \right) = 2 \frac{mD}{kT}. $$

    We may now finally express in a convenient way the force per unit volume which the flowing solute molecules experience as a result of collisions with the solvent. According to (2-143) and (2-142), the force per unit volume exerted by the solvent is

$$\left(\frac{F}{V}\right) = -2\gamma m j.$$

Using now (2-148) for γ in terms of D, we find

$$\left(\frac{F}{V}\right) = -\left(\frac{kT}{D}\right) j. \tag{2-149}$$

This equation will prove to be very important in Section 2.6 on the coupled flow of solvent and solute across membranes.

The discussion just presented can now be simply extended to describe the effect of an external force on the solute molecules.

2.5.B.   Particle Current and the Diffusion Equation in the Presence of a Concentration Gradient and Externally Applied Forces. Drift Velocity

In this section, we generalize the equation for the rate of change dP/dt of the momentum of the solute particles in a volume element V, by adding an *external force* acting on these particles. Our point of departure is (2-142), where we now write kT for mv^2, and kT/D for $2m\gamma$, according to (2-148):

$$\frac{1}{V}\frac{dP}{dt} = -kT\frac{\partial C}{\partial x} - \frac{kT}{D}j(x). \tag{2-150}$$

Although we derived this result in the context of a rather restricted model, it is, in the form of (2-150), much more generally valid; a three-dimensional random walk model, allowing a continuous range of solute velocities, gives exactly the same result.

It is now easy to add the effect of an external force, like gravity, or an electric force, acting on the solute particles. Let f be the force acting on a single particle. In the volume element V there are $N = C(x)V$ particles. The total external force acting on them is therefore

$$fC(x)V$$

and, according to Newton's second law, it contributes to the rate of change of the solute momentum P. Equation (2-150) is therefore generalized to

$$\frac{1}{V}\frac{dP}{dt} = -kT\frac{\partial C}{\partial x} - \frac{kT}{D}j(x) + fC(x). \tag{2-151}$$

Let us restate the physical interpretation of the three terms on the right-hand side:

(i) The first term represents the rate of momentum transport by particles entering or leaving V.

(ii) The second term represents the force exerted by the solvent medium (the "obstacles") on the solute particles in V.

(iii) The third term represents the external force acting on the solute particles.

As explained in the last part of Section 2.5.A, we may assume, for further discussion, that a quasi-stationary situation prevails in which $dP/dt \cong 0$. It is then useful and customary to solve (2-151) for the current j, so as to display how the quasi-stationary current depends upon the concentration gradient and the external force

$$j(x) = -D\left(\frac{\partial C}{\partial x}\right) + \frac{D}{kT}fC(x). \tag{2-152}$$

This then is the desired generalization of Fick's law. It tells us the current density at each point in the medium when both a concentration gradient and an externally applied force is present.

We may convert (2-152) into an expression for the space–time evolution of the concentration. As we mentioned in Section 2.5.A, eventually the current flow will affect the concentration gradients which established it. We can include this effect to obtain the quasi-static evolution of the concentration $C(x, t)$ by using the continuity equation. The continuity equation [(2-91) or (2-137)] is

$$\frac{\partial}{\partial t}C(x, t) = -\frac{\partial}{\partial x}j(x, t).$$

This relates the time rate of change of the concentration at x to the spatial rate of change of the current at the same point. We can incorporate this continuity equation into the equation for the current (2-152) by taking the derivative of that equation relative to x and then replacing $(\partial j/\partial x)$ by $-\partial C/\partial t$. This gives the partial differential equation for forced diffusion, viz:

$$\left(\frac{\partial C}{\partial t}\right) = D\left(\frac{\partial^2 C}{\partial x^2}\right) - \left(\frac{D}{kT}\right) f\left(\frac{\partial C}{\partial x}\right). \qquad (2\text{-}153)$$

This equation tells us how the concentration ($C(x,t)$) evolves in space and time when the solute particles experience, in addition to a concentration gradient, an externally applied force f. Note that when $f = 0$, (2-153) returns to the familiar simple diffusion equation ($\partial C/\partial t) = D(\partial^2 C/\partial x^2)$.

In order to see what the effects of this added force term are, let us consider first the case of a uniform concentration, so that $\partial C/\partial x = 0$. Equation (2-152) then shows that the force f generates a uniform current density j:

$$j = \left(\frac{D}{kT}\right) fC. \qquad (2\text{-}152\text{a})$$

The ratio between current density j, measured in particles/cm^2 s, and the particle density, measured in particles/cm^3, defines a velocity u:

$$j = uC. \qquad (2\text{-}154)$$

We call this velocity the *drift velocity* of the particles produced by the force f acting on them. The proper interpretation of the drift velocity is in the statement that the force f acting on each particle imposes a bias on its random walk, such that on average, the particles undergo a displacement

$$\langle x \rangle_{\text{av}} = ut.$$

From (2-152a) and (2-154), we see that this drift velocity has the value

$$u = \frac{D}{kT} f. \qquad (2\text{-}155)$$

Again, f is the external force that acts on each of the solute particles. We will come back in Section 2.5.C to give a further elaboration of this result. But before doing this, let us see what the effect of this drift is in the case of a *nonuniform* concentration profile, such as the step-profile illustrated in Figure 2.17. We can expect that if a force f is present, this profile would flatten out by diffusion, but at the same time *move* with the drift velocity u. The position of the "boundary," that is, the point of maximum slope, would then be located at

$$x_{\text{boundary}} = ut$$

rather than remain at $x = 0$.

We shall now show that the forced diffusion equation [(2-153)] rigorously predicts a space–time evolution of the concentration profile, $(C(x, t))$ exactly as we guessed above. To show this, we describe the density C in a moving coordinate system by means of a position coordinate x' measured relative to the position of the boundary

$$x' = x - x_{\text{boundary}} = x - ut \tag{2-156}$$

with u given by (2-155). We thus represent the density $C(x, t)$ at position x and time t as a function of x' and t:

$$C(x, t) = \text{function of } x' \text{ and } t \equiv F(x', t).$$

$F(x', t)$ represents the density as "seen" in the moving coordinate system. Let us now see what the diffusion equation for $F(x', t)$ is if $C(x, t)$ obeys (2-153). This transcription is easily achieved if we observe that by the chain rule for derivatives we have

$$\left(\frac{\partial C}{\partial x}\right) = \left(\frac{\partial F}{\partial x'}\right)\left(\frac{\partial x'}{\partial x}\right) = \frac{\partial F}{\partial x'}, \quad \left(\frac{\partial^2 C}{\partial x^2}\right) = \left(\frac{\partial^2 F}{\partial x'^2}\right),$$

$$\frac{\partial C}{\partial t} = \frac{\partial F}{\partial t} + \left(\frac{\partial F}{\partial x'}\right)\left(\frac{\partial x'}{\partial t}\right) = \frac{\partial F}{\partial t} - u\frac{\partial F}{\partial x'}, \tag{2-157}$$

since, from (2-154), we find that $(\partial x'/\partial x) = 1$, $(\partial x'/\partial t) = -u$. Inserting these results into (2-153), we have

$$\frac{\partial F}{\partial t} - u\frac{\partial F}{\partial x'} = D\frac{\partial^2 F}{\partial x'^2} - \frac{D}{kT}f\frac{\partial F}{\partial x'}.$$

Because $u = (D/kT)f$ from (2-155) this equation becomes simply

$$\frac{\partial F(x', t)}{\partial t} = D\frac{\partial F(x', t)}{\partial x'^2}.$$

This result shows that the quantity $F(x', t)$, that represents the density as seen from the moving coordinate system, evolves in space–time *as if no force f were present*: $F(x', t)$ simply undergoes a spread in time due to diffusion. The effect of the force f on the particles lies entirely in a displacement of the profile as a whole, with velocity u, as given by (2-153).

This state of affairs is illustrated in Figure 2.35 for a concentration, which at $t = 0$ has a step-discontinuity at $x = 0$.

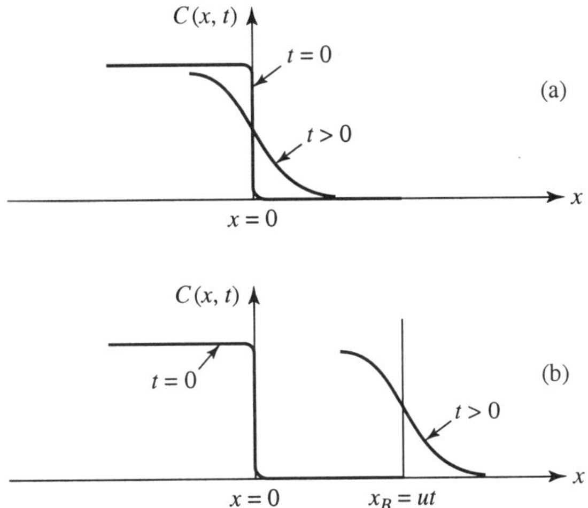

Figure 2.35. Evolution of a step discontinuity in concentration. (a) Diffusive spread with *no* force acting on solute particles. (b) Diffusive spread and drift due to force f on solute particles. Displacement due to drift is $x_B = ut$. Profile (b), as a function of x', is the same as profile (a) as a function of x.

In summary, the solutions to (2-153) for the concentration $C(x, t)$ can be obtained by first ignoring the external force f acting on the particles. We have then a simple diffusive spread of the initial concentration profile. The effect of the force term is then to displace this profile in space by an amount x_B proportional to the elapsed time t:

$$x_B \equiv ut,$$

u being the drift velocity, and whose value is given by (2-156).

In concluding this discussion, we can return to (2-151) for the rate of change of momentum when external forces are also present. Again the term $-(kT/D)j(x)$ represents the average force that collisions exert on the solute particles. In the present case, however, j is given by (2-152). Thus, the force per unit volume associated with *collisions* between solute and solvent is given in general by the equation

$$\left(\frac{F}{V}\right)_{\text{coll}} = -\frac{kT}{D}j(x) = kT\left(\frac{\partial C}{\partial x}\right) - fC(x).$$

The average force on the solute particles, produced by the collisions, exactly balances out in the quasi-static case the flow of momentum associated with diffusion and the externally applied forces per unit volume.

2.5.C.   Mobility and the Stokes–Einstein Relation

The interpretation of the drift velocity u as given by (2-155) is that of a mean velocity of motion of each solute particle produced by the balance between the applied force f and the collision forces betwen solute and solvent molecules. The actual *motion* of a particle is, of course, still a random walk, but it is biased by the applied force. This is illustrated in Figure 2.36. Since the drift velocity u is directly proportional to the applied force f on the molecule according to (2-155) $(u = (D/kT)f)$, it is useful to define a mobility μ which is the ratio of u/f:

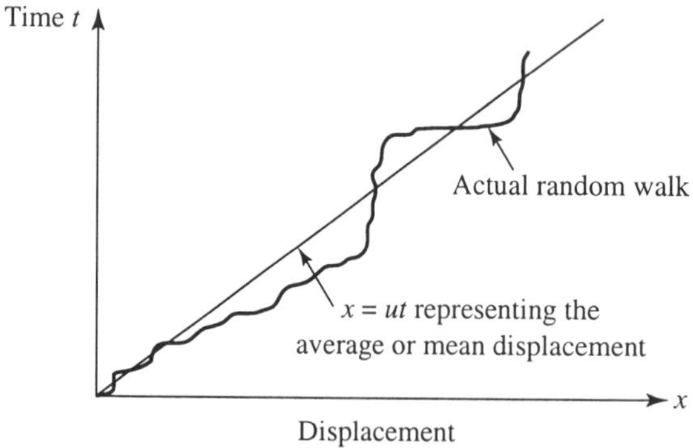

Figure 2.36. Random walk and mean displacement of a particle subject to a constant force f and collisions with solvent molecules.

$$\mu \equiv \left(\frac{u}{f}\right) = \left(\frac{D}{kT}\right).$$

(2-158)

The mobility is independent of the type or magnitude of the force and depends only on the temperature and the diffusion constant D. The expression

$$\mu = \left(\frac{D}{kT}\right)$$

was found by Einstein in 1905 in his investigation of Brownian movement [1]. As a result, this equation for μ is called Einstein's relation.

Einstein realized that it was possible to carry this analysis further when the solute molecule is large compared to the solvent molecule. In such a case, the solute molecule can be regarded as a body moving smoothly through a fluid. In the case that the velocity of the body is small, we saw in Volume I, Section 2.6.B, that the flow through the fluid exerts a viscous drag force F_D that is proportional to the velocity

$$F_D = -\beta v.$$

The coefficient β depends on the viscosity (η) of the fluid, and upon the shape and size of the body in general. In the case, however, that the body is a sphere, Stokes has shown that

$$\beta = 6\pi \eta a,$$

where a is the radius of the sphere. Now suppose such a sphere has applied to it an external force f, then Newton's second law of motion gives us that

$$m\left(\frac{dv}{dt}\right) = f + F_D = f - \beta v.$$

According to this equation the velocity v increases until it reaches a terminal velocity u, which applies when $dv/dt = 0$. From this equation we see that the terminal velocity is that for which $(f - \beta u) = 0$. Thus

$$u = \left(\frac{1}{\beta}\right) f.$$

If we compare this macroscopic expression for the terminal velocity with that obtained above for the average drift velocity of the solute molecule under the action of the same force f, we find

$$u = \left(\frac{D}{kT}\right) f.$$

We see that the microscopic and macroscopic descriptions of the particles average motion will agree provided that

$$D = \left(\frac{kT}{\beta}\right) \qquad (2\text{-}159a)$$

or, in the case of a spherical particle,

$$D = \frac{kT}{6\pi\eta a}. \qquad (2\text{-}159b)$$

The important equations (2-159a, b) are called the Stokes–Einstein relations. These results show that for a large solute molecule, like a globular protein or a virus, the diffusion coefficient is dependent only on the absolute temperature and the viscous drag coefficient β. In the case of a spherical particle, β is proportional to the product of the viscosity η and the radius of the macromolecule. For objects like rods and ellipsoids there are, similarly, formulas relating β to the viscosity, and the size and shape of the object. We see then that measurements of the diffusion coefficient of macromolecules can provide information on their size. Also, if the molecule undergoes a change in size due, for example, to denaturation or to a conformational change, this will affect the diffusion coefficient. Accurate measurements of the diffusion coefficient can therefore be used to detect such conformational changes [15], [16].

Another important application of (2-159) occurs in the measurement of the molecular weight of macromolecules. As we saw in Volume I, Section 3.5.E, the molecular weight can be found by measuring the sedimentation velocity v_s in a centrifuge provided that both the viscous drag coefficient (β) (previously we called this the "friction factor"), and the partial specific volume are known. The general Stokes–Einstein relation [(2-159a)] shows that β can be found from measurements of the diffusion coefficient. Thus, by combining accurate measurements of the diffusion coefficient with those of sedimentation velocity and partial specific volume, one can measure the molecular weight of large macromolecules and viruses quite accurately [17].

2.5.D.   Sedimentation Equilibrium: Scale Heights and the Molecular Weights of Macromolecules. Perrin's Experimental Measurement of Avogadro's Number

In this section we explore the effects of a gravitational force on a solution of macromolecules or a suspension of very small particles in a solvent.

We will show that if such a solution is left to itself, the concentration C of particles will evolve, no matter what the initial concentration profile was, toward a stationary state, the so-called *sedimentation equilibrium.*

To illustrate this, consider a vertical cylinder filled with solvent and an initially uniform concentration of particles. Let ρ (g/cm^3) be the mean density of the *solution.* Each particle has a mass m, and volume v_p. The force on a particle is then the sum of the force of gravity, $-mg$ (choosing the $+x$-axis to point vertically upward), and the force of buoyancy, $+\rho v_p g$:

$$f = -(m - \rho v_p)g. \tag{2-160}$$

As we already mentioned in Volume I, Section 3.5.E, it is useful to express v_p in terms of the particle mass m by defining a "specific volume" $\bar{v}_p$ as

$$\bar{v}_p \equiv \frac{v_p}{m}. \tag{2-161}$$

$\bar{v}_p$ is the volume per unit mass of the particle, and its units are cm^3/g. In terms of this, the force f is then conventionally written as

$$f = -m(1 - \rho \bar{v}_p)g = m^* g. \tag{2-162}$$

The factor $(1 - \rho \bar{v}_p)$ is called the buoyancy correction to mg. For proteins and viruses, $\rho \bar{v}_p$ is typically in the range 0.6–0.75. The quantity m^* in (2-162) is called the effective mass.

The equation we use to find the equilibrium concentration profile is (2-152) for the current j, with f given by (2-162) above, i.e.,

$$j(x) = -D\frac{\partial C}{\partial x} - \frac{D}{kT}mg(1 - \rho \bar{v}_p)C(x). \tag{2-163}$$

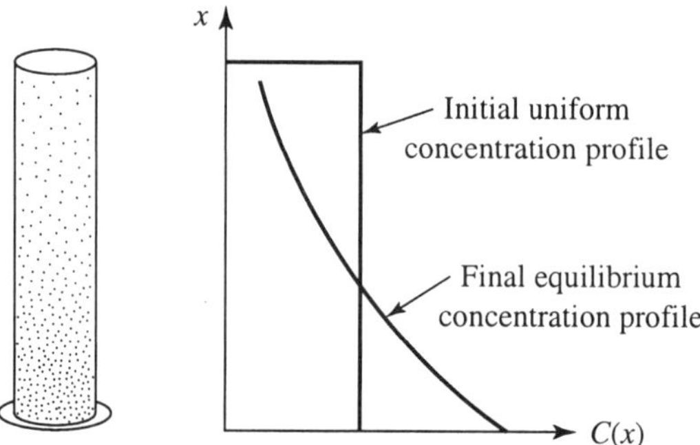

Figure 2.37. Concentration profile in a vertical cylinder. Initial uniform and final exponential profile of concentration.

Let us assume that we start with an initially uniform concentration C. (See Fig 2.37.) We also assume that $(1 - \rho \bar{v}_p)$ is positive, so that the net force f is in the direction of gravity. Since $\partial C/\partial x$ is zero initially, $j(x)$ is negative initially, indicating a downward flow of matter. As this continues, the concentration $C(x)$ becomes larger in the lower parts of the cylinder, smaller in the upper parts. In this way, a gradient $\partial C/\partial x$ arises, which is negative; hence $-D(\partial C/\partial x)$ is positive, and eventually becomes large enough to stop the downward flow. At this point, equilibrium has been reached. At equilibrium, no current flows anymore, anywhere. That is, for all values of x, we now have $j(x) = 0$. Equation (2-163) now tells us that when this equilibrium is reached, $C(x)$ and $\partial C/\partial x$ are related by

$$\frac{\partial C}{\partial x} = -\frac{mg}{kT}(1 - \rho v_p)C(x). \tag{2-164}$$

We recognize this as the equation for a decreasing exponential. The coefficient

$$\frac{mg}{kT}(1 - \rho \bar{v}_p) \equiv \frac{1}{H} \tag{2-165}$$

has the dimension of a reciprocal length, and we write it as $1/H$. H is called the scale height of the equilibrium distribution of concentration. The solution to

$$\frac{dC}{dx} = -\frac{1}{H}C(x)$$

is given by

$$C(x) = C(0)e^{-x/H},$$

as we can easily see by differentiation. $C(0)$ is the value of C at the bottom of the cylinder, $x = 0$.

The interest of (2-165) lies in the fact that a measurement of H allows a direct determination of the mass m of the suspended particles. Conversely, it also tells us the precise form of the density profile when m and $\bar{v}_p$ are known.

The most important particles, whose mass we might thus determine, are macromolecules and virus particles. In order to see whether their mass could indeed be determined experimentally using (2-165), it is important that the density profile be clearly measurable. The form of the profile is of course fixed by the scale height H. Let us then estimate the *mass m* of particles which in sedimentation equilibrium would give a scale height $H = 1$ cm ($\simeq 0.4''$). Using $k = 1.38 \times 10^{-16}$ erg/°K, and $T = 293\,°$K ($\simeq 68°$F), we find from (2-165) with $H = 1$ cm that

$$m^* \equiv m(1 - \rho v_p) \simeq 4 \times 10^{-17}\ \text{g}.$$

If this particle is a giant molecule, its effective molecular weight M^* would be given by

$$M^* = N_0 m^* = 2.5 \times 10^7\ \text{g/mol}.$$

A representative object, which has a scale height of that order, is the Tobacco Mosaic Virus (TMV), whose M is 5×10^7 g/mol (see Table 2.4(b)). The buoyancy correction factor $(1 - \rho\bar{v}_p)$ for a suspension of TMV in water is about 0.25; thus, it has a value M^* of 1.25×10^7 g/mol, and a scale height H of 2 cm. These results also show that *globular proteins*, whose molecular weights are typically in the range of

$$10^4\ \text{g/mol} < M < 10^5\ \text{g/mol}$$

(and have similar buoyancy corrections as TMV) have scale heights of the order

$$100\ \text{m} > H > 10\ \text{m}.$$

This is much too large to be observed. A means to *reduce* the scale height in such cases is to place the solution of macromolecules into an *ultracentrifuge* where the force of gravity is replaced by a centrifugal force. Since centrifugal accelerations up to 10^5 g (10^5 times the gravitational acceleration of 9.81 m/s) are easily obtained, the above-listed scale heights can be reduced by a factor up to 10^5, and then fall into the easily measurable range

$$H < 1 \text{ cm}.$$

This topic of ultracentrifugation will be further developed in the next section, 2.5.E. We conclude this section with a brief discussion of J. Perrin's historic (1914) experiments on the determination of Avogadro's number. In (2-165), the Boltzmann constant k may be written in terms of Avogadro's number N_0 and the well-known gas constant R: $k = R/N_0$, so as to give

$$\frac{1}{H} = N_0 \frac{mg}{RT}(1 - \rho \bar{v}_p). \tag{2-166}$$

We see that if H, m, and $(1 - \rho \bar{v}_p)$ can be measured, N_0 can be determined.

Perrin realized that, under a microscope, he could determine scale heights H as small as 10^{-3} cm, and therefore he could use a suspension of particles whose individual size was large enough to be actually seen and counted. We give the data for one of his experiments:

particles: latex spheres
radius $a = 2.12 \times 10^{-5}$ cm,
density $\rho = 1.194$ g/cm^3,
buoyancy correction: $(1 - \rho \bar{v}_p) = 0.163$,
temperature $T = 293\,°$K $(= 68\,°$F$)$.

The particles were counted in four thin layers at four different heights in the container (Figure 2.38). We leave it as a problem assignment to verify numerically that these data give a value $N_0 \sim 7 \times 10^{23}$. This is clearly not a precision determination of N_0. The main thrust of the experiment, however, was to give a convincing demonstration of the basic validity of the assumptions that go into the derivation of (2-165). These assumptions are the molecular–kinetic picture of gases and liquids, and the equipartition law of kinetic energies. Perrin published his findings in 1914 in the book, *Les Atomes* [2]. At this time, the only other determination of N_0 based purely on the kinetic theory was obtained from an interpretation of the viscosity of ideal gases; Perrin reports a value of $N_0 = 6 \times 10^{23}$ based on that data.

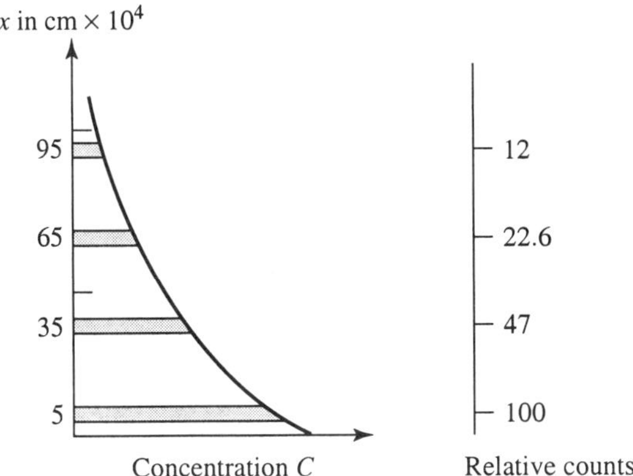

Figure 2.38. Particle concentration as a function of height in Perrin's experiment.

It is interesting to note what, in Perrin's own view, the main thrust of his results was. After having found that the value of N_0 extracted from his data was always the same, whether he replaced water by glycerine, or changed the temperature, or changed the size of particles, he observed: "It becomes thus difficult to deny the objective reality of molecules." One forgets easily that even 80 years ago the atomistic view of matter still needed to be defended!

2.5.E. Ultracentrifugation

The development of the ultracentrifuge and its emergence as one of the major tools of biochemistry began in the 1920s. The pioneering work by Svedberg and his collaborators, both with regard to design and to results, is well documented in the classic text by Svedberg and Pederson [18], and is still worth reading. Much further development has taken place since 1940, both into design, techniques of observation, and in the theory of the sedimentation process. An elementary account of this may be found in the book by Bowen [19], and with more emphasis on theory, in the report by Williams [20].

There are two main types of use of ultracentrifugation in present-day biochemical work: preparative and analytical. Preparative work is qualitative in nature, and consists mainly of the separation by centrifugation of various fractions or components in a solution. Analytical work consists of quantitative determination of sedimentation velocities, or density profiles in sedimentation equilibrium. These measurements are performed by means of optical equipment while the centrifuge is running.

In this section we shall be concerned with the results that such analytical work can yield. We will show in particular how molecular weights, and to an extent, the size and shape of macromolecules and viruses can be determined by ultracentrifugation. So crucial has the role of ultracentrifugation been that modern macromolecular biochemistry is virtually unthinkable without it.

(i)   Design and Performance of the Ultracentrifuge

The main element of the ultracentrifuge is the rotor, a lumpy shaped solid piece of high strength aluminum alloy. It contains two cylindrical slots for the insertion of the sedimentation cells. These cells contain the wedge-shaped sedimentation *chamber*, and are capped at top and bottom by quartz windows, permitting optical measurements. Figure 2.39 gives a sketch of these two components. The rotor is suspended by means of a flexible wire and rotated at high speed, the range of available speeds being from about 25 rev/s to about 10^3 rev/s. As a consequence, the particles in the sedimentation chamber are subject to radially outward-directed, centrifugal force. Because of the high angular velocities of the rotor, this centrifugal force can attain values far in excess of the normal gravitational force. It plays the role of an "artificial" gravity, which can be made large enough to compress the scale height H (see (2-165)) to a value of the order of 1 cm or less.

Before going into the numerical determination of this centrifugal force, let us make sure that there is no misunderstanding about its role and meaning. This force

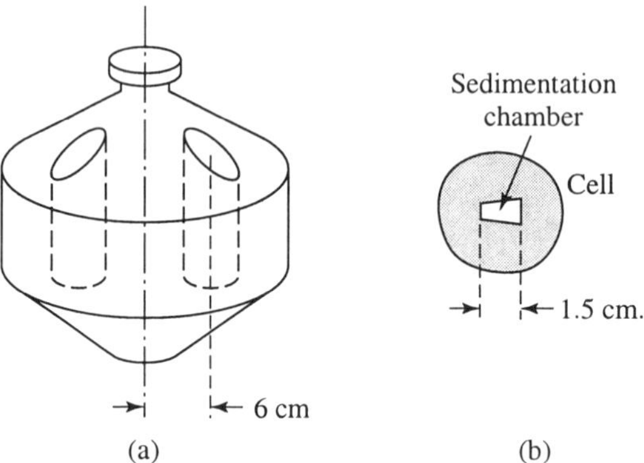

Figure 2.39. Ultracentrifuge components. (a) View of a rotor with slots for sedimentation cells. (b) Top view of a sedimentation cell with a wedge-shaped chamber.

comes into play naturally whenever we deal with circular motion, and describe motion in a rotating system of coordinate axes. Such a system of axes, fixed relative to the sedimentation cell, and participating in its circular motion, is shown in Figure 2.40. Newton's second law for a particle of mass m, subject to a force $\vec{F}$, states that its acceleration $\vec{a} = d\vec{v}/dt$ is given by

$$m\vec{a} = m\frac{d\vec{v}}{dt} = \vec{F}.\qquad(2\text{-}167)$$

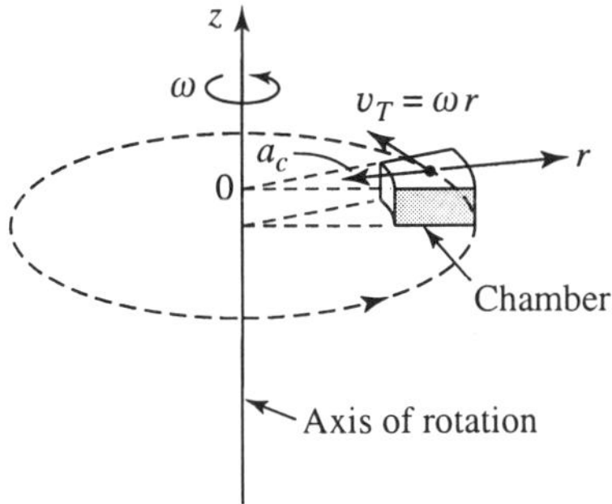

Figure 2.40. Coordinate system with r- and z-axes. The r-axis rotates with the cell at angular velocity ω. A particle *at rest* in the chamber has a tangential velocity $v_T = \omega r$, and a centripetal acceleration $a_C = -\omega^2 r$, directed toward 0.

This is a vector equation, and implies that the *corresponding* vector components, along any axis, of $m\vec{a}$ and $\vec{F}$ must be equal. If we choose the components along the radial (outward) direction, we get from (2-167) the components equation

$$ma_r = F_r\qquad(2\text{-}168)$$

the subscript r indicating the radial component. Elementary kinematics (see Volume I, Section 1.3) shows that the radial component of the acceleration a is not just (dv_r/dt), but rather

$$a_r = \left(\frac{dv_r}{dt}\right) - \omega^2 r,\qquad(2\text{-}169)$$

ω being the angular velocity of the rotor (and hence of the rotating chamber). The term $-\omega^2 r$ is the centripetal acceleration, and points radially inward. Newton's law for the rate of change of the radial velocity $v_r \equiv dr/dt$ is therefore of the form

$$m\left(\frac{dv_r}{dt}\right) = F_r + m\omega^2 r, \tag{2-170}$$

as can be seen from inserting (2-169) into (2-168). The term $+m\omega^2 r$ in (2-170) *is* the "centrifugal force." This equation shows that for a particle to remain stationary or to move at constant radial velocity v_r in the sedimentation chamber, the sum of the force F_r and the centrifugal force $m\omega^2 r$ must be zero. F_r must be directed inward. For a particle suspended in a fluid-filled chamber, F_r is the sum of buoyancy and the drag force. Depending on whether the force of buoyancy is less than $m\omega^2 r$, or exceeds it, the particle will be moving radially outward or inward.

The centrifugal force $m\omega^2 r$ plays the role of "artificial gravity" for the material filling the sedimentation chamber. It is customary, in fact, to express its magnitude by comparing it to the gravitational force $F_g - mg$ and then express the centrifugal acceleration $\omega^2 r$ in units of $g = 9.81 \text{ m/s}^2$. Let us find some numerical values for $\omega^2 r$. We must remember that ω, the angular velocity, is 2π times the number ν of revolutions per second. For r we choose 6 cm. This gives

$$\left(\frac{\omega^2 r}{g}\right) = 4\pi \frac{6 \text{ cm}}{0.981 \times 10^3 \text{ cm/s}^2} \nu^2$$
$$= 0.24 \text{ s}^2 \nu^2$$

or

$$\omega^2 r = (0.24 \text{ s}^2 \nu^2)g. \tag{2-171}$$

Thus, for $\nu = 10$ rev/s, the centrifugal acceleration is 24 g, for $\nu = 100$ rev/s it has become 2400 g, and for $\nu = 10^3$ rev/s it reaches 240,000 g.

These large values make it possible to compress the sedimentation equilibrium curve into a fraction of a centimeter. At the same time, the drift velocities of macromolecules can be made large enough to be observable in reasonable time. These drift velocities, defined in (2-155) ($u = (D/kT)f$), would be intolerably small, if the only force acting on a macromolecule were gravity (corrected for buoyancy, as described in (2-162). Indeed, if we insert $f = m^*g = m(1 - \rho\bar{v}_p)g$ into (2-155) for the drift velocity, we find

$$u = \frac{D}{kT}m(1 - \rho\bar{v}_p)g = \frac{D}{H}, \tag{2-172}$$

H being the scale height of the sedimentation curve for the macromolecule in the fluid. Since for a typical macromolecule, D is of the order of 10^{-6} cm^2/s, and H of the order of 10 m $= 10^3$ cm, we find a drift velocity $u \simeq 10^{-9}$ cm/s! It is only by replacing g in (2-172) by a centrifugal acceleration $\omega^2 r$ of order 10^5 g that we can raise the drift velocity to an observable level of $u \sim 10^{-4}$ cm/s ~ 0.3 cm/h.

(ii)   The Sedimentation Coefficient s. Determination of Molecular Weights

In a typical operation the outer (more distant from the axis) part of the sedimentation chamber is filled with the dilute solution of the macromolecules to be analyzed. The remainder of the chamber is filled with pure buffer. (Protein solutions are generally buffered so as to maintain a pH near the isoelectric point; otherwise, electrostatic effects would come into play which complicate the analysis.) We have thus, at the start of the operation, a concentration profile as shown in Figure 2.41, containing a sharp boundary between the solute and the pure buffer. As the centrifugation proceeds, this boundary moves with the drift velocity u (also called sedimentation velocity in ultracentrifuge work). If several different macromolecules are in solution (in homogeneous solution), the boundary may split into several components if the different species have sufficiently distinct drift velocities.

We shall now show how, by measuring this drift velocity, we can determine the molecular weight M, and then present some data.

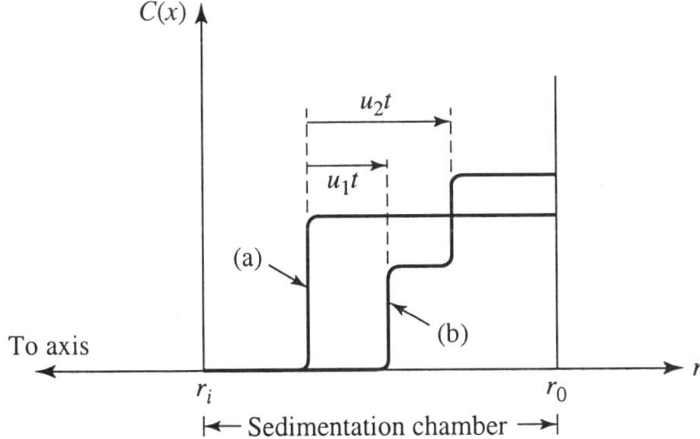

Figure 2.41. Concentration profile in a sedimentation chamber of an ultracentrifuge. (a) At the beginning of centrifugation. (b) At some later time, revealing two solute components drifting with different velocities u_1 and u_2.

The equation describing the drift and diffusive spread of the boundary is (2-153). The force f acting in this case is the centrifugal force corrected for buoyancy

$$f = m\omega^2 r(1 - \rho \bar{v}_p).\tag{2-173}$$

In this expression m is the mass of the macromolecule, ρ the density of the solution (in g/cm^3), and $\bar{v}_p$ the specific volume of the macromolecule, in cm^3/g. This force is not exactly constant since it depends on the distance r of the molecule from the axis of rotation. With this force we get a drift velocity u, according to (2-155), of

$$u = \frac{D}{kT}m(1 - \rho\bar{v}_p)\omega^2 r.\tag{2-174}$$

It is customary to normalize the observed drift velocities against the centrifugal acceleration $\omega^2 r$ used in the experiment, and define a so-called *sedimentation coefficient* s as the ratio

$$s \equiv \frac{u}{\omega^2 r}.\tag{2-175}$$

This sedimentation coefficient depends only on the macromolecule involved, and the density and temperature of the solution

$$s = \frac{D}{kT}m(1 - \rho\bar{v}_p).\tag{2-176}$$

s may also be expressed in terms of the molecular weight $M = N_0 m$ and the gas constant $R = N_0 k = 8.31 \times 10^7$ erg/mol degree:

$$s = \frac{D}{RT}M(1 - \rho\bar{v}_p).\tag{2-176a}$$

The dimension of s is that of time as is easily seen from (2-175). Typical values of s are of the order of 10^{-13} s, as can be seen by using the example of a typical molecule with

$$M = 10^5 \text{ g/mol,}$$

$$D \sim 10^{-7} \text{ cm}^2\text{/s,}$$

$$T = 300\,^\circ\text{K} \to RT = 2.5 \times 10^{10} \text{ erg/mol,}$$

$$(1 - \rho\bar{v}_p) \sim 0.25,$$

giving

$$s = \frac{10^{-7} \text{ cm}^2\text{/s} \times 10^5 \text{ g}}{2.5 \times 10^{10} \text{ g cm}^2 \text{ s}^{-2}} \times .25 \sim 10^{-13} \text{ s.}$$

A value $s = 10^{-13}$ s is called 1 *svedberg unit*, and is given the symbol S. Thus, for example, a sedimentation coefficient of 3.6 svedberg

$$s = 3.6S$$

means

$$s = 3.6 \times 10^{-13} \text{ s.}$$

The relation between the s-value of a molecule and the actual drift velocity u is then easily recovered by means of (2-175). If, for example, $s = 3.6S$, and $\omega^2 r = 10^5$ g, then the drift velocity is

$$u = 3.6 \times 10^{-13} \text{ s} \times 10^8 \text{ cm/s}^2 = 3.6 \times 10^{-5} \text{ cm/s.}$$

A displacement of the boundary by 10^{-1} cm will, in this case, require centrifugation at 10^5 g for about 1 h!

The determination of molecular weights M from the observed sedimentation coefficients s is based on (2-176a). It requires that the diffusion constant be measured separately, and that the buoyancy correction factor be determined by measuring both the density ρ of the solution and the specific volume $\bar{v}_p$ of the solute molecule; this latter, we repeat, may be defined as the volume increment of a dilute solution if 1 g of dry solute is added. This prescription raises some problems if applied to proteins and, in general, the buoyancy correction is the least reliable factor that goes into the determination of M. With all these data determined, we find the molecular weight M (in g/mol) from (2-176a) as

$$M = \frac{RT}{D(1 - \rho\bar{v}_p)}s. \tag{2-177}$$

We shall also present data at the end of the next section 2.5.E(iii) that deals with the determination of M from the concentration profile at sedimentation equilibrium.

(iii)   Determination of Molecular Weights from Sedimentation Equilibrium: Some Data

This method follows the footsteps of Jean Perrin's historic work on measuring the scale height of the equilibrium distribution. As mentioned in Section 2.5.D, these scale heights H would be of the order of 10 to 100 m in the gravitational force field. By using centrifugal accelerations up to a value of 10^5 g, they can be compressed to values of less than 1 cm.

We must here repeat the derivation of the equilibrium concentration profile because, unlike gravity, the centrifugal force is not exactly constant, but increases with distance r from the axis of rotation. We start again with (2-150) for the current, but using now the radial distance r instead of x as the position coordinate, and (2-173) for the force f:

$$j(r) = -D\frac{\partial C}{\partial r} + \frac{D}{kT}m(1 - \rho\bar{v}_p)\omega^2 r C(r). \qquad (2\text{-}178)$$

Whatever the initial concentration profile of the solute particles may be (in ultracentrifuge work it will be uniform), eventually an equilibrium will be reached where the current density j vanishes at every point. Since C is now time-independent, we can write dC/dr for $\partial C/\partial r$, and find

$$\frac{dC}{dr} = \frac{m}{kT}(1 - \rho\bar{v}_p)\omega^2 r C(r). \qquad (2\text{-}179)$$

This may be written, after dividing by $C(r)$, as

$$\frac{1}{C(r)}\frac{dC}{dr} \equiv \frac{d}{dr}\ln C(r) = \frac{m}{kT}(1 - \rho\bar{v}_p)\omega^2 r.$$

Integration of this equation gives

$$\ln C(r) - \ln C(r_0) = \frac{m}{kT}(1 - \rho\bar{v}_p)\frac{\omega^2}{2}(r^2 - r_0^2). \qquad (2\text{-}180)$$

It is useful to work directly with this expression for the concentration, which states that the logarithm $\ln C(r)$ of the concentration increases linearly with r^2:

$$\ln C(r) = \text{constant} + \tfrac{1}{2}\alpha r^2,$$

where α has the value

$$\alpha = \frac{m}{kT}(1 - \rho \bar{v}_p)\omega^2. \tag{2-181}$$

A measurement of this slope $\tfrac{1}{2}\alpha$ therefore gives the *same information* as a measurement of the sedimentation coefficient, and is a useful alternative to determining $M = N_0 m$. We see from (2-180) that the equilibrium profile is not exponential as it would be in the case of a constant force f. From (2-180) we find indeed that

$$C(r) = C(r_0)e^{+\frac{1}{2}\alpha(r^2 - r_0^2)}. \tag{2-182}$$

For r_0, we may choose the position of the bottom (far end) of the sedimentation chamber, as indicated in Figure 2.37. Since the width of the chamber, $(r_0 - r_i)$, is considerably smaller than r_0, we may write for all values of r between r_i and r_0 (i.e., inside the chamber)

$$(r^2 - r_0^2) = (r + r_0)(r - r_0) \simeq 2r_0(r - r_0).$$

In this approximation, the expression (2-182) for $C(r)$ reduces to an exponential

$$C(r) \cong C(r_0)e^{+\alpha r_0(r - r_0)}$$
$$= C(r_0)e^{-(r_0 - r)/H}. \tag{2-182}$$

We see then that $C(r)$ is maximum at the bottom, and decreases approximately exponentially with distance $(r - r_0)$ above the bottom of the chamber. The scale height is given by

$$\frac{1}{H} = \frac{m}{kT}(1 - \rho \bar{v}_p)(\omega^2 r_0)$$

which is the same as (2-165), but with the gravitational acceleration g replaced by the centrifugal acceleration $(\omega^2 r_0)$.

Before giving some data, a word may be said about the manner in which the sedimentation coefficient s, or the equilibrium density profile are actually measured. In both cases, the method is optical, based on the fact that the index of refraction n of the solution increases linearly with the solute concentration C. The shape of the equilibrium profile can be measured by measuring the "optical path length" n for a light beam crossing the chamber, parallel to the rotation axis at a position r. Another method, the so-called "Schlieren" method, can be used to determine the location of the moving boundaries in the procedure based on measuring the sedimentation coefficient s. For details, we refer the reader to [19], or for a very quick survey [21].

We now give a tabulation (Table 2.10) of some data obtained by these two methods. They are taken from a compilation presented in the book by Cohn and

Table 2.10. Molecular Weights from Ultracentrifugation.

Molecule	Specific volume $\bar{v}_p$ in cm^3/g	Sedimentation coefficient s in svedbergs	Diffusion constant D in 10^{-7} cm^2/s	Molecular weight M in g/mol — from sed. velocity	Molecular weight M in g/mol — from sed. equil.	D_0/D
(1)	(2)	(3)	(4)	(5)	(6)	(7)
Myoglobin	0.741	2.04	11.3	16,900	17,200	1.11
Ribonuclease	0.709	1.85	13.6	12,700	13,000	1.04
Lactoglobulin	0.7514	3.12	7.3	41,500	38,000	1.26
Ovalbumin	0.749	3.55	7.8	44,000	40,500	1.16
Hemoglobin (horse)	0.749	4.41	6.3	68,000	68,000	1.24
Serum globulin (horse)	0.745	7.1	4.05	167,000	150,000	1.44
Hemocyanin (Palinurus)	0.740	16.4	3.4	450,000	450,000	1.23
Hemocyanin (Homarus)	0.740	22.6	2.78	760,000	800,000	1.27
Bushy stunt virus	0.739	146			7,600,000	1.09
		132	1.15	10,600,000		1.27
Rabbit papilloma virus	0.756	280	.585	47,100,000		1.49

The data in the second, third, and fourth columns are corrected so as to represent the values of these quantities, in water, at 20 °C.

Edsall [22]. The first column of this table lists the macromolecule or virus particle, the second column its specific volume, in cm^3/g. It is interesting to observe the great uniformity of the values of $\bar{v}_p$. Roughly speaking, we can say that all closely packed macromolecules are equally dense, about 1.3 g/cm^3. Column (3) gives the sedimentation coefficient s, in svedbergs. We see that a wide range of s-values can be successfully observed. D, in column (4), is the diffusion constant derived from separate measurements.

Columns (5) and (6) give calculated values of the molecular weight M, based on sedimentation (drift) velocity (column (5)) and equilibrium density profile (column (6)). We observe that, in general, there is a fair agreement between these two sets of data, the discrepancy rarely exceeding 10%. It is interesting to see that also the "molecular weights" of virus particles can be so determined; it is of the order of 10^7 g/mol for the smallest of them.

Column (7) finally lists the ratio of a calculated diffusion constant D_0 to the one actually observed, D. D_0 is obtained by combining the values of M and $\bar{v}_p$ to calculate an equivalent radius a of the molecule: The quantity $M\bar{v}_p$ is the volume of 1 mol of macromolecules; considering them to be spheres, we can write

$$M v_p = N_0 \left(\frac{4\pi}{3} a^3 \right) \tag{2-183}$$

and get an "equivalent" radius a. This radius may be inserted into the Einstein–Stokes formula [(2-159)] to get a calculated diffusion constant. This calculated quantity is here called D_0 and compared with D in the last column. The values of D and D_0 should agree closely for nearly spherical molecules, and the general agreement gives a sort of consistency check on the data. In individual cases, D may differ from D_0, either because the macromolecule is hydrated, or because its shape is *not* that of a sphere. Perrin has shown (see [22, table, p. 406]) that for a fixed volume the diffusion constant is maximum for a spherical shape. A value D_0/D of 1.1 indicates an ellipsoidal shape with a ratio of 3:1 of the principal diameters, and a value $D_0/D = 1.5$ indicates a ratio of 10:1. In that case, the particle is either a "rod" or a "pancake." Calculation of D_0/D therefore gives additional, though not very precise, information on the shape of the macromolecule or virus particles.

2.6 Flow of Solute and Solvent Across a Membrane in the Presence of Both Pressure and Concentration Gradients

In this section we will generalize the description of the flow of fluid and solute particles across membranes beyond that given in Section 2.4.C(i), (ii), and (iii). In those sections we limited our analysis to the case that either a pressure difference Δp *or* a concentration difference ΔC was established across the membrane. In the former case Δp produced a volume flow per unit area of membrane (J_v) given by equation

$$J_v = L_p \Delta p. \qquad (2\text{-}184)$$

Here J_v is the volume flow measured in units of cm^3/s^2. L_p is the filtration coefficient or hydraulic permeability, and it has the dimensions $(\text{cm/s})/(\text{dyn/cm}^2)$ in cgs units. In case a concentration difference ΔC is present, there is an associated solute flow J_s, as measured in the number of solute particles passing per second across the membrane per unit area. J_s is related to ΔC by

$$J_s = p \Delta C. \qquad (2\text{-}185)$$

Here p is the membrane permeability. It has the units of cm/s.

We must now generalize this description to obtain expressions for J_v and J_s when *both* a pressure difference Δp and a concentration difference ΔC are set up across the membrane. Such a situation occurs commonly among physiological systems. For example, both a pressure and a concentration gradient exist across the wall of a blood capillary. Both affect the flow between the blood plasma inside the capillaries and the interstitial fluid outside the capillary wall. A second example is that of single cells, like red blood cells. If they are placed in pure water, the concentration difference of solutes between the cell interior and the water outside will generate an *inflow* of water into the cell, and thus build up a hydrostatic pressure Δp across the cell membrane. Unless the concentration gradient is reduced, the cell can swell up and finally burst. In fact, this flow of fluid across a membrane from the region of low solute concentration to the region of high solute concentration was omitted in our discussion of Section 2.4,C(i), (ii), and (iii). Such flow is generally described as taking place under the influence of an "osmotic pressure" difference across the membrane.

A dramatic demonstration of the physiological importance of osmotic pressure can be seen in the so-called "edemas of starvation." We have all seen pictures of the swollen bellies of starving children. This swelling or edema actually represents the outflow of fluid from the vascular bed into the surrounding interstitial space under the influence of the hydrostatic pressure provided by the heart. Normally this hydrostatic outflow from the blood capillaries is checked by the osmotic inflow from the interstitial fluid to the blood plasma. In the case of starvation, the proteins in blood plasma have such a low concentration that the osmotic inflow is insufficient to compensate for the hydrostatic outflow and "edema" results.

It is the first objective of this section to describe quantitatively the experimental facts that underly the concept of osmotic pressure, and also to describe the effect of a concentration difference ΔC on the bulk flow across a membrane. That is, we shall show the origin of the term $-\lambda \Delta C$ hypothesized in (2-117a) of Section 2.4.C(iv). We begin our discussion by considering these questions when the membrane is a so-called "semipermeable" membrane. This is a membrane that allows the passage of solvent molecules, but that will not allow the passage of solute particles, i.e., $p = 0$. The walls of blood capillaries are an example of such membranes; they permit passage of water, but not of the globular proteins dissolved in the blood plasma. We will then, in Section 2.6.B, relax this restriction and obtain the general form relating J_v and J_s when both ΔC and Δp are present across a membrane for which $p \neq 0$.

2.6.A.   Hydrostatic Pressure. Semipermeable Membrane. Osmotic Pressure. Van t'Hoff's Law. Volume Flow (J_v) Across a Semipermeable Membrane in the Presence of Both Δp and ΔC

(i)   Hydrostatic Pressure

Let us begin by reviewing very briefly the elements of hydrostatics. (See also Volume I, *Mechanics*, Section 3.5.) The scalar p (the hydrostatic pressure) describes the force $\Delta \vec{F}$ that acts across each element ΔS of area within, or on the surface of a fluid, in accordance with the equation

$$\Delta \vec{F} = -p\hat{n}\Delta S. \tag{2-186}$$

By considering the conditions for static equilibrium as applied to a small pill box of fluid having mass density ρ, it can be shown that for the weight of the fluid to

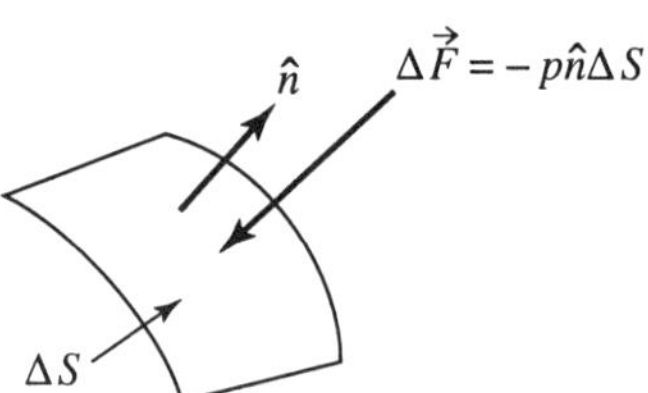

Figure 2.42. Relationship between the force $\Delta\vec{F}$ acting across an area element ΔS and the hydrostatic pressure Δp. $\hat{n}$ is a unit vector pointing outward from ΔS.

be balanced by the internal pressure in the fluid, the pressure must increase with depth (z) beneath the fluid in accordance with Pascal's law

$$\left(\frac{dp}{dz}\right) = \rho g, \tag{2-187}$$

where g is the acceleration of gravity. For the case where the density ρ has a constant value independent of z, we may integrate (2-187) directly to obtain

$$p(z) = p(0) + \rho g z. \tag{2-188}$$

Here $p(0)$ is the pressure at the surface of the fluid, and $p(z)$ is the pressure at the depth z.

A generalized form of this result will now be proven. Suppose that a "body force" $F(x)$, that in general can vary in magnitude and direction for various values of x, acts on the fluid particles. Let the body force act in say the x-direction, and let $(\Delta F_x/\Delta V)$ be the force per unit volume. Then, we shall show that if the fluid is not to experience an acceleration under the action of this force, i.e., if the fluid moves with constant velocity, there must be a pressure gradient in the fluid (dp/dx) whose magnitude is given by

$$\left(\frac{dp}{dx}\right) = \left(\frac{\Delta F_x}{\Delta V}\right). \tag{2-189}$$

That is, the body force per unit volume acting on the fluid is equal to the pressure gradient in the direction of the force when the fluid is in mechanical equilibrium (not accelerating).

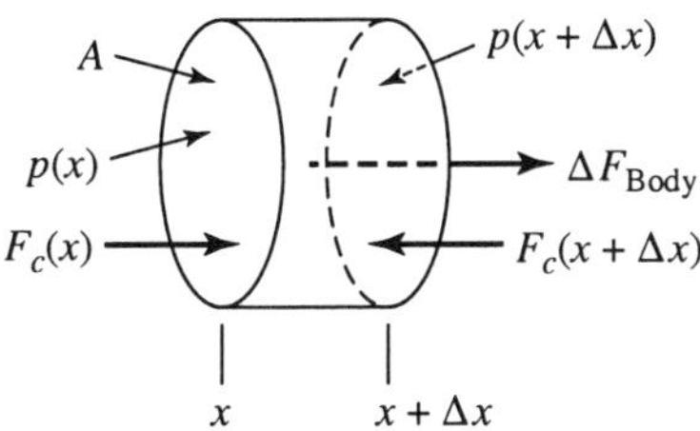

Figure 2.43. Body force F_b, and contact force F_c, acting on bulk and surfaces of a mathematical surface inside a fluid.

This result can be proven quite simply by referring to Figure 2.43. The total force acting on the material inside the mathematical surface shown in Figure 2.43 is

$$F_{\text{tot}} = \Delta F_{\text{body}} + \left(F_c(x) - F_c(x + \Delta x)\right). \qquad (2\text{-}190)$$

Here F_c is the contact force exerted on the pillbox surface by the surrounding liquid. At equilibrium $F_{\text{tot}} = 0$. Also, from the definition of the pressure, we see that (2-190) can be written at equilibrium as

$$0 = \Delta F_{\text{body}} + A p(x) - A p(x + \Delta x)$$

or if Δx is small, we have

$$0 = F_{\text{body}} + A p(x) - A \left(p(x) + \left(\frac{dp}{dx}\right) \Delta x \right)$$

or

$$0 = \Delta F_{\text{Body}} - A \left(\frac{dp}{dx}\right) \Delta x$$

but

$$A \Delta x = \Delta V.$$

So we find

$$\left(\frac{\Delta F_{body}}{\Delta V}\right) = \left(\frac{dp}{dx}\right). \quad \text{Q.E.D.} \tag{2-191}$$

Thus, the pressure gradient (dp/dx) at a point x in the fluid is equal to the external force per unit volume exerted on the fluid.

(ii) Phenomenological Description of Osmotic Pressure and Volume Flow Across a Semipermeable Membrane. Van t'Hoff's Law

We now turn to a discussion of "osmotic pressure." For this purpose we begin by first discussing experimental facts associated with the flow of water across a semipermeable membrane. In Figure 2.44 we show schematically the experimental setup that reveals the principal phenomena. The inverted thistle tube is closed off by a membrane and filled with a solution of solute concentration C_S:

(a) Onset of the experiment: Level of solution z_0 is adjusted so that the same hydrostatic pressure p exists on both sides of the membrane. Water begins to enter the tube.

(b) Intermediate stage (after some elapsed time): Water continues to flow into the tube against a gradient of hydrostatic pressure

$$\Delta p = \rho(\text{solution})g(z_i - z_0).$$

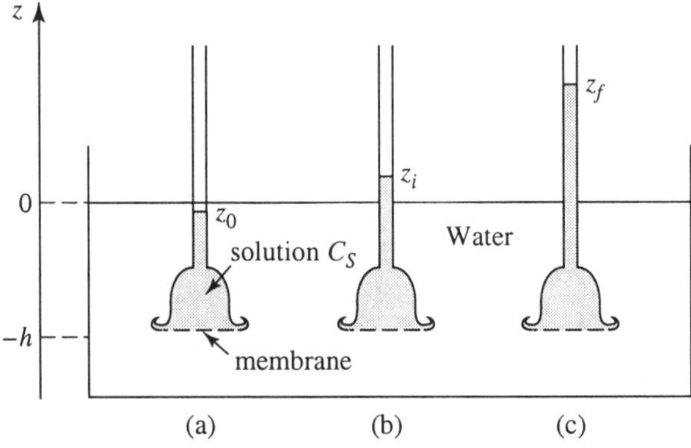

Figure 2.44. The basic osmotic pressure experiment.

(c) Final stage: Water flow stops at some terminal value of hydrostatic pressure difference $\Delta p_f = \rho(\text{solution})g(z_f - z_0)$.

We now describe the observations and the inferences one is led to draw from them:

(a) At the onset of the experiment one observes that the level of fluid in the neck begins to rise, signaling an *inflow* of water into the tube. Clearly, water is *sucked* into the tube across the membrane. The driving force for this flow cannot be the hydrostatic pressure difference Δp, since this is zero at the start. There must therefore be a second agent, in addition to Δp, for the volume flow J_v across the membrane. The qualitative observation that the rate of this inflow increases with solute concentration establishes the fact that this agent resides in the compartment containing the solution. We can express the effectiveness of this agent in terms of a pressure, driving water across the membrane just as a hydrostatic pressure would. This pressure is called the osmotic pressure Π of the solution. Pure water (or other pure solvent) has zero osmotic pressure by definition.

(b) As the experiment proceeds, the water level z_i in the stem of the tube continues to rise, but the rate of rise slows down with time. Inflow now occurs *against* an adverse hydrostatic pressure gradient

$$\Delta p = \rho(\text{solution})g(z_i - z_0)$$

indicating that the osmotic pressure Π of the solution is still larger than Δp.

(c) Water inflow into the tube eventually comes to rest where the level in the tube has risen to a value z_f. At this point the hydrostatic pressure difference

$$\Delta p_f = \rho(\text{solution})g(z_f - z_0) \tag{2-192}$$

checks the effect of the osmotic pressure Π. Recall that Δp_f wants to push the water out, Π wants to suck it into the tube.

Without in any way supplying an explanation of the origin of this mysterious pressure Π, we can describe the observation just made by a generalization of the flow [(2-184)] for the bulk flow J_v. Let us establish a positive sense of flow from inside the thistle tube toward the outside, and let us define Δp as the pressure difference between the inside and outside in agreement with (2-192). Let Π, a positive quantity, represent the osmotic (sucking) pressure of the solution. We may measure Π by its ability to generate volume flow; that is, we view Π as an additional pressure that may either reinforce, or counteract the effect of hydrostatic pressure on volume flow.

The experimental observations described above can be put into quantitative form by generalizing the flow [(2-184)] to the following form:

$$J_v = L_p\{(p(\text{solution}) - p(\text{water}) - \Pi(\text{solution})\}$$

or

$$J_v = L_p(\Delta p - \Pi_s), \qquad (2\text{-}193)$$

where $\Delta p = p(\text{solution}) - p(\text{water})$, and Π_s is the osmotic pressure of the solution. Π_s is by our definition positive, and must therefore be entered with a $(-)$ sign in (2-193), so as to generate an inflow. Let us check that (2-193) indeed *describes* the observed phenomena presented above, and in Figure 2.44.

At the onset of the experiment, $\Delta p = 0$ and $J_v = -L_p\Pi_s$. We have inflow into the tube. This inflow raises the level of fluid in the stem of the tube, and builds up a hydrostatic pressure difference Δp. The volume flow is now

$$J_v = -L_p(\Pi - \Delta p)$$

but Π is still larger than Δp, and inflow continues but at a reduced rate. Eventually, however, Δp reaches a value that exactly compensates Π and all flow stops. At this point, Δp is given by (2-192). Since this value of Δp is equal to Π, we have, in this relation, a method to *measure* the osmotic pressure Π of the solution

$$(\Pi)_{\text{measured}} = \Delta p_f = \rho(\text{solution})g(z_f - z_0). \qquad (2\text{-}193\text{a})$$

In words, we may measure the osmotic pressure of a solution by the hydrostatic pressure difference that must be established across the membrane in order to stop the flow of water.

Having established a means to measure the osmotic pressure Π, with the help of (2-193), we can now review how Π depends on the properties of the solution. It turns out that, for sufficiently dilute solutions, the osmotic pressure of a solution is directly proportional to the concentrations C_s of solute particles, and to the absolute temperature T. This is expressed as Van t'Hoff's law

$$\Pi_s = C_s RT, \qquad (2\text{-}194\text{a})$$

where C_s is the solute concentration in mol/unit volume, T the absolute temperature, and R the universal gas constant. Alternatively, if C_s is measured in units of the number of solute particles per cm^3 (c_s), then Van t'Hoff's law can be written as

$$\Pi = c_s kT, \tag{2-194b}$$

where k is Boltzmann's constant.

Van t'Hoff's law is remarkable by the fact that it is identical in form to the ideal gas law. Indeed, in the case that C_s is expressed in mol/l, we can express C_s in terms of the volume $\overline{V}_s$ occupied by 1 mol of solute, namely,

$$C_s = \frac{1}{\overline{V}_s},$$

so that (2-194a) may be indeed be written as

$$\Pi_s \overline{V}_s = RT. \tag{2-195}$$

This is, of course, recognized as the same form as the ideal gas law.

It is important to recognize that just as the ideal gas law applies accurately only in the low gas density region, so the Van t'Hoff law also only applies for low solute concentration. We will later present data on the range of validity of this Van t'Hoff law for both proteins and small molecules as glucose.

Relation [(2-195)] furnishes the crucial clue to the origin of the osmotic pressure, namely, the kinetic energy of the solute particles as given by the equipartition law of energies.

Before explaining the physical origin of the osmotic pressure, let us extend our phenomenological description to a more general case than that discussed above. The more general case is shown in Figure 2.45. Here a semipermeable membrane

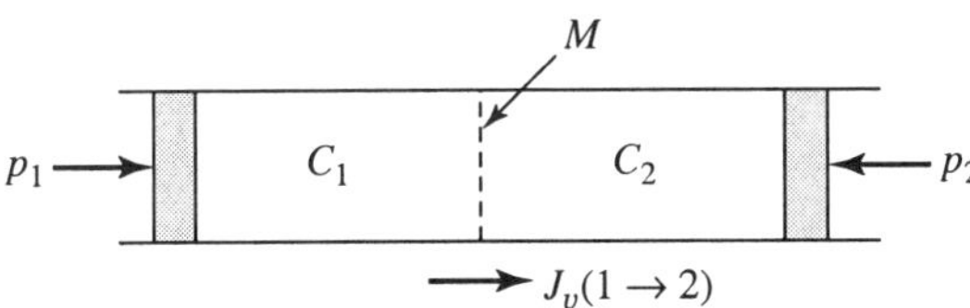

Figure 2.45. Semipermeable (solute impermeable) membrane separating two compartments of solute concentrations C_1 and C_2, with hydrostatic pressures p_1 and p_2 maintained by forces on the pistons.

M separates two compartments (1) and (2). In each compartment the solute concentration is C_1 or C_2, and the hydrostatic pressure is p_1 or p_2. Let J_v again be the volume flow to be counted positive if directed from compartment (1) toward compartment (2), otherwise negative. Let $\Delta p = p_1 - p_2$ the hydrostatic pressure difference across the membrane, and let Π_1 and Π_2 be the osmotic pressures of the solutions (1) and (2) (Π_1 and Π_2 can be separately established by having solutions (1) and (2) facing water across the membrane). One then observes a volume flow J_v that is driven by the difference between the hydrostatic pressure drop $\Delta p = p_1 - p_2$, and the osmotic pressure drop $\Delta \Pi = \Pi_1 - \Pi_2$:

$$J_v = L_p(\Delta p - \Delta \Pi). \tag{2-196a}$$

Also since there is no solute flow in the case of a semipermeable membrane, we have

$$J_s = 0. \tag{2-196b}$$

These two equations describe, for the case of a semipermeable membrane, the relationship between the "fluxes," J_v and J_s, and the "forces," Δp and $\Delta \Pi$. In Section 2.6.B, we will obtain the relationship between the fluxes and the forces in the case of a general permeable membrane.

(iii)   Physical Origin and the Theory for the Osmotic Pressure. Derivation of Van t'Hoff's Law. Poisseuille Flow and the Flow of Solvent Through a Semipermeable Membrane Under the Influence of Both Pressure and Concentration Differences

We now turn to a detailed investigation of the physical origin of the osmotic pressure. We will try to understand theoretically why the water flows from a region of low solute concentration to one of high solute concentration in proportion to the osmotic pressure difference. We first confine ourselves again to the case of a semipermeable membrane; that is, one which does not permit the flow of solute particles. We show in Figure 2.46 two chambers (1) and (2), separated by a semipermeable membrane. At time $t = 0$, let the pressure in chamber (2) be p_2 and that in chamber (1) be p_1. Let the solute concentration be C_s in (1) and 0 in (2). To examine the origin of the osmotic pressure, let us view the situation that exists in the vicinity of a single pore of the membrane. Such a pore is shown schematically in Figure 2.47. Since the membrane excludes the solute molecules from passing through the pores, the average concentration of solute molecules decreases from

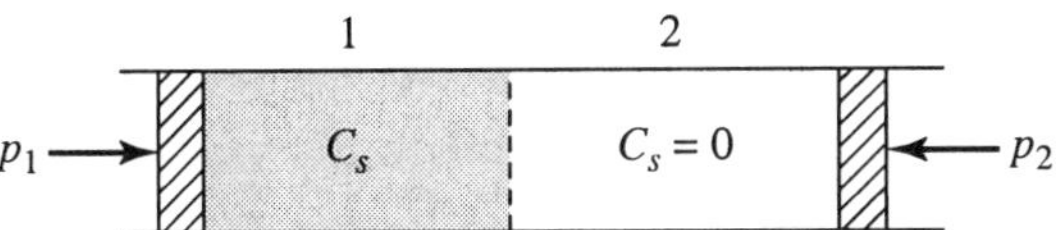

Figure 2.46. Chambers (1) and (2) separated by a semipermeable membrane. Concentration of solute in chamber (1) is C_s, and $C_s = 0$ in chamber (2).

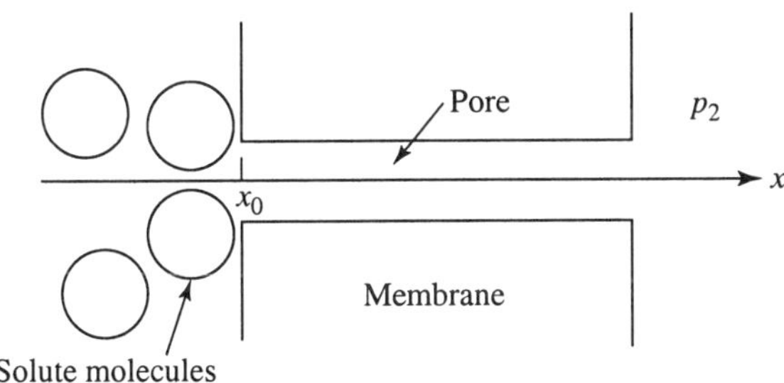

Figure 2.47. Detail of a pore entrance and exit.

C_s to 0 in moving along the pore axis from outside the pore entrance to inside the pore. This is illustrated schematically in Figure 2.48. The solute molecules are in a concentration gradient near the opening of the pore; as a result there should be a diffusion current into the pore. The solute molecules do not, in fact, enter the pore because they hit the edge of the pore and experience a force which *on average* will be in the $-x$ direction. We can find this average force quite simply from (2-152) that relates the solute flux to the concentration gradient and the average force f

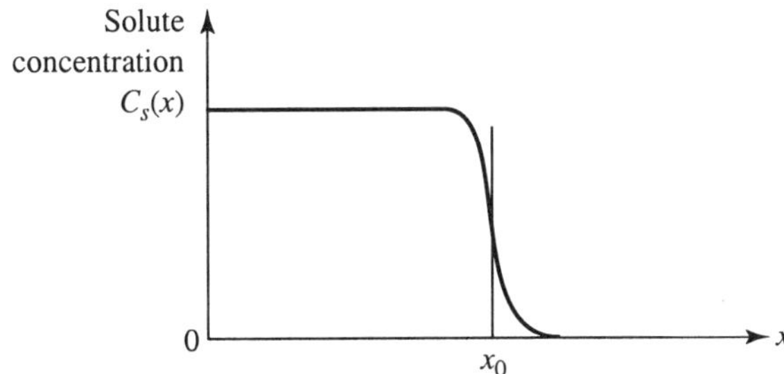

Figure 2.48. Concentration profile for solute molecules in the vicinity of the entrance of a pore of a semipermeable membrane.

exerted on each solute particle, viz:

$$j_s(x) = -D\left(\frac{\partial C_s}{\partial x}\right) + \left(\frac{D}{kT}\right) f C_x(x), \qquad (2\text{-}197)$$

where j_s is the solute current density in particles/cm s. Since the membrane blocks the flow of solute, we have $j_s = 0$, so that the average force f, that the wall exerts on each solute particle, is given by the equation

$$f C_s(x) = kT\left(\frac{\partial C_s}{\partial x}\right). \qquad (2\text{-}198)$$

This formula, in fact, states that the average force per unit volume $f C_s(x)$, that the wall exerts on the solute molecules, is $kT(\partial C_s/\partial x)$. Since the solute molecules are part of each volume element in the fluid as a whole, the semipermeable membrane, in effect, is exerting a body force on a volume element of fluid located directly in front of the pore. Because of the very slow velocity of flow, it is accurate to regard the flow as having essentially zero acceleration. Thus, the fluid may be regarded as being in mechanical equilibrium. The condition of equilibrium in turn implies that there must exist in the fluid, near the pore entrance, a pressure gradient as computed earlier in (2-191). This pressure gradient leads to a drop in pressure at the entrance of the pore. This pressure drop is responsible for the osmotic inflow from compartment (2). Let us now make this argument quantitative. In Figure 2.49 we show a small volume element in the fluid located directly in front of the pore. Let the front surface of area A be at the point x well within the solute. Also, let the

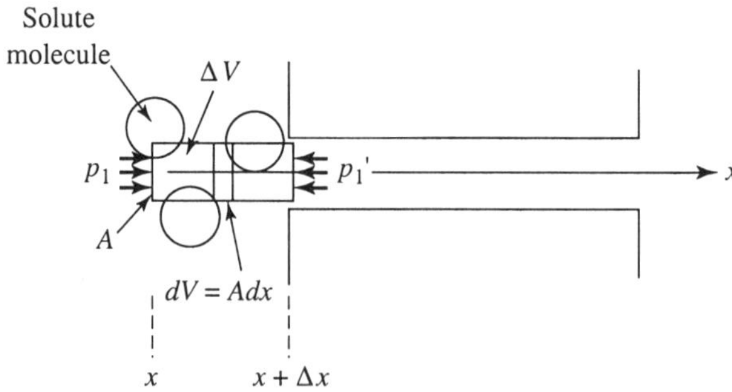

Figure 2.49. Volume element of fluid located directly in front of a membrane pore.

total pressure at x be p_1. Let the near surface of the element be located at $x + \Delta x$, and let the pressure at this point be p_1'. We now apply the condition of equilibrium to the volume element ΔV in Figure 2.49. The total force acting on ΔV consists of the body force $\Delta \vec{F}_B$, plus the net contact force, $\Delta \vec{F}_C$, provided by the surrounding fluid. For equilibrium we have

$$\Delta \vec{F}_B + \vec{F}_C = 0. \tag{2-199}$$

We now compute $\Delta \vec{F}_B$ and $\Delta \vec{F}_C$ separately. The net body force ΔF_B is produced solely by the average force the membrane exerts on the solute particles inside ΔV. Since the average force per unit volume at point x is equal to $f C_s(x) = kT(\partial C_s/\partial x)$, the total body force acting on ΔV is given by

$$\Delta F_B = \int_x^{x+\Delta x} f C_s(x)\, dV = AkT \int_x^{x+\Delta x} \left(\frac{\partial C_s}{\partial x}\right) dx \tag{2-200}$$
$$= AkT[C_s(x + \Delta x) - C_s(x)].$$

Since $C_s(x + \Delta x) = 0$ and $C_s(x) = C_s$, we find

$$\Delta F_B = -AkT C_s. \tag{2-201}$$

The minus sign signifies that the body force on ΔV points to the left as expected.
    We now calculate ΔF_C. This net force is the difference between the forces exerted on the surfaces at x and $x + \Delta x$ by the surrounding fluid. This is

$$\Delta F_C = A p_1 - A p_1'. \tag{2-202}$$

We now can substitute (2-202) and (2-201) into the equilibrium condition, (2-199), and we find at once that

$$p_1 - p_1' = kT C_s.$$

or

$$p_1' = p_1 - kT C_s. \tag{2-203}$$

This is the crucial result. The total fluid pressure *drops* from p_1 to $p_1 - kTC_s$ as one moves from inside compartment (1) into the pore. The drop in pressure kTC_s is that which one obtains when the concentration of solute is so low that the solute molecules move independently of one another; i.e., there is no interaction between the solute molecules. At higher solute concentration there will in general be some drop in pressure that will depart from the formula kTC_s. We denote the pressure drop in general by the osmotic pressure Π, and we will see later that this definition of Π is consistent with our previous phenomenological definition for the osmotic pressure.

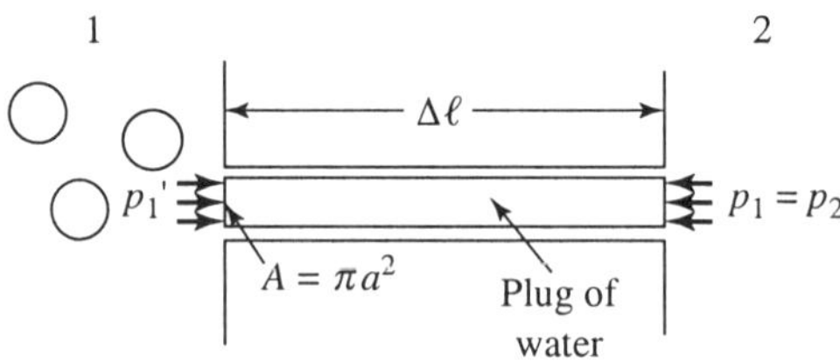

Figure 2.50. Volume element showing a plug of water flowing through the membrane pore.

We now show why this pressure drop acts to drive solvent from compartment (2) to compartment (1). Consider first the case $p_2 = p_1$. We examine the motion of a plug of *water* that extends throughout the pore as shown in Figure 2.50. The pressure profile along the axis of the pore is shown in Figure 2.51. Inside the pore

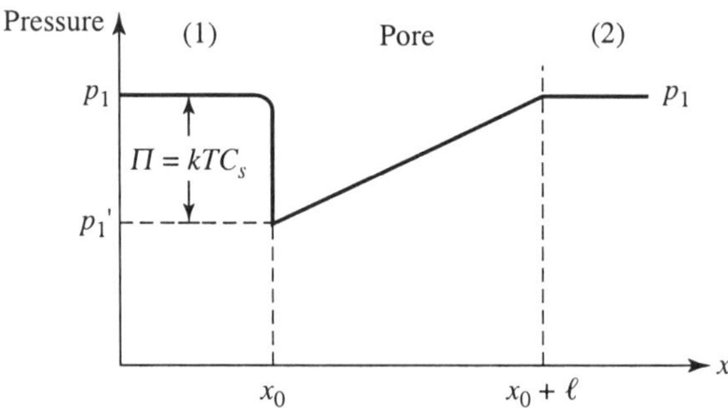

Figure 2.51. Pressure profile in a pore of a uniform cross section semipermeable membrane. In the case $p_1 = p_2$.

the pressure varies linearly with x if the pore cross section is uniform. On entering the channel from compartment (1), the pressure immediately drops from p_1 to $p_1 - kTC_s$ as shown in Figure 2.50. On the other hand, at the right end of the pore the pressure is p_1. In this situation the solvent fluid in the pore experiences a pressure difference between the left and right ends of the pore which is given by $(p_1' - p_1) = -kTC_s = -\Pi$. In addition, there is a frictional drag between the flowing solvent and the pore walls. As a result the solvent flows at a constant rate that is proportional to the constant pressure gradient $(p_1' - p_1)/l$ in accordance with Poiseuille's law

$$J/\text{pore} = \frac{\pi a^4}{8\eta}\left(\frac{-\Pi}{l}\right). \tag{2-204}$$

Alternatively, we can express the volume flow per unit membrane area in terms of Π as

$$J_v = nJ/\text{pore} = \frac{n\pi a^4}{8\eta}\left(\frac{-\Pi}{l}\right). \tag{2-205}$$

In this expression n = number of pores/cm^2, and the coefficient of $(-\Pi)$ in (2-205) is just the filtration coefficient L_p:

$$J_v = -L_p\Pi \tag{2-206}$$

when $p_1 = p_2$. This then is the volume flow associated with the osmotic effect.

The generalization to the case when $p_1 \neq p_2$ is straightforward. If $p_2 \neq p_1$, then the effective pressure head along the pore is

$$\begin{aligned}\Delta p_{\text{eff}} &= (p_1' - p_2)\\ &= (p_1 - \Pi - p_2)\end{aligned} \tag{2-207}$$

or

$$\Delta p_{\text{eff}} = (p_1 - p_2) - \Pi, \tag{2-208}$$

and the volume flow J_v is given by

$$J_v = L_p\Delta p_{\text{eff}} = L_p(\Delta p - \Pi). \tag{2-209}$$

The flow continues from chamber (2) to chamber (1) until the pressure difference $\Delta p = (p_1 - p_2)$ becomes equal to the drop in pressure $(p_1 - p_1') = \Pi$ experienced on entering the pore from the solution. Thus, our definition of $\Pi = (p_1 - p_1')$ is consistent with the phenomenological definition given previously.

In Figure 2.52 we show the pressure profiles that apply as we vary p_1 keeping p_2 fixed. We show the following cases:

$$\left.\begin{array}{l} p_1 - p_2 = 0, \\ 0 < p_1 - p_2 < \Pi, \end{array}\right\} \qquad \text{Flow to left,}$$

$$(p_1 - p_2) = \Pi, \qquad \text{No flow,}$$

$$(p_1 - p_2) > \Pi, \qquad \text{Flow to right.}$$

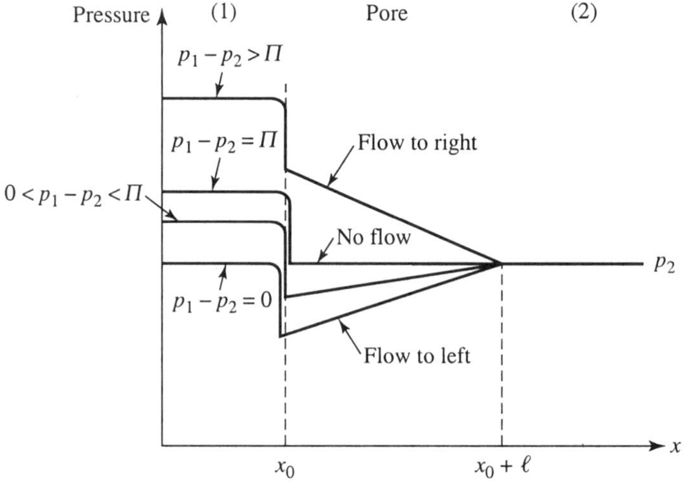

Figure 2.52. The variation of pressure with distance along the pore axis when $C_1 = C_s$, $C_2 = 0$, and p_1 varies as p_2 is held constant.

We can now easily generalize our considerations to the case that applies in Figure 2.45, i.e., chamber (2) has concentration C_2, and chamber (1) has concentration C_1. Let the osmotic pressure associated with chamber (1) be Π_1 and that associated with chamber (2) be Π_2 (Π_1 and Π_2 can beseparately established by having

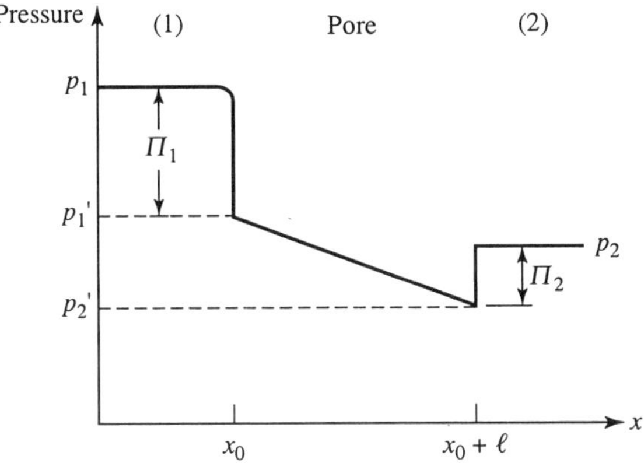

Figure 2.53. The variation of pressure with x when $p_1 \neq p_2$, and $C_1 \neq C_2$.

solutions (1) and (2) facing the water across from the membrane). Since Π_1 and Π_2 are the pressure drops produced on entering the solute-free pore from solution (1) or from solution (2), we see that the pressure profile in general has the form shown in Figure 2.53 (we show the case $p_1 > p_2$). In this case, the volume flow J_v is given by

$$J_v = L_p \Delta p_{\text{eff}} = L_p(p_1' - p_2') \tag{2-210}$$

But

$$p_1' = p_1 - \Pi_1,$$
$$p_2' = p_2 - \Pi_2.$$

Thus

$$J_v = L_p[(p_1 - p_2) - (\Pi_1 - \Pi_2)]$$

or

$$J_v = L_p(\Delta p - \Delta \Pi). \tag{2-211a}$$

To this we add the equation

$$J_s = 0. \tag{2-211b}$$

These equations show that the microscopic description given above for the origin of the osmotic flow is in agreement with the experimentally observed phenomena.

We conclude this discussion of the microscopic basis for the osmotic flow by providing a molecular interpretation for the pressures p_1, Π, and p_1' as shown, for example, in Figure 2.51. The total fluid pressure inside compartment (1) can be thought of as being produced separately by the solute molecules and the solvent molecules. If the solute molecules act as if they have kinetic energy $3kT/2$, and have comparatively small potential energy due to their interaction with the water, they produce a pressure in proportion to their concentration, as an ideal gas; i.e.,

$$p_{\text{solute}} = kTC_s \equiv \Pi.$$

The osmotic pressure then can be thought of as being produced by the collisions of the solute molecules on the membrane wall. Since

$$p_1 = p_{\text{water}} + p_{\text{solute}}$$
$$= p_1' + \Pi = p_{\text{water}} + \Pi \quad \text{(in region 1)},$$

we may interpret p_1' as the pressure (p_{water}) exerted by the water molecules alone. On entering the pore in the membrane, the osmotic pressure produced by the solute molecules disappears, and we have that the pressure p_1' inside the pore is

$$p_1' = p_{\text{water}} \quad \text{(in pore)}.$$

Thus, we find the pressure drops on crossing into the pore by an amount

$$p_1 - p_1' = p_{\text{water}} + \Pi - p_{\text{water}} = \Pi.$$

Thus, if we interpret p_1' as the pressure exerted by the water molecules alone, we see that the pressure drops by Π on crossing the membrane, because the pressure produced by the solute particles disappears as one enters the pore.

2.6.B. Coupled Flow of Solute and Solvent Across a Membrane Subject to Both a Hydrostatic and an Osmotic Pressure Difference. The Three Membrane Parameters

In the previous section, we considered only the case of a semipermeable (= solute impermeable) membrane subject to both a hydrostatic and osmotic pressure difference. In this case of zero membrane permeability, the volume flow J_v (volume of fluid passing the membrane per unit membrane area per second) is given by (2-211a):

$$J_v = L_p(\Delta p - \Delta \Pi), \qquad (2\text{-}211a)$$

whereas the solute flow J_s (= number of solute particles per unit membrane area per second) is zero

$$J_s = 0. \qquad (2\text{-}211b)$$

In this section we generalize these relations to the case of membranes that permit the passage of solvent *and* solute.

This generalization requires that we make a "model" of the permeable membrane. Our model will be a membrane of thickness Δx with straight cylindrical pores of radius a, such as is illustrated in Figure 2.23. In Sections 2.4.C(i) and 2.4.C(ii) we defined the concepts of the filtration coefficient L_p and the permeability p of such a membrane. Then, in Section 2.4.C(iii) we examined the phenomenon of hindered diffusion and introduced the concept of a hindrance factor. We can view a semipermeable membrane as the limiting case of a permeable membrane, where (for a particular solute) the hindrance factor has the value zero, so that the passage of solute particles is totally blocked. Our objective is now to examine the case where there is some hindrance in the passage of solute through a membrane but not total blocking. We will first examine the volume flow (= bulk flow) through a membrane under the joint effect of a pressure difference, $\Delta p = p_1 - p_2$, and a concentration difference, $\Delta C = C_1 - C_2$, between sides 1 and 2 of a membrane. After this, we will examine the solute flow through the membrane.

(i) Volume Flow Through a Permeable, Porous Membrane due to a Pressure and Concentration Gradient

Our starting point will be a result obtained in the investigation of the semipermeable membrane. We showed then that the volume flow J_v (in cm^3/cm^2 s) is determined by the filtration coefficient L_p and the *effective* pressure difference Δp_{eff} that moves the "plug" of fluid through the pore against the viscous drag exerted by the membrane walls. (See (2-209), (2-210).) This effective pressure difference Δp_{eff} is not $p_1 - p_2$, but is rather given by

$$\Delta p_{\text{eff}} = p_1' - p_2',$$

p_1 is the hydrostatic pressure *in the fluid* on side 1, p_1' the pressure *just inside the pore* on side 1. At this location the solute concentration had dropped to zero, and the difference $p_1 - p_1'$ was shown to be due to this drop in concentration from the value $C_s(1) = C_1$ in the solution to 0 in the pore (see (2-203)):

$$p_1' = p_1 - kTC_1. \tag{2-212}$$

A corresponding situation prevailed at side 2 of the membrane.

If we now consider a *permeable* membrane, we may not assume that the solute concentration in the pore is zero. But we will assume that it is smaller than in the solution. Large solute molecules get partly reflected when they attempt to enter the pore. As a result, the equilibrium concentration $C_s'(1)$ just inside the pore on side 1 is smaller than the concentration $C_s(1)$ in the compartment 1. We will express this reduction of concentration by a factor of $(1 - \sigma)$, when σ has a value between 0 and 1: $0 \leq \sigma \leq 1$. We call σ the *reflection coefficient*. It represents the fraction of molecules "rejected" by the pore. We assume that the same situation occurs on side 2 of the pore. In this model then, the concentration profile across the membrane has the form illustrated in Figure 2.54. Fig. 2.54 holds the key to the understanding of the volume flow. Since the solute concentration drops from a value of C_1 in the left compartment to $(1 - \sigma)C_1$ at the pore entrance, there is a net concentration change of (σC_1). In consequence, the hydrostatic pressure p_1' just inside the left pore entrance is now given by

$$p_1' = p_1 - kT(\sigma C_1) \tag{2-213a}$$

rather than by (2-212) which represents the limiting case when the concentration drops to zero inside the pore. In analogous fashion, we have on side 2:

$$p_2' = p_2 - kT(\sigma C_2). \tag{2-213b}$$

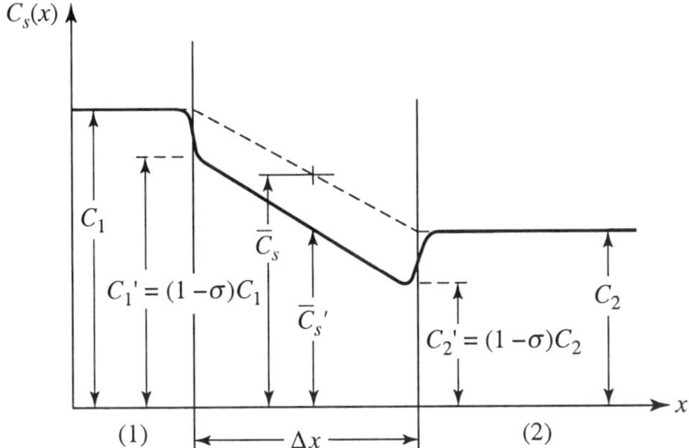

Figure 2.54. Concentration profile of a solute across a permeable membrane pore connecting side 1 with concentration C_1 to side 2 with concentration C_2. The quantities C_1', C_2', and $\overline{C}_s'$ represent the terminal and mean values of the solute concentration *in* the membrane pore.

The difference $p_1' - p_2'$ represents the effective pressure difference moving the plug of fluid through the pore

$$\Delta p_{\text{eff}} = p_1 - p_2 - \sigma kT(C_1 - C_2). \tag{2-214}$$

We will continue to use a definition of the osmotic pressure Π_s of a solution given by the operational definition [(2-193)], using a semipermeable membrane; that is, for dilute solutions we will assume Π_s given by Van t'Hoff's law

$$\Pi_s = kTC_s. \tag{2-215}$$

This terminology is consistent with the view that the osmotic pressure is a property of the solution, and not a property of the membrane. We will call the quantity $\sigma \Pi_s$ the effective osmotic pressure.

The effective pressure Δp_{eff} in (2-214) can therefore be written as

$$\begin{aligned} \Delta p_{\text{eff}} &= (p_1 - p_2) - \sigma(\Pi_1 - \Pi_2) \\ &= \Delta p - \sigma \Delta \Pi. \end{aligned} \tag{2-216}$$

The volume flow generated by this effective pressure is now found from (2-210)

$$J_v = L_p \Delta p_{\text{eff}}$$

or

$$J_v = L_p(\Delta p - \sigma \Delta \Pi). \tag{2-217}$$

This is our first equation for the flow across a permeable membrane. It states that the volume flow is proportional to the applied hydrostatic pressure difference *minus* σ times the osmotic pressure difference. σ, the reflection coefficient, is a pure number between 0 and 1. $\sigma = 1$ characterizes the semipermeable membrane, and $\sigma = 0$, a perfectly permeable membrane in which the pore makes no distinction between the passage of solute or solvent.

(ii)   Solute Flow Through a Permeable Membrane due to Solvent Drag and Diffusion

In this section, we investigate the flow of *solute* in a membrane pore. We assume that fluid flows through the pore with a bulk velocity v. Any dissolved solute of concentration C_s' will be carried along with the same speed.* This "solvent drag" effect, as it is called, thus generates a solute flow that can be represented by the solute current density ($=$ particles/cm^2 s)

$$(j_s)_{\text{drag}} = C_s' v, \tag{2-217a}$$

where C_s' is the solute concentration *in* the pore. In addition, there is a solute flow due to diffusion if there is a concentration gradient along the pore. This diffusion flow is given by Fick's law as

$$(j_s)_{\text{diff}} = -D \frac{\partial C_s'}{\partial x}. \tag{2-218}$$

The two flows together give the total solute flow

---

*This will not be true if there is an incremental resistance between solute particles and the pore walls not included in the hydrodynamic drag. While this effect may occur, we will ignore it here to keep all arguments as simple as possible.

$$j_s = C'_s v - D \frac{\partial C'_s}{\partial x}. \tag{2-219}$$

We will now consider a situation where the bulk velocity v is constant (time independent), and also where the concentrations C_1 and C_2 on both sides of the membrane are constant. In this case, the solute concentration $C'_s(x)$ in the pore does not change with time, and depends on the position x along the pore axis only. The continuity equation [(2-91)]

$$\frac{\partial j_s}{\partial x} = -\frac{\partial C'_s}{\partial t}$$

then requires that the solute current density, j_s, be independent of x; that is, have the same value at every point x along the pore. We will show later that in this case the concentration profile $C'_s(x)$ along the pore is not linear but exponential. For our immediate needs, it is not necessary to determine this profile: We can transform (2-219) into a useful equation by just *averaging* it over all points x, from $x = 0$ to $x = \Delta x$, along the pore. Since j_s, v, and D are constant, this averaging gives us

$$j_s = v \left(\frac{1}{\Delta x} \int_0^{\Delta x} C'_s(x) \, dx \right) - D \left(\frac{1}{\Delta x} \int_0^{\Delta x} dx \frac{dC'_s}{dx} \right).$$

The first term on the right is just v times the average concentration $\overline{C}'_s$ in the pore. The second term can be written as

$$-D \frac{1}{\Delta x} \left(C'_s(\Delta x) - C'_s(0) \right) = -D \frac{C'_2 - C'_1}{\Delta x},$$

where C'_2 and C'_1 are the concentrations, *in the pore*, at the pore terminals 2 and 1. With this, we can write the solute current density j_s in the pore as

$$j_s = v\overline{C}'_s + D \frac{(C'_1 - C'_2)}{\Delta x}. \tag{2-220}$$

Consulting Figure 2.54, we see that the concentrations C'_1 and C'_2 can be related to the concentrations C_1 and C_2 in the fluid compartments by

$$C'_1 = (1 - \sigma)C_1,$$
$$C'_2 = (1 - \sigma)C_2. \tag{2-221}$$

Similarly, we can express $\overline{C}_s'$, the mean value of C_s' in the pore, to a corresponding mean value $\overline{C}_s$ between C_1 and C_2:

$$\overline{C}_s' = (1 - \sigma)\overline{C}_s. \tag{2-222}$$

This gives us the relation

$$j_s = (1 - \sigma)\overline{C}_s v + \frac{(1 - \sigma)D}{\Delta x}(C_1 - C_2). \tag{2-223}$$

This represents the solute current density in a single pore. To obtain from this the solute flow per unit membrane area, we multiply j_s first by the pore area (πa^2), which gives the solute flow (particles/s) per pore. Multiplying this again by n, the number of pores/unit membrane area, gives the solute flow J_s in (particles/s)/unit membrane area

$$J_s = n(\pi a^2)j_s. \tag{2-223a}$$

Multiplying (2-222) with $n(\pi a^2)$, we obtain

$$J_s = (1 - \sigma)\overline{C}_s \left(n(\pi a^2)v\right) + \frac{(1 - \sigma)Dn(\pi a^2)}{\Delta x}(C_1 - C_2). \tag{2-224}$$

The term $n(\pi a^2)v$ represents the total bulk flow per unit membrane area, that we designated by J_v:

$$n(\pi a^2)v = J_v. \tag{2-225}$$

Indeed, if $(\pi a^2)v$ is the volume flow per pore, and multiplied by the number n of pores per unit membrane area, we get the volume flow per unit membrane area (in units $(\text{cm}^3/\text{s})/\text{cm}^2 = \text{cm/s}$). The term

$$(1 - \sigma)\frac{n(\pi a^2)D}{\delta x} = p \tag{2-226}$$

we recognize as the membrane permeability (see (2-110)) with $(1 - \sigma)$ playing the role of the hindrance factor. Inserting (2-225), (2-226) into (2-224), we have the final result

$$J_s = (1 - \sigma)\overline{C}_s J_v + p\Delta C. \qquad (2\text{-}227)$$

This equation expresses the solute flow per unit membrane area as the sum of two terms. The first, the "solvent drag" term, is proportional to the volume flow J_v, the mean concentration between the two compartments on either side of the membrane, and to a hindrance factor $(1 - \sigma)$. The second term is proportional to the concentration difference ΔC between the two sides of the membrane. The coefficient in front of ΔC defines the membrane permeability p.

Equations (2-217) and (2-227) represent the two fundamental equations describing volume flow J_v and solute flow J_s under the joint effect of both a pressure and a concentration gradient across the membrane.

These relations were derived here with the help of a rather special membrane model in the interest of simplicity. It can be shown that the structure of these two equations does not depend on the model used. Every (passive) membrane is characterizable by *three* parameters L_p, σ, and p. Depending on the models, these quantities may have differing interpretations in terms of a detailed description of the membrane. But the relation between the bulk flow J_v and solute flow J_s on the one hand, and the pressure difference Δp and concentration difference (or osmotic pressure difference) $\Delta C = \Delta\Pi/kT$, on the other hand, will *always* be given by (2-217) and (2-227). We will refer to these relations as the coupled flow relations.

(iii)   The Symmetrical Form of the Coupled Flow Relations

The reader may skip this short section which adds nothing new of substance. It deals only with writing the coupled flow relations in a more symmetrical, though practically less useful, form. This form is found in all the modern references on the subject. The equations may be cast into the desired form by introducing the so-called differential flow J_D, defined as

$$J_D = \frac{J_s}{\overline{C}_s} - J_v. \qquad (2\text{-}228)$$

Notice that J_D has the same units as the bulk or volume flow J_v, namely, $(\text{cm}^3/\text{s})/\text{cm}^2 = \text{cm/s}$. At the same time, we also introduce the permeability coefficient ω, defined by

$$p\Delta C = \left(\frac{p}{kT}\right)(kT\,\Delta C) = \frac{p}{kT}\Delta\Pi \equiv \omega\Delta\Pi. \qquad (2\text{-}229)$$

With (2-228) and (2-229), the second flow relation [(2-227)] takes the form

$$J_D = -\sigma J_v + \left(\frac{\omega}{\overline{C}_s}\right)\Delta\Pi,$$

and using (2-217) for J_v, one has

$$J_D = -\sigma L_p \Delta p + \left(\frac{\omega}{\overline{C}_s} + \sigma^2 L_p\right)\Delta\Pi.$$

The two coupled flow relations can thus be written in the symmetric form

$$J_v = L_p \Delta p - (\sigma L_p)\Delta\Pi \tag{2-230}$$

and

$$J_D = -(\sigma L_p)\Delta p + L_D \Delta\Pi, \tag{2-231}$$

where L_D is the sum

$$L_D = \left(\sigma^2 L_p + \frac{\omega}{\overline{C}_s}\right).$$

These relations contain nothing new. They display, however, the interesting feature that the contribution of the *osmotic* pressure to the *volume* flow J_v is equal to the contribution of the *hydrostatic pressure* to the differential flow J_D.

If one only postulates that the flows J_v and J_D are *linearly* related to the pressure differences Δp and $\Delta\Pi$, then the *most general* relation between J_v, J_D, on the one hand, and ΔP, $\Delta\Pi$, on the other hand, would be a set of equations of type

$$J_v = L_p \Delta p + L_{pD}\Delta\Pi,$$
$$J_D = L_{Dp}\Delta p + L_D \Delta\Pi, \tag{2-232}$$

where L_p, L_{pD}, L_{Dp}, and L_D are *four independent* coefficients. It was Onsager who, in 1931, proved generally (without appeal to a specific membrane model) that the two coefficients L_{pD} and L_{Dp} had to be the same [24].

In our following discussions we will always use the coupled flow relations in the form of (2-217) and (2-227) which are much more convenient than the symmetric form of (2-230), (2-231).

(iv)   Measurements of Membrane Parameters on Synthetic Membranes. Data on L_p, σ, p

In the last decade numerous measurements have been carried out on artificial membranes, such as cellulose membranes used for dialysis bags (see Section 2.4.C). The main thrust of the experiments has generally been to gather data on the permeability p and the reflection coefficient σ for a variety of solute molecules of different size. The dependence of both p and σ on molecular radii, and their relation to the pore size, can be used to test various membrane models, or hypotheses on the dynamics of solvent and solute flow through pores. The model that we used to derive the coupled flow equations is somewhat too primitive to warrant such a detailed testing. It will be worthwhile, however, to give a short account of the manner in which some of these data are gathered, to see what the values of membrane parameters are, and whether there is at least general agreement with the coupled flow equations. It may be observed that experiments with membranes are difficult, and precision data on the parameters are not easy to come by.

We will now describe briefly how the membrane parameters L_p, σ, and p are measured, and then report some data on cellophane membranes. This is interesting material, readily available, and used medically in hemodialysis (see Section 2.4.D). Also, it has a complicated pore structure that makes it interesting as an object for testing the flow equations and extracting structural information from the different values of p and σ for different size solute molecules. Cellophane membranes are a thick meshwork of cellulose fibers, having irregular channels of various lengths and widths. Yet data show that a meaningful mean pore radius can be defined. We discussed in Section 2.4.C(iii) the tests made by Bean on fission track etched mica membranes in which the pore radius was determined both from the Poiseuille formula [(2-100)] and the permeability formula [(2-105)]. Similar tests on cellophane membranes give reasonably consistent results.

(a)   Measurement of the Filtration Coefficient L_p.

From (2-217) it appears that the straightforward way to determine L_p is to force fluid through a membrane under hydrostatic pressure alone ($\Delta\Pi = 0$):

$$L_p = \left(\frac{J_v}{\Delta p}\right)_{\Delta\Pi=0} \tag{2-233}$$

Ginzburg and Katchalsky [25] have verified the proportionality of J_v and Δp for dialysis tubing and found a constant value of L_p for pressure differences up to 100 mmHg. Also, L_p appears to be rather independent of solute concentration for a small volume fraction of solute.

(b) Measurement of the Reflection Coefficient σ. Once the filtration coefficient L_p of a membrane is determined, one may again use the flow equation (2-217) and measure the volume flow J_v due to an osmotic pressure difference $\Delta \Pi = kTC_s$ for equal hydrostatic pressure on both sides: $\Delta p = 0$. One then has

$$J_v = L_p(-\sigma \Delta \Pi)$$

and

$$\sigma = -\left(\frac{J_v}{L_p \Delta \Pi}\right)_{\Delta p=0}. \tag{2-234}$$

The values of σ so found depend on the solute concentration; they decrease somewhat as the concentration increases. The values of σ that we will list are extrapolated to zero concentration.

(c) Measurement of the Membrane Permeability p. To measure p, as defined operationally by (2-227), we must measure the solute J_s under conditions that assure zero volume flow J_v. We can therefore not just set up a concentration gradient ΔC across the membrane, and measure J_s. Rather, we must set up a compensating pressure difference Δp so that

$$\Delta p - \sigma \Delta \Pi = \Delta p - \sigma kT \Delta C = 0$$

to ensure that $J_v = 0$. Once this is done, we have from (2-227)

$$p = \left(\frac{J_s}{\Delta C}\right)_{J_v=0}, \tag{2-235}$$

J_s is generally measured by using a radioactively labeled solute.

(d)   Membrane Parameters for Dialysis Tubing.   We now present the results of measurements made on a Visking dialysis membrane that is also used as sausage casing. The structural data available are

$$\text{membrane thickness } \Delta x = 4.5 \times 10^{-3} \text{ cm,}$$
$$\text{water contents (by volume)}$$
$$\text{of wet membrane} = 68\%.$$

This second datum is a measure of the pore volume. It can be used to estimate pore radii, but large errors may occur due to "dead-end" pores.

The authors measure the membrane parameters for a variety of solutes. The results are presented in Table 2.11. The four solutes tested are arranged in order of increasing molecular size. HTO is tritium-labeled water. Column (2) lists estimated molecular radii. These should be compared with the mean pore radius which is of order 25 Å for dialysis tubing. So all molecules have radii small compared to the mean pore radii. This is reflected in the small values of the reflection coefficient σ in column (4). The molecules are too small to be significantly rejected by the pore. Yet, the increase of σ with molecular radius is apparent. In column (3) we list the value of the filtration coefficient. Experimentally no distinction is made in L_p for different solute particles. The value 3.4×10^{-5} (cm/s)/atm represents the value for L_p appropriate for a very dilute solution.

Of greater structural interest is the fact, visible from the data in column (5) on permeabilities, that p decreases much faster with increasing molecular radius than $(1 - \sigma)$. Recall that in the membrane model, used in this section to derive the

Table 2.11. Membrane Parameters for Visking Dialysis Tubing (Ginzburg and Katchalsky [25]).

Solute (1)	Molecular radius (2)	Filtration coefficient L_p (cm/s)/atm (3)	Reflection coefficient σ (4)	Permeability p (cm/s) (5)
HTO†	1.9 Å	3.4×10^{-5}	0	11.1×10^{-4}
Urea	2.7 Å	3.4×10^{-5}	0.013	5.15×10^{-4}
Glucose	4.4 Å	3.4×10^{-5}	0.123	1.78×10^{-4}
Sucrose	5.3 Å	3.4×10^{-5}	0.163	0.971×10^{-4}

†T is the radioactive hydrogen isotope tritium.

Table 2.12. Reflection Coefficients σ as a Function of Molecular Radius/Pore Radius (data from R. Durbin, *J. General. Physiology* **44** 315 (1969) (See Supplem. Reading Section 2.7)).

Solute (1)	Molecular weight M gm (2)	Molecular radius r (Å) (3)	Reflection coefficient σ (4)	r/a (5)	$(1-r/a)^2$ (6)	$1-\sigma$ (7)
D_2O	20	1.9	0.002	0.083	0.84	0.998
Urea	60	2.7	0.024	0.117	0.80	0.976
Glucose	180	4.4	0.20	0.19	0.66	0.80
Sucrose	342	5.3	0.37	0.23	0.59	0.63
Raffinose	594	6.1	0.44	0.26	0.55	0.56
Inulin	991	12	0.76	0.52	0.23	0.24

flow equations, p was proportional to $(1 - \sigma)$ (see (2-224)). Our model neglects the fact that large solute particles experience not only a partial rejection at the pore entrance, but also an additional frictional resistance in the pore channel. This was mentioned in Section 2.4.C, where it was shown that the hindrance factor that describes the reduction of membrane permeability p with solute radius has two components expressed in (2-103). In our membrane model we may identify $(1 - \sigma)$ roughly with the reduction factor $(1 - r/a)^2$ defining the effective pore area, (2-111), and plotted as a solid line in Figure 2.27.

This rough correspondence between $(1 - \sigma)$ and $(1 - r/a)^2$ can be shown from the data on σ collected by Durbin presented in Table 2.12. Table 2.12 lists six solutes in order of increasing molecular weight M (column (2)) and radius (column (3)). Column (4) lists the reflection coefficients σ, based on measurements of osmotic volume flow [(2-234)]. Column (5) lists the ratios r/a of molecular radius to "pore radius" a. This latter was determined by comparing the data for L_p with the permeability p_w for pure radioactively labeled water. Equations (2-100) and (2-105) give

$$L_p = \frac{n(\pi a^2)}{\Delta x} \frac{a^2}{8\eta},$$

$$p_w = \frac{n(\pi a^2)}{\Delta x} D.$$

Since neither the pore density n, nor the actual channel length, can be measured for cellulose membranes, one eliminates $n(\pi a^2)/\Delta x$ between p_w and L_p and finds

$$a = \left(\sqrt{\frac{8\eta D L_p}{p_w}} \right). \tag{2-236}$$

On that basis, a was estimated to be 23 Å. The two last columns of the table compare the values $(1 - r/a)^2$ and $(1 - \sigma)$. These values show a rough agreement, indicating that indeed the reduction of the effective pore area due to the finite size of solute molecules is a major determinant of the value of σ.

2.6.C. Transport of Water and Solute Across Biological Membranes

In applying the results of the previous section to biological membranes, one is faced with a bewildering variety of cases.

On the one side we have the true membranes, such as represented by the cell membrane enclosing a single cell. In animal cells these membranes are typically very thin, of the order of 80–100 Å. Presumably, the nuclear membrane enclosing the cell nucleus and the membranes of organelles within the cell, such as mitochondria, are of similar structure. It is for such structures that the lipid bilayer model was established by Davson and Danielli in 1952. We cannot here enter into a detailed description of membrane models, but refer the reader to the list of Supplementary Reading Section 2.7. Good introductions are provided by the books of Stein, *The Movement of Molecules Across Cell Membranes*, and of Cerreido and Rotunno, *Introduction to the Study of Biological Membranes*.

The word "membrane" in biology is also used to designate much more complex structures, namely, permeable barriers formed by single layers or multiple layers of cells. Examples of this kind of membrane are the blood capillaries, the kidney tubules, the epithelial layer of the intestine, the gills of fish. The concept of "biomembrane" is a *functional* concept, including all permeable or semipermeable barriers that compartmentalize an organism, and serve to regulate and control the flow of water and solutes between compartments.

In what follows, we will give some illustrative examples of typical biomembranes of various types, give values of the membrane parameters, and when possible, relate these values to the function of the membrane.

(i)   The Permeability and Filtration Coefficient of Red Blood Cells

The interior of any cell contains in solution various materials, such as sugars, inorganic salts, and proteins. These materials give the solution an effective osmotic pressure. An external medium in which the cell may be placed may have an osmotic pressure different from that of the cell. An external solution is called isotonic, hypertonic, or hypotonic, depending on whether its effective osmotic pressure is equal, or larger, or smaller than that of the cell. A cell placed into a hypotonic solution will experience an inflow of water into the cell and swell. In a hypertonic solution it will shrink, and in an isotonic solution no volume change will occur.

This rate of volume change can be used to measure the filtration coefficient L_p of the cell membrane.

Consider a cell of volume V and total surface area A. Place this cell into an isotonic solution to which an impermeant solute s of concentration C_s has been added. This addition of solute generates a hypertonic external medium. There will then be an osmotic imbalance

$$\Delta \Pi_s = kT C_s \tag{2-237}$$

between the solution and the cell interior. According to the first flow equation [(2-217)], this will generate a volume flow J_v out of the cell

$$J_v = -L_p \Delta \Pi_s = -L_p(kT C_s). \tag{2-238}$$

The total outflow through the surface area A of the cell will be $A J_v$. This represents the rate of volume loss of the cell

$$\frac{dV}{dt} = A J_v = -L_p A \Delta \Pi_s.$$

From this equation we can determine L_p by measuring the rate of shrinkage at the *onset* of the experiment. (At a later stage the outflow will have reduced the osmotic imbalance.)

$$L_p = \left(\frac{-dV/dt}{A \Delta \Pi_s} \right)_{\text{initial}}. \tag{2-239}$$

Table 2.13. Filtration Coefficients L_p and Water Permeabilities $\mathfrak{p}_w$ for Red Blood Cells (data taken from House [27, Chap. 5]).

Species	Filtration coefficient: L_p (cm/s)/atm	Permeability $\mathfrak{p}_w$ for Water (cm/s)
Human adult	91×10^{-7}	5.3×10^{-3}
Human fetus	86×10^{-7}	3.2×10^{-3}
Horse	127×10^{-7}	—
Cow	120×10^{-7}	5.1×10^{-3}
Cat	250×10^{-7}	—
Dog	147×10^{-7}	6.4×10^{-3}
Eel	24×10^{-7}	—
Chicken	6.1×10^{-7}	—

Based on this relation, values of L_p for red blood cells and eggs of fish, sea urchins, etc. have been determined. We will list the results in Table 2.13.

The permeability $\mathfrak{p}_w$ of red blood cells for radioactively tagged water has also been measured. This is done by measuring the rate of inflow of radioactively labeled (HTO) water that is put into the external, isotonic bathing solutions. The increase of HTO concentration inside the cell is then given by the solution to a two-compartment problem, which we discussed in Section 2.4.C(v): The concentration difference of radioactive water (HTO), $\Delta C(t)$, between outside and inside decreases exponentially from its initial value $\Delta C(0)$:

$$\Delta C(t) = \Delta C(0)e^{-t/\tau_0}. \tag{2-128}$$

Since the outside bathing solution has a volume much larger than the cell volume V, the time constant τ_0 is given by (see (2-126a)):

$$\tau_0 = \frac{V}{A\mathfrak{p}_w}. \tag{2-240}$$

Upon plotting the quantity $\ln(\Delta C(t)/\Delta C(0))$ as a function of the elapsed time t, one should obtain a straight line

$$\ln\left(\frac{\Delta C(t)}{\Delta C(0)}\right) = -\frac{t}{\tau_0}, \tag{2-241}$$

where the slope gives the value τ_0. From (2-240) one then finds the water permeability p_w. We recall the significance of this water permeability. It describes the diffusional flow of radioactive water molecules, which in all other respects behaves as ordinary water molecules. The measurement of p_w forms the basis of determining the "hindrance factor" that must be introduced to account for the permeabilities p_s of larger solute molecules. We now present some data: A striking conclusion can be drawn from these data on L_p and p_w. We notice immediately that the values of L_p for the red cell are about four times smaller than that for the cellulose membrane. On the other hand, the value of p_w for the red cell is five times larger than that for the cellulose membrane. If we use these values of L_p and p in (2-236) for the radius of the membrane pore, we find that the pore radius of the red cell membrane is about 5 Å! This is to be compared with the radius of the water molecules itself which is $\sim$ 2 Å. This is a striking demonstration that biological transport is taking place through structures of molecular dimensions.

One interesting number on which the value of the permeability p_w is based is the time constant τ_0, (2-240), which determines the speed with which the concentration of radioactive water equalizes between the bathing solution and the cell interior. The experimental value of τ_0 found by Paganelli and Solomon [28] is

$$\tau_0 = 6 \times 10^{-3} \text{ s.}$$

Now let us not forget that the red cell stores a large amount of oxygen in the hemoglobin it contains. To be made available, this oxygen must pass into the cell plasma, and then diffuse through the cell membrane into the blood plasma. We may assume that the small oxygen molecule passes the membrane with a permeability not unlike water itself. The above small value of the equlization time constant then shows that the red cell membrane is designed to facilitate a *rapid* delivery of oxygen, if the oxygen pressure in the blood plasma drops, through diffusion into muscular tissue.

(ii)   The Permeability and Filtration Coefficient of Capillary Walls

The walls of capillary blood vessels are made of a single layer of cells. The thickness of this "membrane" is of the order of 10^{-4} cm, that is, about 100 times thicker than a typical cell membrane. It is now assumed that an important property of the blood capillary wall is the existence of flow channels between cells. Water and solute do not actually have to pass through the cells, but rather flow through interstices between cells. In Table 2.14 we reproduce some values of L_p and p for capillary walls in human skeletal muscle. We see from this table that the filtration coefficient L_p is similar to that of the red cell membrane, but the permeabilities are much smaller. This indicates larger pores.

Table 2.14. Values of the Filtration Coefficient L_p and Permeability p for Various Solutes for Capillary Blood Vessels in Human Skeletal Muscle (taken from House [27, Chap. 8]).

Solute	Permeability p (cm/s)	Filtration coefficient L_p (cm/s)/atm
Urea	2.9×10^{-5}	
Glucose	1.3×10^{-5}	
Sucrose	0.9×10^{-5}	6.9×10^{-6}
Raffinose	5×10^{-6}	
Insulin	9×10^{-7}	

As an application, we will now use these data to answer the question whether the permeabilities reported in this last Table 2.14 are adequate to effectively transporting glucose from the bloodstream into the muscular tissue. The average daily glucose requirement of the human body is about 500 g, supplying 2000 kcal of energy. This glucose is provided by carbohydrates (starch, sugar) which are enzymatically broken down into glucose (digestion). Excess glucose is stored, and an elaborate feedback control system (see Volume I, Chapter 6) maintains a rather constant blood glucose level of about 0.8 mg/cm^3. Out of the blood circulating through the body tissues, about 500 g/day or 6 mg/s of glucose must diffuse through the walls of blood capillaries into the body tissues. In Figure 2.55 we give a schematic diagram of a blood capillary and indicate the flow and diffusion of glucose. Blood enters the capillary at A, leaves it at V. During the passage through the capillary a certain fraction of the glucose diffuses through the walls into the interstitial space.

We can find what this fraction is by a simple argument. The total blood flow is about 75 (cm^3/s) at a normal heart rate. The pumping rate of glucose is therefore

$$75 \text{ cm}^3/\text{s} \times 0.8 \text{ mg/cm}^3 = 60 \text{ mg/s}.$$

But we have just stated that about 6 mg/s must be absorbed. This is 10% of the pumped rate. It indicates that in passing through the capillary, the blood will lose

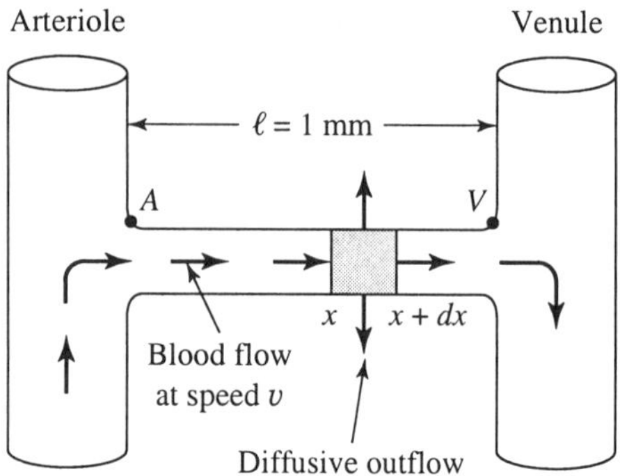

Figure 2.55. Flow of blood through a capillary. Blood enters the capillary at A and leaves it at V. Glucose leaves capillaries by diffusion through the walls.

about 10% of its glucose. The glucose concentration at the exit point V must be 10% less than at the entrance point A of the capillary.

The loss of glucose due to diffusion through the capillary walls depends on the flow rate of blood, on the length and surface of the capillary, and on its permeability for glucose. We now proceed to show that the "design parameters" of the system are adequate for the task.

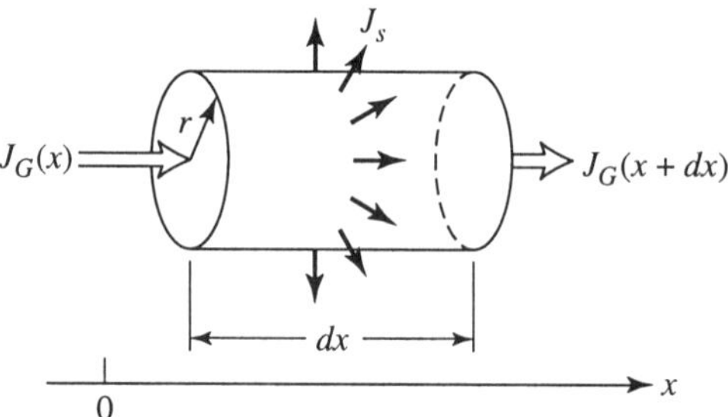

Figure 2.56. Segment of capillary showing glucose inflow at x, glucose outflow at $x + dx$, and diffusional outflow through the walls of the capillary.

To establish this, we present in Figure 2.56, an enlarged view of the small segment of the capillary, shaded in Figure 2.55. Let v be the flow velocity of blood through the capillary, and r its radius. Let us designate by $C_G(x)$ the glucose concentration in the blood, and by C_G^0 the glucose concentration in the interstitial fluid surrounding the capillary. C_G^0 will be lower than $C_G(x)$. In considering Figure 2.56, we see that we have the following "bookkeeping" relation for glucose:

inflow at x = outflow at $(x + dx)$ + diffusional flow through walls.

The inflow of glucose at x is the inflow of fluid volume times the glucose concentration at x:

$$J_G(x) = C_G(x)v(\pi r^2). \tag{2-242a}$$

Similarly, the outflow at $x + dx$ is

$$J_G(x + dx) = C_G(x + dx)v(\pi r^2). \tag{2-242b}$$

Finally, the diffusional outflow through the walls is given by the product of surface area $2\pi r\, dx$ and the diffusional solute flow J_s as determined by (2-227). (NB, the average volume flow J_v through the capillary walls is zero!) So we have

$$\text{diffusional outflow} = 2\pi r\, dx\, J_s = 2\pi r\, dx\, p_G(C_G(x) - C_G^0), \tag{2-243}$$

where p_G is the permeability for glucose, and $C_G(x) - C_G^0$, the concentration difference across the capillary wall. From (2-243) and (2-242) and the previously stated relation

$$J_G(x) = J_G(x + dx) + 2\pi r\, dx\, J_s$$

we find

$$(C_G(x) - C_G(x + dx))\, v\pi r^2 = 2\pi r\, dx\, p_G \left(C_G(x) - C_G^0\right).$$

The term in parentheses at the left can be written as

$$-\left(\frac{dC_G}{dx}\right) dx$$

and the equation restated as

$$\frac{dC_G(x)}{dx} = -\left(\frac{2p_G}{vr}\right)\left(C_G(x) - C_G^0\right). \tag{2-244}$$

We can solve this equation if we assume that the outside glucose concentration C_G^0 is essentially constant. This enables us to write, with $\Delta C(x) = C_G(x) - C_G^0$:

$$\frac{d(\Delta C)}{dx} = -\left(\frac{2p_G}{vr}\right)\Delta C(x). \tag{2-245}$$

This is the familiar equation for a decaying exponential with the solution

$$\Delta C(x) = \Delta C(0)e^{-(2p_G/vr)x}, \tag{2-246}$$

$\Delta C(0)$ is the concentration difference at the capillary entrance A. At the exit, V, a distance $x = l$ downstream, the concentration difference will have dropped to

$$\Delta C(l) = \Delta C(0)e^{-(2p_G/vr)l}.$$

We can write this in the more useful form

$$\frac{\Delta C(0) - \Delta C(l)}{\Delta C(0)} = 1 - e^{-(2p_Gl/vr)}. \tag{2-247}$$

At the left we have in the numerator, the concentration drop *in* the capillary between points A and V:

$$\Delta C(0) - \Delta C(l) = C_G(0) - C_G(l),$$

and in the denominator the concentration difference $C_G(0) - C_G^0$ between the *inside* and *outside* at point A. In terms of these values, which are illustrated in Figure 2.57, we can write our result [(2-247)] as

$$\left(\frac{C_G(0) - C_G(l)}{C_G(0)}\right) = \left(\frac{C_G(0) - C_G^0}{C_G(0)}\right)\left(1 - e^{-(2p_Gl/rv)}\right). \tag{2-248}$$

At the left we now have the *fractional concentration drop* of glucose in the capillary. We know this must be of the order of 10%, or 0.10. On the right we have a factor containing the (unknown) glucose concentration C_s^0 outside the capillary, which should be smaller than $C_G(0)$, but not zero. All the parameters in the second

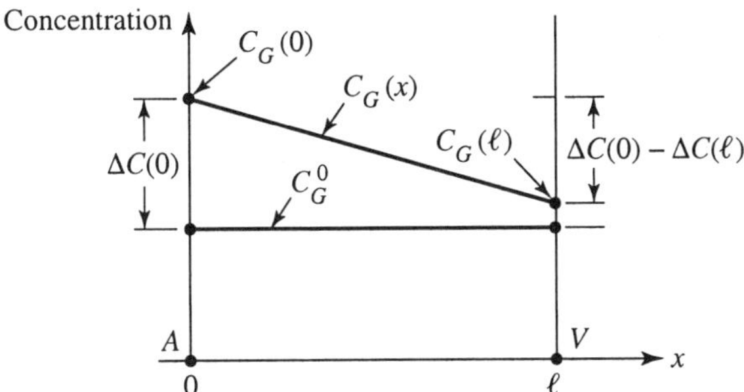

Figure 2.57. Concentration profile for glucose inside capillary $(C_G(x))$ and in interstitial fluid outside (C_G^0).

exponential factor, however, are known. Let us now insert the known numerical values of l, r, v, p_G:

average length l of capillary: $l = 0.1$ cm,

average capillary radius: $r = 4 \times 10^{-4}$ cm,

average flow velocity: $v = 0.04$ cm/s,

glucose permeability: $p_G = 2.6 \times 10^{-5}$ cm/s.

These data give, for $(2p_G l/rv)$, the value

$$\frac{2p_G l}{rv} = 0.325$$

and

$$1 - e^{-(2p_G l/vr)} = 0.278.$$

Using this value in (2-248) with a value of (0.1) for the left-hand side, we see that this equation can be satisfied with a value

$$\frac{C_G(0) - C_G^0}{C_G(0)} \cong \frac{(0.1)}{0.278} = 0.36.$$

This indicates that with a trans-capillary concentration drop of a bit more than 50%, the diffusional flow of glucose across capillary walls can supply the glucose needs of the body tissues.

Appendices

2.A1   Derivation of the Relation (2-6):
Total Kinetic Energy $= \frac{3}{2}pV$

We will give here a simplified derivation, using elementary geometric and kinetic arguments only.

Consider N molecules inside a sphere of radius R. We assume that molecules collide only rarely with each other. In between collisions, their trajectory is then a plane polygon, as shown in Figure 2.58. Let us now follow a molecule that has a velocity v. Its polygonal path is such that consecutive segments of the polygon all subtend the same angle α with the center of the sphere. The impact at the wall takes place at an angle $(\pi - \alpha)/2$ with the normal to the spherical surface at the point of impact. In the collision with the wall, the particle undergoes a momentum change $\Delta \vec{p}_c$:

$$\Delta \vec{p}_c = (m\vec{v})_{\text{after impact}} - (m\vec{v})_{\text{before impact}}.$$

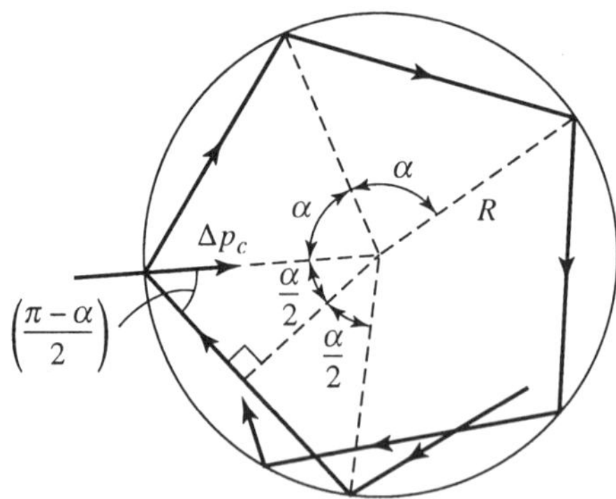

Figure 2.58. Trajectory of a molecule inside a spherical container of radius R.

This p_c is directed toward the center of the sphere, and has a magnitude

$$\Delta p_c = 2mv \cos\left(\frac{\pi - \alpha}{2}\right) = 2mv \sin\left(\frac{\alpha}{2}\right). \qquad (2\text{-}249)$$

Now we must find how often this happens. The time between impacts with the wall is

$$\tau = \frac{\text{distance between impacts}}{\text{velocity of molecule}}$$
$$= \frac{2R \sin \alpha/2}{v}. \qquad (2\text{-}250)$$

The momentum transfer/second is therefore given by the momentum transfer per impact divided by time between impacts

$$\frac{\Delta p_c}{\tau} = \text{momentum transfer/s} = \frac{mv^2}{R}. \qquad (2\text{-}251)$$

We see to our amazement that this does *not* depend on the type of polygonal track any more; it is the *same* for every kind of track, and only dependent on the kinetic energy of the molecule. We now add the contributions of all N molecules in the sphere. With (2-251) this gives

$$\sum_{\text{all molecules}} \frac{\Delta p_c}{\tau} = \frac{1}{R} \sum mv^2. \qquad (2\text{-}252)$$

The left-hand side of this equation can now be interpreted as follows: The sum of all momentum transfers/unit time represents the total inward force exerted by the wall of the sphere on the molecules inside, and is equal to pressure $\times$ surface area:

$$\sum \frac{\Delta p_c}{\tau} = \text{pressure} \times \text{surface area}$$
$$= p \times 4\pi R^2. \qquad (2\text{-}253)$$

Inserting this into (2-252), we find

$$p \times 4\pi R^2 = \frac{2}{R} \sum \left(\frac{mv^2}{2} \right).$$

Since the volume V of the sphere is given by $V = (4\pi/3)R^3$, we can write this as

$$pV = \frac{2}{3} \sum \left(\frac{mv^2}{2} \right). \tag{2-254}$$

By means of a skillful use of Newton's second and third laws, it is possible to show that (2-254) or (2-6) of the text holds for a container of any shape. The simple case used here, however, offers more insight into the way this result comes about.

2.A2   Proof of the Equipartition Law for a "Test Particle" of Mass M in a Gas at Temperature T

Consider a gas of molecules of mass m at temperature T. We assume that a foreign object of mass m_t, subsequently called the "test particle," is suspended in that gas. The test particle may be a molecule of a different kind, or a grain of dust. We will prove that whatever the mass m_t of this object is, its mean kinetic energy will be $(3/2)kT$. The test particle therefore "tests" (measures), so to speak, the temperature of the gas.

To establish this result, we must consider a collision of the test particle with a gas molecule. Let

$$\vec{v}_t, \vec{v}_t' \quad \text{be the velocities of the test particles,}$$

and

$$\vec{v}, \vec{v}' \quad \text{the velocities of the gas molecules before and after the collision.}$$

We also introduce the corresponding *momenta*

$$\vec{p}_t = m_t \vec{v}_t, \qquad \vec{p}_t' = m_t \vec{v}_t',$$

and $\qquad$ (2-255)

$$\vec{p} = m\vec{v}, \qquad \vec{p}' = m\vec{v}'.$$

In Figure 2.59, a collision between a gas molecule and the test particle is sketched. In this collision both momentum and energy are conserved; that is, they are the *same* before and after the collision

$$\vec{P} \equiv \vec{p} + \vec{p}_t = \vec{p}' + \vec{p}'_t \equiv \vec{P}', \tag{2-256}$$

$$E = \frac{p^2}{2m} + \frac{p_t^2}{2m_t} = \frac{p'^2}{2m} + \frac{p_t'^2}{2m} = E'. \tag{2-257}$$

Let $M = m + m_t$ be the total mass of the two collision partners, and let us define a *relative* momentum $\vec{\pi}$ *before* the collision by the relations

$$\vec{p} = \frac{m}{M}\vec{P} + \vec{\pi},$$

$$\vec{p}_t = \frac{m_t}{M}\vec{P} - \vec{\pi}, \tag{2-258}$$

when $\vec{P}$ is the total momentum defined in (2-256). Using $\vec{p}'$ and $\vec{p}'_t$ in (2-258) defines the relative momentum π' *after* collision. In terms of π, the total energy E, (2-257) can be written as

$$E = \frac{1}{2M}\vec{P}^2 + \left(\frac{1}{2m} + \frac{1}{2m_t}\right)\vec{\pi}^2.$$

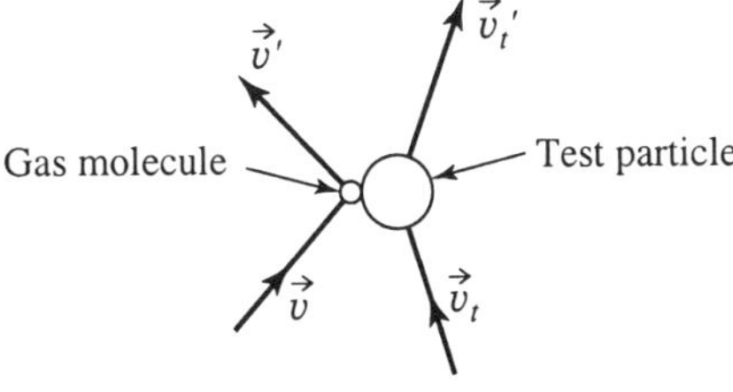

Figure 2.59. Collision of a gas molecule with a "test particle."

Similarly, the energy after collision is

$$E' = \frac{1}{2M}P'^2 + \left(\frac{1}{2m} + \frac{1}{2m_t}\right)\vec{\pi}'^2.$$

Since $E = E'$, and $\vec{P} = \vec{P}'$, it follows from these two equations that

$$\vec{\pi}^2 = \vec{\pi}'^2, \tag{2-259}$$

that is, the square magnitude of the relative momentum is the same before and after the collision. Let us now calculate the *energy change* of the test particle in this collision

$$\Delta E_t \equiv E_t' - E_t = \frac{1}{2m_t}(p_t'^2 - p_t^2).$$

Inserting (2-258) to express p_t in terms of P and π, and the corresponding equation for p_t' in terms of p and π', and using (2-259), we find

$$\Delta E_t = \frac{1}{2m_t}\left\{\left(\frac{m_t}{M}\vec{P} - \vec{\pi}'\right)^2 - \left(\frac{m_t}{M}\vec{P} - \vec{\pi}\right)^2\right\} = +\frac{1}{M}\vec{P}\cdot(\vec{\pi} - \vec{\pi}'). \tag{2-260}$$

This is the energy change in one particular collision. We must now find what the *average* value of ΔE_t is, averaged over very many collisions of the test particle. This averaging contains *two* steps:

(i) For given initial momenta $\vec{p}$ and $\vec{p}_t$ average over all possible values of $\vec{\pi}'$.

(ii) Average over all possible initial momenta $\vec{p}$ and $\vec{p}_t$ of a gas molecule and test particle.

Consider the first step. From (2-258) we can see that $\vec{\pi}$ and $-\vec{\pi}$ are the momenta of a gas molecule and test particle, as seen in the center of a mass frame of reference, where $\vec{P} = 0$. In this frame the collision looks as described in Figure 2.60.

Figure 2.60(a) shows the two momenta before the collision, and Figure 2.60(b), (c) show two possible outcomes of the collision, in which $\vec{\pi}'$ forms an angle θ with $\vec{\pi}$. Remember also that $|\vec{\pi}'| = |\vec{\pi}|$. Since the outcomes of Figure 2.60(b) and (c) are equally probable, the mean value of $\vec{\pi}'$ for Figure 2.60(b) and (c) is

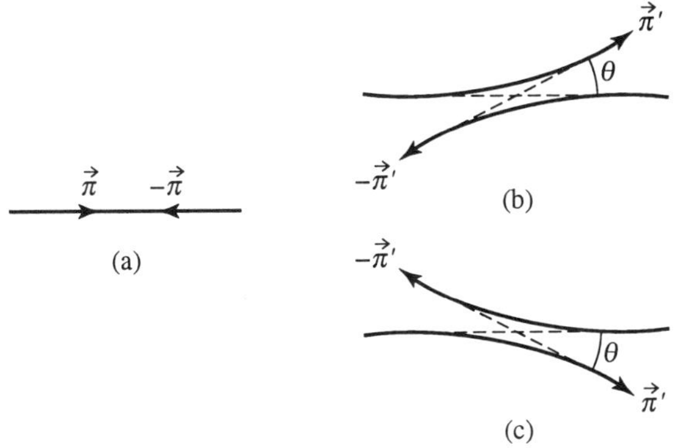

Figure 2.60. Momenta of a gas molecule and test particle as seen in the center of mass frame: (a) momenta before collision, and (b) and (c) momenta after collision. Cases (b) and (c) are equally probable.

$\vec{\pi}\cos\theta$. In eq. (2-260), we need the value of $\vec{P}\cdot(\vec{\pi}-\vec{\pi}')$. Figure 2.60 shows that $(\vec{P}\cdot\vec{\pi}')=(\vec{P}\cdot\vec{\pi})\cos\theta$. Hence

$$\vec{P}\cdot(\vec{\pi}-\vec{\pi}')=(\vec{P}\cdot\vec{\pi})(1-\cos\theta). \tag{2-261}$$

This *first* averaging therefore leads to

$$\overline{\Delta E_t}=\frac{1}{M}(\vec{P}\cdot\vec{\pi})(1-\overline{\cos\theta}). \tag{2-262}$$

We must now average this over all possible *initial* values p_t and p of the momenta of the two collision partners. For this purpose, we invert the relation [(2-258)]:

$$\vec{P}=\vec{p}_t+\vec{p}$$

$$\vec{\pi}=\frac{m_t}{M}\vec{p}-\frac{m}{M}\vec{p}_t. \tag{2-263}$$

With this we can rewrite (2-262) as

$$\overline{\Delta E_t} = \frac{1}{M}(\vec{p}_t + \vec{p}) \cdot \left(\frac{m_t}{M}\vec{p} - \frac{m}{M}\vec{p}_t\right)(1 - \overline{\cos\theta})$$

$$= \frac{1 - \overline{\cos\theta}}{M} \cdot \left\{\frac{2m_t m}{M}\left(\frac{p^2}{2m} - \frac{p_t^2}{2m_t}\right) + \frac{m_t - m}{M}(\vec{p} \cdot \vec{p}_t)\right\}. \quad (2\text{-}264)$$

At this point, we are ready to average over all values of $\vec{p}$ and $\vec{p}_t$: The average of $p^2/2m$ is $\frac{3}{2}kT$. The average of $p_t^2/2m_t$ is $(m_t/2)v_t^2 = \overline{E_t}$. The average of $(\vec{p} \cdot \vec{p}_t)$ must be *zero* since the relative orientation of $\vec{p}$ and $\vec{p}_t$, the initial momenta in the collision, are *random*. For each collision with a gas molecule of momentum $\vec{p}$, the test particle will also experience a collision with a gas molecule of momentum $-\vec{p}$.

Inserting this into (2-264), we find that the fully averaged value $\overline{\overline{\Delta E_t}}$ of the energy change of the test particle is given by

$$\overline{\overline{\Delta E_t}} = \frac{2m_t m}{M^2}(1 - \overline{\cos\theta})\left(\tfrac{3}{2}kT - \overline{E_t}\right). \quad (2\text{-}265)$$

This is the result to be proven. It shows that on average the test particle *gains* energy in a collision if its energy is *less* than $\frac{3}{2}kT$, and *loses* energy if its energy is larger than $\frac{3}{2}kT$. A situation must therefore evolve where the mean kinetic energy of the test particle is exactly $\frac{3}{2}kT$, whatever the mass of the particle is.

2.A3   Gaussian Integrals

In Section 2.2 we have found the distribution function

$$P(x, t) = \sqrt{\frac{1}{4\pi Dt}}e^{-x^2/4Dt}. \quad (2\text{-}266)$$

We will show in this appendix that the *normalization* integral

$$\int_{-\infty}^{+\infty} dx\, P(x, t),$$

which represents the total probability that the particle is somewhere between $x = -\infty$ and $x = +\infty$ is indeed equal to 1. We shall also show, by direct integration, that the *second* moment

$$x^2 = \int_{-\infty}^{+\infty} dx\, x^2 P(x, t) \tag{2-267}$$

is equal to $2Dt$.

The basic result needed for these proofs is the formula, valid for every positive *constant* α, that

$$I = \int_{-\infty}^{+\infty} dx\, e^{-\alpha x^2} = \sqrt{\frac{\pi}{\alpha}}. \tag{2-268}$$

The proof of this is most easily given by calculating I^2, which may be written as the double integral

$$I^2 = \int_{-\infty}^{+\infty} dx\, e^{-\alpha x^2} \cdot \int_{-\infty}^{+\infty} dy\, e^{-\alpha y^2} = \iint_{-\infty}^{+\infty} dx\, dy\, e^{-\alpha(x^2+y^2)}.$$

So to calculate I^2, we have to sum $e^{-\alpha(x^2+y^2)} = e^{-\alpha r^2}$, once the whole area of the x–y plane. Since the integrand depends only on $r^2 = x^2 + y^2$, this can be done by using polar coordinates to write the surface element in the plane. Using the polar surface element as shown in Figure 2.61, we have

$$I^2 = \int_0^\infty r\, dr \int_0^{2\pi} d\phi\, e^{-\alpha r^2} = 2\pi \int_0^\infty r\, dr\, e^{-\alpha r^2}.$$

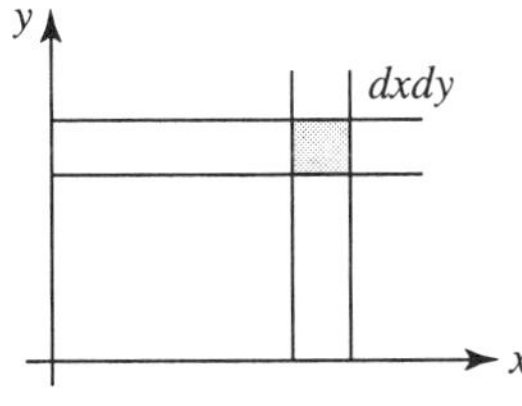

(a) Cartesian surface element
$dA = dxdy$

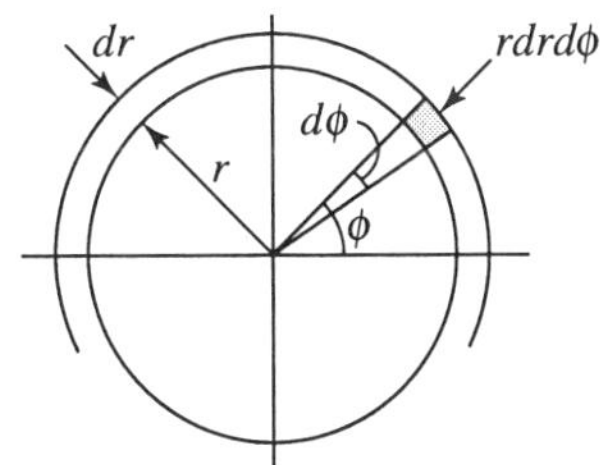

(b) Polar surface element
$dA = rdrd\phi$

Figure 2.61. Surface elements in Cartesian and polar coordinates. (a) Cartesian surface element $dA = dx\, dy$. (b) Polar surface element $dA = r\, dr\, d\phi$.

We now have a single integral. To evaluate it, we change variables. Call $\alpha r^2 = s$. Then,

$$ds = 2\alpha r\, dr \qquad \text{or} \qquad r\, dr = \frac{1}{2\alpha} ds.$$

I^2 is then given by the integral

$$I^2 = 2\pi \frac{1}{2\alpha} \int_0^\infty ds\, e^{-s} = \left(\frac{\pi}{\alpha}\right) \int_0^\infty ds\, e^{-s}.$$

The integral of the exponential has the value 1, so

$$I = \sqrt{\frac{\pi}{\alpha}}.$$

With this, we can show that $P(x, t)$ is indeed normalized. In (2-266), α has the value

$$\alpha = \frac{1}{4Dt},$$

therefore

$$\int_{-\infty}^{+\infty} dx\, P(x, t) = \sqrt{\frac{\alpha}{\pi}} \int_{-\infty}^{+\infty} dx\, e^{-\alpha x^2} = \sqrt{\frac{\alpha}{\pi}} \sqrt{\frac{\pi}{\alpha}} = 1.$$

We also calculate the *second moment* of P:

$$\overline{x^2} \equiv \int dx\, x^2 P(x, t).$$

To do this, we make use of the device known as "differentiation with respect to a parameter." It amounts to writing $x^2 e^{-\alpha x^2}$ as $-(\partial/\partial\alpha)(e^{-\alpha x^2})$ and postponing the α-derivative until after the x-integral is computed

$$\overline{x^2} = \sqrt{\frac{\alpha}{\pi}} \int dx\, x^2 e^{-\alpha x^2}$$

$$= \sqrt{\frac{\alpha}{\pi}} \int dx \left(-\frac{\partial}{\partial\alpha}\right) e^{-\alpha x^2}$$

$$= \sqrt{\frac{\alpha}{\pi}} \left(-\frac{\partial}{\partial \alpha} \right) \int dx\, e^{-\alpha x^2}$$

$$= \sqrt{\frac{\alpha}{\pi}} \left(-\frac{\partial}{\partial \alpha} \right) \sqrt{\frac{\pi}{\alpha}}.$$

The task is now reduced to take the α-derivative of $\sqrt{\pi/\alpha} = \sqrt{\pi}\, \alpha^{-1/2}$. Since

$$\frac{\partial}{\partial \alpha} (\alpha^{-1/2}) = -\tfrac{1}{2}\alpha^{-3/2} = -\frac{1}{2\alpha}\alpha^{-1/2}$$

we find

$$\sqrt{\frac{\alpha}{\pi}} \left(-\frac{\partial}{\partial \alpha} \sqrt{\frac{\pi}{\alpha}} \right) = \sqrt{\frac{\alpha}{\pi}} \left(\frac{1}{2\alpha} \right) \sqrt{\frac{\pi}{\alpha}} = \frac{1}{2\alpha}.$$

Therefore

$$\overline{x^2} = \frac{1}{2\alpha}, \qquad \alpha = \frac{1}{4Dt}, \qquad \text{hence} \quad x^2 = 2Dt.$$

2.7 References and Supplementary Reading

1. *Investigations on the Theory of the Brownian Movement*, by A. Einstein; edited by R. Furth. E. P. Dutton, New York (1956).

2. *Atoms*, by J. Perrin, translated by D. L. Hammick. Constable, London (1923).

3. Observations of the Spectrum of Light Scattered by Solutions of Biological Macromolecules, by S. B. Dubin, J. H. Lunacek, and G. B. Benedek. *Proc. National Academy of Science USA* **57**, 1164–1171 (1967).

4. The Determination of Diffusion Constants of Proteins by a Refractometric Method, by O. Lamm and A. Polson. *Biochem. Journal* **30**, 528 (1936).

5. Ueber Diffusion, by A. Fick. *Annalen der Physik* **94**, 59 (1855).

6. Filtration, Diffusion and Molecular Sieving through Porous Cellulose Membranes, by E. M. Renkin. *Journal of General Physiology* **38**, 225 (1954).

7. Hindred Diffusion in Microporous Membranes with Known Pore Geometry, by R. E. Beck and J. S. Schultz. *Science* **170**, 1302 (1970).

8. Physics of Porous Membranes—Neutral Pores, by C. P. Bean. In *Membranes, Vol. 1: Microscopic Systems and Models*, edited by G. Eisenman. Marcel Dekker, New York (1972).

9. *Biomembranes: Structural and Functional Aspects* (2 vols.), by M. Shinitzky. Weinheim, New York (1994).

10. *Biological Membranes: A Molecular Perspective*, edited by Kenneth M. Merz and B. Roux. Birkhäuser, Boston (1996).

11. *Human Physiology: The Mechanism of Body Function*, 3rd ed., by A. J. Vander, J. H. Sherman, and D. S. Luciano. McGraw-Hill, New York (1980).

12. *The Kidney*, by H. W. Smith. *Scientific American*, January (1953).

13. C. Colton, PhD Thesis, Massachusetts Institute of Technology, Cambridge, MA (1969).

14. *Renal Pathophysiology*, 3rd ed., by A. Leaf and R. Cotran. Oxford University Press, Oxford, UK (1985).

15. Study of the Chemical Denaturation of Lysozyme by Optical Mixing Spectroscopy, by S. B. Dubin, G. Feher, and G. B. Benedek. *Biochemistry* **12**, 714 (1973).

16. Dissociation of E. Coli Ribosomes by Optical Mixing Spectroscopy, by L. Hocker, J. Krupp, G. Benedek, and J. Vournakis. *Biopolymers*, **12**, 1677 (1973).

17. Coliphages and Coliphage DNA, by S. B. Dubin, G. B. Benedek, F. C. Bancroft, and D. Freifelder. *Journal of Molecular Biology*, **54**, 547 (1970).

18. *The Ultracentrifuge*, by T. Svedberg and K. O. Pedersen. Oxford University Press, Oxford, UK (1940).

19. *An Introduction to Ultracentrifugation*, by T. J. Bowen. Wiley-Interscience, New York (1970).

20. *Ultracentrifugal Analysis in Theory and Experiment*, edited by J. W. Williams. Academic Press, New York (1963).

21. *Physical Biochemistry*, by K. E. van Holde (Foundation of Modern Biochemistry Series). Prentice Hall, Englewood Cliffs, NJ (1971).

22. *Proteins, Amino Acids and Peptides*, by E. J. Cohn and J. T. Edsall. Hafner, New York (1963).

23. Thermodynamic Analysis of Permeability of Biological Membranes, by O. Kedem and A. Katchalsky. *Biochem. Biophys. Acta* **27** 229–246 (1958).

24. There is no elementary proof of Onsager's theorem. For a discussion, see *Nonequilibrium Thermodynamics in Biophysics*, by A. Katchalsky and P. F. Curran. Harvard University Press, Cambridge, MA (1967). See also: Onsager Reciprocal Relations in Complex Membranes, by F. M. Snell and B. Stein. *Journal of Theoretical Biology* **10**, 77–183 (1966).

25. The Frictional Coefficients of the Flows of Non-Electrolytes Through Artificial Membranes, by B. Z. Ginzburg and A. Katchalsky. *Journal of General Physiology* **47**, 403 (1963).

26. Exchange of Substances Through Capillary Walls, by E. M. Landis and J. R. Pappenheimer. In *Handbook of Physiology*, Vol. II, Chap. 29, Sect. 2. American Physiological Society, Washington, DC (1963).

27. *Water Transport in Cells and Tissues*, by C. R. House. Edward Arnold, London (1974).

28. The Rate of Exchange of Tritiated Water Across the Human Red Cell Membrane, by C. V. Paganelli and A. K. Solomon. *Journal of General Physiology* **41**, 259 (1957).

29. The Gaussian Error function $\Phi(x)$ can be found tabulated in the following works:

   (a) *A Short Table of Integrals* (3rd edition), by B. O. Pierce. Ginn and Company, Boston, MA (1929).

   (b) *Tables of Functions with Formulae and Curves* (4th edition), by Eugene Jahnke and Fritz Emde. Dover Publications, New York (1945).

   (c) *Handbook of Chemistry and Physics*; Section on Mathematical Tables. The Chemical Rubber Publishing Co., Cleveland, OH. (This Handbook is republished annually. The required information is found only in early editions, prior to 1974!)

30. The topics of ultracentrifugation (along with electrophoresis) in Chapter 7 and Cell Membrane Transport in Chapter 15 are treated in some detail in the text: *Molecular Cell Biology*, by Darnell, Lodish, and Baltimore. Scientific American Books (1986).

There is an enormous literature on the properties of membranes. The following titles may be of general interest:

Exchange of Substances Through Capillary Walls, by E. M. Landis and J. R. Pappenheimer. In *Handbook of Physiology*, Vol. II, Chap. 29, Sec. 2. American Physiological Society, Washington, DC (1963).

The Movement of Molecules Across Cell Membranes, by W. D. Stein. Academic Press, New York (1967).

Osmotic Flow of Water Across Permeable Cellulose Membranes, by R. P. Durbin. *Journal of General Physiology* **44**, 315 (1969).

Introduction to the Study of Biological Membranes, by M. Cerreido and C. A. Rotunno. Gordon and Breach, New York (1970).

Cell Membrane Transport: Principles and Techniques, by A. Kotyk and K. Janaceik. Plenum Press, New York (1970).

The Role of Membranes in Metabolic Regulation, edited by M. A. Mellman and R. W. Hanson. Academic Press, New York (1972).

Water Transport in Cells and Tissues, by C. R. House. Edward Arnold, London (1974).

Biochemistry of Cell Membranes: A Compendium of Selected Topics, edited by S. Papa and J. M. Tager. Birkhäuser Verlag, Basel (1995).

2.8   Problems

(The problems are grouped in sections; and in each section ordered roughly in order of increasing difficulty.)

Molecular Movement and the Physical Properties of Gases

1. If the temperature of the atmosphere were the same at all heights, the density of air ($n(z)$ in molecules/cm^3) will decrease exponentially with the height in accordance with the law

$$n(z) = n_0 e^{-z/H},$$

where the scale height $H \simeq 10$ km.

   (a) From the definition of the mean free path l, write a formula relating the mean free path to the number density $n(z)$.

(b) Assume that at $z = 0$, the mean free path is 10^{-5} cm. Find the height z for which l is equal to 1 cm.

2. Gas pressure and molecular collisions against the containing vessel.

(a) Show that the *mean* momentum (Δp_x) transferred to the containing walls of a box per colliding gas molecule is $\Delta p_x = \sqrt{4mkT}$. (Use the equipartition theorem.)

(b) Express the pressure on the wall in terms of the mean momentum transfer Δp_x and the number of collisions/cm^2 s (dN/dt).

(c) Assume that oxygen is in a container at $300\,°K$ and exerts a pressure of 1 atm. Compute numerically (dN/dt) the number of collisions/cm^2 s between the oxygen molecules and the wall. Use $k = 1.38 \times 10^{-16}$ erg/°K, 1 atm $= 10^6$ dyn/cm^2, and $m_{O_2} = 5.31 \times 10^{-23}$ g.

Random Walk

3. In the one-dimensional random walk the probability $(P_N(R, L))$ that, after N steps there will be R steps to the right and L steps to the left, is given by

$$P_N(R, L) = \frac{N!}{R!\,L!}p^R q^L.$$

Here p and q are the a priori probabilities for a right and a left step. The mean value for R is $\overline{R} = Np$, while the mean value for L is $\overline{L} = Nq$. In Chapter 1 we saw that $\overline{R^2} = (\overline{R})^2 + Npq$.

(a) Show that $\overline{L^2}$, the mean square value of the number of left-hand steps, is
$\overline{L^2} = N^2 q^2 + Npq$.

(b) If we define the variable h as

$$R = \overline{R} + \frac{h}{2}, \qquad L = \overline{L} - \frac{h}{2},$$

then the displacement after N steps is $(R - L) = (\overline{R} - \overline{L}) + h$. Show that $\overline{h^2} = 4Npq$.

4. In the symmetrical one-dimensional random walk $(p = q = \frac{1}{2})$, the probability of a displacement $m = (R - L)$ after N steps is given by

$$P_N(m) = \sqrt{\frac{2}{\pi N}}\, e^{-m^2/2N}.$$

(a) Show, by starting from the general ($p \neq q$) form of the Bernoulli distribution, that in the case $N \gg 1$ the probability $P_N(m)$ of a displacement $m = (R - L)$ after N steps is given by

$$P_N(m) = \sqrt{\frac{1}{2\pi Npq}}\, e^{-m^2/8Npq}$$

or

$$P_N(m) = \sqrt{\frac{1}{2\pi \Delta^2}}\, e^{-m^2/8\Delta^2}.$$

Here Δ is the half-width of the original distribution for $P_N(R)$. $\Delta = \sqrt{Npq}$.

Mathematical hints:

$$P_N(R, L) = \frac{N!}{R!\, L!} p^R q^L \cong p^R q^L \sqrt{\frac{N}{2\pi RL}}\, e^{R\ln(N/R)+L\ln(N/L)}.$$

Also, since $p^R = e^{R\ln p}$, $q^L = e^{L\ln q}$, we can also write

$$P_N(R, L) \cong \sqrt{\frac{N}{2\pi RL}}\, e^{R\ln(Np/R)+L\ln(Nq/L)}$$

Finally, define m as $R = \overline{R} + m/2$, $L = \overline{L} - m/2$.

(b) Show from your result in (a) that $\overline{m^2} = 4Npq$. Compare this with the solution you found in Problem 3(b).

5. A crude, but useful approximation to the one-dimensional random walk probability distribution $P(x, t)$ is a rectangle of height $P_0 = 1/\sqrt{10Dt}$ and width $2\sqrt{(5/2)Dt}$. In this problem we will examine the time–space evolution for $P(x, t)$ using this approximation. Let $D = 5 \times 10^{-7}$ cm^2/s, as is appropriate for a protein of molecular weight $\sim 100{,}000$ diffusing in water.

(a) Make plots of $P(x, t)$ as a function of x for times $t = 1, 4, 10, 40, 100, 400, 1000$ millisec, using the rectangular approximation.

(b) How long does it take until the particle has a reasonable chance to diffuse a distance equal to the wavelength of light $\sim 6 \times 10^{-5}$ cm.

(c) For $t = 40$ millisec, make a graph showing both the accurate form for $P(x, t)$, $(P(x, t) = \sqrt{1/4\pi Dt}\, e^{-x^2/4Dt})$ and the rectangular approximation given above.

(d) Suppose that at $t = 0$, 10,000 protein molecules are introduced near $x = 0$. Write a formula for the one-dimensional number density (n) of these molecules as a function of the time. Make a graph showing n as a function of time. (Use the rectangular approximation to $P(x, t)$.)

(e) How long does it take before the particle density is less than one per wavelength of light?

6. Suppose that a random walk consisting of N steps is broken into two stages. The first a walk of N_1 steps, the second a walk of N_2 steps where $N_1 + N_2 = N$. Show that $P_N(m)$, the probability of a displacement m after N steps, can be written as

$$P_N(m) = \sum_{\text{all } m_1} P_{N_1}(m_1) P_{N_2}(m - m_1).$$

Here m_1 is the displacement after N_1 steps. This result is called the "addition theorem" for displacements in a random walk.

Hints:

(i) Use the multiplication rule for the joint occurrence of independent events, or

(ii) more mathematically, you may use the addition theorem coefficients found in Chapter 1, Problem 20(b).

7. The addition theorem of Problem 6 permits one to solve the following problems.

(a) Suppose two drunkards start a one-dimensional random walk at the origin. If both take N steps, find the probability that on the Nth step they are separated by k steps.

(b) Find the probability that the drunkards *meet* on the Nth step. Check the correctness of your result by considering the *relative* motion of the two random walkers.

8. In this problem we shall examine, more carefully than we did in Section 2.2.D, how to generalize the one-dimensional random walk probability distribution to obtain the two- (or three-dimensional) random walk probability distribution.

Consider a two-dimensional random walk in which the steps have length l, and can occur with equal probability in the four directions $\pm x$, $\pm y$.

(a) After a total of N steps show that the probability of the particle being at $m_x l$ and $m_y l$ is given by

$$P_N(m_x, m_y) = \sum_{\substack{N_x, N_y \\ (N_x+N_y=N)}} \frac{N!}{N_x! \, N_y!} \frac{1}{2^N} P_{N_x}(m_x) P_{N_y}(m_y),$$

where $P_N(m)$ is the one-dimensional random walk distribution.

(b) When N is large, show that the term $N!/N_x! \, N_y!$ is very sharply peaked around the values of $N_x = N_y = N/2$.

(c) Show that as a result of this sharp peaking, one in effect can regard P_{N_x} and P_{N_y} as equal to $P_{N/2}$, and the sum evaluated accurately to yield

$$P_N(m_x, m_y) \xrightarrow[N\gg1]{} P_{N/2}(m_x) P_{N/2}(m_y) \sum_{N_x} \frac{N!}{N_x! \, (N - N_x)!} \frac{1}{2^N}$$

or

$$P_N(m_x, m_y) = P_{N/2}(m_x) P_{N/2}(m_y).$$

By carrying out a perfectly similar argument for three dimensions, we have the result used in Section 2.2.D.

Molecular Dimensions and the Diffusion Coefficient

9. Use the data given in Table 2.4(b) in connection with the following questions.

   (a) Show that the product Da, where D is the diffusion coefficient and a is the molecular radius, is a constant for the large globular proteins.

   (b) Using $\eta = 1 \times 10^{-2}$ g/(cm/s), examine the degree to which $kT/6\pi\eta a$ (the Einstein–Stokes equation) predicts the diffusion coefficient.

   (c) Write a formula for the radius of a globular protein in terms of its molecular weight and the effective protein density.

   (d) Compute the value of the effective protein density that is required to give the radii of globular proteins listed in Table 2.4(b) using the formula obtained in (c) above. Compare these densities with a typical value of 0.7 cc/g for the partial specific volume of proteins.

10. Determine the radius of the Bushy Stunt Virus assuming that it is a sphere. Use the data in Table 2.4(b).

11. In this problem you will see how one may empirically estimate molecular weights from a measurement of its diffusion coefficient.

   (a) If all macromolecules have approximately the same mean density $\bar{\rho}$, show that the molecular weight can be related to the diffusion coefficient D approximately using the equation

   $$\text{molecular weight} = N_0\left(\tfrac{4}{3}\pi\right)\bar{\rho}\left(\frac{kT}{6\pi\eta}\right)^3\frac{1}{D^3}. \tag{1}$$

   (b) Use the following approximate experimental data for the molecular weight and diffusion coefficient of various macromolecules to examine the validity of (1) above:

Molecule	Molecular weight g/mol	Diffusion coefficient D in units of 10^{-7} cm^2/s ($T = 20\,^\circ$C)
Lysozyme	14,500	10.6
Bovine serum albumin	60,000	6
Catalase	250,000	4
70 s ribosomes	3,000,000	2
T7 bacteriophage virus	50×10^6	0.603

   Make a graph of molecular weight versus $1/D^3$ and obtain experimentally the magnitude of the coefficient relating molecular weight to $1/D^3$.

   (c) Use $\eta = 1 \times 10^{-2}$ g/(cm/s), what value of $\bar{\rho}$ is needed to bring the prediction of (1) for the coefficient relating molecular weight to $1/D^3$ into agreement with the value found from the data in (b) above. Is this value of $\bar{\rho}$ reasonable?

Space–Time Evolution of Concentration Variations: The Diffusion Equation

12. By substitution into the diffusion equation, show that

$$C(x, t) = \left(\frac{N_0}{\sqrt{4\pi Dt}} \right) e^{-x^2/4Dt}$$

is a solution of the diffusion equation, subject to the initial condition that all (N_0) of the particles are placed inside a small region of space $|\Delta x| \ll \sqrt{4\pi D \Delta t}$ in a short period of time Δt ($\Delta t \ll t$).

13. Suppose that at $t = 0$ the concentration $C(x', 0)$ variation in the x-direction is as shown in the diagram below:

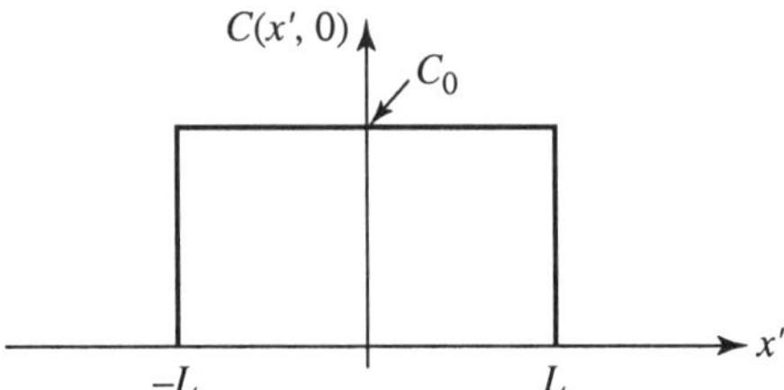

(a) Using the integral form of solution of the diffusion equation, show that $C(x, t)$ at any time–space point later is given by

$$C(x, t) = \tfrac{1}{2} C_0 \left\{ \phi \left(\frac{L - x}{\sqrt{4Dt}} \right) + \phi \left(\frac{L + x}{\sqrt{4Dt}} \right) \right\}$$

where ϕ is the Gauss error function defined on pp. 132 [(2-69)].

(b) Make approximate graphs of $C(x, t)$ for $t = \frac{1}{10}(L^2/4D)$, $t = \frac{1}{2}(L^2/4D)$, $t = 2(L^2/4D)$. Use values of ϕ obtained either from mathematical tables, or from the graph on page 132.

14. Suppose that at $t = 0$, the concentration profile $C(x', 0)$ is a Gaussian distribution of half-width L. That is, it is of the form

$$C(x', 0) = \frac{C_0}{\sqrt{2\pi L^2}} e^{-x'^2/2L^2}.$$

(a) Use the integral form of the solution of the diffusion equation to show that at any space–time point later

$$C(x, t) = \frac{C_0}{\sqrt{2\pi(L^2 + 2Dt)}} e^{-x^2/2(L^2 + 2Dt)}.$$

Thus, we see that at each instant of time later the distribution is still Gaussian in shape, except that its half-width occurs at the point $x_{1/2} = (L^2 + 2Dt)$.

Hint: First show that

$$-\frac{(x - x')^2}{4Dt} - \frac{x'^2}{2L^2}$$

$$= -\frac{x^2}{(4Dt + 2L^2)} - \left(x'\sqrt{\frac{1}{4Dt} + \frac{1}{2L^2}} - \frac{x}{4Dt}\frac{1}{\sqrt{\frac{1}{4}Dt + \frac{1}{2}L^2}} \right)^2.$$

Then integrate over x' keeping x constant.

(b) Sketch the form of $C(x, t)$ as a function of x/L for $t_1 = \frac{1}{10}(L^2/2D)$, $t_2 = \frac{1}{2}(L^2/2D)$, and $t_3 = 2(L^2/2D)$.

15. In this problem you will see how a "Fourier analysis" permits one simply to compute the smoothing out of a concentration gradient. Suppose that at $t = 0$ the concentration profile varies with x' as shown in the following diagram:

$$C(x') = \overline{C} + C_1\left(1 - \frac{2x'}{L}\right)$$

This formula applies to $C(x')$ except at the edges $x' = 0$, $x' = L$ where it is slightly rounded off. At $t = 0$, we *assert* that the profile can be described as

$$C(x') = C_0 + \frac{8}{\pi^2}C_1\left(\cos\frac{\pi x}{L} + \tfrac{1}{9}\cos\left(\frac{3\pi x'}{L}\right) + \tfrac{1}{25}\cos\left(\frac{5\pi x'}{L}\right)\right).$$

This represents the first few terms in the "Fourier series" for the concentration profile shown.

(a) By computing $C(x')$ for a few values of x', examine how well the formula above actually approximates the figure shown.

(b) For $t \geq 0$, the amplitude of each of the cosine factors decays exponentially in time just as was discussed in Section 2.3.D. That is, we may write

$$C(x,t) = C_0 + \frac{8C_1}{\pi^2}\left(a_1(t)\cos\frac{\pi x}{L} + a_3(t)\cos\frac{3\pi x}{L} + a_5(t)\cos\frac{5\pi x}{L}\right),$$

where

$$a_1 = e^{-t/t_0}, \qquad a_3 = \tfrac{1}{9}e^{-9t/t_0}, \qquad a_5 = \tfrac{1}{25}e^{-25t/t_0},$$

when $1/t_0 = D\pi^2/L^2$. Using these results, compute the form of $C(x,t)$ for several values of x at $t = 0.5t_0$, $t = t_0$, $t = 2t_0$.

Particle Conservation, Particle Current, and Diffusion

16. This problem is useful in helping to understand the continuity equation. On a superhighway two observation posts A and B, a few miles apart are set up. There are no side exits or entrances in between.

There is steady, dense traffic with all lanes occupied. At a given time the following observation is made:

At A: Cars move at a mean speed of 50 ft/s. Cars are about 25 ft apart (front of car to front of next car).

At B: Cars move at a mean speed of 75 ft/s. Cars are about 50 ft apart.

Find, for both A and B:

(a) Mean concentration c of cars, in cars/ft per lane.

(b) Mean current (flow) density j, in cars/s per lane.

(c) Number of cars/s crossing the line at A, and at B.

(d) Find the rate of change (car/s) in the *number* of cars located *between* A and B (shaded area).

17. Consider the Lam–Polson experiment we discussed in Section 2.3.B.

(a) Obtain a formula for the current density of solute particles crossing the plane at $x = 0$. (For short times, i.e., $t < L^2/4D$.)

(b) From the concentration gradient $(\partial C/\partial x)$ at $x = 0$, compute the rate at which the number of particles N in the lower portion of the tube decreases with time. Show that

$$\left(\frac{dN}{dt}\right) = -N_0 \frac{1}{L}\sqrt{\frac{D}{4\pi t}}.$$

Here N_0 is the number of particles originally in the lower portion length L of the tube. D is the diffusion coefficient of the particles.

(c) Use the result found in (b) above to show that in the time interval $t \ll L^2/4D$ that the fractional decrease in the number N is equal to

$$\left(\frac{\Delta N}{N_0}\right) = -\frac{2}{L}\sqrt{\frac{Dt}{4\pi}}.$$

(d) For what range of values of $(\Delta N/N_0)$ will this formula accurately apply?

(e) Suppose $L \simeq 5$ cm, how long will it take for the concentration of a glucose solution in the lower portion of the tube to decrease by 20%? Use the data for D given in Table 2.4(b). Repeat the calculation for the Bushy Stunt Virus.

18. In this problem you will learn how to determine the effect of an absorbing wall on the smoothing out of a concentration variation. At $t = 0$, a large number N_0 of particles begin a random walk starting at $x = 0$. At some time later the particle concentration in particles/cm is given by

$$C(x, t) = \frac{N_0}{\sqrt{4\pi\,Dt}} e^{-x^2/4Dt}.$$

(a) Give an expression for the particle current $j(x, t)$ at each position along the x-axis.

(b) Indicate graphically how $j(x, t)$ varies in space for various times. Suppose that at the position $x = L$, there is a wall that absorbs or binds all the particles that touch it. This wall has the effect that $C(L, t)$, the concentration of particles in solution at the position of the wall, is zero. The effect of this wall can be obtained by subtracting from $C(x, t)$ its mirror image whose center is located at $x = 2L$. This is indicated in the figure below:

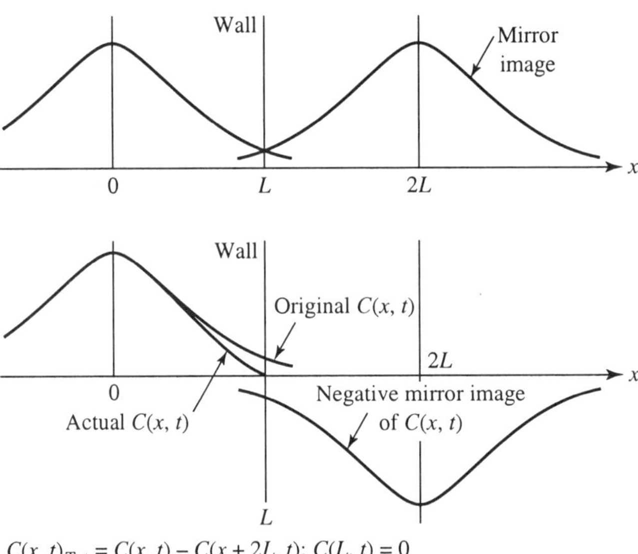

$$C(x, t)_{\text{Tot}} = C(x, t) - C(x + 2L, t); \quad C(L, t) = 0$$

(c) Show that the current density $j(x, t)$ at $x = L$ is now twice what it would be in the absence of the wall.

(d) Can you give a physical reason why the absorber increases the value of j?

(e) Give a formula for the number of particles absorbed per second at $x = L$, and the resulting fractional decrease in the total number of diffusing particles.

19. Suppose that at some time t the concentration $C(x, t)$ in (molecules/cm^3) in a tube of cross-sectional area A fluctuates sinusoidally in space in accordance with the formula

$$C(x, t) = \overline{C} + \Delta C(t) \sin \frac{2\pi x}{L}.$$

Here $\Delta C(t)$ is the amplitude of the concentration fluctuation:

(a) Calculate the additional number of particles $(\Delta N(t))$ in the region $(0 \leq x \leq L/2)$ over and above the average number $\overline{C} A L/2$.

(b) Calculate the current of solute particles at the points $x = 0$ and $x = L/2$.

(c) Use the conservation of numbers of particles to calculate $(d\Delta N(t)/dt)$, the rate of change of ΔN between $0 \leq x \leq L/2$.

(d) From your result in (c) above obtain the decay time for the relaxation of the concentration fluctuation $\Delta C(t)$.

(e) Compute the decay time if $L = 6000$ Å and $D = 1 \times 10^{-7}$ cm^2/s.

**Flow Across Porous Membranes Under a Pressure
or Concentration Gradient**

20. An experimenter is studying the conformational change in an enzyme (lysozyme, for example). He induces this conformational change by altering the concentration of the "denaturant" that is guanidine hydrochloride (GuCl). He starts with some concentration (C) of GuCl in the solution and reduces the concentration by dialysis. That is, he places the solution in a container of volume V whose surface (area A) is a semipermeable membrane with a permeability p that is the same for both the Gu$^+$ and Cl$^-$ ions. This dialysis "bag" is placed in a tank containing flowing water so that the concentration of GuCl outside the bag is always nearly zero.

(a) Write a differential equation that describes the rate of change of the GuCl concentration in the bag. Assume that the GuCl is always uniformly distributed within the bag.

(b) Solve the equation and obtain an expression in terms of the parameters given above for the time required to reduce the GuCl concentration to about one-tenth of its initial value.

(c) Suppose that the solution has a volume of 50 cm^3, and that the permeability is 0.5×10^{-3} cm/s, what should the surface area of the bag be so that the concentration of GuCl falls to one-tenth its initial value in 15 h?

21. Filtration of the blood plasma in the kidney.

(a) Calculate the filtration coefficient L_p for the basement membrane of the kidney glomeruli, using the following data:

pore radius: $a = 35$ Å,

pore length: $l = 600$ Å,

fractional pore area: 5%,

blood plasma viscosity: $\eta = 0.2$ poises (dyn s/cm^2).

(b) Calculate the volume of filtrate per day (24 h) using a total basement membrane area of 1.5 m^2 and an effective pressure across the membrane of

$$\Delta p \cong 20 \text{ mmHg.}$$

Compare your result with the known physiological fact that the total volume of filtrate is of the order of 180 l/day.

22. Suppose the body fluids, plus the blood, have a burden of 100 g of urea. How long would it take for a hemodialyzer, with pumped dialyzant that always has zero concentration of urea, to reduce the body burden to 10 g of urea? Assume that the membrane permeability is 0.5×10^{-3} cm/s, membrane area $= 1$ m^2, and body and blood fluid volume $= 40$ l.

23. In the previous problem, suppose the dialyzant is not pumped, but that the concentration of urea in it can increase in time from an initial value of zero urea concentration. If the volume of dialyzant is 400 l, can one reduce the burden of urea to 10 g? How long will it take to reduce the body burden to 15 g urea? How long for 20 g of urea?

24. Two membranes A and B have permeabilities p_A and p_B, respectively. They are joined together to form a single two-layered membrane. The questions following relate to the properties of this composite membrane.

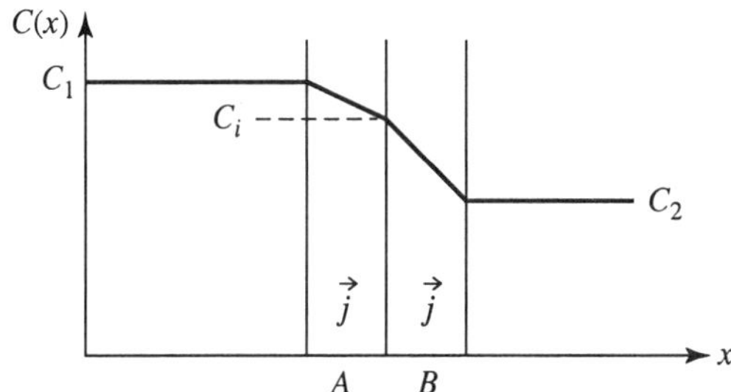

(a) How would you *define* the *overall* permeability p_{AB} of the two-layered membrane?

(b) Why is the current density j of particles diffusing through the membrane the same in layer A as in layer B?

(c) What are the equations that, for *given* concentrations c_1 and c_2, determine the value of c_i, and the concentration at the inner boundary of A and B?

(d) Find an expression for p_{AB} in terms of p_A and p_B.

Forces, Flow, and Diffusion

25. In this problem you will use Perrin's data given in Figure 2.38 to calculate Avogadro's number from the equilibrium variation of concentration of solution with height. Calculate first the scale height H from the graph. Then use the measured values of the properties of the individual particles to compute Avogadro's number. Compare your result with the current accepted value of N_0.

26. Stokes–Einstein relation. Tobacco Mosaic Virus particles of mass m and diffusion constant D are uniformly dispersed at low concentration in water. The density of water is seven-tenths of the density of each virus particle. At first, let us imagine that the mass of each particle is small enough so that no appreciable settling out or precipitation occurs. Suppose then that the pH of the solution is changed, and that as a result, the molecules coalesce to form spherical aggregates containing n molecules each.

(a) Obtain a formula for the speed with which each of the aggregates moves as it sinks, because of the pull of gravity, to the bottom of the container.

(b) Suppose the Tobacco Mosaic Virus has a molecular weight of 4×10^6 g/mol and a diffusion coefficient of $D \cong 0.4 \times 10^{-7} \mathrm{cm}^2/\mathrm{s}$. Examination of the aggregates shows they drop to the bottom with the speed of 2×10^{-3} cm/s. How many virus particles are in each aggregate?

(c) After all the settling has occurred, one will observe a gradient in the density of the aggregated particles. Give a formula for the scale height Z_0, at which the number of aggregates/cm^3 is reduced by a factor $(1/e)$ from the density found at the bottom. (In this part of the question the concentration gradient cannot be neglected.)

(d) Compute Z_0 numerically for the case $n = 1$ and the value of n obtained above in part (c).

27. **Sedimentation of insulin.** Table 11.8 in R. A. Alberty's *Physical Chemistry* gives the following data for the sedimentation of insulin (at $t = 20\,^\circ\text{C}$):

$$\text{sedimentation coefficient: } s = 1.7 \times 10^{-13}\ \text{s},$$

$$\text{buoyancy correction: } (1 - \rho\bar{v}_p) = 0.28,$$

$$\text{diffusion constant: } D = 1.5 \times 10^{-6}\ \text{cm}^2/\text{s}.$$

(a) Use these data to determine the molecular weight M of insulin.

(b) Assuming that the insulin molecule is spherical, use (2-183) to determine its radius, a.

(c) Using the Stokes–Einstein relation [(2-59)], *calculate* the diffusion coefficient of the molecule, using the radius determined in (b) above. Use

$$\eta = 10^{-2}\ \text{dyn s/cm}^2,$$

$$k = 1.38 \times 10^{-16}\ \text{dyn cm/degree},$$

$$T = 293\,^\circ\text{K}.$$

(This calculated coefficient is called D_0 in Table 2.10.) Is the result you obtained in (c) consistent with the given value of D?

28. **Density gradient centrifugation.** We have generally assumed the density ρ of the solution in the sedimentation chamber to be constant, and closely equal to that of water (dilute solutions are used only).

It is possible to introduce a *tapered* density $\rho(r)$ by skillfully adding a heavy salt, like cesium chloride, or sucrose, in layers. (See Van Holde [21, Section 5.5].) In this case, it is possible to make the buoyancy correction factor $(1 - \rho\bar{v}_p)$ to be zero at some value r_c in the chamber. For $r < r_c$ the suspended proteins are then denser, for $r > r_c$ less dense than the solution, and the buoyancy drives them *toward* r_c, where they accumulate in a narrow band. Near $r = r_c$ one has for the buoyancy correction

$$\left(1 - \rho(r)\bar{v}_p\right) \cong 1 - \left(\rho(r_c) + (r - r_c)\left(\frac{d\rho}{dr}\right)_{r_c}\right)\bar{v}_p$$

and since $\rho(r_c)\bar{v}_p = 1$ by definition, one has

$$1 - \rho(r)\bar{v}_p \cong (r - r_c)\left(\frac{d\rho}{dr}\right)\bar{v}_p.$$

(a) Using (2-179), and the subsequent arguments, show that near $r = r_c$, the steady state protein concentration $C(r)$ is given by the Gaussian

$$C(r) = C(r_c)e^{-(r-r_c)^2/L^2},$$

where

$$L = \sqrt{\frac{2RT}{M(\omega r_c)\bar{v}_p(d\rho/dr)_{r_c}}},$$

M being the molecular weight of the protein, $\bar{v}_p$ its specific volume.

(b) The width L can be quite small, and in this lies the main interest of the method. Using $M = 10^6$ gm, $(\omega^2 r_c) = 10^6$ g, $\bar{v}_p = 0.75$ cm^3/gm, and $(d\rho/dr)_{r_c} = 0.5$ gm/cm^4, show that L is of the order of 10^{-2} cm!

(c) Assume that the solution contains two different proteins of similar molecular weight M, but slightly different specific volumes, $\bar{v}_p$ and $\bar{v}'_p$. They will accumulate at slightly different positions r_c and r'_c. Show that the separation $r'_c - r_c$ is related to the difference in specific volumes by

$$(r'_c - r_c) = (\bar{v}'_p - \bar{v}_p)\frac{\rho(r_c)}{v_p(d\rho/dr)_{r_c}}.$$

(d) Comparing this with the width L of each band, show that the condition for the separability of two proteins of different specific volumes $\bar{v}_p$ and $\bar{v}'_p$ is

$$\rho(r_c)|\bar{v}'_p - \bar{v}_p| > \sqrt{\frac{2RT}{M}\frac{v_p(d\rho/dr)_{r_c}}{(\omega^2 r_c)}}.$$

(e) A successful separation of nearly identical macromolecules by the density gradient method was achieved with DNA. DNA contains about 30% nitrogen by weight. Nitrogen occurs in two isotopes ^{14}N and ^{15}N. DNA from bacteria grown in media containing only the one or the other isotope have a different density. Find the fractional difference $|\bar{v}_p - \bar{v}'_p|/\bar{v}_p$ of the specific volumes of the two kinds of DNA. Decide whether or not the two types of DNA can be separated by the density gradient method, assuming that the data of part (b) do apply to this case.

Coupled Flow of Solute and Solvent. Osmotic Pressure

29. Consider a thistletube osmometer as shown below. Let L_p be the filtration coefficient of the membrane.

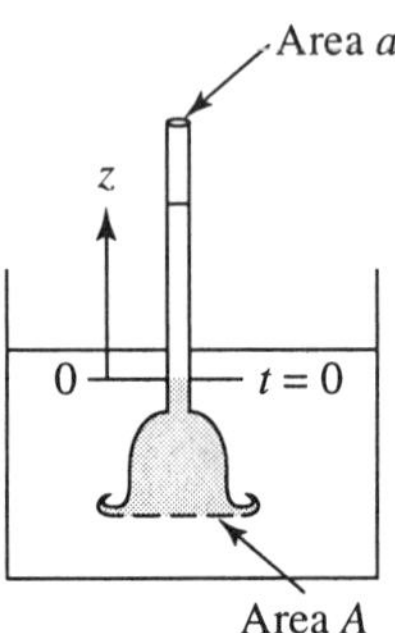

(a) At time $t = 0$, the pressure difference across the semipermeable membrane is zero. Show that the height z of the solution in the tube rises exponentially to its final value z_f:

$$z(t) = z_f(1 - e^{-t/\tau}).$$

To find this result, relate the rate of rise (dz/dt) to the volume inflow across the membrane. Remember that the pressure difference Δp across the membrane is $\Delta p = \rho_s g z$, ρ_s being the density of the solution. Show that the time constant τ is given by

$$\tau = \frac{a}{A\rho_s g L_p}.$$

(b) Suppose the solute concentration c_s is 10^{-3} mol/l, and the density ρ_s of the solution is 1 g/cm^3. Calculate the final height z_f of the fluid in the column.

(c) Suppose the filtration coefficient has a value

$$L_p = 3 \times 10^{-11} \text{ cm}^3/\text{dyn s}$$

(typical for cellophane, for example), and that

$$A = 10 \text{ cm}^2, \qquad a = 10^{-2} \text{ cm}^2.$$

Calculate the numerical value of the time constant τ. From this calculation you can see that the thistletube geometry is not a practical one for a rapid measurement of osmotic pressure.

(d) Suggest an improved osmometer design that would reduce the value of τ to minutes.

30. Molecular weight determination from osmotic pressure. The osmotic pressure of blood plasma proteins has been measured to be about 28 mmHg at physiological temperature ($T = 310\,°\text{K}$). The quantity of plasma proteins present has been measured to be about 60 g/l. Use these data to get a mean molecular weight M (in g/mol) for these plasma proteins, assuming the validity of Van t'Hoff's law.

31. The following scheme has been proposed as a means of obtaining fresh water from the ocean. A well, consisting of a tube closed at the bottom end by a semipermeable membrane, is lowered into the ocean to a depth of H_2.

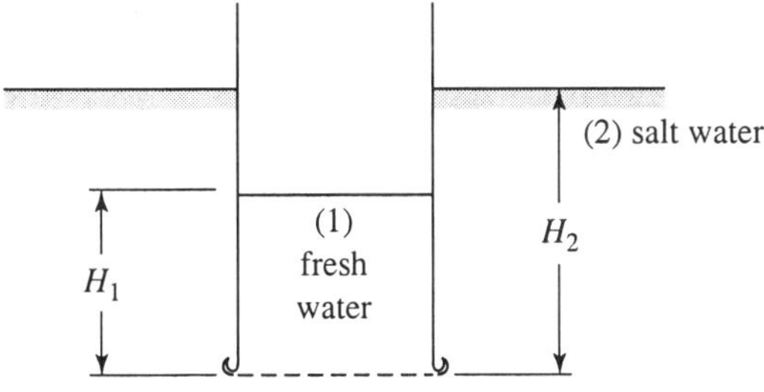

The osmotic pressure Π_s of salt water is 22 atm (1 atm = 760 mmHg = 10 m water.)

(a) Considering only the osmotic pressure difference across the membrane, what is the magnitude and direction of the volume flow of fresh water? (Express your result in terms of L_p, Π_s, and A.)

(b) Considering only the effect of the hydrostatic pressure difference across the membrane, what is the magnitude and direction of the volume flow of fresh water? (Express your result in terms of H_2, H_1, L_p, A, g and the density (ρ_0) of water. You may here neglect the density difference between salt and fresh water.)

(c) What is the minimum depth $(H_2)_{min}$ required so that at equilibrium fresh water just fills the bottom of the well?

(d) Numerically, how deep is $(H_2)_{min}$?

(e) If the well is inserted deeper $H_2 > H_{2_{min}}$ into the ocean, what is the location of the fresh water level relative to the surface of the ocean? Explain your result.

(f) If H_2 gets very large, the density difference between the heavier sea water and the lighter fresh water begins to make itself felt. Describe qualitatively what happens to the water level in the well as the tube goes deeper into the ocean. The density difference between salt and fresh water is 3%.

32. A cylinder of total volume $2V_0$ is divided into two parts by a perfect, semi-permeable piston. This piston is free to move along the axis of the cylinder, has area A, and a filtration coefficient L_p.

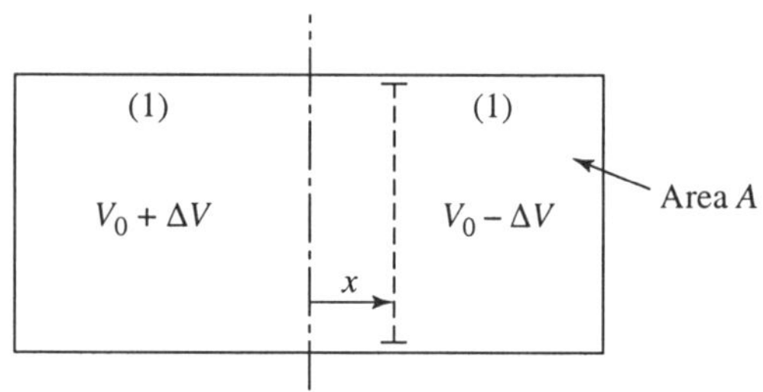

Each compartment contains an equal number N of solute particles in water. At time $t = 0$, the piston is in a position somewhat to the right of the central position, so that volume (1) is $V_1 = V_0 + \Delta V$, and volume (2) is $V_0 - \Delta V$.

(a) Describe qualitatively what will happen to the position of the piston, and why.

(b) Explain why the net fluid pressure p will be the same in both compartments at all times.

(c) Give an expression for the volume flow $A J_v$ of water across the membrane, in terms of L_p and the concentrations C_1 and C_2 of solute in the two compartments.

(d) Find a differential equation for the time-dependence of ΔV.

(e) Obtain an expression for the time constant τ_0, that characterizes the rate at which the system approaches equilibrium.

33. Volume flow through capillary walls. On average, there is no net out- or inflow of fluid from blood capillaries into interstitial space. But since the pressure difference across the capillary wall is not constant along the capillary, there is *localized* in- and outflow of fluid. The two figures below illustrate this (see also Vander, Sherman, and Luciano, *Human Physiology* [11, p. 275], and Landis and Pappenheimer [26] in the list of references).

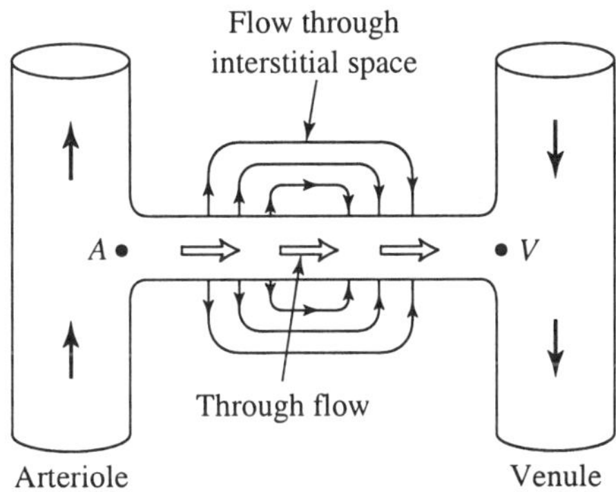

Flow pattern around capillary linking an arteriole and venule.

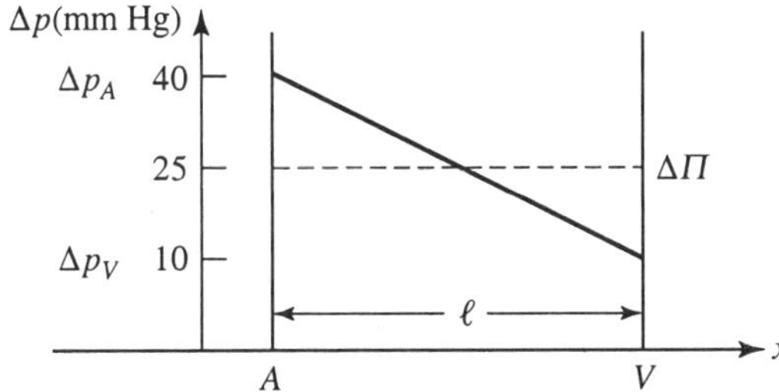

Pressure difference Δp across the capillary wall between blood plasma and interstitial fluid. $\Delta \Pi$ is the osmotic pressure difference, due to plasma proteins.

(a) Explain qualitatively the flow pattern sketched in the first figure.

(b) The difference $p_A - p_V = \Delta p_A - \Delta p_V$ is responsible for driving blood through the capillary. Calculate the flow rate J (in cm^3/s) of blood through the capillary, using the values

$$a = \text{capillary radius} = 5 \times 10^{-4} \text{ cm},$$
$$v = \text{flow velocity} = 0.05 \text{ cm/s}.$$

(c) The outflow through the capillary wall, per unit wall area, is given by

$$J_v = L_p(\Delta p - \Delta \Pi)$$

Calculate the total outflow of fluid through the left half section of the capillary wall, using a mean value of $\Delta p - \Delta \Pi$ of 7.5 mmHg. The length of the capillary is 0.1 cm, and L_p is given by Table 2.14 as 6.9×10^{-6} (cm/s)/atm.

(d) Find the fraction of the fluid that moves from A to V by going into the interstitial space and back into the capillary.

(e) Assume that the blood plasma osmotic pressure is lowered by 5 mmHg due to a deficiency of plasma proteins. What would be the total accumulation of fluid in interstitial space per day, given that the total area of open capillaries is about 250 m^2?

34. Transfer of oxygen from blood capillaries to body tissue. Blood flows through a capillary of length l with a velocity v. It carries oxygen which is free to diffuse out of the blood under the influence of a concentration gradient. The capillary wall has a radius r_0 and a permeability for oxygen equal to p.

(a) If the blood were not flowing, show that the time constant τ for the depletion of oxygen from the blood through the capillary wall is equal to

$$\tau = \left(\frac{r_0}{2p}\right).$$

(b) For efficient oxygen transport, the time that the flowing blood remains in the capillary should be $\gtrsim 2\tau$.

(i) Obtain a formula for the maximum speed of blood flow in the capillary.

(ii) Using $l \sim 0.1$ cm, $r_0 = 0.4 \times 10^{-3}$ cm, and $p = 3 \times 10^{-4}$ cm/s, compute numerically the expected speed of blood flow. (The value of p is only an estimate.)

(c) In order for oxygen to effectively nourish tissues, the distance δ between blood capillaries must be about equal to the distance that an oxygen molecule can diffuse in time τ. Use this criterion to obtain a formula for δ in terms of the diffusion coefficient (D) of oxygen in tissue, r_0 and p.

(d) Use $D = 2 \times 10^{-5}$ cm^2/s and the parameters given in part (b)(ii) above to calculate numerically the expected spacing of capillaries in the vascular bed. NB. Experimentally $v \simeq 0.04$ cm/s, $\delta \approx 0.025$ cm.

35. Bean's experiment. (solute flow for $\sigma = 0$). In Bean's experiment, the results of which are reproduced in Fig 2.25, a volume flow J_v is generated by a hydrostatic pressure difference Δp across the porous membrane. On side (1) of the membrane, radioactive water (HTO) is admixed in a small concentration C_1. The HTO concentration on the other side (2) is kept close to zero by stirring.

(a) Assuming fluid flows with velocity v through the pore from side (1) to side (2), the flow density j_s (particles/cm^2 s) of HTO particles is

$$j_s = C_s(x)v - D\frac{\partial C_s}{\partial x}.$$

In stationary flow, j_s and v are independent of x. Show that in this case, the HTO concentration $C_s(x)$ has the profile

$$C_s(x) = \frac{j_s}{v}\left(1 - e^{v(x-l)/D}\right),$$

$l = \Delta x$ being the length of the pore, and $C_s(l) \equiv C_2 = 0$ is assumed.

(b) j_s may be expressed in terms of $C_1 = C_s(x = 0)$ and v. Show that this relation is

$$j_s = C_1\left(\frac{v}{1 - e^{-vl/D}}\right).$$

(c) Plot j_s as a function of the bulk velocity v. v is proportional to the pressure difference Δp across the membrane. Show that this curve has the form drawn in Figure 2.25.

(d) Consider the limiting cases:

(i) $\dfrac{vl}{D} \gg l$; and

(ii) $\dfrac{vl}{D} \to 0$.

Show that in (i):

$$j_s \simeq C_1 v$$

and in (ii):

$$j_x \simeq D\frac{C_1}{l} = D\left(\frac{\Delta C}{\Delta x}\right).$$

This substantiates the statements contained in the text in equations (2-106) and (2-107).

36. Renkin's ultrafiltration experiment (Renkin [6]). This is an experiment similar to Bean's (Problem 35), except that Renkin uses a solute with a finite reflection coefficient σ; Bean uses HTO, for which $\sigma = 0$. The problem is: Assuming that a solution, containing a solute at concentration C_1, is driven by hydrostatic pressure through a membrane, with reflection coefficient $\sigma \neq 0$ for the solute particle, what is the concentration of the fluid collected on the other side of the membrane? The hydrostatic pressure is adjusted so as to provide a constant flow velocity v in the pore. We break down this question into three parts. Consult Figure 2.54 of the text, for the notation used.

(a) Let $C'(x)$ be the solute concentration in the pore. If fluid flows at speed v, the solute flow density (particles/cm^2 s) j_s is then given by

$$j_s = C'_s(x)v - D\frac{\partial C'_s}{\partial x},$$

j_s and v are constants independent of x in a stationary flow. $C'(x)$ is therefore not a linear function of x, but given by the equation

$$\frac{dC'_s}{dx} - \frac{v}{D}C'_s(x) = -\frac{j_s}{D}.$$

Verify that this equation has the solution

$$C_s'(x) = C_1' - (C_2' - C_1')\frac{(e^{vx/D} - 1)}{(e^{vl/D} - 1)},$$

C_1' and C_2' being the concentrations in the pore, at the left and right terminals, respectively, and l being the length of the pore ($l = \Delta x$).

(b) Verify that with this formula for C_s', j_s is indeed independent of x, and may be written as

$$j_s = v\left(C_1' + \frac{C_1' - C_2'}{e^{vl/D} - 1}\right).$$

(c) The relation between C_1', C_2' and the concentrations C_1, C_2 in the compartments are given by

$$C_1' = (1 - \sigma)C_1, \qquad\qquad (\alpha)$$
$$C_2' = (1 - \sigma)C_2, \qquad\qquad (\beta)$$

C_2 is the concentration of the *filtrate collected* on side (2). This concentration must therefore be given by

$$\frac{\text{number of particles passing/cm}^2\text{s}}{\text{volume of fluid passing/cm}^2 \text{ s}} = \left(\frac{j_s}{v}\right).$$

So we have

$$C_2 = \frac{j_s}{v}. \qquad\qquad (\gamma)$$

Insert (α), (β), (γ) into the expressions for j_s found in (b). Show that you obtain, for the ratio C_2/C_1 of the concentrations of filtrand (C_1) and filtrate (C_2):

$$\frac{C_2}{C_1} = \frac{1 - \sigma}{1 - \sigma e^{-vl/D}}.$$

Plot this curve as a function of the bulk speed v. (Renkin's data agree with this form, but do not reach large enough values of v to show the asymptotic value $C_2/C_1 = 1 - \sigma$.)

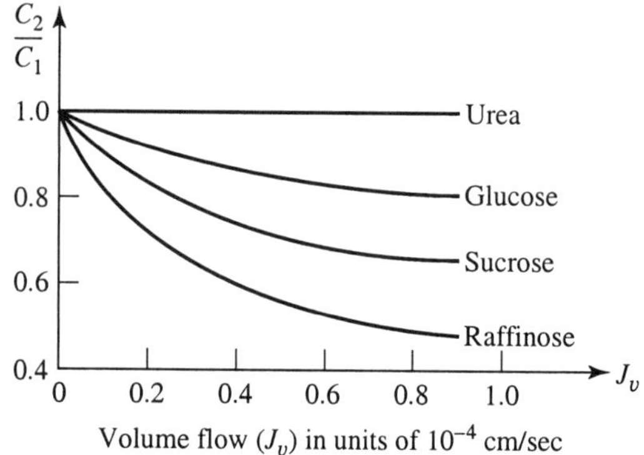

Qualitative plot of Renkin's data for Visking cellulose.

CHAPTER 3

Poisson Statistics

3.1 Introduction

The probability distribution we will discuss in this chapter was first found by Poisson in 1837 [1]. It found little scientific application until 1907 when W. G. Gosset [2] obtained it as a description of the numbers of corpuscles to be found in repeated samplings of drops of liquid taken from a much larger container having some average density of corpuscles. In 1910, E. Rutherford and H. Geiger were the first to discover [3] that the numbers of particles emitted during some fixed sampling time by radioactive atomic nuclei could be described using the Poisson distribution. In this case the Poisson distribution describes the probability that in a fixed interval of time, during which, on average, say, $\bar{n}$ particles are emitted, the actual number observed will be n, where n is any number from 0 to $n \gg \bar{n}$.

Since the beginning of this century the Poisson distribution has been found to describe a growing number of important phenomena in fields as diverse as physics, biology, engineering, operations research, epidemiology, and economics.

In physics the Poisson distribution describes, for example, such phenomena as:

(a) The thermionic emission of electrons from a heated metal surface (the cathode of an electronic vacuum tube).

(b) The photoelectric effect: the emission of electrons from the illuminated surface of a metal (e.g., the photo cathode of a photo tube).

(c) The intensity of sunlight scattered by fluctuations in the number density of molecules in the atmosphere—and the associated "blueness" of the sky.

(d) Radioactive emission of particles from unstable atomic nuclei.

In biology Poisson statistics describes such phenomena as:

(a) The detection of light by the photo receptors of the eye. The fluctuations associated with the number of photo responses determines the "glare sensitivity" of the eye and sets the minimum level of light that the eye is able to "see."

(b) The intensity of light scattered from solutions of biological macromolecules can be related to the number density of molecules in the solution by Poisson statistics. This in turn permits the measurement of the molecular weight of the scattering molecules [4].

(c) Poisson statistics are needed in the analysis of the numbers of resistant bacteria that survive attack by bacteriophage virus. By examining in detail the fluctuations in the numbers of survivors, it was conclusively demonstrated by S. Luria and M. Delbruck [5] that survival occurs by bacterial mutation and not by the inheritance of a characteristic acquired by exposure to the virus. The analysis of the variance of the distribution of surviving bacteria also permits a measurement of the rate of mutations.

In this chapter we will discuss in detail a number of these examples. Let us begin by obtaining the form of the Poisson distribution in a number of different ways.

3.2   Derivation of the Poisson Probability Distribution

The Poisson distribution initially emerged as a limiting case of the Bernoulli distribution applicable when the probability of a success (p) is a very small number. It should be pointed out at once that in certain physical situations the Poisson distribution emerges exactly as the description of the system, and does not emerge as a limiting case of the Bernoulli or binomial distribution at all. Keeping this in mind, let us first proceed by obtaining the Poisson distribution as a limiting case of the Bernoulli distribution applicable when p, the a priori probability of a success in a single trial, is very small.

3.2.A.   The Poisson Distribution and the Sampling of Particles from a Solution

A physical situation in which the Poisson distribution emerges from the Bernoulli distribution is in the problem [2] of counting particles in a small "sample volume" v, taken from a much larger volume V. Suppose that we have N particles in solution in volume V. The mean number density ρ is defined as

$$\rho \equiv \left(\frac{N}{V} \right). \tag{3-1}$$

Let us now focus our attention on the sample volume (v), and examine the number of particles that can be found in this small volume. We can find the probability, that any number k is to be found in v, by imagining that the N particles are placed, at random locations, one at a time, into the total volume V. Each time a particle appears in the sample region (v) a "success" is registered. This corresponds to a Bernoulli process with N trials. The probability of success in a single trial (p) is given by

$$p = (\bar{k}/N). \tag{3-2}$$

Here $\bar{k}$ is the average number of bacteria in v:

$$\bar{k} = \rho v. \tag{3-3}$$

Thus p, which is often called the "a priori probability" is given by

$$p = v/V. \tag{3-4}$$

The probability that there will be exactly k particles in the sampling volume when there are N particles altogether (i.e., N trials) is given by the Bernoulli probability distribution

$$P_N(k) = \binom{N}{k} p^k (1 - p)^{N-k}. \tag{3-5}$$

Here $\binom{N}{k}$ represents the binomial coefficients discussed in Chapter 1.

$$\binom{N}{k} = \frac{N(N-1)\cdots(N-k+1)}{k!}. \tag{3-6}$$

Using $p = \bar{k}/N$ from (3-2), we can reexpress (3-5) as

$$P_N(k) = \frac{1}{k!}(\bar{k})^k \left(1 - \frac{\bar{k}}{N}\right)^N \left\{\left(1 - \frac{\bar{k}}{N}\right)^{-k} \frac{N(N-1)\cdots(N-k+1)}{N^k}\right\}. \tag{3-7}$$

We now can easily see the form of this when N is very large compared to k or $\bar{k}$. In such a case the term in braces is approximately equal to unity. We also recall that, according to the definition of the exponential function,

$$\lim_{N \to \infty} \left(1 - \frac{\overline{k}}{N} \right)^N = e^{-\overline{k}}. \tag{3-8}$$

Using the approximation

$$\left(1 - \frac{\overline{k}}{N} \right)^N \cong e^{-\overline{k}} \tag{3-9}$$

for $N \gg 1$ and the condition $N \gg k, \overline{k}$, we see that $P_N(k)$ is, to a high approximation, independent of N, and depends only on k and $\overline{k}$. Calling $P_N(k)$ in this limit $P(k, \overline{k})$, we have

$$P_N(k) = P(k, \overline{k}) \cong \frac{(\overline{k})^k e^{-\overline{k}}}{k!}. \tag{3-10}$$

This is the Poisson probability distribution. It gives the probability that k particles will be counted in a particular sample if the average number to be expected in the sample is $\overline{k} = \rho v = N(v/V)$. The distribution given above will be accurate provided that both k and $\overline{k}$ are very small compared to the total number of particles in the entire solution. A simple immediate application of the Poisson statistics is an estimate of how concentrated the parent solution must be, to make it likely that a particle will in fact be found when examined. The probability that no particle will appear in a small sample is

$$p(0, \overline{k}) = e^{-\overline{k}}. \tag{3-11}$$

And the probability that one or more particles will appear on examination of a sample is

$$(1 - p(0, \overline{k})) = (1 - e^{-\overline{k}}).$$

In order to be likely to see one or more particles in a single sample, it is clearly necessary that $\overline{k}$ be at least of the order 1. Even in this case though there is a probability of $1/2.718 = 0.37$ that no particle will be seen. Thus on average about one sample in three will not show a particle even if on average there is one particle in each sample.

3.2.B. The Poisson Distribution and Radioactive Decay

The Poisson distribution also emerges as a limiting case of the Bernoulli distribution in the description of the numbers of particles emitted by radioactive nuclei. Suppose we have a box containing N radioactive nuclei. Each nucleus can undergo a spontaneous radioactive "decay" in which a particle is emitted from the nucleus. The actual process of emission is random in the sense that it can occur at any time. However, the probability q that any given nucleus has *not* decayed after a time t has elapsed is found to be given by the formula (see also homework Problem 2)

$$q = e^{-t/\tau_0}. \tag{3-12}$$

The parameter τ_0 is the "lifetime" of the particular nucleus. For example, a nucleus of radium has a "lifetime" $\tau_0 \sim 2300$ years against the emission of an α particle (a particle containing two protons and two neutrons). The a priori probability (p) that a nucleus will have emitted a particle at some time during the interval t is

$$p = (1 - q) = (1 - e^{-t/\tau_0}). \tag{3-13}$$

The probability distribution for the emission of k particles from the system of N nuclei is clearly a Bernoulli distribution involving N trials, i.e.,

$$P_N(k) = \binom{N}{k} p^k q^{N-k}.$$

On using (3-12) and (3-13), we find

$$P_N(k) = \binom{N}{k} (1 - e^{-t/\tau_0})^k (e^{-t/\tau_0})^{-k} e^{-Nt/\tau_0}$$

$$= \binom{N}{k} (e^{t/\tau_0} - 1)^k e^{-Nt/\tau_0}. \tag{3-14}$$

Now, if the observation time t is short compared to the lifetime τ_0, then t/τ_0 is a small number and

$$(e^{t/\tau_0} - 1) \cong \left(\frac{t}{\tau_0}\right). \tag{3-15}$$

Also the mean number $\bar{k}$ of counts in the interval t is

$$\bar{k} = Np \cong \frac{Nt}{\tau_0}. \tag{3-16}$$

Using (3-15) and (3-16) in (3-14), we find

$$P_N(k) = \binom{N}{k} \left(\frac{t}{\tau_0}\right)^k e^{-\bar{k}}. \tag{3-17}$$

Now, if the number of nuclei is large compared to the mean number of counts and to k, i.e., $N \gg k, \bar{k}$, we have $\binom{N}{k} \sim N^k/k!$ and hence (3-17) becomes, on using (3-16),

$$P_N(k) = \frac{(\bar{k})^k e^{-\bar{k}}}{k!} \equiv P(k, \bar{k}). \tag{3-18}$$

Thus the probability distribution for k counts if the mean number of counts in $\bar{k}$ is just the Poisson distribution. In the final equation above, we use the symbol $P(k, \bar{k})$ as equivalent to $P_N(k)$ to denote the fact that the probability distribution for k can be expressed solely as a function of $\bar{k}$, the mean number of counts in the sample interval.

Equation (3-18) simply describes the probability that exactly k particles will be emitted by the decaying nuclei during an observation period in which, on average, $\bar{k}$ decays occur, provided that the sampling time is short compared to the lifetime τ_0.

In this example we see once again that when the total number of elements are distributed with a low a priori probability (p) in space or time, the Poisson distribution emerges as a rather accurate approximation to the Bernoulli distribution. The Poisson distribution predicts the probability that exactly k events occur in the sample provided that the a priori probability that the event will occur for $N = 1$ is very small.

3.2.C. The Poisson Distribution and the Photoelectric Effect

In the previous examples the Poisson distribution emerged as a limiting case of the Bernoulli distribution. This occurs when the process under consideration consists of fixed, large numbers (N) of trials (particles) in which the a priori probability (p) of success in a single trial is very small compared to unity.

The Poisson distribution also emerges rigorously and independently of the Bernoulli statistics when we consider a random process in which individual events occur with no correlation relative to one another, apart from the fact that *on average* they appear at some constant rate. In such situations the total number of events is not defined, the events just keep coming, at an average rate that is constant in time.

One physical realization of this situation occurs in the "photoelectric effect": If a monochromatic light beam falls on the surface of a metal, the metal will emit electrons from the surface at a rate proportional to the intensity (I) of the light. It is possible to examine experimentally the statistical character of the time sequence of the emission of electrons by making the metal surface the cathode of a vacuum tube. The anode of the tube collects the emitted electrons, and the emission of each electron can be recorded as a function of time, as a sequence of current pulses as shown in Figure 3.1.

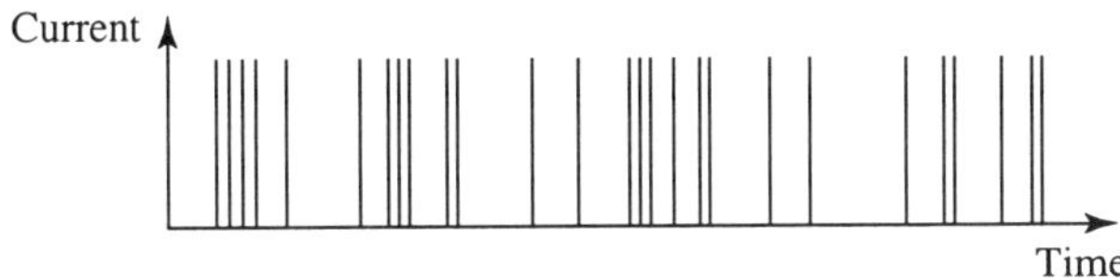

Figure 3.1. Sequence of photoelectrons emitted from a metal illuminated by a light beam.

In this sequence of electron emissions there is no correlation in the emission of successive electrons, except that *on average* the electrons are emitted at a rate $\overline{dn/dt}$ that is proportional to the light intensity. If one actually measures the number of "photoelectrons" (n_τ) emitted during some *sampling interval* of length τ, one will find that n_τ will fluctuate from sample to sample about the mean value $\overline{n}_\tau$ where

$$\overline{n}_\tau = \overline{(dn/dt)}\tau. \tag{3-19}$$

We will show that the Poisson statistics rigorously describes the probability distribution $P(n, \overline{n}_\tau)$ for the number n of photoelectrons observed in many measurements of n over a time interval τ.

It may be of interest to digress briefly from our analysis of the statistics of this problem to see how $\overline{dn/dt}$ is related to the incident light intensity. This relation was given by A. Einstein in 1905 as part of the theory of the photoelectric effect. (In that same year he also produced the special theory of relativity, and theory of Brownian movement!)

By extending M. Planck's idea that matter could *exchange* energy with radiation only in quanta of size $h\nu$, Einstein proposed that the energy in the light beam is *carried*, in fact, in quanta called "photons," each of whose energy is $h\nu$. (Here h is Planck's constant: 6.62×10^{-27} erg/s, and ν is the light frequency (in cycles/s).) The intensity I of a light beam is the energy flux of the beam, i.e., the energy carried per second across a unit cross section of the beam. Hence, a light beam of intensity I and cross-sectional area A brings $(IA/h\nu)$ photons onto the metal surface each second. The process of emission of electrons from the metal consists in the transfer of energy from the photon to the electron. If the photon energy $(h\nu)$ is large enough, the electron can obtain sufficient kinetic energy to escape the binding of the surrounding atoms and emerge from the surface of the metal. If each incident photon can succeed in emitting an electron, the emission rate would be on average $\overline{dn/dt} = (IA/h\nu)$. In actual fact, each photon is not successful in producing an emitted electron. On average only a fraction (ε) of the maximum emission rate actually occurs (e.g., the electron can absorb the photon but not be moving in a direction toward the surface). Typical values of ε are $\varepsilon \sim 0.05$. Thus we expect, on average, that

$$\overline{dn/dt} = (\varepsilon I A / h\nu), \tag{3-20}$$

where I is the intensity of the light beam (erg/cm^2 s), A is the cross-sectional area of the beam, ε the quantum efficiency, and $h\nu$ the energy of the photon. It is important to stress, however, that while (3-20) describes the average rate of emission, the detailed temporal sequence in the emission of electrons will be random even if I, the light intensity, is constant.

We now proceed to a calculation of $P(n, \tau)$, the probability that n pulses appear in a random sequence of events having a constant average rate $\overline{dn/dt}$. At the conclusion of our analysis we will see that $P(n, \tau)$ depends only on n and $\overline{n}_\tau$, the latter being equal to $(\overline{dn/dt})\tau$. Thus at the end of the calculation we will denote $P(n, \tau)$ as $P(n, \overline{n}_\tau)$.

One possible sequence of pulses is shown in Figure 3.2. To find the probability distribution desired, let us first consider the slightly longer time interval $(\tau + \Delta\tau)$. The probability that n particles appear in $\tau + \Delta\tau$ is denoted as $P(n, \tau + \Delta\tau)$. n particles can appear in $(\tau + \Delta\tau)$ by having either n in τ and none in $\Delta\tau$, or by having $n - 1$ in τ and 1 in $\Delta\tau$, or by having $n - 2$ in τ and 2 in $\Delta\tau$, etc. If we denote by $P(n - m, \tau/m, \Delta\tau)$ the probability that $n - m$ events occur in τ and m

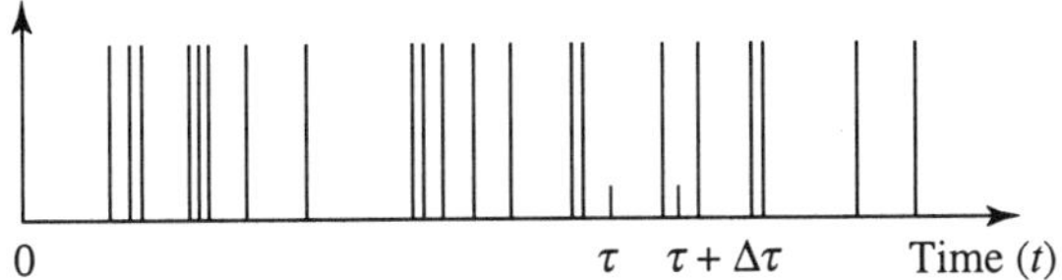

Figure 3.2. One possible random sequence of pulses.

in $\Delta\tau$, we see that

$$P(n, \tau + \Delta\tau) = P(n, \tau/0, \Delta\tau) + P(n - 1, \tau/1, \Delta\tau)$$
$$+ P(n - 2, \tau/2, \Delta\tau) + \cdots. \tag{3-21}$$

We assume from the outset that the arrival of m pulses in $\Delta\tau$ is completely independent of the occurrence of $(n - m)$ pulses in τ. Thus, by the multiplication rule of probabilities

$$P(n - m, \tau/m, \Delta\tau) = P(n - m, \tau)P(m, \Delta\tau). \tag{3-22}$$

Thus (3-21) can be written as

$$P(n, \tau + \Delta\tau) = P(n, \tau)P(0, \Delta\tau) + P(n - 1, \tau)P(1, \Delta\tau)$$
$$+ P(n - 2, \tau)P(2, \Delta\tau) + \cdots. \tag{3-23}$$

Let us now make the time interval $\Delta\tau$ very small so that on average we can expect that one particle is quite unlikely to appear in $\Delta\tau$. This can always be done if $\overline{dn/dt}$ is finite by making $\Delta\tau$ arbitrarily small. In this case

$$P(1, \Delta\tau) \ll 1. \tag{3-24}$$

Furthermore, by its definition, $P(1, \Delta\tau)$ is related to the average emission rate by

$$P(1, \Delta\tau) = \left(\overline{\frac{dn}{dt}}\right) \Delta\tau \tag{3-25}$$

and

$$P(0, \Delta\tau) = (1 - P(1, \Delta\tau)). \tag{3-26}$$

Also $P(2, \Delta\tau)$ is proportional to $(\Delta\tau)^2$. For convenience, we will define

$$\alpha \equiv \overline{dn/dt}. \tag{3-27}$$

Upon inserting (3-25)–(3-27) into (3-23), and retaining only terms linear in $\Delta\tau$, we find for small $\Delta\tau$ that

$$P(n, \tau + \Delta\tau) \cong P(n, \tau)(1 - \alpha\Delta\tau) + P(n - 1, \tau)\alpha\Delta\tau$$

or

$$\frac{P(n, \tau + \Delta\tau) - P(n, \tau)}{\Delta\tau} = \alpha P(n - 1, \tau) - \alpha P(n, \tau). \tag{3-28}$$

Thus in the limit $\Delta\tau \to 0$ we observe that the left-hand side in (3-28) is simply the derivative of $P(n, \tau)$ relative to τ, and we obtain a "differential recursion relation"

$$\left(\frac{dP(n, \tau)}{d\tau}\right) + \alpha P(n, \tau) = \alpha P(n - 1, \tau). \tag{3-29}$$

This equation permits us to deduce $P(n, \tau)$ if $P(n - 1, \tau)$ is known. This is a frequently useful mathematical procedure: one builds up the solution for any value of n, step by step, using a "recursion relation" starting with a known value of the solution for $P(0, \tau)$. The probability $(P(0, \tau))$ that no pulses occur in time τ is found easily by breaking τ into $\tau/\Delta\tau$ intervals of length $(\Delta\tau)$. The probability that no pulses occur in τ is the product of the probabilities that no pulses occur in each of the intervals of length $\Delta\tau$, i.e.,

$$P(0, \tau) = (P(0, \Delta\tau))^{\tau/\Delta\tau}. \tag{3-30}$$

But

$$P(0, \Delta\tau) = (1 - \alpha\Delta\tau), \tag{3-31}$$

$$P(0, \tau) = \lim_{\Delta\tau \to 0} \left(1 - \alpha\tau\left(\frac{\Delta\tau}{\tau}\right)\right)^{\tau/\Delta\tau}. \tag{3-32}$$

But the exponential e^{-x} is

$$e^{-x} = \lim_{N \to \infty} \left(1 - \frac{x}{N}\right)^N.$$

Thus, by making the correspondence $(\tau/\Delta\tau) = N$, we find

$$P(0, \tau) = e^{-\alpha\tau}. \tag{3-33}$$

Note that since

$$\alpha\tau = \frac{\overline{dn}}{dt}\tau \equiv \bar{n}_\tau, \tag{3-34}$$

where $\bar{n}_\tau$ is the average number of pulses in time τ, we see that the probability that no events occur in τ can be expressed as

$$P(0, \tau) = e^{-\bar{n}_\tau}. \tag{3-35}$$

The probability of no events diminishes as the negative exponential of the number of expected events.

With this result we can now deduce, using the recursion relation (3-29) for $P(n, \tau)$. For this purpose, it is convenient to recognize that (3-29) can be written as

$$\frac{d}{d\tau}(e^{\alpha\tau} P(n, \tau)) = \alpha e^{\alpha\tau} P(n - 1, \tau) \tag{3-36}$$

or, calling

$$f(n, \tau) \equiv e^{\alpha\tau} P(n, \tau), \tag{3-37}$$

we find that (3-36) is equal to

$$\frac{d}{d\tau}(f(n, \tau)) = \alpha f(n - 1, \tau). \tag{3-38}$$

Now when $n = 1$, the right-hand side is

$$\alpha f(0, \tau) = \alpha e^{\alpha\tau} P(0, \tau) = \alpha e^{\alpha\tau}(e^{-\alpha t}) = \alpha, \tag{3-39}$$

thus, we have for $f(1, \tau)$:

$$\frac{d}{d\tau} f(1, \tau) = \alpha$$

and

$$f(1, \tau) = \alpha \tau. \tag{3-40}$$

Note that the constant of integration is zero, since as $\tau \to 0$, $P(n, \tau) \to 0$ (for $n \geq 1$); the probability of observing one or more pulses is zero when $\tau \to 0$.

We now proceed to find $f_2(\tau)$: Using (3-38) and (3-40), we have

$$\frac{d}{d\tau} f(2, \tau) = \alpha f(1, \tau) = \alpha^2 \tau,$$

$$f(2, \tau) = \frac{(\alpha \tau)^2}{2!}. \tag{3-41}$$

Similarly

$$f(3, \tau) = \frac{(\alpha \tau)^3}{3!} \tag{3-42}$$

and

$$f(4, \tau) = \frac{(\alpha \tau)^4}{4!}.$$

In general, then

$$f(n, \tau) = \frac{(\alpha \tau)^n}{n!}. \tag{3-43}$$

To get $P(n, \tau)$, we remember from (3-37) that

$$f(n, \tau) \equiv e^{\alpha \tau} P(n, \tau), \tag{3-44}$$

thus, we find the desired general result for $P(n, \tau)$:

$$P(n, \tau) = \frac{(\alpha \tau)^n}{n!} e^{-\alpha \tau}. \tag{3-45}$$

This is the Poisson distribution. We can express this result in terms of $\bar{n}_\tau$, the mean number of counts in time τ, using the fact that

$$\alpha \tau = \left(\overline{\frac{dn}{dt}} \right) \tau = \bar{n}_\tau.$$

Finally, we can emphasize that $P(n, \tau)$ is the probability that n pulses will be counted when there are, on average, $\overline{n}_\tau$ pulses in the counting interval by defining

$$P(n, \tau) \equiv P(n, \overline{n}_\tau).$$

Thus, we have

$$P(n, \overline{n}_\tau) = \frac{(\overline{n}_\tau)^n e^{-\overline{n}_\tau}}{n!}. \tag{3-46}$$

The analysis which gave this result shows that the Poisson distribution did not arise as a limit of the Bernoulli distribution. It is the rigorous description, valid for any finite n, of the probability that n events will occur provided only that successive events occur independently and that there is some fixed, finite average rate of events.

3.3   Properties of the Poisson Distribution

Having obtained a closed mathematical formula for the Poisson distribution, we now examine it more carefully so as to acquaint ourselves more fully with its detailed properties. We begin by showing that the distribution is in fact normalized. We can compute the average value of the distribution. Following this, we will show how in actual experiments one can measure, with any desired accuracy, the average number of events even though the number measured in each counting interval fluctuates around the average.

The graphical form of $P(n, \overline{n}_\tau)$ will be shown for various values of $\overline{n}_\tau$, and we will demonstrate that for large values of $\overline{n}_\tau$ the Poisson distribution becomes the same as the "normal" or Gaussian distribution.

3.3.A.   Normalization and Average Value of the Poisson Distribution

The probability that any event (i.e., $n = 0$, or 1, or 2, or 3...) will occur must be equal to unity. We examine if this is the case. The probability that any event occurs is

$$P(0, \overline{n}_\tau) + P(1, \overline{n}_\tau) + P(2, \overline{n}_\tau) + \cdots = \sum_{n=0}^{\infty} P(n, \overline{n}_\tau) = e^{-\overline{n}_\tau} \sum_{n=0}^{\infty} \frac{(\overline{n}_\tau)^n}{n!} = 1.$$

Since the series representation of the exponential function is

$$e^{\bar{n}_\tau} = 1 + \bar{n}_\tau + \frac{(\bar{n}_\tau)^2}{2!} + \frac{(\bar{n}_\tau)^3}{3!} + \cdots.$$

Thus we see that the Poisson distribution is indeed normalized correctly.

Let us compute the average value of n in the Poisson distribution. If our previous treatment was correct, this average should be exactly the same as the parameter $\bar{n}_\tau = \overline{(dn/dt)}\tau$:

$$\text{average }(n) \equiv \sum n P(n, n_\tau) = \sum_{n=0}^{\infty} \frac{n(n_\tau)^n e^{-n_\tau}}{n!},$$

$$= e^{-\bar{n}_\tau} \sum \frac{n(\bar{n}_\tau)^n}{n!} = e^{-\bar{n}_\tau}\left(\bar{n}_\tau + \frac{(\bar{n}_\tau)^2}{1} + \frac{(\bar{n}_\tau)^3}{2!} + \frac{(\bar{n}_\tau)^4}{3!} + \cdots\right)$$

$$= \bar{n}_\tau e^{-\bar{n}_\tau}\left(1 + \bar{n}_\tau + \frac{(\bar{n}_\tau)^2}{2!} + \frac{(\bar{n}_\tau)^3}{3!} + \cdots\right) = n_\tau e^{-\bar{n}_\tau} e^{\bar{n}_\tau}$$

$$= \bar{n}_\tau.$$

3.3.B.   Fluctuations of n Around the Average: Accurate Measurement of the Average Number of Events

Suppose we actually measure the number of events in some interval τ. The actual number n observed in each sample will fluctuate up and down around the average value. We now compute how large this deviation is from the mean, on average. Since we can expect on average that n is as likely to be greater than $\bar{n}_\tau$ as less than $\bar{n}_\tau$, we see that $\overline{(n - \bar{n}_\tau)} = 0$, and so this is not a useful measure of the magnitude of the average deviation from the average. We need a measure of the absolute magnitude of the difference between n and $\bar{n}_\tau$. A convenient, easily computed, measure of the magnitude of this deviation is the square root of the average square deviation. In short, this is called the root mean square (rms) deviation, viz:

$$\text{rms deviation} = \sqrt{\overline{(n - \bar{n}_\tau)^2}}. \tag{3-47}$$

This tells us how much a typical measured value of n will deviate from the average $\bar{n}_\tau$. We now show how this root mean square deviation depends upon $\bar{n}_\tau$, the average number of counts in the interval. For this purpose we consider first the mean squared deviation

$$\overline{(n - \overline{n}_\tau)^2} = \overline{n^2 - 2n\overline{n}_\tau + (\overline{n}_\tau)^2} = \overline{n^2} - 2(\overline{n}_\tau)^2 + (\overline{n}_\tau)^2.$$

Thus

$$\overline{(n - \overline{n}_\tau)^2} = \overline{n^2} - (\overline{n}_\tau)^2 \tag{3-48}$$

and we need only to compute the average of n^2 to find the mean square deviation. The mean square value of n in the Poisson distribution is

$$\overline{n^2} = \sum_{n=0}^{\infty} n^2 P(n, \overline{n}_\tau) = \sum_{n=0}^{\infty} n^2 \frac{(\overline{n}_\tau)^n e^{-\overline{n}_\tau}}{n!},$$

$$\overline{n^2} = e^{-\overline{n}_\tau} \sum_{n=0}^{\infty} \frac{n^2 (\overline{n}_\tau)^n}{n!}. \tag{3-49}$$

Writing this out gives

$$\overline{n^2} = e^{-\overline{n}_\tau} \left\{ \overline{n}_\tau + \frac{2^2}{2!}(\overline{n}_\tau)^2 + \frac{3^2}{3!}(\overline{n}_\tau)^3 + \frac{4^2}{4!}(\overline{n}_\tau)^4 + \cdots \right\}$$

$$= e^{-\overline{n}_\tau}\overline{n}_\tau \left\{ 1 + 2\overline{n}_\tau + \frac{3(\overline{n}_\tau)^2}{2!} + \frac{4(\overline{n}_\tau)^3}{3!} + \cdots \right\}$$

Now note that the term inside the brackets is equal to the sum $(e^{\overline{n}_\tau} + \overline{n}_\tau e^{\overline{n}_\tau})$. We see this by rewriting the brackets as

$$\left\{ \right\} = 1 + \overline{n}_\tau + \frac{(\overline{n}_\tau)^2}{2!} + \frac{(\overline{n}_\tau)^3}{3!} + \frac{(\overline{n}_\tau)^4}{4!} + \cdots$$

$$+ \overline{n}_\tau + \frac{2(\overline{n}_\tau)^2}{2!} + \frac{3(\overline{n}_\tau)^3}{3!} + \frac{4(\overline{n}_\tau)^4}{4!} + \cdots,$$

$$\left\{ \right\} = e^{\overline{n}_\tau} + \overline{n}_\tau e^{\overline{n}_\tau}.$$

Thus we find

$$\overline{n^2} = e^{-\overline{n}_\tau}\overline{n}_\tau(e^{\overline{n}_\tau} + \overline{n}_\tau e^{\overline{n}_\tau}) = (\overline{n}_\tau + (\overline{n}_\tau)^2) \tag{3-50}$$

Thus, on using (3-48) and (3-50), we find

$$\overline{(n - \overline{n}_\tau)^2} = \left[\overline{n}_\tau + (\overline{n}_\tau)^2\right] - (\overline{n}_\tau)^2$$

or, finally,

$$\overline{(n - n_\tau)^2} = \overline{n}_\tau. \tag{3-51}$$

The mean square deviation of n from $\overline{n}_\tau$ is $\overline{n}_\tau$ itself, and the root mean square deviation of n from $\overline{n}_\tau$ is therefore the square root of $\overline{n}_\tau$, i.e.,

$$\sqrt{\overline{(n - \overline{n}_\tau)^2}} = \sqrt{\overline{n}_\tau}. \tag{3-52}$$

Thus, suppose, for example, the counting interval contains, on average, $\overline{n}_\tau = 100$ events. Successive measurements of n in this interval will give values of n that, on average, differ from 100 by about ± 10. Thus, successive measurements will give numbers like 105, 93, 115, 89, 92.... How can we obtain from such a fluctuating series of numbers an accurate determination of the average? Intuitively, we sense that by taking the average of many samples we will more accurately determine the $\overline{n}_\tau$. Let us analyze this sense carefully.

Suppose we accumulate M measurements of n in interval τ. This corresponds in essence to counting for an interval of length $M\tau$. If we add together the values of n obtained in each of the M intervals, we will get a number that will depart from the actual average $M\overline{n}_\tau$ by an amount $\sim \pm\sqrt{M\overline{n}_\tau}$, the root mean square deviation.

The average value of n_τ obtained after M samplings $(\overline{n}_{\tau M})$ is then a number that is likely to be in the range

$$\overline{n}_{\tau M} = \left(\frac{M\overline{n}_\tau \pm \sim \sqrt{M\overline{n}_\tau}}{M}\right) = n_\tau \pm \sim \sqrt{\frac{\overline{n}_\tau}{M}}. \tag{3-53}$$

The fractional uncertainty in the determination of $\overline{n}_\tau$ is defined as

$$\text{fractional uncertainty} \equiv \frac{\sqrt{\overline{n}_\tau / M}}{\overline{n}_\tau} = \frac{1}{\sqrt{M}}\left(\frac{1}{\sqrt{n_\tau}}\right). \tag{3-54}$$

Since the fractional uncertainty in a single measurement of n_τ is $\sim 1/\sqrt{n_\tau}$, we see that by making M samplings we diminish the uncertainty by a factor of $1/\sqrt{M}$.

Conversely, the accuracy of the determination of $\bar{n}_\tau$ (or, equivalently, (dn/dt)) increases only as the square root of the total number of samples measured.

Thus, in a process in which $\bar{n}_\tau = 100$, if we wish experimentally to determine $\bar{n}_\tau$ to an accuracy of say $\pm 1\%$, we must have

$$\frac{1}{\sqrt{M}} \frac{1}{\sqrt{\bar{n}_\tau}} = 0.01,$$

i.e., $M = 100$; one hundred samples of the number in interval $\bar{n}_\tau$ must be made even though the accuracy of a single sample is $\pm 10\%$. As we can see, obtaining more accurate information can be quite time consuming.

3.3.C.   Graphs of $P(n, \bar{n}_\tau)$

We present, in Figures 3.3–3.7, graphs of

$$P(n, \bar{n}_\tau) = \frac{1}{n!} (\bar{n}_\tau)^n e^{-\bar{n}_\tau}$$

as plots of P, as a function of n for values of $\bar{n}_\tau = 1, 2, 5, 10, 500$.

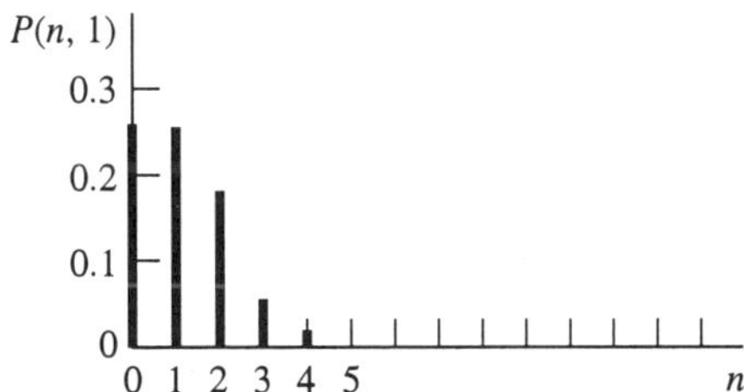

Figure 3.3. The Poisson distribution $P(n, \bar{n}_\tau)$ for $\bar{n}_\tau = 1$.

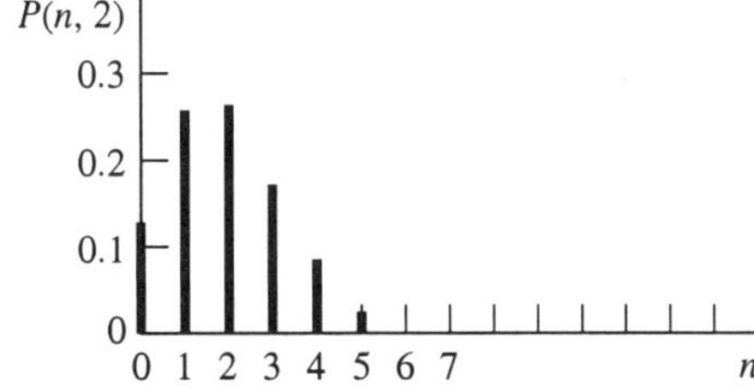

Figure 3.4. The Poisson distribution $P(n, \bar{n}_\tau)$ for $\bar{n}_\tau = 2$.

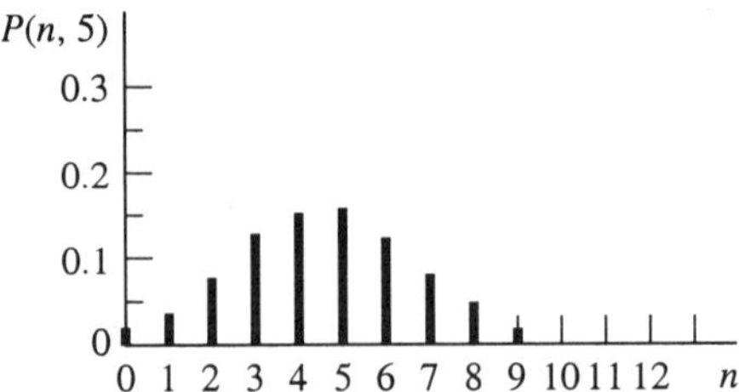

Figure 3.5. The Poisson distribution $P(n, \bar{n}_\tau)$ for $\bar{n}_\tau = 5$.

We observe that the peak value of $P(n, \bar{n}_\tau)$ occurs near $\bar{n}_\tau$. As $\bar{n}_\tau$ becomes greater than $\sim \bar{n}_\tau \sim 10$, the distribution becomes quite symmetrical about the average value $\bar{n}_\tau$. We may examine how the peak value of the probability changes with $\bar{n}_\tau$ by evaluating $P(n, \bar{n}_\tau)$ for $n = \bar{n}_\tau$:

$$P(\bar{n}_\tau) = \frac{(\bar{n}_\tau)^{\bar{n}_\tau}}{(\bar{n}_\tau)!} e^{-\bar{n}_\tau}.$$

But if n_τ is large, say greater than 10, we can use Stirling's approximation, namely,

$$(\bar{n}_\tau)! \cong \left(\frac{\bar{n}_\tau}{e}\right)^{\bar{n}_\tau} \sqrt{2\pi \bar{n}_\tau}. \tag{3-55}$$

Thus, approximately, we find

$$P(\bar{n}_\tau) \simeq \frac{1}{\sqrt{2\pi \bar{n}_\tau}}. \tag{3-56}$$

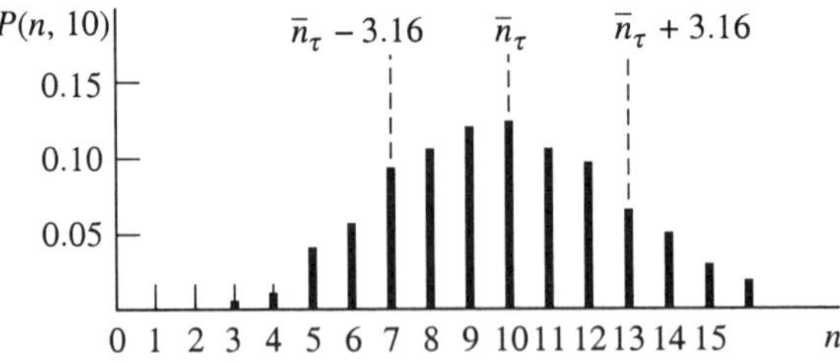

Figure 3.6. The Poisson distribution $P(n, \bar{n}_\tau)$ for $\bar{n}_\tau = 10$.

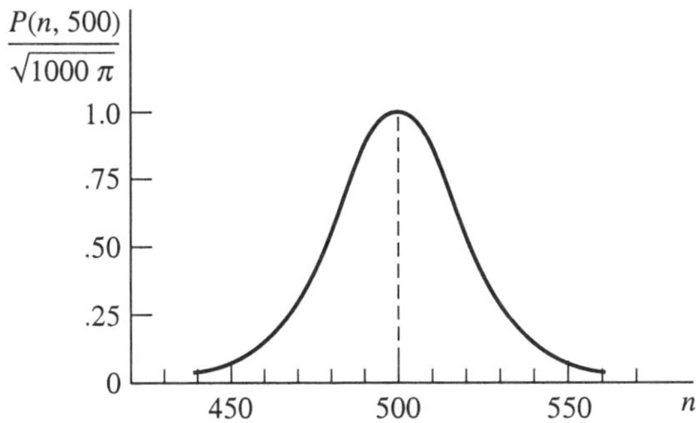

Figure 3.7. The Poisson distribution $P(n, \bar{n}_\tau)$ for $\bar{n}_\tau = 500$.

Thus, we see that the amplitude of the probability decreases as $1/\sqrt{2\pi \bar{n}_\tau}$ for $\bar{n}_\tau$ larger than, say, 10. The root mean square fluctuation in n that we computed above measures the *width* of the distribution about the mean. Thus, for example, for $\bar{n}_\tau = 10$ the root mean square width is ~ 3.16. This is indicated in Figure 3.6.

We observe that, in general, when $\bar{n}_\tau$ is the average number of events expected in the sample, the actual number that do occur can range widely about the average. As the graphs show, the probable range for measured values of n are between $\bar{n}_\tau - \sqrt{n_\tau}$ and $\bar{n}_\tau + \sqrt{n_\tau}$, at least for $\bar{n}_\tau \gg 1$.

When $\bar{n}_\tau$ is large in comparison with unity, say $\bar{n}_\tau > 10$, we can write the Poisson distribution in a different form—a form in which the symmetry of the distribution around $\bar{n}_\tau$ is explicitly clear, and the position of the points $\pm\sqrt{n_\tau}$ are easily located. This form is the so-called Gaussian or normal distribution. This distribution occurs very frequently in statistical physics, and it is appropriate to obtain it at this point.

3.3.D.   The Form of the Poisson Distribution for Large $\bar{n}_\tau$: The Normal, or Gaussian Distribution

In Chapter 2, Section 2.2.B, we saw that the Bernoulli distribution

$$P_N(n) = \binom{N}{n} p^n q^{N-n}$$

can be written accurately as

$$P_N(n) = \frac{1}{\sqrt{2\pi\, Npq}}\, e^{-(n-\bar{n})^2/2Npq}, \tag{3-57}$$

provided that

$$Np \gg 1$$

and

$$N \gg n, \qquad \bar{n} \gg 1.$$

In (3-57), $\bar{n} = Np$ is the average number of successes. This is the same as $\bar{n}_\tau$ used above, but here we drop the subscript τ. These conditions are consistent with those under which the Poisson distribution emerges from the Bernoulli distribution, i.e., $(\bar{n}/N) = p \ll 1$ and $N \gg n, \bar{n}$. The only restriction that must be placed on the Poisson distribution for it to be described by (3-57) is that the average number of

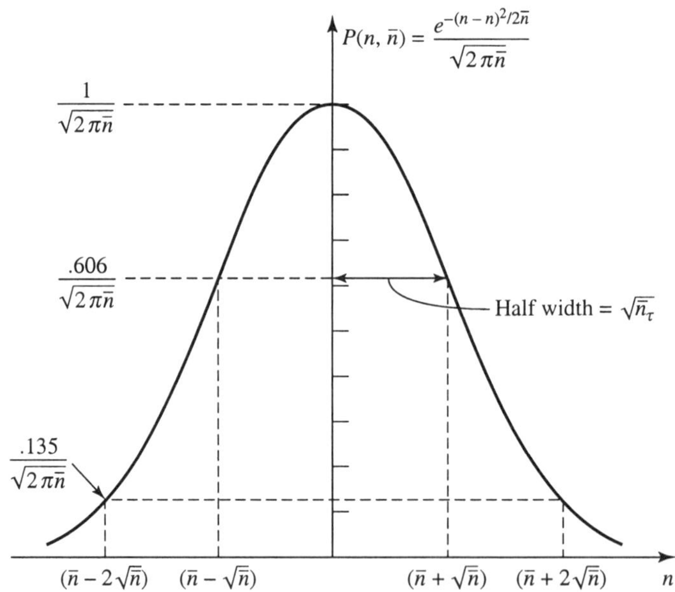

Figure 3.8. The Gaussian distribution plotted as a function of n in the vicinity of $\bar{n}$. This is also the form of the Poisson probability distribution in the case that $\bar{n}$ is large compared to unity.

successes $\bar{n} = pN$ must be large compared to unity. Thus, we see at once, since $Np = \bar{n}$ and $q \cong 1$, that the Poisson distribution, in the limit $\bar{n} \gg 1$, is given by the equation

$$P_N(n) \equiv P(n, \bar{n}) = \frac{1}{\sqrt{2\pi\bar{n}}} e^{-(n-\bar{n})^2/2\bar{n}}.$$

This distribution is called the Gaussian or the normal distribution. This is a symmetrical, bell-shaped curve centered around n. The area of the distribution is unity. Its height at maximum is $1/\sqrt{2\pi\bar{n}}$ and its width is $2\sqrt{n}$. The location of the points $\bar{n} \pm \sqrt{n}$ are shown in the schematic diagram of Figure 3.8.

3.4   Poisson Statistics and the Detection of Light by the Eye

The first step in the detection of light by the eye takes place on a molecular level. Photons in the incident light are absorbed by rhodopsin molecules (visual purple) located in the receptor cells in the retina. The detection of light at this first stage is a probabilistic process. The number of molecules that absorb the incident photons, during some sampling interval (τ), fluctuate in accordance with the Poisson statistics about the mean absorption number $\bar{n}_\tau$.

This fluctuation in the number of primary events in the light detection process plays a central role in the detectability of light at very low intensity: near the very threshold of vision. This fluctuation is also closely connected with our inability to "see" in the presence of light of high background intensity. By the word "see" we mean the ability to detect differences in light intensity coming from different points in the visual field. For example, we cannot "see" a slide photograph projected onto a screen when the room light is very bright. The explanation for this phenomenon and the associated "glare sensitivity" of patients who suffer from certain kinds of eye pathology is to be found in the statistical and fluctuating character of the first step in the light detection process: the absorption of light by the photoreceptor molecules.

The study of the statistical properties of the visual process, both by psychophysical measurements and by direct neurophysical measurements of the discharge pulses along the retinal neurons continues to play a vital role in the elucidation of the way in which the eye detects light and transmits the visual information to the brain.

In the present section we will examine how the Poisson statistics helps us to explain measurements of the detectability of light at the very threshold of vision.

We will also see how the Poisson statistics sets the "visual contrast threshold" required for "seeing" as a function of the intensity of the background light. Finally, we will discuss the character of the actual neuronal discharges that issue from the retinal neurons as a function of light intensity.

3.4.A.   The Detection of Light at the Threshold of Vision

In 1942, S. Hecht, S. Shlaer, and M. Pirenne published [6] an investigation of the detectability of light at the very threshold of vision. Their paper is a model of scientific clarity and completeness. It is a classic demonstration of the importance of statistical physics in analyzing a fundamental problem in physiology.

Hecht et al. were looking for the "minimum" number of photons that are capable of eliciting a visual response. In essence, their experiment consisted of allowing a measured number of photons in a short flash of light to fall upon the eye. The observer was asked to register "yes" or "no," depending on whether or not he saw the flash of light. If the number of photons is very small, the observer will register no light in each test flash. On the other hand, for flashes containing a number of photons much greater than the minimum detectable number, he will nearly always register a "yes." In the region between these two extremes we can construct a probability-of-seeing curve, also called [6] a "frequency-of-seeing curve." At each value of flash intensity, one measures the number (n) of yes responses in M trials. The probability-of-seeing curve is a graph of (n/M) as a function of $\bar{n}$, the mean number of photons in a flash. If the detectability of light was a perfectly deterministic process, then we would expect that the probability of seeing $(P(\bar{n})) = (n/M)$ would be a step function: zero for $\bar{n}$ less than the threshold and unity for $\bar{n}$ greater than the threshold minimum. On the other hand, if, as we expect from the probabilistic character of the absorption of a photon, there are fluctuations in the actual absorption of photons, even when the same light intensity is incident on the receptor molecules, then we will expect that the frequency of seeing curve will rise gradually from zero to unity as we pass through the value of $\bar{n}$ at which the detection of light starts to become definite. The Hecht et al. experiment determined this frequency-of-seeing curve and fixed the value of $\bar{n}$ at which the curve rises substantially from zero to unity.

We will shortly give their experimental results and their theoretical analysis of this frequency-of-seeing curve. However, before doing this, it is useful to examine how they arranged the viewing conditions of the observer so as to provide the most favorable conditions for the detection of light. An understanding of the conditions they established for their experiment will give us useful insight into important features of the anatomy and physiology of the eye. A very detailed discussion of these conditions is to be found in a book by T. Cornsweet [7].

(i)   Anatomical and Physiological Conditions for Maximum Sensitivity

Dark Adaptation.   It is a familiar experience that on turning off the light in one's bedroom, it is at first very difficult to see anything in the room. As time goes by, however, it becomes possible to "see" even if there is relatively little stray light coming in through the windows. This improvement in the ability to see in the dark is called "adaptation." Measurements of this effect show that during adaptation the sensitivity of the eye increases linearly at first with time. After about 10 min in the dark the sensitivity increases to a level about 500 times that which applied at the start ($t = 0$) of the adaptation period. After 10 min the increase in sensitivity becomes more gradual, and 20 min after the start of adaptation the sensitivity becomes constant at a level about 2000 times greater than at $t = 0$. To insure that the adaptation was as complete as possible, Hecht et al. arranged to have their subjects adapt in the dark for 40 min before beginning the experiment.

Location of Light Flash on the Retina.   The light receptor cells in the retina are described anatomically as rod cells and cone cells. The rod cells are sensitive at low levels of light intensity. The rod cells do not produce a color sensation, and at low light levels one cannot detect colors in the visual field.

The cone cells are color sensitive and are effective at high levels of light intensity. The cone cells are packed together at very high density at the fovea of the eye. This spot, the fovea, is on the optic axis of the cornea and the lens, and is the place where objects of interest are normally focused. (See Figure 3.9.) The high density of cone cells at the fovea permits a high resolution examination of the colored image.

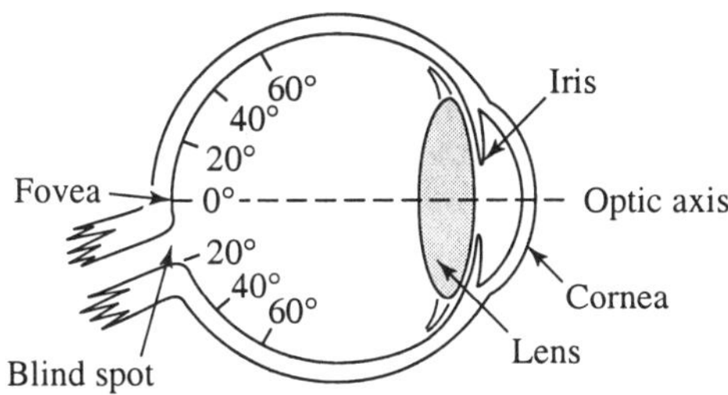

Figure 3.9. Elements of the anatomy of the eye.

The rod cell density is very low at the fovea, but peaks up to a maximum density at a location about 20° away from the fovea. In Figure 3.9 we see the elementary anatomy of the left eye showing the retinal positions marked off in degrees.

In Figure 3.10 we see the density of the rod and cone cells in units of thousands/mm^2 as a function of the angular position on the retina. This figure shows that to obtain maximum sensitivity for the detection of light at low intensity levels, it is important to focus the light in the left eye at the position of maximum rod density, i.e., at a point 20° toward the temporal direction on the retina. Thus to see faint lights at night, it is best not to look directly "at" the light for then its image will fall on the fovea where the cones are relatively insensitive to low light levels. Instead one should look away from the source. With the left eye one should have the source about 20° to the right of the optic axis, and with the right eye one should have the source about 20° to the left of the optic axis.

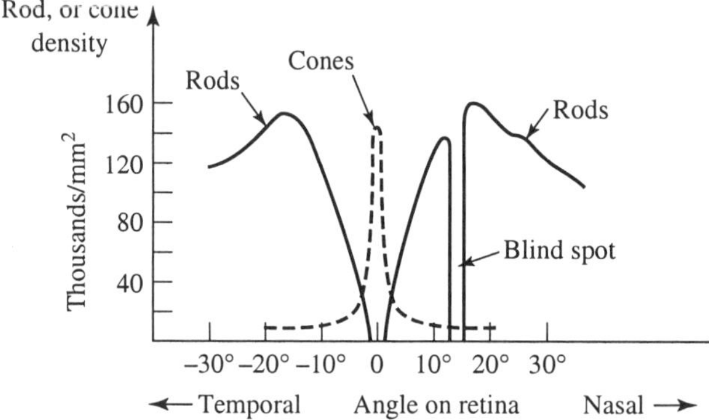

Figure 3.10. Density of rods and cones in the retina of the left eye (after M. Pirenne [9]).

In their experiments Hecht et al. arranged to "fixate" the eye with a weak red light. That is, the subject looks directly at a weak steady point of light. This establishes the orientation of the optical axis of the eye. The position of the test flash source was then arranged to be along a line 20° off this axis so that its image on the retina falls at the point of maximum rod density.

Size of Image of the Light Source on the Retina.   The rod cells can be thought of as being organized into regions called "receptive fields." The angular size of a receptive field is roughly 10 min of arc. If light is incident on the retina in a circular image that is smaller than this, the visual response remains the same, as the size of the image is reduced. That is, the visual response is independent of the size of the image and is dependent only on the total energy in the flash provided that the light is confined within a "receptive field." If the light is spread out into an image that overlaps several receptive fields, light in one field can inhibit the response of neighboring fields. This is called lateral inhibition.

The limitation on angular size is expressed in "Ricco's law": for equal amounts of radiant energy the visual response is the same independent of the illuminated area provided that the area is ~ 10 min of arc or smaller. In their experiment, Hecht et al. were careful to keep the retinal image of the source less than this size.

Temporal Duration of the Flash.   In Section 3.2.C we saw that the number of photons that cross an area A during a light flash of duration t is $\bar{n}_t = (IAt/h\nu)$. Here I is the light intensity, i.e., the power per unit area, h is Planck's constant, and ν is the frequency of the light. It is found experimentally (Bloch's law) that within a receptive field, if the flash duration (t) and the intensity change in such a way that it is a constant, then the visual response will be identical—provided that the flash time is less than 0.1 s [8]. If the flash duration time exceeds ~ 0.1 s, more light intensity is needed to evoke the same visual response. Thus, the eye appears to integrate and respond to the total number of incident photons provided that they occur within the integration time of 0.1 s. Photons that are detected after this characteristic integration period are not counted effectively. It is interesting to note that 0.1 s is about the same as the flicker-fusion time. This flicker-fusion effect permits us to see, as a continuously changing picture, individual pictures that appear one after another in times less than 0.1 s.

Hecht et al. arranged their flash duration time to be 0.001 s. Thus, they were well within the time domain in which the eye can temporally integrate the number of absorptions.

Color of the Test Flash.   The eye is most sensitive to light of wavelength 5100 Å: the blue–green. Hecht et al. used this as the color of the test light.

(ii)   The Frequency-of-Seeing Curve

Having arranged for the maximum sensitivity in the detection of light, Hecht et al. presented the observer with repeated test flashes, each of which contained some number $\bar{n}$ of quanta that entered the eye. For each value of $\bar{n}$ the number (n) of successes in seeing the flash was recorded out of a total of M trials at that light level. The quantity $(n/M) = P(\bar{n})$ is the probability that the observer will see a flash containing $\bar{n}$ photons at the cornea. In Table 3.1 we give the values of $P(\bar{n})$ versus $\bar{n}$ obtained by one of the observers, S. Shlaer [6].

Table 3.1. Frequency of Seeing ($P(\bar{n})$) Versus $\bar{n}$, the Number of Photons at the Cornea (from Hecht et al. [6]).

$P(\bar{n})$	$\bar{n}$
0	37
.12	59
.44	93
.94	149
1.0	239

The frequency-of-seeing curve is zero for small values of $\bar{n}$. However, in Figure 3.11, around $\bar{n} \sim 100$, the frequency-of-seeing curve rises and becomes about equal to unity when $\bar{n} \sim 200$. For convenience, Hecht et al. regarded the value of $\bar{n}$ for which $P(\bar{n}) = 0.60$ as the "minimum" number of quanta necessary for vision. Of course, the curve of $P(\bar{n})$ shows that, below this value, light can still be seen, and that $P(\bar{n})$ does not have a sharp step function character as one would

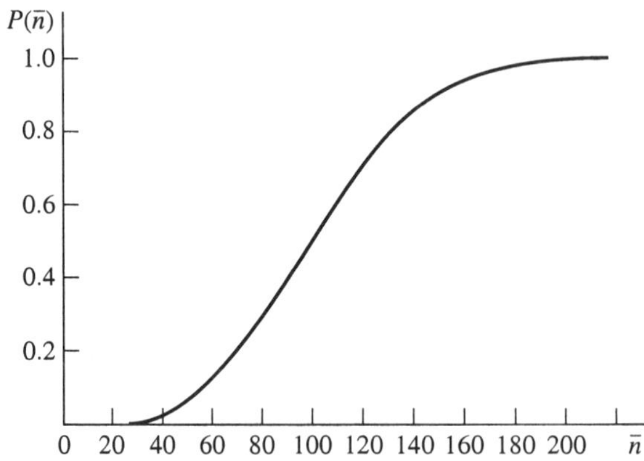

Figure 3.11. $P(\bar{n})$ the probability of seeing versus $\bar{n}$ at the threshold of vision (from Hecht et al. [6]).

expect for a go–no go situation. We shall, in fact, shortly examine the theory for the precise shape of $P(\bar{n})$, but at the moment let us examine the meaning of this loose idea of the minimum number of quanta necessary for vision. By obtaining frequency-of-seeing curves many times for each of the subjects (which were, by the way, the authors themselves), they found that the "minimum" number of quanta at the cornea, capable of eliciting the visual response, ranged from 50 to 150. How many molecules of the visual pigment are actually activated by these quanta?

On entering the cornea a small amount ($\sim 4\%$) of the incident light is reflected backward because of the difference between the index of refraction of air and the cornea. After entering the cornea only one-half of the 5100 Å light actually reaches the retina. The other half is either absorbed (primarily by the lens) or scattered by irregularities in the optical tissues of the eye. Having reached the retina, we may inquire now what percentage of the light is actually absorbed by the photosensitive molecules. Hecht et al. determined, using optical absorption measurements, that the absorption efficiency was actually about 20%, which is quite high in comparison with the quantum efficiency of the photoelectric effect in metals.

Thus, if $\bar{m}$ is the number of molecular absorptions (or, equivalently, the number of "excited" molecules) produced in the retina, we may relate $\bar{m}$ to $\bar{n}$, the number of photons arriving at the cornea, by

$$\bar{m} \cong \tfrac{1}{2} \times \tfrac{1}{5}\bar{n} \cong 0.1\bar{n}. \tag{3-58}$$

Thus, if one takes crudely the value of $\bar{n}$ at $P(\bar{n}) = 0.60$ as an indication of the threshold of vision, one finds that between 5 and 15 molecules excited in the visual receptors suffices to produce a visual response.

(iii)   Theory for the Shape of the Frequency-of-Seeing Curve

Now let us look more carefully at the precise shape of the frequency-of-seeing curve. By this analysis we will be able, independently, to establish a figure for the minimum detectable number of molecular absorptions in the retina required for vision. Let m_0 be the minimum number of excited molecules of rhodopsin required in order that the visual sensation be registered. Then, if $\bar{m}$ is the average number of excitations associated with the arrival of $\bar{n}$ photons, the probability that exactly m molecules are excited is given by the Poisson distribution $P(m, \bar{m})$:

$$P(m, \bar{m}) = \frac{(\bar{m})^m}{m!} e^{-\bar{m}}. \tag{3-59}$$

Since vision is presumed to be associated with the excitation of m_0 or more molecules, we see that the probability $(\mathcal{P}(m_0, \overline{m}))$ of seeing is

$$\mathcal{P}(m_0, \overline{m}) = P(m_0, \overline{m}) + P(m_0 + 1, \overline{m}) + P(m_0 + 2, \overline{m}) + \cdots \qquad (3\text{-}60)$$

or

$$\mathcal{P}(m_0, \overline{m}) = \sum_{m=m_0}^{\infty} P(m, \overline{m}). \qquad (3\text{-}61)$$

$\mathcal{P}(m_0, \overline{m})$ tells us the probability that m_0 or more molecules are excited by a flash that on average excites $\overline{m}$ molecules. By fitting this expression to the frequency-of-seeing curve, one should be able to determine whether this probabilistic view of the visual response is correct, and also one should be able to obtain the value of m_0.

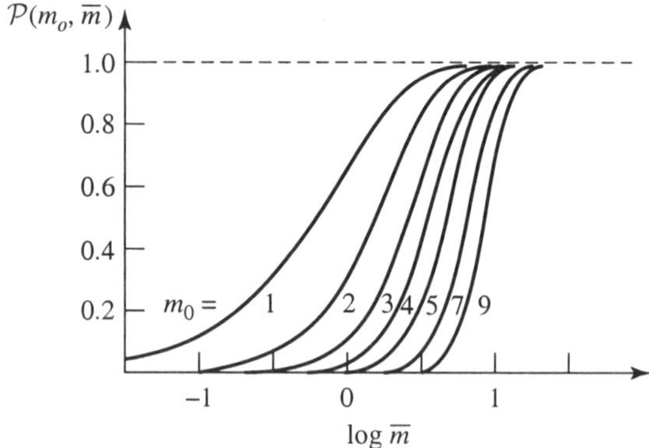

Figure 3.12. Graphs of $\mathcal{P}(m_0, \overline{m})$ for various values of m_0.

In Figure 3.12 we show a graph of $\mathcal{P}(m_0, \overline{m})$ versus $\overline{m}$ for various values of m_0. From these graphs we see that while $\mathcal{P}(m_0, \overline{m})$ varies qualitatively just like the frequency-of-seeing curve, the steepness of the rise between 0 and 1 depends markedly on the value of m_0. The curve of $\mathcal{P}$ for $m_0 = 9$, for example, rises much more markedly than does that for $m_0 = 1$.

To obtain the value of m_0, Hecht et al. compared the experimental data on the frequency-of-seeing curve plotted as a function of $\log \overline{n}$, with graphs of $\mathcal{P}(m_0, \overline{m})$ plotted as a function of $\log \overline{m}$. Since $\overline{m} = c\overline{n}$, the two curves can be compared regardless of the size of the coefficient c that relates the mean number of incident photons to the mean number of molecular absorptions $\overline{m}$. The comparison is achieved simply by displacing the two graphs horizontally relative to one another. In this way they obtained the fit shown in Figure 3.13. By carrying out this comparison for each of the subjects, they found that m_0, the minimum number of molecular absorptions required for vision, was in the range $5 < m_0 < 8$!!

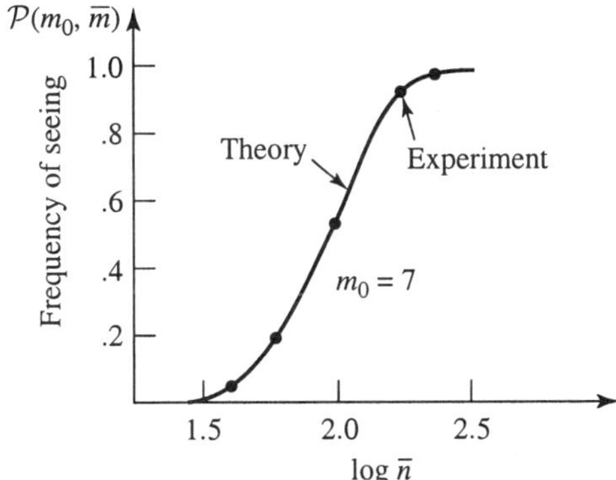

Figure 3.13. Comparison between the theoretical result for $\mathcal{P}(7, \overline{m})$ and the frequency-of-seeing data (after Hecht et al. [6]).

The extraordinary agreement between the data and theory shown in Figure 3.13 is a beautiful demonstration of the appropriateness of a quantitative analysis in the elucidation of the elements of a biological system.

Hecht et al. conclude their paper with the following remark [6].

The fact that for the absolute visual threshold the number of quanta is small makes one realize the limitation set on vision by the quantum structure of light. Obviously the amount of energy required to stimulate any eye must be large enough so as to supply at least one quantum to the photosensitive material. No eye need be so sensitive as this, but it is a tribute to the excellence of natural selection that our own eye comes so remarkably close to the lowest limit.

The fluctuations in the number of excited photosensitive molecules that are responsible for the shape of the frequency-of-seeing curve are also responsible for the uncertainty of vision at high light levels. In the following section we will explain this phenomenon and will then be able to understand the observed value for the minimum number of molecular absorptions (m_0). We will in fact show that m_0 must be ~ 6 in order to overcome latent spontaneous molecular excitations called "dark current noise" in the retinal photoreceptors.

3.4.B.　"Seeing" in the Presence of Background Light. Visual Contrast Thresholds and the Detection of Signals in the Presence of Noise

In order to "see," it is necessary that there be differences in intensity in the light emanating from different points in the field of view. It is also known that our ability to "see" can be greatly impaired if the eye is subjected to additional light of uniform illumination falling on the retinal image of this field. Thus, for instance, it may be perfectly possible to see through a slightly dirty window, but as soon as the sun shines on the window, our ability to see through it is greatly reduced.

The present section is devoted to a quantitative understanding of the phenomena of seeing in the presence of such "veiling lights." Let us begin by citing some additional examples: Perhaps the most dramatic and familiar example is that of viewing slides projected upon a screen in a well-lighted room. We know intuitively that under such conditions one cannot see the projection on the screen. It is necessary to darken the room so that light only comes from the projected slide. Despite its familiarity this is nevertheless a rather curious fact, because if we were to measure the *difference* in intensity of light emanating from two points on the screen, we would find that this difference is the same in the presence or in the absence of room lighting. Thus, the detectability of differences in intensity in the visual field does not depend solely on the absolute magnitude of this difference. Somehow, the *general level* of illumination is also important in the detection process.

A second example of the role of background light is our inability to see stars during the daytime. At nighttime, when the Sun has set and the sky is black, we can see stars. Clearly the stars do not "come out" at night and "go in" during the day. They are there shining with the same brightness, day and night. We are unable to see stars during the day because the general illumination of the sky, produced by the scattering of sunlight by the air molecules, inhibits their detectability. Again, intensity differences become undetectable in the presence of a high level of background or general ambient light.

Another simple example of this phenomenon is familiar to those of us who have to attend committee meetings in rooms that have large glass windows. It is a common experience to find oneself seated so as to face the light coming through the windows. One then discovers that he cannot see clearly the faces of colleagues whose backs are to the windows. In a similar way, at home, we may have a bright light behind a chair which makes it difficult to see the face of a person seated in the chair. We see, in these situations, the same phenomenon as before: the presence of a light source in the same part of the visual field as the object to be seen will seriously inhibit our ability to see the object.

Finally, it should be pointed out that patients who have pathologic eye conditions, such as early cataracts or corneal swelling that *scatter* large amounts of light onto the fovea, suffer from a condition called "glare sensitivity." They have great difficulty seeing in bright outdoor light. The objects in their visual field seem "washed out." However, indoors or in a darkened room they can see, fairly clearly, objects that are illuminated by a well-directed beam of light. Such persons also have great difficulties when driving a car at night. The headlights of oncoming cars scatter so much light onto the fovea that these patients cannot see the fainter illumination coming from the road ahead of them. They simply cannot see the road under these conditions and driving becomes a frightening and dangerous experience. Again, we see that a high level of light acts to limit the detectability of intensity differences in the visual field.

How may we quantitatively characterize the essential features of these phenomena experimentally? How is this phenomenon related to the fact that the eye appears capable of detecting as few as 5–8 photons absorbed in the visual pigment, and yet cannot detect intensity differences in the presence of a large background rate of absorption of photons? We will see in the following sections how to answer these questions, and we will learn that the actual detection of intensity differences corresponds to the detection of a signal in the presence of noise—just as in the detection of a signal in a noisy radio communication.

(i)   Experimental Measurement of the Visual Contrast Threshold

The essential features of the phenomena described above can be measured by the following experiment. Present to an observer an object consisting of two concentric illuminated circles (Figure 3.14). Let (I) be the intensity of light collected by the eye from the outer ring of the test source. Let ($I + \Delta I$) be the intensity of light collected by the eye from the inner circle. The experiment consists of setting I to some fixed level and then increasing ΔI from zero until it reaches a level at which the observer can reliably state that the inner circle is brighter than the light in the surrounding circle. The quantity (ΔI) is called the threshold intensity difference.

All the phenomena described previously are the result of the fact that ΔI, the threshold intensity difference, must *increase* as I, the background illumination, increases. A graph of the dependence of ΔI on I is called the visual contrast threshold curve.

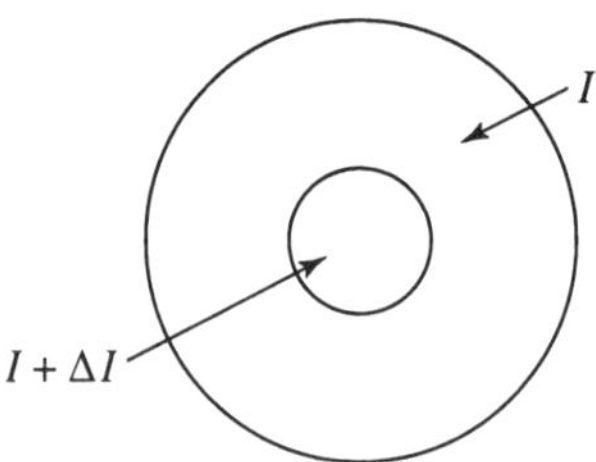

Figure 3.14. Essential target elements in the measurement of the visual contrast threshold.

In conducting a measurement of the visual contrast threshold curve, we must state the conditions under which the experiment is conducted, just as Hecht et al. did in their threshold measurements in which the background intensity was zero. The two most important factors involved in the visual contrast threshold measurements are the proper choices of the area of the source as it appears on the retina and the time during which the object is presented to the viewer. Our previous discussion of the Hecht et al. experiment suggests that the minimum threshold value of ΔI occurs when the size of the inner circle is small; i.e., less than 10 min of arc when viewing occurs in the region of maximum rod density. Also, to obtain the minimum ΔI, we expect that the change in intensity should be generated in the form of flashes whose time duration is short compared to the integration time (0.1 s) described in our discussion of the Hecht et al. experiment. On the other hand, it is clear that under *normal* viewing conditions the visual field is large, and the display or observation time is long. We will therefore show later the results of experiments in which large-area, long-time sources are used.

We now present the results of measurements in which the experimental conditions are adjusted to provide the minimum threshold values of ΔI. These experiments used short-time, small-area sources.

In 1957, H. Barlow measured the visual contrast threshold curve for a human eye [10]. In his experiment the outer circle of Figure 3.14 was large. It subtended an angle of 13° at the cornea. The light emanating from this circle and from the area of the small circle inside it had a *constant* intensity I as measured at the cornea. In the inner circle, of area 5 $(\text{min})^2$, the intensity was raised by means of intermittent flashes of 8.6 ms duration. This experiment is therefore analogous to the Hecht et al. experiment, except that the flashes occur against a background of steady illumination I, rather than against darkness. The wavelength of the light used was $\sim$ 5070 Å.

In Figure 3.15 we show Barlow's results for this short-time, small-area test flash stimulus. ΔI is presented in terms of ΔN, the number of threshold quanta in the flash incident on the cornea. The background intensity I is measured in units of the number of quanta incident on the cornea per second per (degree)^2. The logarithm of this quantity is plotted on the horizontal axis.

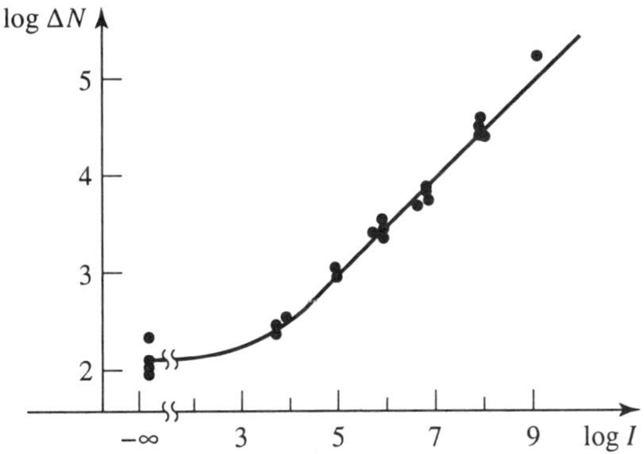

Figure 3.15. The visual increment contrast threshold curve. (ΔN, the minimum detectable number of photons in a flash is plotted versus the logarithm of the background light intensity for a short-time, small-area test source. I is measured in units of number of quanta/s(deg)^2 (from Barlow [10]).

To give the reader a feeling for the range of light intensities I with which one deals in this experiment, we present values of I that apply when the eye examines various sources through a 2 mm diameter entrance pupil. $I \sim 5 \times 10^8$ quanta/s(deg)^2 for light from a white sheet of paper illuminated by a 100 W bulb 1 m from the paper. Also, $I \sim 2.5 \times 10^5$ quanta/s(deg)^2 for light falling on a white paper illuminated by the light of the full Moon.

The data shown in Figure 3.15 can be fitted accurately by the equation

$$\Delta N = C(I + X)^{1/2}. \tag{3-62}$$

Once again, ΔN is the minimum number of photons in a flash that can be detected when the background intensity is I. The intensity I of the light from the background source is measured at the cornea, and represents the number of quanta/s(deg)^2 entering the iris of the eye.

Figure 3.15 and equation (3-62) show that at very low light intensities ($\log I < 3$) the number of incident quanta that can be detected is independent of I and is determined by the so-called "dark light" (X). The numerical value of X is

$$X = 1260 \text{ quanta/s (deg)}^2. \tag{3-63}$$

At this point it suffices to say that this quantity, the "dark light" X, represents the fact that the eye, in detecting increments of light, behaves *as if* on average 1260 quanta/s(deg)^2 were incident on the cornea, even when the actual background has zero intensity.

C is a numerical constant whose units are deg $\sqrt{s}$:

$$C \cong 3.3 \text{ deg (s)}^{1/2}. \tag{3-64}$$

It is important to note that (3-62) applies accurately to measurements of ΔN that range over about three orders of magnitude. Over this entire range the detectability of the photons in the test flash is proportional to the square root of the effective incident light intensity ($I + X$).

(ii)   The Noise Theory of the Visual Contrast Threshold Curve: For Short-Time, Small-Area Test Sources

The basic ideas which permit us to understand Barlow's measurements of 1957 were presented much earlier by A. Rose [11] (in 1948) and by H. De Vries [12] (in 1943). These ideas in turn are natural extensions of those established by Hecht et al. They are based on the conception that the absorption of light by the photopigment is a probabilistic process that is accurately described using the Poisson statistics.

In Barlow's experiment the incremental number of photons (ΔN) falling on the cornea must be detected in the presence of the continual arrival of many photons coming from the background source. Let us specify clearly what one means by the last phrase; i.e., detected *in the presence* of the continual arrival of many background photons. The actual detection of the increment ΔN occurs within one "receptive field" on the retina. Only those photons incident *within that receptive field* play a role in the detectability of the incremental quanta ΔN. If Ω is the angular area (in (degrees)^2) subtended at the cornea by a receptive field, then the number of quanta per second falling within the receptive field is

$$\left(\frac{d\bar{n}}{dt}\right) = I\Omega. \tag{3-65}$$

Our earlier discussion (Section 3.4.A(i)) of conditions required for maximum sensitivity in the Hecht et al. experiment showed that associated with the receptive field there is a characteristic integration time (τ) within which the receptive field in effect can sum up the number of photoabsorptions produced by the incident light. The total number of incident photons that affect the detection of the flash are photons which are contained in this integration time τ. We see then that the detection of ΔN test flash photons takes place effectively in the presence of a number $(\bar{n}_\tau)$ of background photons where

$$\bar{n}_\tau = (d\bar{n}/dt)\tau = I\Omega\tau. \tag{3-66}$$

These background photons are absorbed by the photopigment molecules in a random manner. On average, the number (m'_τ) of excited, i.e., absorbing, photopigment molecules associated with n_τ photons is

$$\overline{m'_\tau} = f\bar{n}_\tau = I\Omega\tau f. \tag{3-67}$$

Here f is the fractional number of corneal photons that actually excites the photoreceptive pigment molecules. (We saw from our discussion of the Hecht et al. experiment that $f \sim 0.1$. [See (3-58).] The process of photoabsorption results in a conformational change in the photoreceptor molecules under the action of the absorbed photon. Such a change can also occur spontaneously due to thermal agitation. Thus, we may expect that during the integration time there may be a small, constant number $\overline{m^s_\tau}$ of "spontaneous" excitations of the sensitive photopigment molecules even in the absence of light. Thus, on average, we may express the average number $\overline{m}_\tau$ of molecular excitations, in the receptive field and within the integration time, as the sum of $\overline{m'\tau}$ and $\overline{m^s_\tau}$:

$$\overline{m_\tau} = \overline{m'_\tau} + \overline{m^s_\tau} = I\Omega\tau f + \overline{m^s_\tau}. \tag{3-68}$$

For convenience we will express $\overline{m^s_\tau}$ in terms of x, a fictitious intensity of light needed to produce the spontaneous absorptions, by the equation

$$\overline{m^s_\tau} \equiv x\Omega\tau f. \tag{3-69}$$

Thus, we find $\overline{m_\tau}$ can be written as

$$\overline{m_\tau} = (I + x)\Omega\tau f. \tag{3-70}$$

The flash of light containing ΔN photons falls within one receptor field and on average excites $\overline{\Delta m}$ photoabsorptions, where

$$\overline{\Delta m} = f(\Delta N). \tag{3-71}$$

And now we come to the nub of the theory: In effect, the eye must detect $(\overline{\Delta m})$ incremental absorptions in the presence of $\overline{m_\tau}$ background absorptions. This can be a difficult process because the actual number (m_τ) of photoabsorptions is constantly fluctuating due to the random character of the absorption process. (3-70) only tells us the *mean value* of the number of photoabsorptions connected with the background light and the spontaneous molecular excitations. In fact, if one actually sampled the number of molecular excitations over intervals of length τ, one would find a graph like that shown in Figure 3.16. m_τ fluctuates up and down about the mean value of $\overline{m_\tau}$. This fluctuation about the mean is governed by the Poisson statistics. The mean square amplitude of the fluctuation (Δm_τ) is therefore given by

$$\overline{(\Delta m_\tau)^2} = \overline{(m_\tau - \overline{m_\tau})^2} = \overline{m_\tau}. \tag{3-72}$$

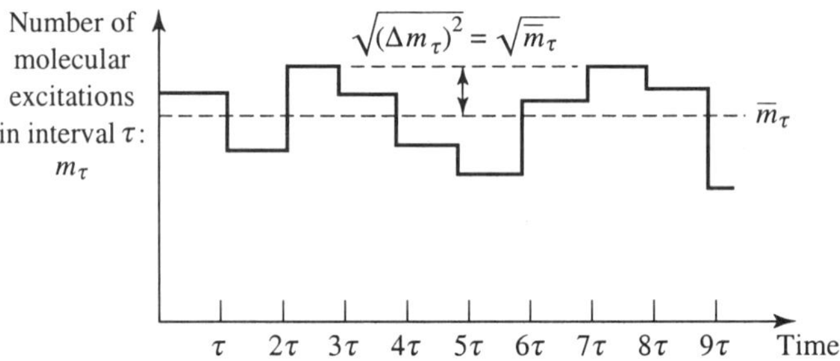

Figure 3.16. Schematic diagram showing the fluctuation in the number of molecules excited either "spontaneously" or by background light in one receptive field during one integration time τ.

The continual fluctuation of m_τ about its average represents a "noise" against which the eye is required to detect the "signal" consisting of the incremental photoabsorptions $\overline{\Delta m}$ associated with the flash. The "seeing" of the flash is thus a matter of detecting a signal in the presence of noise. This is only possible if the signal is clearly larger than the noise.

Let us call the signal ($\mathcal{S}$) the incremental number of photoabsorptions associated with the flash

$$S = \overline{\Delta m} = f\Delta N \tag{3-73}$$

and let us call the noise ($\mathcal{N}$) the root mean square fluctuation in the number of molecular excitations associated with the intensity I plus the spontaneous excitations, i.e.,

$$\mathcal{N} = \sqrt{\overline{m_\tau}} = \sqrt{(I+x)\tau\Omega f}. \tag{3-74}$$

With this notation, the incremental threshold needed for the detection of the flash in the presence of the background occurs when $\mathcal{S}$ exceeds $\mathcal{N}$ by some fixed factor K called the "threshold signal-to-noise ratio"

$$K \equiv (\mathcal{S}/\mathcal{N}). \tag{3-75}$$

K is a pure number whose magnitude can be expected to be greater than unity and smaller than ten.

Using (3-73) and (3-74) in (3-75), we obtain the prediction of the Rose–De Vries noise theory for the visual incremental threshold curve, namely,

$$K = \frac{(f\Delta N)}{\sqrt{(I+x)\tau\Omega f}}.$$

We may rearrange this equation to express ΔN as a function of I. This gives

$$\Delta N = \left(K\sqrt{\frac{\tau\Omega}{f}}\right)\sqrt{(I+x)}. \tag{3-76}$$

Comparison of this result with the experimental finding in (3-62) shows that the theory has the correct form required by the experiment.

Let us further examine whether the theory is capable of predicting the observed value of the coefficient C given by Barlow's experiment. We will then consider the case $I \to 0$, and will spell out the interpretation of X, the "dark light." Finally, we will apply this theory to the Hecht et al. experiment, and will come to understand why between five and eight molecular absorptions are required for the detection of light at the absolute threshold of vision.

The experimentally observed value of the coefficient C found in Barlow's experiment is $C = 3.3 \text{ deg (s)}^{1/2}$. The theoretical value of C can be found from (3-76) to be

$$C = K\sqrt{\frac{\tau\Omega}{f}}. \tag{3-77}$$

We can obtain a theoretical value for C from values of the parameters τ, Ω, f, and K. The quantities τ and f are given by the values

$$\tau = 0.1 \text{ s}, \tag{3-78}$$
$$f \cong 0.1. \tag{3-79}$$

These values have already been presented in previous discussions of the Hecht et al. experiment. The quantity Ω, which is the square of the angular diameter of the receptive field, must now be considered. Barlow [13] has investigated the size of Ω and has shown that the effective diameter of a receptive field is not a sharply defined number, but varies in different regions of the retina. He concludes [13] that the upper limit for the size of the receptive field is about

$$\Omega \sim 0.4 \text{ (deg)}^2. \tag{3-80}$$

The value of K, the threshold signal-to-noise ratio is a bit tricky to pin down unambiguously because it depends in part on the observer's own individual judgment as to when he is "certain" that he has seen the flash. It would appear possible that an observer would be just barely capable of seeing the test flash when $K = \mathcal{S}/\mathcal{N}$ is equal to 2. In order to have some confidence, however, in his judgment, the observer would very well need to have $K \cong 4$. A value of $K \sim 4$ would thus appear intuitively to be reasonable. We could substitute this value of K, and the values above, for Ω, τ, and f into (3-76) to predict C. Instead, let us alter this procedure slightly; let us use (3-76) to *compute* the value of K appropriate to the observer in Barlow's experiment. We will then examine whether the theoretically required value of K is, or is not, reasonable. In this way, we find

$$K = (C)_{\mathrm{exp}\,t} \Big/ \sqrt{\frac{\tau\Omega}{f}}, \tag{3-81}$$

$$K = 3.3 \text{ deg (s)}^{1/2} \Big/ \sqrt{\frac{(0.1)(0.4)}{(0.1)}} \text{ deg (s)}^{1/2},$$

$$K = 5.2.$$

This value of K (~ 5) is in fact in reasonable agreement with what we expected. We may therefore conclude that the noise theory of Rose and De Vries is indeed capable of explaining the observed size of the coefficient C in Barlow's experiment.

Let us now examine the term $(I + x)^{1/2}$ in (3-75). This has the same form as Barlow's experimental result $(I + X)^{1/2}$. We may thus interpret Barlow's constant X as being equal to our quantity x defined in (3-69). Barlow's "dark light" X then represents the number of photons/s(deg)^2 incident on the cornea that would be needed to produce the rate of *spontaneous* molecular excitations. The photopigment molecules are in fact *unstable* and can be excited even in the absence of light, albeit with low probability, by their thermal motion.

These spontaneous molecular excitations are responsible for limiting the number of detectable photons at the very threshold of vision, when $I = 0$. We can see this simply by examining the implications of the numerical value of X found experimentally: namely, $X = 1260$ quanta/s(deg)^2. Using this in (3-69), we can obtain the mean number $\overline{m^s_\tau}$ of *spontaneous* absorptions that occur within a receptive field and one integration time. The value of $\overline{m^s_\tau}$ is

$$\overline{m^s_\tau} = X\Omega\tau f$$

or

$$\overline{m^s_\tau} = 1260 \text{ quanta/s (deg)}^2 (0.4) \text{ (deg)}^2 (0.1) \text{ s } (0.1),$$

$$\overline{m^s_\tau} = 5.$$

Thus the mean number of "dark" or spontaneous molecular excitations is about 5. These dark pulses themselves produce noise because their actual number fluctuates from one integration period to the next. The root mean square fluctuation in the dark excitations is, in accordance with the Poisson statistics, given by

$$\left[\overline{(\Delta m_\tau^s)^2}\right]^{1/2} = \sqrt{\overline{m_\tau^s}}$$

Thus, the noise $\mathcal{N}$ associated with the existence of spontaneous molecular excitations in the dark is

$$\mathcal{N} = \sqrt{\overline{m_\tau^s}} = \sqrt{5}$$

In order to detect a flash *at the threshold of vision*, the mean number (m_0) of photoexcited absorptions (that is equal to the signal $\mathcal{S}$) must clearly exceed the noise by some threshold factor K:

$$\frac{\mathcal{S}}{\mathcal{N}} \geq K.$$

Since $\mathcal{S} = m_0, \mathcal{N} = \sqrt{5} = 2.2$, we see that at the threshold of vision the minimum number (m_0) of photoabsorptions that is distinguishable above the spontaneous noise is

$$m_0 = 2.2K.$$

In the Hecht et al. experiment the quantity m_0 was discovered to be a number between 5 and 8. We may, therefore, look upon the measurement of the minimum number of detectable photoabsorptions as a signal-to-noise discrimination in which the detection criterion was one in which the threshold signal-to-noise ratio (K) ranged between

$$2 < K < 4.$$

In the language of the present theory, the reason that between five and eight photoabsorptions are needed, at the minimum, for the detection of light in complete darkness, is that with this number the signal produced by the flash of light is between two and four times greater than the noise associated with spontaneous thermal excitations of the photopigment molecules.

We now conclude our discussion of the limiting threshold for the detection of short-time, small-area light sources, both in the presence of background light and in utter darkness, by summarizing our principal findings. For those readers whose interest in this subject has been aroused, we will also mention briefly, and give references to, recent neurophysiological measurements of the actual electrical response of the retinal nerve cells to the test flash and the background light.

Barlow's experiment and its analysis using the Rose–De Vries noise theory makes it clear that the ability to "see" short-time, small-area light flashes is in reality a *process* of detecting a signal in the presence of noise. The *signal* is the number $\overline{\Delta m}$ of excited photopigment molecules produced by the test flash. The noise is the continual random fluctuation in the number (m_τ) of photopigment molecules excited by the background light and the "dark light" during one integration time τ and one receptive field (Ω). The root mean square fluctuation in m_τ characterizes the *noise*. Its magnitude is given according to the Poisson statistics by $\sqrt{\overline{m_\tau}}$. The flash signal can be seen only if Δm clearly is larger than $\sqrt{\overline{m_\tau}}$. The ratio $(\Delta m / \sqrt{\overline{m_\tau}})$ is called the threshold signal-to-noise ratio (K), and experimentally it is found to be in the range between 2 and 5 depending on the observer. In sum then we cannot "see" a weak test flash in the presence of a background light because the background produces noise! Only when the test flash light signal is substantially larger than the noise, do we see it! The quantitative characterization of τ and Ω are very important in setting the precise limits for light detection. In fact, these psychophysical experiments lead us directly to asking important neurophysiological questions such as: What is the neuronal mechanism by which the retina *temporally integrates* the number of photoabsorptions over the time τ? Why is the value of $\tau \sim 0.1$ s? What is the physiologic and anatomic mechanism by which the retina *spatially integrates* photopigment absorptions over the area (Ω) of the receptive field? Can one detect experimentally the photopigment absorptions or the electrical impulses they produce in the retinal neural network? What are the statistics of these pulses? The beauty of the quantitative analysis, so far provided, lies in large part in the stimulus it gives to the framing of new questions whose answers lead to a deeper and richer understanding of the transduction of light into a neural response in the brain.

We close this section by giving a glimpse of the neurophysiological measurements of the electrical impulses produced by photopigment excitations in the retina. By using very thin microelectrodes, it is possible to detect and record the sequence of action potential spikes produced in the retinal ganglion cells of the cat by the absorption of light. (See [14], [15], [16], and [17].) These experiments permit one to measure the visual contrast threshold curve directly by observations of the minimum detectable change in the neural discharge rate. These neurophysiological measurements confirm the psychophysical measurements. They also show, among other things, that the steady discharge rate associated with light of intensity I is proportional not to I, but more closely to $\log I$. Thus, though the light intensity may change by a factor of 10^7, the rate of appearance of action potential spikes increases only by a factor of about 7! Quite similar results are found in the receptor cell of the eye of the horseshoe crab [18]. Functionally, this dependence of discharge rate on $\log I$ is very advantageous since the need for the large signal-carrying capacity of the nerve cells is thereby markedly reduced. If the neuronal discharge rate were directly proportional to the rate of photopigment excitations, the nerve cell would have to be capable of transmitting as few as ~ 5 pulses/s and as many as 10^7 pulses/s or more. The precise mechanism for the logarithmic relationship between the rate of action potential spikes, and the rate of photopigment excitations is one of fundamental importance in neurobiology.

(iii)　Visual Contrast Thresholds for Long-Time, Large-Area Test Sources

Barlow's crucial experiment of 1957 using small-area, short-time test sources of light showed [10] that the threshold (ΔI) for the detection of light varied as $(I + X)^{1/2}$. This square root dependence was the essential "tip-off" for an interpretation of these results in terms of noise theory. Before Barlow's experiments the insights afforded by the noise theory were largely ignored because neurophysiologists were convinced that the threshold intensity ΔI varied in direct proportion to I. This relationship $\Delta I \propto I$ is called the Weber–Fechner law. Since its publication in 1860 in Fechner's *Elemente der Psychophysik*, this experimental finding gradually grew to assume the proportions of a general law of psychophysics despite the fact that later experiments showed that at low and high light levels ΔI was, in fact, not linearly proportional to I.

In his 1957 paper, Barlow also demonstrated [10] clearly that the functional dependence of ΔI upon I depends critically upon the size of the test source and on the length of time of the test flash. Almost all of the many psychophysical studies of the threshold intensity levels, in fact, used *long-time, large-area* test sources. Such experiments do in fact demonstrate that over an appreciable range of light intensities ΔI is proportional to I, and further that the absolute magnitude of ΔI is considerably higher than that set by the fundamental noise limitation considerations presented previously. In Figure 3.17 we present an example of such measurements by Aguilar and Stiles [19] where we plot $\log \Delta I$ versus $\log I$ using a large-area, long-time (0.2 s) flash test source.

In order to obtain a qualitative appreciation in everyday terms for the magnitude of the background light intensity I represented in Figure 3.17, it is of interest to note that the brightness of a white sheet of paper illuminated by direct sunlight is such as to produce an intensity of $\sim 1 \times 10^6$ scotopic trolands in passing through

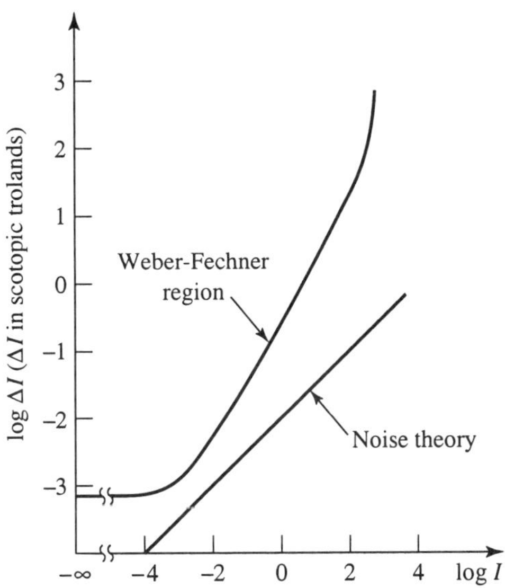

Figure 3.17. Increment threshold curve of large-area, long-time test sources. This curve is the mean result of five observers. We indicate the Weber–Fechner region $\Delta I \propto I$, and also show the prediction of the noise theory without dark light. After Aguilar and Stiles [19] and after Barlow [20]. I is measured in units called "scotopic trolands." A light intensity of one scotopic troland corresponds to 4.46×10^5 quanta/s $(\deg)^2$ of light having wavelength 5070 Å.

a 2 mm diameter pupil into the eye. Using these units of intensity, $\log I \sim 6$. Similarly, $I \sim 1 \times 10^3$ scotopic trolands for the light produced in the eye by a sheet of white paper illuminated by a 100 W bulb 1 m from the paper. In this case, $\log I \sim 3$. The data shown in Figure 3.17 is useful in describing the detectability of the sort of objects that come into our visual field under normal viewing conditions. As such, this data is helpful in quantifying clinically important phenomena, such as the debilitating "glare sensitivity" of patients suffering from pathologic intro-ocular light scattering [21].

From a fundamental point of view, however, if we wish to understand the data shown in Figure 3.17, it is necessary to understand more fully the temporal and spatial *interaction* between receptive fields. It is well known that the illumination of adjacent receptive fields *inhibits* the visual response of the individual fields. This phenomenon is known as lateral inhibition [22]. Lateral inhibition surely plays an important role in determining the magnitude and shape of the curve of Figure 3.17, and must be considered *in addition* to the noise fluctuations in an effort to explain the Weber–Fechner law. From this point of view, we see that the *fundamental* limits on the detection of light bring us to the functional relation $\Delta I \propto (I + X)^{1/2}$ applicable to individual receptive fields. The Weber–Fechner law $\Delta I \propto I$, so deceptively simple from the mathematical point of view, in fact includes the complicated effects of temporal and spatial interaction between receptive fields, and so far has not found a quantitative explanation.

It is important to appreciate that even with large-area, long-time sources, for which the Weber–Fechner law applies, the crucial element in "seeing" is still the detection of signal in the presence of noise. In the Weber–Fechner regime ΔI is no longer simply proportional to $I^{1/2}$. Instead *more* signal is needed to overcome noise than the minimum theoretically required. This indicates that the interaction between receptive fields reduces detectability of the source below that which is optimally possible with a small-area, short-time test source.

Let us conclude our discussion of this topic by relating the experiments described above to one of the visual phenomena described earlier. For example, consider the visibility of a star in the presence of skylight. In particular, let us imagine the conditions that apply at twilight as the sun sinks below the horizon. If we search the sky as the light wanes, the brightest star suddenly becomes visible. What actually happens in our retina? We detect the star when the number of photopigment excitation it produces in one integration time exceeds by about a factor of 5 the fluctuations in the number of absorptions produced by the surrounding skylight. The appearance of the star in the fading skylight is a detection of signal in the presence of noise! In a bright sky the noise far exceeds the signal and the star is not visible. At night the signal exceeds the noise, and we can see the stars.

3.4.C.   Phototransduction

The analysis presented above naturally leads to the quest for an understanding of the molecular mechanisms that connect the absorption of a photon by the rhodopsin molecules, that reside in discs in the retinal rod and cone cells, to the consequent neural response of these cells. Since the publication of the first edition of this book, considerable progress has been made in unraveling this vital first step of phototransduction. This first step determines the absolute sensitivity, and the temporal resolution of the overall visual response in the brain.

The progress that has been made in understanding the molecular basis for phototransduction was reviewed at a colloquium entitled "Vision: From Photon to Perception" held at the National Academy of Sciences in Irvine, California, May 20–22, 1995. The papers presented at this colloquium are to be found in the

Proceedings of the National Academy of Sciences (USA), Vol. 93, pp. 557–639, 1996.

Below, we briefly discuss some of the new findings and include at the end of the list of references for this chapter, papers published since 1970, that will assist the interested reader in following the exceptional scientific advances that have been made in understanding the biophysical and biochemical basis of the phototransduction process.

The earliest steps in phototransduction occur at the individual rod and cone photoreceptor cells in the retina. In these cells, the input signals are the photons of visible radiation. The output of these cells is the secretion of a chemical transmitter, probably glutamate, released from the cells' synaptic endings (Baylor [25]). These transmitters connect to the retinal bipolar and horizontal cells. The rate of release of this transmitter is controlled by the electrostatic potential difference between the inside and outside of the photoreceptor cell wall, viz., the transmembrane potential.

The question naturally arises as to the role of light in controlling this potential difference. In the absence of light, this potential difference amounts to about -40 mV. This difference is controlled by the permeability of cationic channels in the membrane through which flow sodium ions. The permeability of these channels is reduced when the receptor cells absorb light, and as a result, the transmembrane potential shows hyperpolarization, i.e., the voltage difference can increase to $\sim$ -65 mV. The hyperpolarization in turn reduces the rate of transmitter release at the synaptic terminal (see Baylor [25]). Modulation of the transmembrane potential results in a modulation of the rate of transmitter secretion.

Recent work (see Baylor [25]) has elucidated the molecular basis for the connection between light absorption by rhodopsin and the consequent reduction in channel permeability and transmembrane voltage in the vertebrate photoreceptor. This connection may be summarized as follows (see also Lamb [27], and Molday [28]). In darkness, the concentration in the cytoplasm of free cyclic guanosine monophosphate (cGMP) is relatively high. cGMP binds to the cationic sodium channels in such a way as to increase the permeability of each channel. The concentration of free cGMP, however, is controlled by a series of proteins that are bound to the disc membranes in the rod outer segment. These discs are in intimate contact with the cytoplasm. The membrane proteins are rhodopsin (the sensor molecule R), the G-protein, transducin (the transducer G), and finally, the protein phosphodiesterase PDE (the effector E).

The initial molecular step in phototransduction is the absorption of light by a ligand (11-*cis*-retinal) bound to rhodopsin (R), which results in the conversion of rhodopsin into an activated molecular (R^*) metarhodopsin II. Such activated R^* molecules are free to diffuse on the surface of the disc where they and the G

proteins, transducin, are located. Each activated R^* molecule, on encountering a G-protein, activates it catalytically. The activation process consists in removing from the transducin a GDP molecule and its replacement with a GTP molecule from the cytoplasm, thereby producing a molecule G^*. The two-dimensional diffusion of this activated G^* transducin protein brings it into contact with the effector molecule phosphodiesterase, and activates this effector molecule into a form that is capable of hydrolyzing the cytoplasmic cGMP. This hydrolysis very effectively reduces the level of cGMP in the cytoplasm, and as a result, the light sensitive channels close, leading to a drop in the rate of transmitter release.

Detailed quantitative studies have been made of the amplification and the kinetics of the G-protein cascade described above (see Pugh and Lamb [29] and Baylor [26]]. In this connection, it has been found that a single photoabsorption results in the blockage of about one million elementary charges from passage through the sodium channels. Furthermore, these findings provide a molecular explanation for the important integration time of the order of 0.1 s, that we have used above in our statistical analysis of the Hecht-Shlaer-Pirenne experiments and the Barlow experiments. Indeed, the characteristic times associated with the diffusion and activation of the proteins in the G-protein cascade has been shown to be ~ 0.2 s. Thus, the slow temporal response associated with human rod vision is directly attributable to the phototransduction process in the retinal photoreceptors. Finally, dark shot noise, which we have deduced from the psychophysical experiments, appears to be consistent with thermally driven spontaneous excitation of rhodopsin to metarhodopsin II.

3.5   The Luria–Delbrück Experiment: Mutation as the Source of Bacterial Immunity to Virus Attack

3.5.A.   Introduction

If a bacterial culture is brought into contact with bacteriophage virus particles, the viruses will attack the bacteria and kill them in a matter of hours. However, a small number of bacteria do survive the attack. These survivors will reproduce and pass on to their descendants their resistance to the virus. The form of resistance of the offspring of the surviving bacteria is that their surface does not adsorb the attacking virus. Bacterial strains can also be resistant to metabolic inhibitors, such as streptomycin, penicillin, and sulphonamide. If a bacterial culture is subjected to attack by these antibiotics, the resistant strain will emerge just as in the case of the phage resistant bacteria.

In the early 1940s, Luria and Delbrück were working on "mixed infection" experiments in which the bacteriophage resistant strain of *E. coli* bacteria were used as indicators in studies they were making on T_1 and T_2 virus particles. Starting in the Fall of 1942, they began to put aside the mixed infection experiment and asked themselves: What is the origin of those resistant bacterial strains that they were using as indicators [23]?

They vacillated between the following two ideas:

(1) that the resistant bacteria were spontaneous mutants present in the colony before exposure to the virus; or

(2) that the resistant bacterial strain came into being as a result of exposure to the virus, and then transmitted this *acquired* characteristic (resistance) to subsequent generations.

These two possibilities represent, in the field of bacterial genetics, the age-old problem as to whether evolutionary development is the result of mutation, or if it is the inheritance of a characteristic acquired by the organism on exposure to the external environment. The appearance of the resistant bacterial species raises these questions quite sharply: Are the resistant bacteria descendants of mutants, or have they inherited an acquired characteristic, as Lamarck believed?

In 1943 Luria and Delbrück performed an experiment that answered this question decisively [5]. Their experiment and its analysis made three fundamental contributions to bacterial genetics. First, it proved quite dramatically that the phage resistant bacteria were the result of *mutations*, and did not arise from the phage attack. Second, it provided a means for measuring mutation rates from statistical fluctuations in the number of mutants present in an ensemble of separate bacterial cultures. Third, it permitted measurement of mutation rates that were lower by several orders of magnitude than was possible previously. By providing this quantitative means of studying very small mutation rates, their experiment came to constitute one of the main roots in the development of molecular biology [23].

The Luria–Delbrück experiment, in essence, is a measurement of the probability distribution for the number of resistant mutants that appear in an ensemble of identically prepared, but separate bacterial cultures after a growth period t. To be specific, let us imagine an ensemble of C growth cultures. At $t = 0$, an inoculum of say ~ 100 bacteria (*E. coli*) is placed in each of the cultures.

The bacteria are then allowed to reproduce and increase their numbers for a period of time t. At time t, each culture is exposed to an excess number of bacteriophage particles ($T1$ virus). This exposure quite rapidly kills all but a few bacteria. These surviving bacteria transmit their resistance to the offspring. By allowing the surviving bacteria to grow into colonies, the experimenter can count colonies and determine the original number r_i of survivors that existed in each of the bacterial cultures immediately following exposure to the phage. From these numbers one can construct a probability distribution $P(r_i, t)$, that describes the probability that r_i resistant bacteria will appear following a growth period t and exposure to the phage

$$P(r_i, t) = \frac{\mathcal{N}(r_i)}{C}.$$

(3-82)

Here $\mathcal{N}(r_i)$ is the number of cultures having r_i surviving bacteria.

In January, 1943, S. Luria realized that the form of the distribution for $P(r_i)$ would be quite different depending on whether the resistant bacteria resulted from adaptation on exposure to the phage, or were mutations. Consider, first, the adaptation possibility. In this case, we can expect that each bacterium has some small probability ε of developing resistance on exposure to the phage. Further, it is reasonable to suppose that the conversion of one bacterium to a resistant strain is independent of whether or not some other bacterium in the culture also becomes converted. Under these conditions, in a culture containing a total of $N(t)$ bacteria at time t, the actual number of resistant bacteria produced by exposure to the phage will fluctuate from culture to culture according to a Poisson distribution whose mean is $\bar{r} = \varepsilon N(t)$. That is, the adaptation hypothesis, combined with the assumption that the bacteria convert independently of one another, leads to the prediction that $P(r_i)$ would be a Poisson distribution with mean $\bar{r} = \varepsilon N(t)$. One could examine whether this actually occurred by computing the mean square fluctuation in r_i, the number of resistant bacteria. If the probability distribution was Poisson, the variance

$$\overline{(r - r_i)^2} \equiv \sum_{i=1}^{C} (\bar{r} - r_i)^2 P(r_i)$$

(3-83)

is equal to the mean number of resistant bacteria; i.e.,

$$\overline{(\bar{r} - r_i)^2} = \bar{r}.$$

(3-84)

Thus, on the adaptation hypothesis, if $\bar{r} = \varepsilon N(t)$ is much larger than unity, the relative root mean square fluctuation, the number of resistant bacteria from culture to culture, will be small; i.e.,

$$\frac{\sqrt{(\bar{r} - r_i^2)}}{\bar{r}} = \sqrt{\frac{1}{\bar{r}}} = \sqrt{\frac{1}{\varepsilon N(t)}}.$$
(3-85)

When the experiment was performed, Luria found, in fact, that there were great fluctuations in the actual number of resistant bacteria that appeared in the different cultures in the ensemble. Luria and Delbrück put it this way [5]:

> In an attempt to determine accurately the proportion of resistant bacteria, great variations of the proportions were found, and the results did not seem to be reproducible from day to day.

They soon came to understand that the large fluctuations in r_i about $\bar{r}$ was not an experimental artifact, but was the inevitable consequence of the fact that the resistant bacteria were mutants that appeared in each of the cultures well before the introduction of the phage.

Let us examine qualitatively at first how mutations can produce a probability distribution $P(r_i)$ with a variance much larger than the Poisson distribution. It is consistent with the mutation hypothesis to assume that the mutant bacteria appear randomly in time with some average mutation rate. In fact, the mutation rates are so small that, on average, it is very unlikely that a mutation will appear in the early generations of bacterial division. In fact, most of the mutants appear toward the latter part of the growth period because of the exponential increase in the number wild-type bacteria. *On occasion*, however, a mutant appears early, and this can produce the same number of mutant offspring in its clone, as do all the later appearing mutants. As a result of these factors the actual number of resistant bacteria that appear in a finite number of samples can deviate very markedly from the mean expected number. The probability distribution $P(r_i)$, under the hypothesis of mutation, will be one with a long and significant tail representing numbers of resistant bacteria much larger than the average number, i.e., a distribution with a very large variance.

It is interesting to observe that Luria came to suspect that this was what was happening in his bacterial cultures

> ...while watching the fluctuating returns obtained by various colleagues of mine *gambling* on a slot machine at the Bloomington Country Club, where faculty dances were then held one Saturday a month [23].

The bacteria and the slot machine are analogous in the following sense. In a fair slot machine, on average, the jackpot will occur after, say, R trials. If the player plays only a fraction fR of trials ($f < 1$), he will probably lose the money he puts into the machine. Thus, the amount of money won will be less than that expected on average. On the other hand, there can also be a certain number of trials in which the jackpot is hit, and his return far exceeds what would be expected on average. The slot machine player sees a big *variance* in his fortunes around the average in a finite number of trials.

We will calculate below, in detail, the variance of $P(r)$ to be expected under the hypothesis of mutation, but it is amusing to realize that the connection with gambling played a germinal role in coming to understand bacterial genetics, just as three hundred years earlier the gambling of the Chevalier de Méré provided the spur that led Pascal and Fermat to found the mathematical theory of probability.

3.5.B. Theory of the Probability Distribution for Phage Resistant Bacteria Under the Hypothesis of Mutation

(i) Growth of Bacterial Population. Division Time

Let us imagine that at $t = 0$ each of the tubes illustrated in Figure 3.18 are inoculated with a small number (N_0) (between 50 and 500) of *E. coli* bacteria. The nutrient media in each tube enables growth in the number of bacteria. After a period of roughly 25 min each bacterium will divide producing two offspring. By this process of cell division the number of bacteria $N(t)$ present at time t is related to the number Q of bacterial generations by the equation

$$N(t) = 2^Q N_0. \tag{3-86}$$

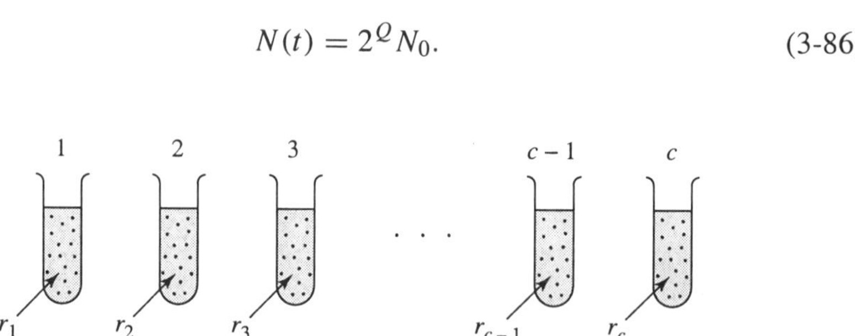

Figure 3.18. An ensemble of C identically prepared bacterial cultures, indicating the number $r_i(t)$ of resistant bacteria to be found in each culture after a growth period (t) and exposure to the bacteriophage particles at time t.

We can express Q in terms of the elapsed time t and the division time t_D by the equation

$$Q = t/t_D. \tag{3-87}$$

Alternatively, we can describe the increase in bacterial numbers using the exponential function $e^{t/\theta}$ in the following way. Write 2^Q as

$$2^Q = e^{\ln 2^Q} = e^{Q \ln 2} = e^{t \ln 2/t_D}.$$

Now we may define an exponential growth time θ as

$$\theta = t_D/\ln 2 = t_D/(0.693). \tag{3-88}$$

and we see that $N(t)$ can be written as

$$N(t) = N_0 e^{t/\theta}. \tag{3-89}$$

Typically the initial bacterial inoculum was ~ 100 bacteria/cm^3, and the final number $N(t)$ of bacteria present at the time of plating with bacteriophage virus was $\sim 3 \times 10^8$. Thus

$$\frac{N(t)}{N(0)} = 3 \times 10^6 = 2^Q = e^{t/\theta}.$$

So we find that

$$Q \approx 21 - 22.$$

Thus, roughly 21 bacterial generations take place during the period of their experiments. The actual time needed for these divisions to occur depends in detail on the growth medium. For example, $t_D = 19$ min in a nutrient broth and $t_D = 35$ min in an asparagin–glucose synthetic medium. To eliminate special reference to the particular growth medium employed, in the forthcoming discussion we measure time in units of the exponential growth time θ. That is, we will use the symbol t to denote the ratio of the actual time to the characteristic time θ:

$$t \equiv \text{time}/\theta. \tag{3-90a}$$

Hence we now write (3-89) as

$$N(t) = N_0 e^t. \tag{3-90b}$$

(ii) Probability Distribution for Clones of Resistant Bacteria

Let us now examine the evolution of the population of resistant bacteria according to the hypothesis of mutation. The crucial element of the mutation hypothesis is that there is a fixed, small chance per unit time for a bacterium to undergo a mutation to resistance. The time scale that is appropriate here is clearly the time as measured in units of the bacterial lifetime, i.e., t as deferred above. If we define as (a) the probability per unit time (t) that a bacterium will mutate, then it is clear that in a population containing $N(t)$ wild-type bacteria the number $d\overline{m}$ of mutant bacteria that appear on average in time dt will be

$$d\overline{m} = a\, dt\, N(t). \tag{3-91}$$

The mutation rate a is then given by

$$a = \frac{1}{N(t)}\left(\frac{d\overline{m}}{dt}\right). \tag{3-92}$$

(Here t is measured in units of actual time/θ as we agreed above.) A mutant bacterium will also divide, and we will assume that it is so like the wild type that its division time is essentially the same as the wild type. On each division the property of resistance is transmitted to the progeny, and so each mutant will give rise to a "clone" of resistant bacteria. Let us first examine how many such clones will *on average* appear in time t.

Since

$$d\overline{m} = a N(t)\, dt,$$

we may use (3-90) to obtain

$$\overline{m} = \int_0^\tau d\overline{m} = \int_{t'=0}^t a N_0 e^{t'}\, dt',$$

$$\overline{m} = a(N(t) - N_0). \tag{3-93}$$

Thus, on average, the number of mutant clones is the mutation rate (a) times the increase in the number of wild-type *E. coli* bacteria.

Now if we could actually examine the number of resistant *clones* in each of the C different cultures, we could tabulate $\mathcal{N}(m)$, the number of cultures containing m clones, and measure the probability distribution of clones

$$P_c(m, \overline{m}) = \lim_{C \to \infty} \left(\frac{\mathcal{N}(m)}{C} \right). \tag{3-94}$$

Since the appearance of a mutant bacterium as the conversion of a wild-type bacterium is perfectly random, and since the mean rate of conversion is constant, we expect $P_c(m, \overline{m})$ to be a Poisson distribution. Under these conditions the probability that no mutants arise is given by

$$P_c(0, \overline{m}) = e^{-\overline{m}}. \tag{3-95}$$

Now using (3-93), we find

$$P_c(0, \overline{m}) = e^{-a(N(t) - N_0)}. \tag{3-96}$$

Thus, by measuring the number of cultures that *do not* contain mutants, one can obtain a value for the mutation rate (a). This assumes, of course, that the mutation hypothesis is correct. We shall soon show how Luria and Delbrück demonstrated this quite conclusively. At this point, let us use this knowledge to gain an estimate of the mutation rate: In an ensemble of 87 cultures $(C = 87)$ a total of 29 were found for which no resistant bacteria existed. Thus, for this limited sample

$$P_c(0, \overline{m}) = \frac{29}{87} = 0.33.$$

The average number $N(t)$ of bacteria in each culture was measured to be

$$N(t) \cong 2.4 \times 10^8, \qquad N_0 \sim 100$$

Thus, we have

$$e^{-a(2.4 \times 10^6)} = 0.33.$$

This leads to the value

$$a = 0.47 \times 10^{-8}\,\text{mutations/bacteria} \times \text{``time''} \tag{a}$$

or

$$a = 0.47 \times 10^{-8}(0.693) = 0.32 \times 10^{-8}\,\text{mutations/bacteria} \times \text{division time.} \tag{b}$$

This procedure represents one means by which the bacterial mutation rate (a) can be found. Of course, the present value of (a) must be regarded as only approximate because of the limited number of cultures (C) used in determining $P_c(0, \overline{m})$.

(iii)   The Mean Value and Variance of the Probability Distribution for the Number of Resistant Bacteria

Luria and Delbrück did not measure the probability distribution of *clones*. They measured, for a finite number of cultures, the probability distribution for the *total number of resistant bacteria*. Each clone arising from a single mutant will contain $e^{\Delta t}$ mutant bacteria in which Δt is the time between the appearance of the mutant and the exposure of the culture to the phage. Although the distribution of clones is Poissonian, the distribution for the total number of resistant bacteria will be quite different from Poisson because the number of bacteria in a clone depends in detail on when it began to grow. We will not attempt to present a theory for the full probability distribution $P(r)$ for the number r of resistant bacteria. Instead, we will confine ourselves to characterizing the distribution by computing its mean and its variance. This can then be compared with the prediction for the mean and variance on the hypothesis of adaptation.

Let us first focus our attention on the average number $(\overline{r})$ of resistant bacteria to be found in the ensemble of C separate cultures. Experimentally, one finds $\overline{r}$ in the following way. One first counts the number of cultures $\mathcal{N}(r_i)$ containing r_i resistant bacteria. Here $r_i = 1, 2, 3, \ldots$. (The number of resistant bacteria in each culture is, of course, measured from the number of colonies of resistant bacteria that appear after phage attack at time t.) Experimentally, one then obtains $P(r_i)$ as

$$P(r_i) = \left(\frac{\mathcal{N}(r_i)}{C}\right). \tag{3-97}$$

Only when C is very large do we, by this means, approximate accurately the limiting probability $P(r)$. The average $\overline{r}$ is, of course,

$$\bar{r} = \frac{1}{C} \sum_i (r_i \mathcal{N}(r_i)). \tag{3-98}$$

We now calculate $\bar{r}$ theoretically. Let us consider the effect of mutations occurring in some time interval $\Delta t'$ around t', where t' is any time between 0 and t.

How many mutant bacteria on average arise in the period Δt? According to the definition of the mutation rate, we have

$$\Delta \overline{m} = a N(t') \Delta t'.$$

Now how many mutant bacteria do these new variants produce? Each one gives rise to a clone containing $e^{t-t'}$ bacteria. Thus, the number of resistant bacteria existing at time t produced by these new mutants is

$$\Delta \bar{r} = \Delta \overline{m} e^{t-t'} \tag{3-99}$$

or

$$\Delta \bar{r} = a N(t') \Delta t' e^{t-t'}, \tag{3-100}$$

and the total number of resistant bacteria at time t, as a result of mutations that first arose at any time t' between 0 and t, is just the sum

$$\bar{r} = \int \Delta \bar{r} = \int_{t'=0}^{t} a N(t') e^{t-t'} \Delta t'. \tag{3-101}$$

Let us now examine the dependence of the integrand on the time t'. Since the number $N(t')$ of wild-type bacteria at t' is related to the final number at time t by $N(t') = N(t) e^{-(t-t')}$, we see that

$$a N(t') e^{(t-t')} = a N(t) e^{-(t-t')} e^{+(t-t')}$$

or

$$a N(t') e^{t-t'} = a N(t). \tag{3-102}$$

This shows that the contribution of each time interval to the mean number of mutant bacteria is the same. Early appearing mutants produce large numbers of bacteria in their clones, but the number of such early mutants is down by exactly the growth factor $e^{-(t-t')}$. Later mutants have not the time to grow large clones, but more appear later in the time interval $0 \to t$, because of the exponential growth of the wild-type bacteria. Again the growth factor $e^{t-t'}$ cancels exactly against the appearance factor $e^{-(t-t')}$, and each interval $(\Delta t')$ contributes equally to the final average number of resistant bacteria. On integrating (3-101) using (3-102), we find

$$\bar{r} = aN(t)t. \tag{3-103}$$

The mean number of resistant bacteria increases relatively more rapidly than does the number of wild-type bacteria. In principle, if $\bar{r}$ could be determined with sufficient accuracy, one could use (3-103) to determine the mutation rate (a). For this to be a practical procedure the root mean square fluctuation in r must be sufficiently small in comparison to $\bar{r}$. We will presently show that, in fact, the mean square fluctuation in r can be very large. In fact, this large value of the fluctuation in f proved to be the decisive evidence in favor of the mutation hypothesis.

At present, however, we wish to examine rather more carefully the meaning and form of (3-103), especially keeping in mind the fact that the actual numbers of resistant bacteria are not large. In obtaining (3-103), we treated the number of new mutants as if this was a continuous variable that increased quite smoothly in time. When one deals with a limited number of mutant bacteria, this is not an accurate procedure. For example, on carrying out the integration over the time $0 \to t$, it is clearly not sensible to include in the integration that time period from 0 to t_0 during which it is unlikely that even one mutant will appear in the *entire ensemble* of C cultures. We can estimate the time t_0 quite simply as follows. According to (3-93), if we have a total number of $CN(t)$ bacteria, the average number of mutants that are produced from the wild-type during time t is

$$\overline{m} = aC(N(t) - N_0).$$

We now seek the time t_0 for which $\overline{m} = 1$. This is found by setting $t = t_0$ above and neglecting N_0 compared to $N(t)$, i.e.,

$$1 = aCN(t_0). \tag{3-104}$$

We can express $N(t_0)$ in terms of $N(t)$, a known quantity, as follows

$$N(t_0) = N(t)e^{-(t-t_0)}.$$

Thus, the equation for the relevant time interval $(t - t_0)$ is

$$1 = aCN(t)e^{-(t-t_0)} \qquad \text{(3-105a)}$$

or

$$(t - t_0) = \ln(aCN(t)). \qquad \text{(3-105b)}$$

We can estimate $(t-t_0)$ and t from the Luria–Delbrück data. In their most favorable experiment they had $C = 100$. If we use the preliminary estimate of (a) obtained previously [see (3-97)] $a \simeq 0.5 \times 10^{-8}/\text{bacteria}\cdot\text{time}$, and the value $N(t) = 3 \times 10^8$ bacteria/culture, then we find

$$(t - t_0) = \ln(0.5 \times 10^{-8})(100)(3 \times 10^8)$$

or

$$(t - t_0) \simeq \ln 150 \approx 5.$$

On the other hand, the total elapsed time t required to produce the 3×10^8 bacteria in each culture was $e^t = (N(t)/N(0)) \simeq \left((3 \times 10^8)/100\right)$, assuming $N_0 \sim 100$. This gives

$$t \approx 15.$$

Thus we see that $t_0 \sim 10$. Thus, even for 100 cultures, the number of clones $(aCN(t) \sim 150)$ is so small that during an appreciable fraction of the growth period $(t_0/t \approx 10/15)$, there are likely to be no mutants at all present in any of the cultures. To correct for this, it is necessary to carry out the integration of (3-101) over the domain $t_0 < t' < t$. This is a much more accurate estimate of the growth period for the resistant bacteria. We then have, as a more accurate value of $\bar{r}$,

$$\bar{r} \cong aN(t)(t - t_0). \qquad \text{(3-106)}$$

This result will be applicable when the number of cultures C is limited. We may conclude our discussion of the value for the mean number of resistant bacteria by noting that (3-106) can be written entirely without reference to the time $(t - t_0)$ if we use the expression (3-105b) for $(t - t_0)$ in (3-106). This gives

$$\bar{r} = aN(t)\ln(aCN(t)). \tag{3-107}$$

If $\bar{r}$ can be determined accurately, then, according to the mutation hypothesis, this value can be used to determine the mutation rate (a) according to (3-107), since $N(t)$, the number of bacteria per culture, can be simply measured. Equation (3-107) then represents an alternative means for determining the mutation rate (a), provided that the resistant bacteria occur as a result of mutation.

Let us now calculate the variance $\overline{(r - \bar{r})^2}$ of the probability distribution for the number of resistant bacteria. In the time interval $\Delta t'$ the average number $\overline{\Delta m}$ of resistant bacteria that appear by mutation is given by (3-91) as

$$\overline{\Delta m} = aN(t')\Delta t'.$$

While this represents the average value of Δm, the actual number that appear in the C cultures will be distributed in accordance to the Poisson distribution

$$P(\Delta m, \overline{\Delta m}) = \lim_{C \to \infty} \frac{\mathcal{N}(\Delta m)}{C} = \frac{(\overline{\Delta m})^{\Delta m} e^{\overline{\Delta m}}}{(\Delta m)!}.$$

As the time goes on beyond the interval $\Delta t'$, this distribution evolves in time because of the growth into clones of each of the mutants. The distribution changes in time as follows; the mean value $\overline{\Delta m}$ grows exponentially with time as

$$\overline{\Delta m(t)} = \overline{\Delta m(t')}e^{t-t'}.$$

Thus, if we imagine a graph of $P(\Delta m, \overline{\Delta m})$ plotted as a function of Δm, we must imagine that the distribution $P(\Delta m, \overline{\Delta m})$ spreads out in time in proportion to $e^{t-t'}$. Because of this scaling factor the *variance* of each of these partial distributions $(P(\Delta m))$ grows as $\overline{\Delta m^2}e^{2(t-t')}$ where $\overline{\Delta m^2} = \overline{\Delta m} =$ mean square width of the distribution at $t = t'$. At the measurement time t, the mean square width $\overline{(r - \bar{r})^2}$ of $P(r, \bar{r})$ is the sum of the mean square widths of the partial distributions that emerged at any time t' between 0 and t, i.e.,

$$\overline{(r - \bar{r})^2} = \sum_{\Delta t'} \overline{\Delta m^2}(t')e^{2(t-t')} \tag{3-108}$$

$$= \sum_{\Delta t'} \overline{\Delta m}(t')e^{2(t-t')},$$

$$\overline{(r - \bar{r})^2} = \int aN(t')e^{2(t-t')}\,dt'. \tag{3-109}$$

In (3-109) we have replaced the sum over all time intervals $\Delta t'$ by an integration. In computing this integral over t' we must examine carefully, once again, the proper limits of integration. In the limit $C \to \infty$ clearly $0 \le t' \le t$. However, for a finite number of cultures (C) there is a period $0 \to t_0$ in which not a single mutant will appear in any of the C cultures. Under these circumstances, to get a reasonable approximation to $\overline{(r - \bar{r})^2}$ for a finite number of cultures, one should in fact carry out the integration only between t_0 and t, where t_0 is given by (3-104) or (3-105). With this consideration in mind we find that

$$\overline{(r - \bar{r})^2} = \int_{t_0}^{t} aN(t')e^{2(t-t')}\,dt'. \tag{3-110}$$

But

$$N(t') = N(t)e^{-(t-t')},$$

according to (3-90a,b), so that

$$\overline{(r - \bar{r})^2} = \int_{t_0}^{t} aN(t)e^{(t-t')}\,dt'. \tag{3-111}$$

On integrating, we find

$$\overline{(r - \bar{r})^2} = aN(t)(e^{(t-t_0)} - 1). \tag{3-112}$$

This equation expresses the mean square fluctuation in the number of resistant bacteria to be found in the various cultures. It is useful to express this result explicitly in terms of $t - t_0$ by using our expression for $\bar{r}$ in (3-106):

$$\bar{r} = aN(t)(t - t_0).$$

This equation shows that we can express $aN(t)$ as equal to $\bar{r}/(t - t_0)$, so that (3-112) can be written as

$$(r - \bar{r})^2 = \bar{r}\left[\frac{e^{t-t_0} - 1}{t - t_0}\right]. \tag{3-113}$$

This equation shows very dramatically that according to the mutation hypothesis the mean square fluctuation in the number of resistant bacteria found in the various cultures will be vastly larger than that which applies for a Poisson distribution. If $P(r, \bar{r})$ was a Poisson distribution, as would be the case under the adaptation hypothesis, then we would expect

$$\overline{(r - \bar{r})^2} = \bar{r} \qquad \text{(acquired immunity hypothesis)}.$$

On the other hand, if the resistant bacteria arose by mutation and then exponential growth of resistant bacteria in clones during the time interval between 0 and t, then $\overline{(r - \bar{r})^2}$ is given by

$$\overline{(r - \bar{r})^2} = \bar{r}\frac{(e^{t-t_0} - 1)}{(t - t_0)} \qquad \text{(mutation hypothesis)}.$$

We saw in our discussion of $\bar{r}$ that the factor $t - t_0 \sim 5$. Thus, we expect that the mean square fluctuation will be enormously larger than $\bar{r}$. In fact $(e^{t-t_0}/(t - t_0)) \sim e^5/5 \sim 30$. The form of (3-113) also illustrates quite clearly how important it is to exclude the time interval $0 - t_0$ in computing the mean square fluctuation $(r - \bar{r})^2$. If we had integrated (3-109) from $0 \to t$, we would have found $(r - \bar{r})^2 = \bar{r}e^t/t$. But we saw above that $t \sim 15$, and $(e^t/t) \sim (e^{15}/15) \sim 2 \times 10^5$. This is to be compared with $e^{t-t_0}/(t - t_0) \sim 30$. This shows that the exclusion of the time period in which no mutations are likely is very important in obtaining a reasonable estimate for $(r - \bar{r})^2$.

We will examine the Luria–Delbrück measurements in greater detail later. At the present point, however, we shall simply give the result they found on measuring an ensemble of 100 cultures. They found $\bar{r} \cong 40$, and $\overline{(r - \bar{r})^2} = 6270$. Clearly, the observed variance is consistent with the mutation hypothesis, and is not consistent with the hypothesis of acquired immunity. It is interesting to quote directly from the Luria–Delbrück paper regarding this large fluctuation in the number of resistant bacteria found in similarly prepared cultures

In the attempt to determine accurately the proportion of resistant bacteria, great variations of the proportions were found, and the results did not seem to be reproducible from day to day.

Eventually it was realized that these fluctuations are a necessary consequence of the mutation hypothesis and that the quantitative study of the fluctuations may serve to test the hypothesis [5].

To close this discussion of the variance of the distribution of resistant bacteria, we can use the expression for the variance to obtain yet another means of estimating experimentally the mutation rate (a). In fact, if we use (3-105a, b) to eliminate $(t - t_0)$ from (3-112) we find, on neglecting 1 in comparison with e^{t-t_0}, that the variance may be related to (a) by

$$\overline{(r - \bar{r})^2} = a^2 N^2(t)C.\tag{3-114}$$

3.5.C. The Experimental Data of Luria and Delbrück. Comparison Between Theory and Experiment. Determination of the Bacterial Mutation Rate

In Table 3.2 we show Luria and Delbrück's experimental measurements of the probability distribution of numbers of resistant bacteria to be found in two ensembles containing 100 and 87 separate cultures. Each culture was inoculated with about ~ 100 $E.\ coli$ bacteria. The cultures grew until $N(t)$, the total number of bacteria, was $\sim 3 \times 10^8$. The cultures were then exposed to phage attack. This destroyed all the bacteria except the resistant bacteria in each culture. The resistant bacteria could be counted because they produced colonies as they developed in the growth medium. The number r_i of resistant bacteria in each culture was counted. Using this, they tabulated the number of cultures $\mathcal{N}(r_i)$ containing r_i resistant bacteria. Since $\sum_i \mathcal{N}(r_i) = C$, where C is the total number of cultures, they were able to calculate $P(r_i)$, $\bar{r}$, and $(r - \bar{r})^2$ in accordance with the following formulas

$$P(r_i) = \frac{\mathcal{N}(r_i)}{C},\tag{3-115}$$

$$\bar{r} = \sum_i r_i \frac{\mathcal{N}(r_i)}{C},\tag{3-116}$$

$$\overline{(r - \bar{r})^2} = \sum_i (r_i - \bar{r})^2 \frac{\mathcal{N}(r_i)}{C}.\tag{3-117}$$

Table 3.2 on the following page lists the results they found.

Table 3.2. Number Distribution of Phage-Resistant Bacteria (from Luria and Delbrück [5]).

	Experiment #22 $C = 100$ cultures Culture volume $= 0.2$ cm^3		Experiment #23 $C = 87$ cultures Culture volume $= 0.2$ cm^3	
r_i Number of resistant bacteria	$\mathcal{N}(r_i)$ Number of cultures with (r_i) resistant bacteria		(r_i)	$\mathcal{N}(r_i)$
0	57		0	29
1	20		1	17
2	5		2	4
3	2		3	3
4	3		4	3
5	1		5	2
6–10	7		6–10	5
11–20	2		11–20	6
21–50	2		21–50	7
51–100	0		51–100	5
101–200	0		101–200	2
201–500	0		201–500	4
501–1000	1		501–1000	0
$\bar{r}$	40.48		$\bar{r}$	28.6
$\overline{(r - \bar{r})^2}$	6270		$\overline{(r - \bar{r})^2}$	6431
$\overline{(r - \bar{r})^2}/\bar{r}$	150		$\overline{(r - \bar{r})^2}/\bar{r}$	220
$N(t)$	2.8×10^8		$N(t)$	2.4×10^8
a [from (3-114)]	2.8×10^{-8}		a [from (3-114)]	3.6×10^{-8}
a [from (3-107)]	2.3×10^{-8}		a [from (3-107)]	2.0×10^{-8}

For each ensemble, the mean $\bar{r}$ and variance $\overline{(r - \bar{r})^2}$ of the distribution is listed. We note first of all that the ratio of the variance to the mean is vastly larger than unity. In fact, for $C = 100$, this ratio is 150, while for $C = 87$, this is 220. We then see that $\bar{r}$, the mean number of resistant bacteria, is, in fact, rather poorly determined. The relative root mean square fluctuation in the two experiments is given by

$$\frac{\sqrt{(r-\bar{r})^2}}{\bar{r}} = \frac{79}{40.5} \cong 1.9 \quad \text{(for Experiment 22)}$$

$$= \frac{80}{29} \simeq 2.8 \quad \text{(for Experiment 23)}.$$

The very large variance in the values of r shows that $\bar{r}$, the mean number of resistant bacteria, is uncertain by a factor of about 2–3. Despite this, Luria and Delbrück used their measured values of $\bar{r}$ to obtain the mutation rate (a) from (3-107). We will not follow their procedure, but will leave it as a problem for the student at the end of this chapter. The values for (a) found in this way are listed in the final row in Table 3.2.

Instead it is much simpler to calculate a using (3-114) that relates the variance $(r-\bar{r})^2$ to a, $N(t)$, and C. According to this equation, a is given by

$$a = \frac{\sqrt{(r-\bar{r})^2}}{N(t)C^{1/2}} = \frac{\sqrt{6270}}{2.8 \times 10^8 (10)} = 2.8 \times 10^{-8} \quad \text{for} \quad C = 100$$

$$= \frac{\sqrt{6431}}{2.4 \times 10^8 \sqrt{87}} = 3.6 \times 10^{-8} \quad \text{for} \quad C = 87.$$

These values are listed in the second row from the bottom of Table 3.2.

It is important to observe that these calculated values of a are much larger than those estimated in Section 3.5.B(ii) using only the data on the number of cultures that have no resistant bacteria. We will regard the values of a given in Table 3.2 as the more reliable ones, and use them to reexamine the magnitude of the quantity $(t-t_0)$. Since (3-115) is equivalent to (3-114), and since $\overline{(r-\bar{r})^2}/\bar{r} \sim 180$ for the ensembles they studied, we have

$$\frac{e^{t-t_0}}{t-t_0} \sim 180.$$

This implies $(t-t_0) \sim 7.2$. This is to be compared with the value of $t \sim 15$ that comes from the total number of bacteria. Thus, we expect $t_0 \sim 8$. Our previous estimate of t_0, based on the bacterial mutation rate deduced from the number of cultures showing no resistant bacteria, was $t_0 \sim 10$. That value of t_0 indicated that during $\frac{10}{15} = \frac{2}{3}$ of the growth period no mutant bacteria, on average, were to be found. The present more accurate estimate for the bacterial mutation rate implies that during $\frac{8}{15} \sim \frac{1}{2}$ of the growth period no mutant bacteria appeared. A value of $t_0 = 8$ implies, alternatively, that $(8/0.693) \cong 11$ bacterial generations are required, on average, before the appearance of a single bacterial mutant in their ensemble of C cultures.

3.6   References and Supplementary Reading

1. *Recherches sur la Probabilité des Jugements en Matière Criminelle et en Matière Civile*, by S. D. Poisson. Bachelier , Paris (1837).

2. On the Error of Counting with a Hemocytometer, by W. S. Gosset (pseud. "student"). *Biometrika* **5**, 351 (1907).

3. The Probability Variations in the Distribution of α-Particles, by E. Rutherford and H. Geiger. *Philosophical Magazine* **20**, 700 (1910).

4. Light Scattering in Solutions, by P. J. W. Debye. *Journal of Applied Physics* **15**, 338 (1844).

5. Mutations of Bacteria from Virus Sensitivity to Virus Resistance, by S. E. Luria and M. Delbrück. *Genetics* **28**, 491 (1943).

6. Energy Quanta and Vision, by S. Hecht, S. Shlaer and M. Pirenne. *Journal of General Physiology* **25**, 819 (1942).

7. *Visual Perception*, by T. Cornsweet. Academic Press, New York (1970).

8. Area and Intensity—Time Relation in the Peripheral Retina, by C. H. Graham and R. Margeria. *American Journal of Physiology* **113**, 299 (1935).

9. *Vision and the Eye*, 2nd ed., by M. Pirenne. Associated Books, London (1967).

10. Increment Thresholds at Low Intensities Considered as Signal/Noise Discriminations, by H. B. Barlow. *Journal of General Physiology* **136**, 469–488 (1957).

11. On the Sensitivity Performance of the Human Eye on an Absolute Scale, by A. Rose. *Journal of the Optical Society of America* **38**, 196 (1948).

12. The Quantum Character of Light and its Bearing on: The Threshold of Vision, The Differential Sensitivity, and The Acuity of the Eye, by H. I. De Vries. *Physica* **10**, 553 (1943).

13. Temporal and Spatial Summation in Human Vision at Different Background Intensities, by H. Barlow, *Journal of Physiology* (London) **141**, 337 (1958).

14. Coding of Light Intensity by the Cat Retina, by H. B. Barlow and W. R. Levick. In *Rendiconti della Schuola Internazionale di Fisica E. Fermi*, XLIII course (1969).

15. Three Factors Limiting the Reliable Detection of LIght by Retinal Ganglion Cells of the Cat, by H. Barlow and W. R. Levick. *Journal of Physiology* **200**, 1 (1969).

16. Changes in the Maintained Discharge with Adaptation Level in the Cat Retina, by H. Barlow and W. R. Levick. *Journal of Physiology* **202**, 699 (1969).

17. *Responses to Single Quanta of Light in the Retinal Ganglion Cells of the Cat*, by H. B. Barlow, W. R. Levick and M. Yoon. Vision Research Supplement 3, pp. 87–101. Pergamon Press, London (1971).

18. Voltage Noise in Limulus Visual Cells, by F. A. Dodge, B. W. Kinght, and J. Toyoda. *Science* **160**, 88–90 (1968).

19. Saturation of the Rod Mechanism of the Retina at High Levels of Stimulation, by M. Aguilar and W. S. Stiles. *Optica Acta (Paris)* **1**, 59–65 (1954).

20. The Physical Limits of Visual Discrimination, by H. B. Barlow, in *Photo-Physiology*, Vol. II, Chap. 16. Academic Press, New York (1964).

21. *Intraocular Light Scattering: Theory and Clinical Application*, by D. Miller and G. B. Benedek. C. C. Thomas, Springfield, IL (1973).

22. *Mach Bands: Quantitative Studies on Neural Networks in the Retina*, by F. Ratliff. Holden-Day, New York (1965).

23. Mutations of Bacteria and Bacteriophage, by S. E. Luria, In *Phage and the Origins of Molecular Biology*, edited by J. Cairns, G. Stent, and J. Watson. Cold Spring Harbor Laboratory of Quantitative Biology (1966).

24. Vision: From Photon to Perception. Papers from a colloquium organized by J. Dowling, L. Stryer, and T. Wiesel. In *Proc. Natl. Acad. Sci. USA* **93**, 557–639 (1996).

25. Baylor, D., Photoreceptors and Vision (Proctor Lecture). *Invest. Ophth. and Vis. Sci.* **28**, 34–49 (1987).

26. Baylor, D., How Photons Start Vision, *Proc. Natl. Acad. Sci. USA* **93**, 560–565 (1996).

27. Lamb, T. D., "Gain and Kinetics of Activation in the G-Protein Cascade of Phototransduction, *Proc. Natl. Acad. Sci. USA* **93**, 566–570 (1996).

28. Molday, R. S., Photoreceptor Membrane Proteins, Phototransduction, and Retinal Degenerative Diseases. *Invest. Ophth. and Vis. Sci.* **39**, 2493–2513 (1998).

29. Pugh, E. N. Jr. and Lamb, T. D., Amplification and Kinetics of the Activation Steps in Phototransduction, *Biochem. et Biophys. Acta* **1141**, 111–149 (1993).

3.7   Problems

(These problems are grouped topically, and in each group they are arranged roughly in order of difficulty.)

Poisson Statistics: Bacterial Counts and Radioactive Decay

1. Suppose that a broth culture of volume 0.2 cm^3 contains 10^8 bacteria. Show that N, the total number of bacteria in the culture, can be measured with an accuracy of 10% by first diluting the culture in a volume of 1 l, and then counting the number of bacteria in 1 mg. How many 1 mg samples should be counted to increase the accuracy in the determination of N to 1%?

2. The probability that a radioactive nucleus will emit an α-particle in a short time interval Δt is proportional to Δt, i.e.,

$$p(1, \Delta t) = \left(\frac{1}{\tau_0} \right) \Delta t.$$

   Here $(1/\tau_0)$ is the proportionality coefficient. By dividing a finite time interval into a large number of small intervals Δt, show that the probability q, that no α-particles have been emitted in time t, is given by the formula

$$q = e^{-t/\tau_0}.$$

   This result was used in the discussion in Section 3.2.B, p. 291.

3. The probability $p(1, \Delta t)$ that a radioactive nucleus will emit an α-particle in a short time Δt is proportional to Δt:

$$p(1, \Delta t) = \frac{1}{\tau_0} \Delta t$$

   Here $(1/\tau_0)$ is the constant of proportionality. Consider an ensemble of nuclei.

   (a) Designate by $n(t)$ the number of nuclei that have *not* yet decayed at time t. Show that the average number $\bar{n}(t)$ of such nuclei decreases with time in accordance with the formula

$$\frac{d\bar{n}}{dt} = -\frac{\bar{n}(t)}{\tau_0}.$$

(b) Integrate this formula to show that the mean rate of emission of α-particles decreases exponentially with time.

(c) Assume that the time constant τ_0 is sufficiently small so that a change in decay rate $(d\bar{n}/dt)$ can be observed during two reasonably spaced time intervals Δt_1 and Δt_2 around t_2. Show that the time constant τ_0 can be measured from the values of the decay rates according to the formula

$$\tau_0 = \frac{t_2 - t_1}{\ln(d\bar{n}/dt)_{t_1} - \ln(d\bar{n}/dt)_{t_2}}.$$

4. The nucleus of an atom of U^{234} can spontaneously emit an α-particle with a characteristic time constant of $\tau_0 = 3.61 \times 10^5$ years. The mass of one such U^{234} atom is 3.88×10^{-22} g. For such a long-lived nucleus it is not possible to detect a change in emission rate during a reasonable period of time. Instead, one can determine τ_0 by a measurement of the mean number of decays $\bar{k}$ produced in a sampling time Δt in accordance with the formula

$$\bar{k} = \left(\frac{\Delta t}{\tau_0}\right) N.$$

Here N is the number of U^{234} atoms in the sample right after chemical preparation, and $\bar{k}$ is the mean number of decays observed after repeated samplings during intervals of length Δt.

(a) Suppose 1 μg of U^{234} is available, and suppose that the counting time t is about 1 min. Estimate the mean number of counts that one expects to observe.

(b) If one actually counts the number of α-particles emitted in the time interval of 1 min, how accurately can one expect to estimate τ_0 using the 1 μg sample of U^{234}?

5. A beam of X-rays is used to make a two-dimensional ($\ell \times \ell$) scan of the brain in an effort to locate a tumor. The geometry of the scan is shown below.

   After passage through the brain the X-ray intensity is measured by counting the number of X-ray photons (n) in each grid square. Suppose that the "signature" of the tumor is that it preferentially absorbs X-rays by an amount that is 0.1% greater than for healthy tissue.

(a) What is the minimum number n of X-ray photons that should be used in scanning each grid square?

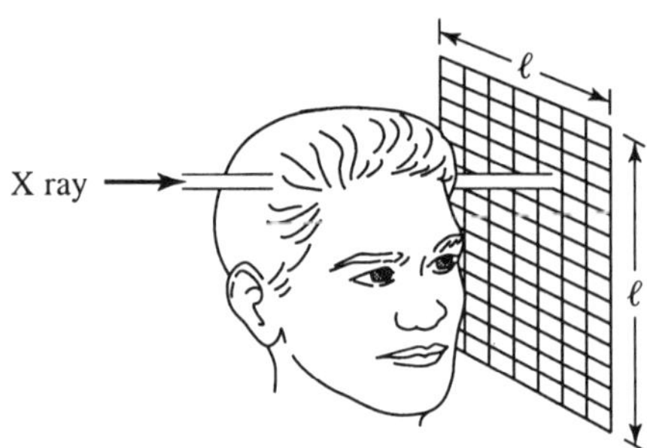

(b) Suppose that the maximum dosage permitted to the brain consists of a total of $\mathcal{N}$ X-ray photons. Obtain a formula for the minimum size of the grid square in terms of $\mathcal{N}$ and ℓ. This, in fact, determines the minimum tumor size that can be detected.

6. Consider a Poisson process in which only every third pulse is counted. (Such selective counting is called "scaling.")

(a) Given that the original pulses are Poisson distributed according to $P(n, \bar{n}) = [(\bar{n})^n / n!]e^{-\bar{n}}$, show that after scaling the probability $P_s(m, \bar{m})$ of counting m pulses in an interval, that on average contains $\bar{m}$ of them, is given by

$$P_s(m, \bar{m}) = \text{constant}\ \frac{(3\bar{m})^{3m}}{(3m)!}.$$

(b) In the case of large $\bar{n}$, the Poisson distribution goes over into the Gaussian

$$P(n, \bar{n}) \to \frac{e^{-(n-\bar{n})^2/2\bar{n}}}{\sqrt{2\pi\bar{n}}}$$

Show that the corresponding limit of the scaled distribution is

$$P_s(m, \bar{m}) \to \sqrt{\frac{3}{2\pi\bar{m}}}\, e^{-(m-\bar{m})^2/\frac{2}{3}\bar{m}}.$$

(c) Show that the variance $(m - \bar{m})^2$ of the scaled distribution has the value $\frac{1}{3}\bar{m}$. How much smaller is this value than that of a Poisson distribution with $\bar{n} = \bar{m}$?

(d) Show that the accuracy with which $\overline{m}_\tau$ can be determined is $\sqrt{\overline{\Delta m_\tau^2}}/\overline{m}_\tau$ $= 1/\sqrt{3\overline{m}_\tau}$, and that as a result, the accuracy with which $\overline{n}_\tau$ can be determined is still $1/\sqrt{\overline{n}_\tau}$. Thus, the use of the scaler as described does not degrade the accuracy of the measurement of $\overline{n}_\tau$.

(e) The narrowing of the variance of the distribution in m, that you found above, can be demonstrated quite dramatically in the following way. Construct, graphically, a "random" sequence of ~ 50 pulses with the same "average" pulse rate. Now, select from this sequence every fifth pulse, and plot these pulses directly underneath the initial sequence. You will see at once that the scaled sequence is much more regularly spaced than the original sequence you first drew.

Poisson Statistics Applied to Epidemiology, Sports, Business, Politics, and War

7. In the United States each year 400,000 people are killed or injured in automobile accidents out of a total population of 200 million people. Suppose that you have 100 friends and acquaintances. What is the statistical probability that during this coming year one or more of them will be injured in an auto accident? What is the probability that one or more will be injured over the next 5 years?

8. The incidence of poliomyelitis during the years 1949–1954 was approximately 25 per 100,000 people. In a city of 40,000 people, what is the probability that there are five cases or less? What is the probability that there are 20 or more cases? In a city of 1×10^6 persons what is the probability that there are five or less cases? What is the probability that there are more than 250 cases?

9. In baseball a good hitter will on average get three hits for each 10 times at bat. Assume that each game represents 12 times at bat. Suppose that we examine the batter's performance in any consecutive series of three games:

    (a) What is the probability that he gets zero hits in such a three-game series?

    (b) What is the probability that he will get three hits or less in a three-game series?

    (c) What odds would you offer if someone was willing to bet that the batter would get five hits or more in his next three games?

10. The number of goals scored in a game (per team) in the 1966 World Cup soccer matches is given in the table below. There were 32 games played:

Goals per team	Observed frequency
0	18
1	20
2	15
3	7
4	2
5	2
> 5	0

Find the average number of goals per team per game. Using this average, show that the distribution of goals follows closely a Poisson distribution; that is, compute the expected number of goals per team per game using a Poisson distribution with this mean, and compare with the table above.

(a) What is that probability that a game between two equally matched teams will be a tie? What then is the probability that an injustice (i.e., no tie) is perpetrated if a single game is used to decide between two teams?

(b) How many games would two equally matched teams have to play until the probability of a draw is 0.5?

(c) Suppose that two teams, with scoring averages of 1.0 and 1.5 goals per game, meet in a single match. What is the probability the inferior team will win?

11. Despite his best efforts a printer makes typesetting errors randomly at a rate of one error every two pages. Suppose he selects four pages of his work at random to demonstrate his skill. What is the probability that no error appears in these four pages? How many customers need he have until he can expect to have one successful demonstration? What is the probability that no error will occur if only one page is sampled? What is the probability that two errors occur on one page? What is the probability that five errors occur on four pages?

12. In a large fleet of delivery trucks the average number inoperative on any day, because of repairs, is two. Two standby trucks are available. What is the probability that on any day

 (i) no standby trucks will be needed, and

 (ii) the number of standby trucks is inadequate?

13. Major motor failures occur among the buses of a large bus company at the rate of two a day. Assuming that each motor failure requires the services of

one mechanic for a whole day, how many mechanics should the bus company employ to insure that the probability is at least 0.95 that a mechanic will be available to repair each motor as it fails? (More precisely, find the smallest integer K such that the probability is greater than or equal to 0.95 that K or fewer motor failures will occur in a day.)

14. (a) During the 96 years from 1837 to 1932, 48 resignations or deaths resulted in vacancies in the Supreme Court. What is the probability that in any single 4-year term the President will have the opportunity to make one appointment to the court? What is the probability that he will be able to make one or two?

    (b) Compute the probabilities that in any one year 0, 1, 2, and 3 vacancies, respectively, can occur. In the 96 years involved compute the number of times that 3, 2, 1, and 0 vacancies have occurred in a year. Below is given a table that gives the actual historical data. Compare your results with the table.

Number of vacancies	Number of times these occurred in one year (Expt.)	Your calculation
0	59	
1	27	
2	9	
3	1	

15. The following data (R. D. Clarke, An Application of the Poisson Distribution, *Journal of the Institute of Actuaries*, **72** (1946)) gives the number of flying bomb (V-2) hits recorded in each of 576 small areas of size $\frac{1}{4}$ km^2 in the south of London during World War II:

k = Number of flying bomb hits per area	N_k = number of areas with k hits
0	229
1	211
2	93
3	35
4	7
5 or over	1

Make a theoretical prediction of the probability that any one area will be hit by 1 bomb, 2 bombs, etc. Compute the total number of areas hit by 1, 2, 3, etc., bombs, and compare the results with those given in the table. Do you think the bombs were aimed at any of the areas in particular?

Seeing in the Presence of Background Light and at the Threshold of Vision

16. Sunglasses impair intensity discrimination at low light levels, but improve it at high light levels. In this problem you will see why. In the case of long-time, large-area observations, the noise associated with the retinal discharges varies with light intensity I as

$$(\Delta I)_{\text{noise}} = \text{constant } I^n.$$

The exponent n itself is a function of the light intensity I. At low light levels $n = 0$. In the Weber–Fechner regime $n = 1$, and at high light levels $n \geq 1$ (see Figure 3.17). The sunglasses reduce intensity differences $(\Delta I)_f$ in the visual field by some factor R $(R \gg 1)$, and also reduce the background intensity I by the same factor.

(a) Show that the signal-to-noise ratio $(S/\mathcal{N})$, that applies when sunglasses are used, is related to the signal-to-noise ratio in their absence $(S/\mathcal{N})_{\text{no glasses}}$ by the formula

$$\left(\frac{S}{\mathcal{N}}\right) = \left(\frac{S}{\mathcal{N}}\right)_{\text{no glasses}} \times R^{n-1}.$$

(b) Use Figure 3.17 to show that when $\log I \leq 2$ the use of sunglasses impairs visual discrimination. Show that only at high light levels: i.e. $\log I \geq 2$ is visual discrimination improved by the use of sunglasses.

17. Visibility of faint stars. The visual brightness B of a star is characterized by astronomers in terms of an index called the *visual magnitude m*. The relation between B and m is

$$m = 2.5 \log_{10}\left(\frac{B_0}{B}\right).$$

where B_0 is the brightness of a reference star with magnitude 0. The star α-Centauri, one of the brightest stars in the southern sky, has a magnitude $m \cong 0$. The brightness B_0 represents a flux of about 10^6 photons/cm s in the range of wavelengths for which the eye is sensitive.

(a) By what factor does the brightness of a star *decrease* as the visual magnitude *increases* by 1?

(b) Compute the visual magnitude m of the faintest star that should be visible to the unaided eye. Remember that the pupil of the eye represents an entrance aperture of about $\frac{1}{4}$ cm^2. Also, remember that the sampling or integration time for detection of light by the eye is $\frac{1}{10}$ s.

The Luria-Delbrück Experiment

18. If time is measured in units of the characteristic time θ (see (3-90)), the bacterial mutation rate a is determined to be $a = 2 \times 10^{-8}$. What is the probability per bacterium for a mutation in a single division time?

19. The mean number $\bar{r}$ of resistant bacteria in an ensemble of C cultures can be used to compute the bacterial mutation rate (a). Use the data given for c, $N(t)$, and $\bar{r}$ given in Table 3.2 to calculate a for the two ensembles $C = 100$ and $C = 87$.

   Hint: The equation for $\bar{r}$ [(3-107)] is a transcendental equation. You may find it convenient to find a by putting this equation in the form

$$C\bar{r} = x \ln x.$$

20. Here we will show that the uncertainty in the experimental value of the mean number $\bar{r}$ of bacteriophage-resistant bacteria cannot be reduced by increasing the number C of cultures in the test ensemble.

(a) Show that the root mean square fluctuation in r is related to $\bar{r}$ by the equation

$$\frac{\sqrt{(r - \bar{r})^2}}{\bar{r}} = \frac{\sqrt{C}}{\ln(aCN(t))}.$$

(b) Use the values $a \simeq 2 \times 10^{-8}$ and $N(t) \sim 3 \times 10^8$ to show that the uncertainty in $\bar{r}$ increases as C increases. Compute $\sqrt{\overline{\Delta r^2}}/\bar{r}$ for $C = 10^2$, 10^3, 10^4.

(c) Use the values $C = 10^2$, $a \approx 2 \times 10^{-8}$ to calculate how big $N(t)$ must be to provide, say, a value of $\bar{r}$ that has an accuracy of $\sim 10\%$. If $N_0 \sim 100$, how many bacterial generations are needed to get this value of $N(t)$?

21. We here investigate the number C of cultures required so that $t_0 \ll t$, i.e., that bacterial mutations are likely to appear in the first bacterial generation.

(a) From (3-106a), we have that

$$1 = aCN_0e^{t_0}.$$

Show that the condition $t_0 \ll t$ is satisfied when

$$1 \cong aCN_0. \tag{1}$$

(b) From the experiments of Luria and Delbrück, determine how large C must be to satisfy this condition ($N_0 \sim 100$).

(c) Show that the physical significance of the value of t_0 in the Luria–Delbrück experiment is just this: At t_0 the number of wild-type bacteria is so large that in the next generation following t_0, it is quite likely that there will be at least one mutation. Thus, instead of increasing C, Luria and Delbrück simply let time pass until $aCN(t_0) = 1$, and in essence they count time starting from t_0.

Thermal Equilibrium. The Boltzmann Factor. Entropy and Free Energy. The Second Law of Thermodynamics. Application to Physics, Chemistry, and Biology

4.1 The Statistical Nature of Thermal Equilibrium

4.1.A. Introduction: Thermal Equilibrium in Gases, Solids and Fluids. Equilibrium Between Phases. Chemical Reaction Equilibrium. Statistical Physics Versus Thermodynamics

In this introductory section we will discuss the concept of thermal equilibrium. Taken in the most general sense, this also includes the equilibrium between different phases of a given material (e.g., gas–liquid), as well as the equilibrium between the partners partaking in a chemical reaction.

Some aspects of thermal equilibrium are familiar, and can be established by simple, everyday observations: A cup of tea, left to itself, will cool and eventually acquire the same temperature as its environment. If the tea is stirred, the turbulent motion of the fluid will eventually die out, and the fluid will come to rest. If a few drops of sweetener are put into the cup, it will eventually spread uniformly over the tea. In all three cases we see the natural evolution of an initially "unbalanced" situation toward a final, quiescent state characterized by what we may call maximum *uniformity*: Temperature differences, differences of motion, and differences of concentration are wiped out. This final stage *is* thermal equilibrium.

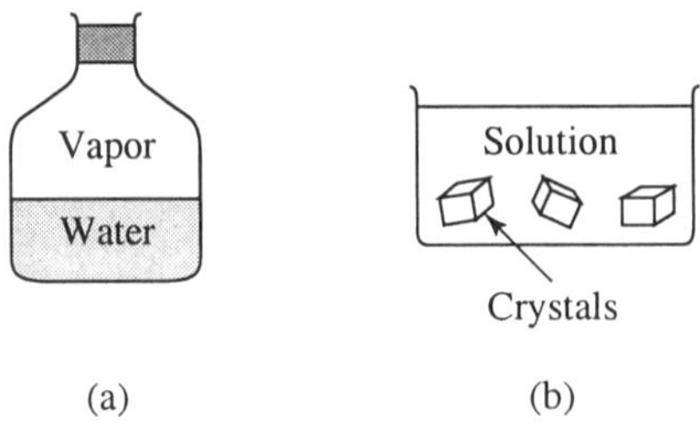

Figure 4.1. Phase equilibria.

But, as we have already remarked, thermal equilibrium encompasses a wider range of phenomena than are contained in the above-mentioned teacup. We will now give a few additional examples, in order to see this concept demonstrated in its full generality:

(a) We all know that if water is placed in an open, shallow dish, it slowly evaporates until no water is left. If, however, water (or any other fluid, for that matter) is placed in a closed bottle, this evaporation stops after a while, and the fluid level remains constant thereafter. What happens is that the evaporating fluid molecules are now confined to a finite volume, and increasing evaporation must raise the vapor concentration above the fluid. The vapor *concentration* finally reaches an *equilibrium* value, after which further evaporation ceases.

(b) In a dish filled with water we place some crystals of salt. The salt begins to dissolve, but if enough salt is present, this process comes to an end, once a specific salt concentration is reached. The solution of salt is now in *equilibrium* with the solid crystalline salt phase, and at this point the process of dissolving comes to an end.

(c) As an example of a chemical reaction equilibrium, consider the mixing of a certain amount of iodine vapor (I_2-molecules in gas phase) with molecular hydrogen gas (H_2). Upon mixing, a chemical reaction begins to take place, leading to the formation of hydrogen–iodide

$$H_2 + I_2 \rightarrow 2\,HI.$$

This process does *not* proceed all the way, i.e., until all the H_2 or I_2 are used up. It stops once a certain quantity of HI is formed. At this point, the initial reaction partners, H_2 and I_2, are in *equilibrium* with the reaction product HI, and from then on, the relative abundances of H_2, I_2, and HI remain constant.

(d) Another similar example is given by the dissociation equilibrium of a weak acid. If a small amount of concentrated acetic acid, CH_3-COOH, is diluted with plenty of water, partial dissociation takes place

$$CH_3-COOH \rightarrow CH_3OO^- + H^+.$$

Again, this process does not proceed to completion, but stops once a characteristic concentration of hydrogen ions (H^+) has been reached.

These examples give us enough variety so that we may begin to investigate what they have in common. First, let us observe that all these equilibria can be described macroscopically, i.e., by bulk properties readily measured by the chemist and physicist. The macroscopic parameters needed for such properties are temperature, volume (or pressure), and then the amounts of material in the various phases or chemical forms, expressed in moles, molarities, or partial volumes, as the case may be.

But, let us now ask the question of the significance and characteristics of these equilibria on the *molecular* level. As a first example, let us choose the case of the dissociation of acetic acid diluted in water. A similar case was already discussed in Chapter 1, Section 1.4.C, on the distribution of electric charges on the hemoglobin molecule. We saw there (Appendix to Section 1.4.C) that the dissociation equilibrium must be understood as the *dynamic* balance between two opposite processes:

(i) the dissociation of the protonated acid:

$$CH_3-COOH \rightarrow CH_3-COO^- + H^+;$$

(ii) the recombination of the dissociation products:

$$CH_3-COO^- + H^+ \rightarrow CH_3-COOH.$$

Both processes, which we can summarize in the relation;

$$AH \rightleftharpoons A^- + H^+,$$

are constantly occurring in the solution. To describe them, we introduce the quantity α, the degree of dissociation. It is defined by

$$\alpha = \frac{\mathcal{N}(H^+)}{\mathcal{N}_0} = \frac{\mathcal{N}(A^-)}{\mathcal{N}_0},$$

where $\mathcal{N}(H^+) = \mathcal{N}(A^-)$ represents the number of H^+ or A^- ions in the solution, and $\mathcal{N}_0$ the total number of AH molecules *before* dissociation. After partial dissociation has occurred, $\mathcal{N}_0$ is given by

$$\mathcal{N}_0 = \mathcal{N}(AH) + \mathcal{N}(H^+) = \mathcal{N}(AH) + \mathcal{N}(A^-).$$

In terms of α, we thus have the relations

$$\mathcal{N}(H^+) = \mathcal{N}(A^-) = \alpha \mathcal{N}_0,$$
$$\mathcal{N}(AH) = (1 - \alpha)\mathcal{N}_0.$$

Let us now see what we can say about the rates at which these two processes occur. We expect then, the *mean* number of dissociations occurring per second to be proportional to the number of undissociated molecules, which is $\mathcal{N}_0(1 - \alpha)$:

$$\text{dissociation rate} = c_1 \mathcal{N}_0(1 - \alpha) = k_1(1 - \alpha). \tag{4-1}$$

The *mean* number of recombinations per second will be proportional to the product of the number of partners that must meet to recombine, which is therefore proportional to α^2:

$$\text{recombination rate} = k_2 \alpha^2. \tag{4-2}$$

In Figure 4.2, these rates are plotted as a function of α. Assume that at a given time the actual degree of dissociation is $\alpha_1 < \alpha_{eq}$. Then Figure 4.2 shows that the dissociation rate is larger than the recombination rate, and hence α increases, moving it toward α_{eq}. If, on the other hand, the system at a given time has a degree of dissociation $\alpha_2 > \alpha_{eq}$, then the recombination rate is the larger, α decreases, and again moves toward α_{eq}. α_{eq} is therefore the stable, average value of the degree of dissociation. This argument clearly illustrates the dynamic character of this equilibrium. Notice well that the rates described in (4-1) and (4-2) are *average* rates, which are meaningful because the solution contains an enormous number of molecules. On the level of the individual molecule or ion, these processes are *random* events: Recombination is based on a chance encounter of the two ions CH_3—COO and H^+, and the dissociation of a given CH_3—COOH molecule results from fluctuations of its energy.

Upon reflection, it is readily seen that *all* equilibria mentioned above have this same dynamic aspect in common. In example (a) on the equilibrium vapor pressure, we have again two opposite processes going on all the time:

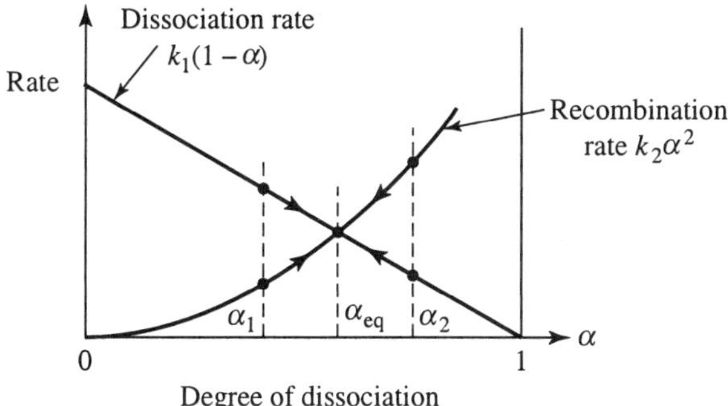

Figure 4.2. Plot of dissociation and recombination rates in a weak acid, as a function of the degree of dissociation α.

(i) the escape of water molecules from the surface of the fluid; and

(ii) the collision of water molecules in the vapor phase with the fluid surface and ensuing capture.

The rate process (ii) clearly increases with increasing concentration (pressure) of the water vapor, and at some sufficiently large value of this concentration, the rates for (i) and (ii) become equal. At this point, equilibrium is reached. Examples (b) and (c) can be analyzed along similar lines.

Even in the case of the tea in the teacup reaching equilibrium, we deal with an analogous situation: The cooling of the tea represents a net transfer of energy to the surrounding environment. This is, of course, partly achieved by heat loss due to evaporation, but in part also through heat transfer from the tea to the cup, and from the cup to the air. The recipients of this energy are the air molecules, which collide with the teacup. We have seen in Appendix 2.A.2 of Chapter 2 how this energy exchange can be described statistically: In any individual collision, an air molecule, i.e., nitrogen or oxygen, may either pick up or loose energy. On *average*, however, it will pick up energy if the teacup is hotter than air. Again then, we have two opposite processes, molecular collisions when the molecules gain energy, and collisions when they loose energy. The final equilibrium is not characterized by a termination of this energy exchange process, but by a dynamic "truce" in which energy gain by some molecules is balanced by energy loss by other molecules. When this point is reached, we say (and observe) that the temperatures of the air and the teacup have become equal.

Let us finally look at the spread of the sweetener put into the tea. The cyclamate molecules spread over the fluid by diffusion, i.e., by random walk. This is another reminder that on the molecular level the processes at work, which drive the system toward equilibrium, are random processes. The uniformity of the sweetener concentration is the ultimate average result of the spreading by random walk.

We therefore take the general view of the evolution of the systems mentioned in the above examples as a random process. Therefore, many different outcomes are possible in principle. The apparent inevitability of the evolution toward an equilibrium indicates that this equilibrium represents a class of outcomes that are overwhelmingly more probable than others. To state it in other words: The evolution of a system toward thermal equilibrium is the most probable result of the enormous number of random events occurring on the molecular level.

We may, at this point, bring in an analogy with a case previously discussed. It is a fact of experience that in a large population the ratio between males and females has a rather fixed value within narrow limits. But on the "molecular" level of the individual, the birth of a male or female offspring is a random event. But it is overwhelmingly more probable that the ratio of boys to girls born in a large population is between say 0.512 and 0.515 than it is outside this range.

A necessary consequence of this view of thermal equilibrium is that there will be fluctuations in the mean values of the parameters that describe this equilibrium. For instance, in a small volume element of air, there will be fluctuations in the number of molecules about its mean value. Their total energy will fluctuate too. Or, in a volume element containing different chemical species in a reaction equilibrium, there will be fluctuations of the relative number of reacting partners about their mean relative values.

These fluctuations are quite fundamental, theoretically. Recently, it has become possible to measure these fluctuations experimentally. One technique is to measure the magnitude and time-dependence of the intensity of light scattered by the fluctuations [1]. From the amplitude of the fluctuations one can obtain important information on the equation-of-state of the system. Thus in a single component system, as in a gas or liquid, one can measure the compressibility from the amplitude of the density fluctuations. In a mixture of two fluids the amplitude of the concentration fluctuations is proportional to the osmotic compressibility $(\partial C/\partial \pi)_T$. The time-dependence of the fluctuations, in turn, gives information on the "transport coefficients" that govern the rate at which fluctuations return to the equilibrium state. For example, in a solution of macromolecules in water, the time-dependence of the fluctuations permits one to measure the diffusion coefficient of the macromolecules [2]. On the other hand, from the time-dependence of the fluctuation of the number of reaction partners in a chemical reaction, it has been proven possible to determine chemical reaction rate constants [3].

Statistical Physics Versus Thermodynamics.   It will be the main theme of this chapter to show how thermal equilibrium in its various aspects can be understood as the "most probable outcome" of the random processes operating on the atomic–molecular level.

This way of looking at thermal equilibrium has evolved into a highly developed discipline called Statistical Mechanics. Its main power and attractiveness derives from the very detailed molecular description it provides for an equilibrium situation. This detailed description makes it possible to give a microscopic interpretation of the observed bulk properties of systems.

The predictive power of statistical mechanics is limited mainly by mathematical difficulties.

The statistical–mechanical method is not the only way to arrive at an understanding of thermal equilibria. We mentioned already that such equilibria are described entirely in terms of macroscopic quantities, temperature, pressure, partial volumes, mole number, etc. Historically, such equilibria were first analyzed by a method that entirely avoids a *microscopic* (atomic–molecular) description of the systems. This is the so-called *thermodynamic* method (a misnomer, by the way; it should be called the thermostatic method since it deals with equilibria). Thermodynamics is based on a semiempirical approach. The bulk properties of the materials that compose the systems are described by the empirical *equations of state*. These equations relate, for instance, the volume of a gas, liquid, or solid to pressure and temperature, or describe its heat capacity as a function of these same variables.

In addition to these equations of state, the thermodynamic method relies on two general laws, the so-called first and second law of thermodynamics.

We have already encountered the first law in Volume I, Chapter 5. The first law of thermodynamics is essentially the general statement of energy conservation in the process of energy exchange of a system with its environment. The essential point is that in such processes we must recognize *heat exchange* as a form of energy exchange, in addition to the more usual exchange of energy in the form of mechanical, or electromagnetic, or chemical work. So, if E is the total (internal) energy of a system, then the change ΔE of E is related to the heat input ΔQ minus the energy transfer ΔW from the system to its environment in the form of work

$$\Delta E = \Delta Q - \Delta W. \tag{4-3}$$

Thus, the first law takes care of the general bookkeeping for the acquisitions and expenditures of energy in its various guises. But, clearly, this is not enough to tell us about equilibria. For these, we need an additional law or rule equivalent in content to the statistical statement of the most probable set of outcomes of a random process.

Here, thermodynamics relies on the observational fact of the existence of manifestly *irreversible* processes, leading the system inexorably toward the equilibrium state. The cooling of a cup of tea is an example. The word *irreversible* here has the connotation that the opposite process—a cup of tea getting warmer or colder than its environment—*does not happen*. In the light of our statistical understanding of this process of cooling, this amounts to saying that "the most probable is inevitable." This, in fact, is the point of departure for the formulation of the so-called *second law*: This law is essentially a sharpening up of the simple observational fact that a cup of tea will not spontaneously get hotter or colder than its environment.

The second law generalizes this observational fact, and states categorically *that there exists no conceivable process, whose only effect would be that a certain amount of heat has been transferred from a colder to a hotter body.*

The emphasis is on the word *only*. It *is* possible to transfer heat from a colder to a hotter body. This is done on a large scale in air conditioners and refrigerators. The second law requires that to operate such a "heat pump," a power input is required.

4.1.B. Elements of Quantum Physics

Chapter 4 has a very ambitious objective, namely, to describe the macroscopic properties of systems consisting of many individual atoms and molecules. To do this, we will draw upon profound ideas from three fundamental fields of physics, viz, quantum mechanics, statistical mechanics, and thermodynamics.

In the following three subsections we summarize and present, *without proof*, some fundamental ideas and facts from quantum mechanics. These are needed to characterize correctly the microscopic statistical properties of the assembly of individual atoms and molecules. We ask the student to accept these facts of quantum mechanics without proof, leaving until later in his/her studies a more complete analysis of their theoretical and experimental basis. The field of quantum mechanics represents the juncture between classical mechanics on the one hand, and chemistry on the other. The language of chemistry, that is so different from that of classical mechanics, is in fact the same as the language of applied quantum mechanics.

(i) Energy States in Atoms, Molecules, Macromolecules, and Solids

The outstanding *fact* of quantum physics is that the *bound* structures of two or more particles have their internal energy *quantized*. A bound system is characterized by a negative total internal energy; i.e., the partners in the system cannot escape from each other. This internal energy is simply the energy of the system when its center of mass is at rest. The total energy of such a structure is the sum of total internal energy and the kinetic energy of motion through space.

Most of the detailed evidence on the quantization of internal energy comes from spectroscopic data, that is, from the observation of absorption and emission of light (from infrared to ultraviolet). Light of frequency ν consists of energy packages, "photons," of energy $h\nu$ where h is Planck's constant. The elementary process of absorption or emission of light by an atom or molecule consists of picking up or emitting one photon. In that process, the energy of the atom or molecule changes by

$$\Delta E = \pm h\nu \quad \begin{cases} +: & \text{absorption,} \\ -: & \text{emission.} \end{cases}$$

The fact of energy quantization then follows directly from the *observation* that atoms and molecules absorb and emit only light of well-defined, characteristic frequencies. On the basis of such observation, an *energy level diagram* can be constructed for each atom or molecule. In Figure 4.3, we give a few examples. In the case of the hydrogen atom the energy of the various energy levels can be

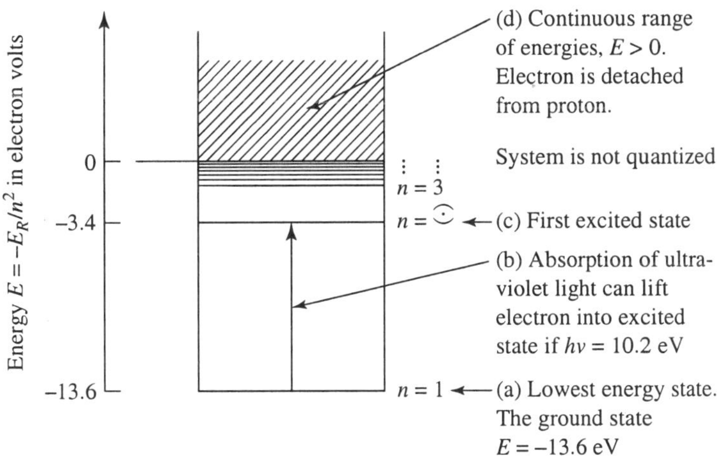

Figure 4.3. Energy levels of a hydrogen atom. The negative energy states, i.e., bound states, are quantized. At positive energy ($E > 0$) one has a continuous distribution of energy. The electron is not bound to the proton, and can fly away.

expressed by a formula

$$E_n = -E_R \frac{1}{n^2}, \qquad n = 1, 2, 3, \ldots \qquad (4\text{-}4)$$

The integer n is a number running from 1 toward ∞, and both identify the level and its energy. Such numbers are called *quantum numbers*. In Figure 4.3, and in later discussions, we use, as a convenient unit of energy, the "electron volt (eV)." 1 eV is equal to 1.60×10^{-12} erg. (This is equal to the kinetic energy that an electron gains on being accelerated through an electric potential difference equal to 1 V.)

Figure 4.4 lists the energy levels of the alkali atom sodium. In this slightly more complicated level spectrum (spectrum = set of all levels), the energy levels are grouped in series. These different series correspond to different values of the angular momentum of the orbiting valence electron. Angular momentum is also quantized, and comes in portions $(h/2\pi)\sqrt{l(l+1)}$ with l—the angular momentum quantum number—taking on the integer values $0, 1, 2, 3 \ldots$. If we turn to even the *simplest molecules*, such as hydrogen, the level spectrum becomes much more complex. A molecule has energy states corresponding to a given arrangement of the orbiting electron. These different arrangements define the *electronic energy states*. For *each* such electronic state, the molecule as a whole can also vibrate and rotate, and the energies of rotation and vibration are quantized too.

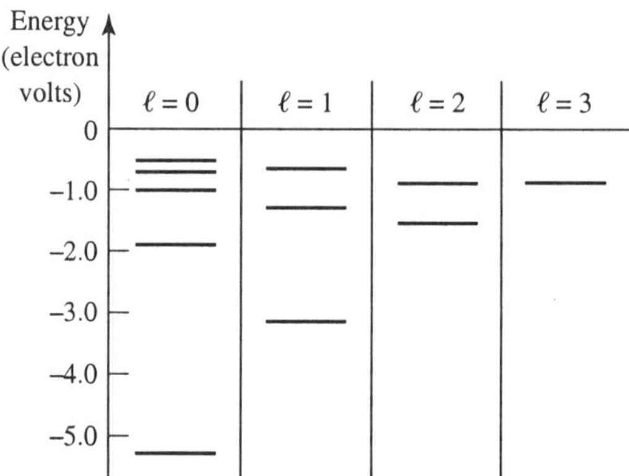

Figure 4.4. Energy levels of the sodium atom. Levels are grouped into series; all levels in a series have a common value of the angular momentum quantum number l.

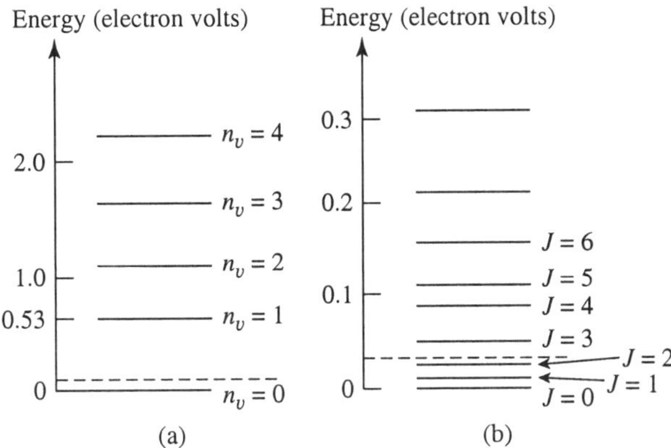

Figure 4.5. Molecular energy states. (a) The vibrational energy states of the hydrogen molecule. They are labeled by a vibrational quantum number $n_v = 0, 1, 2, \ldots$. (b) The rotational energy states of the hydrogen molecule. They are labeled by a quantum number $J = 0, 1, 2, \ldots$. (The dashed lines represent the energy $E = kT$ for $T = 300\,°$K.)

In Figure 4.5 we display the quantized energy states of the *vibration* and *rotation* of the hydrogen molecule (H_2). In this figure the energy levels plotted represent the *excitation* energies of rotation and vibration. The total energy of the molecule is the sum of this *positive* excitation energy and the *negative* binding energy. The binding energy represents the total potential and kinetic energy of the electrons and nuclei when the latter assume their stationary equilibrium position. The molecule as a whole remains bound, as long as the total vibrational and rotational excitation energy does not exceed the binding energy. Also, in Figure 4.5, note the numerical values on the energy scale. The vibrational levels are about $\frac{1}{2}$ eV apart, which is considerably closer than the lower atomic energy level spacings in sodium and hydrogen. The rotational levels are even more closely spaced, the level $J = 1$ being only about 7.3×10^{-3} eV above the lowest state $J = 0$. In Figure 4.5 we have also drawn two dashed lines representing the energy kT for $T = 300\,°$K ($80\,°$F). For the vibrational levels, the spacing of levels is large compared with kT, which is 2.6×10^{-2} eV. By contrast, there are several rotational levels below the kT line. The importance of this distinction will be made clear shortly.

Since the internal energy of a molecule is the sum of the energy of the electron configuration and of the energies of vibration and rotation, the complete energy level spectrum of the entire molecule is quite complex. Each energy level is represented by at least three quantum numbers, specifying the value of the electronic, vibrational, and rotational energy, respectively.

Let us pay some attention to larger, so-called polyatomic molecules. We have then two new phenomena. The first is the occurrence of several patterns of vibrations (also called normal modes). The molecule can invest energy simultaneously in the various patterns of vibration. These energies are quantized for each pattern, and additive, if excitation occurs simultaneously in more than one pattern. We may illustrate this by means of the *water* molecule (H_2O). This molecule has three patterns of vibration as shown in Figure 4.6. The quantized energies of each mode are given by the oscillator formula

$$E_n = n\varepsilon_a, \qquad n = 0, 1, 2, \ldots, \tag{4-5}$$

and the actual state of vibrational excitation of the molecule is

$$E(n_a, n_b, n_c) = \varepsilon_a n_a + \varepsilon_a n_b + \varepsilon_a n_c, \tag{4-6}$$

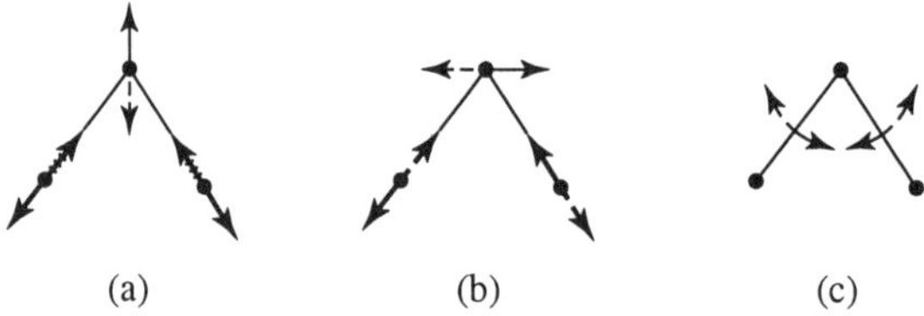

Figure 4.6. The three modes (patterns) of vibration of the water molecule. (a) Symmetric stretch mode, (b) asymmetric stretch mode, and (c) bending mode.

where n_a, n_b, n_c may each assume the values $0, 1, 2, \ldots$. It thus takes a set of three quantum numbers to specify a given vibrational energy level of this molecule. The lowest few levels, and their quantum numbers, are plotted in Figure 4.7. Proceeding to molecules more complex than water, a new feature of importance in its consequence for statistical physics is the occurrence of *conformational isomers* for a given molecule. A simple illustration of this is given by the molecule 1,2-dichloroethylene, shown in Figure 4.8. These are two isomers commonly called the *cis-* and *trans-*form. These two isomers do not have exactly the same energy in their lowest state. Depending on the height of the *potential barrier* separating the two isomeric forms of the molecule, the molecule may jump from one to the other form at a certain rate.

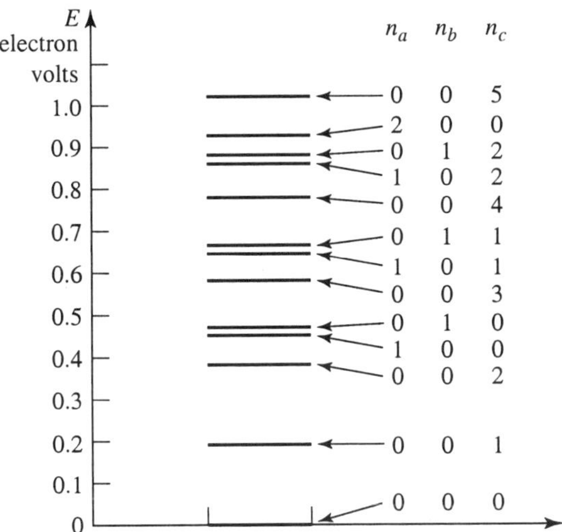

Figure 4.7. Vibrational energy levels of the water molecules and associated quantum numbers n_a, n_b, n_c. The energies ε_a, ε_b, ε_c are $\varepsilon_a = 0.192$ eV, $\varepsilon_b = 0.452$ eV, and $\varepsilon_c = 0.466$ eV.

Figure 4.8. Two conformational isomers of the molecule 1,2-dichloroethylene.

(ii)   Quantum States. Stability of Atoms and Molecules

In 1925, Schroedinger demonstrated that the quantization of energy could be understood, and is a consequence of the *wave nature* of matter. In Schroedinger's *wave mechanics* (which is a version of quantum mechanics), a particle such as an electron is not described as a *point* moving through space and time, but rather as a spatially spread-out probability distribution. This probability distribution $P(\vec{r})$ is given by the square modulus of a *wave function* $\psi(r)$:

$$P(\vec{r}) = |\psi(r)|^2, \tag{4-7}$$

and $\psi(r)$ itself must obey a certain wave equation, the so-called *Schroedinger equation*. Schroedinger showed that for bound systems of electrons and nuclei, such as atoms and molecules, there were no meaningful solutions to this wave equation except for certain discrete or quantized values of the energy of the system. The acceptable solutions form a denumerable ("countable") set, and the numbers used to tag each solution are the quantum numbers we referred to previously.

Each solution ψ_n (n being the quantum numbers) represents a possible "state of existence" of the atom or molecule. These states, labeled and identified by the quantum numbers n, are called *quantum states*.

Different quantum states in general have different energies; occasionally, however, a group of distinct quantum states have the same energy. Such an energy value (energy level) is then called *degenerate*.

By far the most important consequences of this quantum mechanical description of atoms and molecules is that it offers an explanation of the *structural stability* of the different chemical species as we know them. An oxygen molecule is stable because the quantization of its internal energy requires that a finite (and large) amount of energy be added before *any* change *at all* can occur in the structure of the molecule. The quantization of energy also explains why all members of a chemical species are *alike*, in fact, it explains why sharply defined chemical species exist at all. For instance, any two helium atoms in helium gas at room temperature are exactly alike, simply because they represent the *same* quantum state (the state of lowest energy) of the system of two electrons and a nucleus of charge $+2e$. Also, the helium atom in its lowest quantum state is *stable*; the next lowest energy state is 20 eV higher; no energy anywhere near this can be supplied by collision in the gas, where the typical kinetic energy is $\frac{3}{2}kT = 4 \times 10^{-2}$ eV. In such collisions, therefore, *nothing at all* can happen to the electronic structure of the helium atom.

On the other hand, collisions between hydrogen *molecules* can cause the molecule to jump from one low-lying rotational energy state to another. This is seen from Figure 4.5, where the energy kT is comparable with the rotational energies for small values of the quantum number J. The vibrational states of the hydrogen molecules are unaffected by collisions since kT is much smaller than the spacing between the vibrational states.

(iii)   Free Particles. De Broglie Wavelength and Uncertainty Relations

The energy of an atom or molecule is the sum of the internal energy and the energy of its center of mass motion. This latter is not quantized. In quantum mechanics this motion is described by a *moving wave packet*, such as is illustrated in Figure 4.9.

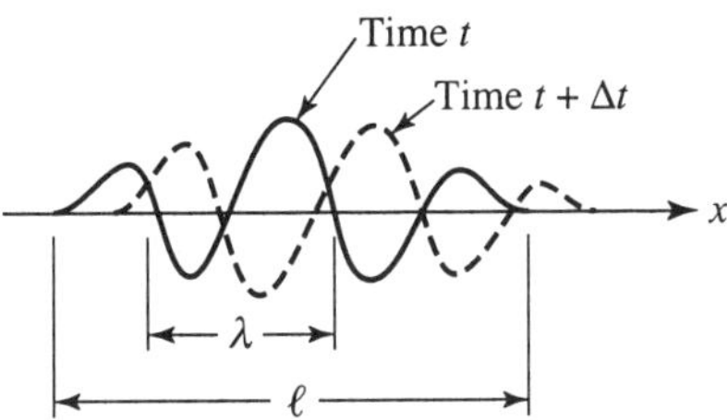

Figure 4.9. One-dimensional wave packet. The solid and dotted lines show the packet at two different times, indicating displacement of the packet.

What interpretation can we attach to this illustration? The wave equation prescribes that this packet moves with a velocity $v = h/m\lambda$, where h is Planck's constant, m the mass of the particle, and λ the wavelength of the packet. (This is the famous de Broglie relation $p = mv = h/\lambda$, relating the wavelength λ to the momentum p.) This displacement of the wave packet represents the motion of the particle through space.

Any such wave packet represents a *quantum state* of the moving particle. Since the wave packet has a finite spatial extension l, the quantum state does not specify the location of the particle with accuracy. This location has an *uncertainty* δx about equal to l, the length of the packet

$$\delta x \cong l. \tag{4-8}$$

Because of the finite size of the packet, also its wavelength, λ cannot be accurately defined. The precision with which λ can be specified is determined by the number of wavelengths (l/λ) in the wave packet. We have the estimate

$$\left(\frac{\delta\lambda}{\lambda}\right) \cong \frac{\lambda}{l}. \tag{4-9}$$

This, in turn, implies an indeterminacy of the momentum $p = mv$ of the particle in this quantum state of an amount

$$\delta p = m\,\delta v = h\frac{\delta\lambda}{\lambda^2}$$

$$= h/l. \tag{4-10}$$

Irrespective of the detailed structure of the wave packet, we have therefore the general relation connecting the uncertainty δx in location and the uncertainty δp in momentum

$$\delta p \, \delta x \cong h. \tag{4-11}$$

This is the famous *Heisenberg uncertainty relation.* This relation points out an important contrast between the classical and quantum mechanical description of a free particle. The state of motion of a particle (in one dimension) can be specified by the values $x(t)$ and $p(t)$ of position and momentum as a function of time. Graphically, this can be represented as a *moving point* in the so-called *phase space* (see Figure 4.10).

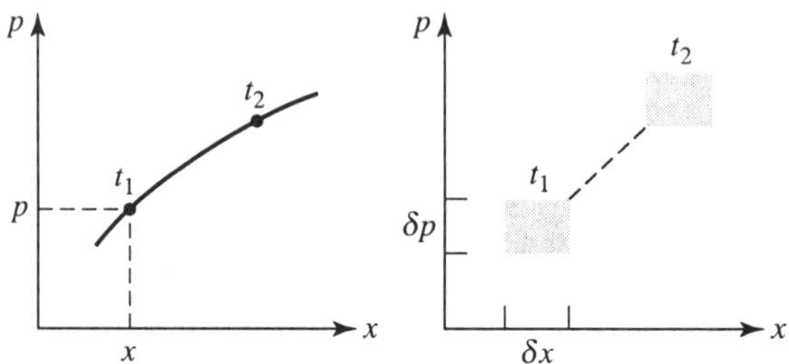

Figure 4.10. Classical and quantum mechanical description of motion in phase space.

We will have more to say about phase space in Section 4.2 of this chapter in the discussion of the thermal equilibrium of a gas. We will see there that a key question is to find the number of *distinct* distributions of N particles in phase space. If these particles are represented by *points* in phase space, there is clearly an infinite number of distinct distributions; this says that there is really no way to *count* the number of distinct distributions. In the quantum mechanical description there is no such difficulty. Each particle is described by a "patch" of area h.* The number of *distinct* distributions of N particles in phase space is then given by the

---

*In the description of the three-dimensional motion of a particle in a six-dimensional phase space, this "patch" becomes a small volume element of size h^3.

number of arrangements of N nonoverlapping patches,[†] each patch having the size h (or h^3; see footnote). This number is well defined and can easily be determined.

(iv)   The Importance of Quantization for Statistical Physics. The Principle of Detailed Balance

In statistical physics we describe the behavior of systems containing a large number N of elements in probabilistic terms. In such a large system the elements can arrange themselves in a great many ways. In a probabilistic approach we must assign a definite probability to each way. A prerequisite for this is that we should be able to identify exactly what we mean by two different "ways" of arranging the elements in a large system, and that we should be able to *count* all distinct arrangements.

Quantum mechanics makes this possible. Any particular "way" of arranging the elements of a large system is represented by a specific "microstate" in which each element of the system is in a specific quantum state. If just one element of the system moves into a different quantum state, then a new, distinct microstate of the system is generated. On this basis, the number of distinct microstates of the system can be determined and counted.

There remains the task of attaching a probability to each microstate. The basis for this is a dynamic principle, the so-called *principle of detailed balance*. We best illustrate it by means of an example: This example consists of three identical, but distinct, oscillators, each of which has an energy spectrum

$$\varepsilon_n = n\varepsilon_0, \qquad n = 0, 1, 2, 3, \ldots .$$

A quantum state of this system ($\equiv$ a "microstate") is defined by specifying the quantum numbers n_1, n_2, and n_3 of each oscillator. The energy of a microstate $\{n_1, n_2, n_3\}$ will then have the value

$$E = (n_1 + n_2 + n_3)\varepsilon_0.$$

Now we will assume that the three oscillators are capable of exchanging energy. This can happen for a variety of reasons. One may be that the oscillations are

---

[†] In a more precise, mathematical language, two distinct wave packets $\psi_a(x)$ and $\psi_b(x)$ are characterized by the property of "orthogonality": $\int dx\, \psi_a(x)\psi_b^*(x) = 0$, ψ^* being the complex conjugate of ψ. The graphic statement of nonoverlapping patches in phase space is equivalent to the property of "orthogonality" of the wave packets.

coupled to some degree; that is, the motion of one oscillator has a small effect on its two neighbors. Or, some external agent, a neighboring atom or molecule, can *perturb* the motion of the three oscillators and induce energy exchange.

In the language of quantum physics, such an energy exchange is called a *transition*. An example of a transition is the process

$$\{n_1, n_2, n_3\} \rightarrow \{n_1, n_2 + 1, n_3 - 1\}. \tag{4-12}$$

The process

$$\{n_1, n_2 + 1, n_3 - 1\} \rightarrow \{n_1, n_2, n_3\} \tag{4-13}$$

is called the *inverse* transition. (See Figure 4.11.)

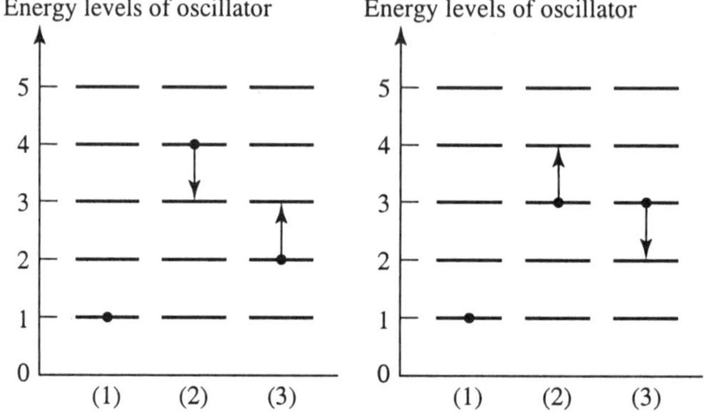

Figure 4.11. Energy states of three oscillators (1), (2), (3). On the left the transition $\{n_1 = 1, n_2 = 4, n_3 = 2\} \rightarrow \{n_1 = 1, n_2 = 3, n_3 = 3\}$ is shown. On the right, the *inverse* transition is presented.

The general methods of quantum physics make it possible to calculate the so-called *transition probability*; that is, the probability that the process [(4-12)] will occur in a certain time interval. We need not go into this, but we must take notice of an extremely important result that comes out of such a calculation.

Whatever the nature of the agent responsible for the transition, and irrespective of how large or small the transition probability is, the *transition [(4-12)] and its inverse [(4-3)], are equally probable*. In general, the transition from any microstate *of the system* to a different microstate *of the same total energy*, and the *inverse* process, are equally probable. This is called *the principle of detailed balance*. It is the basis for the very existence of thermal equilibrium. Let us use the example of the three oscillators to illustrate the implications of this principle.

Let us assume we "start" the system in the state $\{3, 2, 2\}$, which stands for $\{n_1 = 3, n_2 = 2, n_3 = 2\}$, and that there is a large transition probability to the state $\{2, 3, 2\}$, but a very small one to the state $\{5, 0, 2\}$. One might then think that it is much more likely to find the system in the state $\{2, 3, 2\}$ than in the state $\{5, 0, 2\}$. This is a totally false conclusion. If state $\{2, 3, 2\}$ is easy to get into, it is also easy to get *out* of. Conversely, if the system has a low probability to move into state $\{5, 0, 2\}$, it will stay there longer because it has an equally low probability to move *out* of this state. The net effect of this manifestation of the principle of detailed balance is that in the long run the system is found with equal probability in any of the microstates $\{n_1, n_2, n_3\}$ that have a common total energy. Another way of saying it is that the system will spend an equal fraction of its time in any of the states $\{n_1, n_2, n_3\}$ of common energy.

In summary, we see that quantum physics introduces two ingredients of cardinal importance into statistical physics:

(a) First, by means of the concept of the quantum state, it permits a clear-cut, unique way to *define* and *count* all the distinct "states" of a system of N elements. Only quantum physics gives a precise meaning to the concepts of the "state of the system" on the microscopic level.

(b) The quantum mechanical principle of detailed balance leads us to assign equal probability to all microstates ($\equiv$ quantum states of the system) that have a common energy.

With these two results we have at hand the basic means to proceed to a statistical, probabilistic description of the properties of systems composed of a large number of elements.

4.2   The Probability Distribution of Energy. The Boltzmann Factor

In this section we will establish the probability distribution for the energy of the constituent particles of a large system.

We will start with two simple cases to illustrate both the form and meaning of this energy distribution. The first case to be discussed is an idealized crystal of identical atoms, the second an ideal monoatomic gas, such as helium. In both cases, we will assume that the total energy of the entire crystal, or the gas as a whole, has some constant value E. We may then ask: How is this energy distributed over the constituent particles that make up the system? Underlying this question is the view that the constituent particles can *exchange* energy between each other, and that this energy exchange occurs in a random fashion. This random nature of the exchange is clearly apparent in the gas, where it occurs during the collision of two gas atoms. In the crystal, lattice and surface imperfections are the agents of energy exchange. As a result of these exchanges of energy, the various atoms of the system will, at any given time, have widely different energies. This raises the question of the *probability distribution* for the energy of an individual particle of the systems.

In the case of the crystal the vibrational energies of the individual atoms are quantized (discrete). An atom may be found in any of the possible quantum states (n) of energy ε_n. We can therefore raise the fundamental question: What is the *probability distribution* $P(n)$ of finding an atom in any of the possible *energy states* (n)? In the case of the monoatomic gas the energy of an atom is its translational kinetic energy, and possibly a potential energy, and has therefore a continuous range. Since we may, in this case, specify the energy of a particle by its momentum, $\vec{p}$, we may phrase the fundamental question again in the following manner: What is the continuous distribution that describes the probability that the particle can be found within a certain range of momenta $\overrightarrow{\Delta p}$? At the outset it is by no means clear how this probability $P(n)$ will depend on the particle energy. Certainly it will be determined in part by the condition that the total energy of all N particles must add up to a fixed amount E. One might guess therefore that the probability distribution would, for example, be *peaked* about the mean energy per particle, (E/N). In fact, this naive expectation is wide off the mark. The correct distribution, as we will demonstrate, always assigns the largest probability to the lowest energy state of a particle, and falls off *exponentially* with increasing energy of the particle. The factor in the probability distribution that describes this exponential fall off is called the *Boltzmann factor*.

The objective of this section is the derivation of the above-mentioned probability distributions, and to show the origin of the Boltzmann factor.

The route that we will have to travel to obtain the probability distribution for the energy of a single particle may appear rather tortuous at first reading. Just because energy sharing is a cooperative enterprise, the search for the probability distribution for the energy of a single particle requires that we first investigate an apparently much more complex quantity, the so-called joint probability $P(N_0, N_1, \ldots)$. In the case of the crystal this is the probability that a number N_0 of atoms are in the energy state $n = 0$, N_1 of them in the energy state $n = 1$, etc. (A similar joint probability is defined for the gas.) We will show, however, that another key concept evolves from this joint probability—that of the entropy S of a system. It is this concept that will later enable us to relate the statistical description of thermal equilibrium, elaborated in this Section 4.2, to a purely macroscopic ($=$ thermodynamic) description. This connection will be made in Section 4.3.

4.2.A.   Probability Distribution for the Energy of Vibrating Atoms in a Crystalline Solid

As mentioned in the introduction, we begin the discussion of the probability distribution of energy with the example of a system with *quantized* energy states. As we will see, our main task is then reduced to a proper counting of states.

(i)   The Einstein Crystal as a Model

We are all familiar with crystals, a regular array or lattice of either atoms, ions, or molecules. Each atom in the crystal has a well-defined mean position about which it may oscillate. The energy of this oscillation is the thermal energy of the crystal; it increases with temperature and, in the limit of high temperature, this energy is $3kT$ per atom (or ion) in the crystal lattice (see Section 2.1.C). The factor three arises from the fact that *each atom* can oscillate in three perpendicular directions.

Now, in a *real* crystal, these oscillations of individual atoms are strongly coupled to each other. This leads to certain technical complications if one wants to find the quantized energy states of the crystal. We shall, therefore, study here the energy distribution not for a real crystal, but for a simplified *model* of a crystal, the so-called Einstein crystal. In this model each atom oscillates about its mean position in the lattice without being much influenced by the motion of its neighbors. As a result, the energy states of *each atom* separately are quantized, and all atoms have the same set of energy values. We will designate these quantized vibrational energies by ε_n, with the quantum number n assuming the integer values $n = 0, 1, 2, \ldots$. To obtain the *form* of the probability distribution, including the Boltzmann factor, it does not matter what the values of ε_n are. We will, however, in the end, assume that the atoms perform a *harmonic* oscillation, in which case the energy levels ε_n are equidistant

$$\varepsilon_n = n\varepsilon_0.$$

This simple relation will allow us to push the calculation a bit further, and begin to make the connection between the mean energy per particle, as determined from the probability distribution of energy and temperature.

(ii)　Definition of the Probability Distribution $P(n)$ for the Energy ε_n of an Atom in the Crystal

Let us recall that the question we are trying to answer is: Given that the crystal as a whole has a total energy E of thermal excitation, an individual atom on average has an energy (E/N). Nevertheless, its actual energy at a given time may be any one of the quantized energies $\varepsilon_n : \varepsilon_1, \ldots, \varepsilon_n, \ldots$. We wish to find the probability $P(n)$ that an atom is found in any of these states n. To obtain this probability, we may imagine making a total of $\mathcal{N}$ measurements of the energy of a single atom. We then tabulate the number $\mathcal{N}(\varepsilon_n)$ of times that the atom was found in the energy state ε_n. The probability $P(n)$ is then, by definition, the limit of the ratio $\mathcal{N}(\varepsilon_n)/\mathcal{N}$, as $\mathcal{N}$ becomes very large

$$P(n) = \left(\frac{\mathcal{N}(\varepsilon_n)}{\mathcal{N}}\right)_{\mathcal{N}\to\infty}. \tag{4-14}$$

Alternatively, $P(n)$ represents the fractional amount of time that an individual atom spends in energy state n. This probability is the *same* for each atom of the crystal. Therefore, the mean number $\overline{N}_n$ of atoms to be found in state n may also be viewed as the total number N of atoms times the probability that any particular atom is in state n:

$$\overline{N}_n = N P(n). \tag{4-15}$$

This gives us the alternative relation that the probability $P(n)$ may also be described by means of

$$P(n) = \frac{\overline{N}_n}{N}. \tag{4-16}$$

In words: The probability $P(n)$ of finding an atom in the quantum state n is equal to the mean fraction $\overline{N}_n/N$ of atoms in the crystal that are in the quantum state n. Our immediate objective has thus become the determination of the mean numbers $\overline{N}_n$ of atoms that are in the quantum state n.

　　The number N_n of atoms of the crystal that at a given time are in the quantum state n is constantly changing in a random manner. This is a direct consequence of the random exchange of energy between neighbors. An atom in state n may acquire energy from a neighbor, or lose energy to a neighbor, changing its energy state from n to say $n + 1$ or to $n - 1$. In both cases, the number N_n will have decreased to $N_n - 1$.

We are thus led to look at the probability distribution for the numbers N_n. This distribution will forcibly have to be a so-called *joint* probability distribution of *all* numbers N_n, since these numbers cannot change in complete independence. They are interlocked by two conditions. The first expresses the fact that the total number of atoms in the crystal is fixed, which gives

$$\sum_n N_n = N. \qquad (4\text{-}17)$$

The second expresses the fact that the total energy E of the crystal is fixed. Since the N_n atoms, that are in energy state n, contribute the amount $\varepsilon_n N_n$ toward the total energy, this second condition reads

$$\sum_n \varepsilon_n N_n = E. \qquad (4\text{-}18)$$

(iii)   Microstates and Macrostates of the Einstein Crystal. Weight of a Macrostate

In the previous subsection we introduced the concept of the mean number $\overline{N}_n$ of atoms in quantum state n and its relation to the probability $P(n)$ that a given single atom was to be found in quantum state n.

In this subsection, we proceed to calculate these numbers. For this purpose it is necessary to begin with the previously mentioned joint probability distribution

$$P_N(N_0, N_1, \ldots, N_n, \ldots),$$

which gives the joint probability that N_0 of the atoms are in state $n = 0$, N_1 of them in state $n = 1$, N_2 of them in state $n = 2$, etc. Having found an expression for this distribution, we will calculate the mean values $\overline{N}_0, \overline{N}_1, \ldots, \overline{N}_n, \ldots$.

In order to find the expression for the probability distribution $P(N_0, N_1, \ldots, N_n, \ldots)$, it is useful to introduce two concepts that will occur again and again in our statistical considerations—that of *microstates* and *macrostates*. These concepts are really quite simple, and without using these names we have encountered them already in the example of coin tossing.

Assume our "system" is a set of N coins numbered 1 to N. Each toss of this set of coins produces a certain *outcome*, that we may also call a "state" of the set of N coins, and which is described by a specific sequence of heads and tails. For instance,

Coin No.		1	2	3	4	$\cdots$	$(N-1)$	N
	(a)	h	t	t	h	$\cdots$	h	t
Sample	(b)	t	h	t	h	$\cdots$	h	t
Outcomes	(c)	t	t	h	t	$\cdots$	t	h

Two outcomes are considered distinct unless the corresponding sequences of h's and t's are congruent. (In the table above, outcome (a) is distinct from outcome (b).) We know that for N coins there are 2^N distinct outcomes. They are all equally probable, and have, therefore, a probability of $\left(\frac{1}{2}\right)^N$ each. Any such possible outcome of the toss of N coins (as illustrated by the samples (a), (b), (c), above) we will call a *microstate* of the system of N coins.

When we discussed coin tossing previously, in Chapter 1, we were also addressing the question: What is the probability that in a toss of N coins, a number H of heads and a number $T = N - H$ of tails appear? To answer that question, we grouped together all the outcomes (microstates) that had a *common* number H of heads and T of tails. The specification of an outcome by the two numbers H and T only, is much less detailed than the indication of the exact sequence of h's and t's. We call an outcome that is described by means of H and T *only*, a *macrostate* of the system of N coins. Two distinct macrostates must differ by the values of H and T.

It follows right away that two *distinct macrostates* are *not* equally probable as the outcome of a toss of N coins. However, we learned how to compute the probability of a macrostate (H, T) in Chapter 1: The probability $P(H, T)$ is given by

$$P(H, T) = \text{probability of a microstate}$$
$$\times \text{ number of microstates with common values of } H \text{ and } T.$$

The first factor in this equation is $\left(\frac{1}{2}\right)^N$, since there are 2^N microstates altogether, and they are equally probable. The second factor is the binomial coefficient

$$W(H, T) = \frac{N!}{H!\,T!} \tag{4-19}$$

We see that the probability $P(H, T)$ of the macrostate (H, T) has the form of a normalization constant $\left(\frac{1}{2}\right)^N$ times the binomial coefficient $W(H, T)$. The crucial dependence of the probability P on H and T is entirely contained in this latter factor W, which we shall call the *weight* of the macrostate.

After these preliminaries, we are now ready to define microstates and macrostates of the Einstein crystal. To avoid all irrelevant complexity, we shall assume that this crystal is made up of N atoms with their mean positions forming a lattice, and that each atom is capable of oscillating in one direction only. In this way, a single quantum number n, with values $n = 0, 1, 2, \ldots$, defines the quantum state and the energy ε_n of each atom.

A microstate of this system is now defined by indicating the quantum number n of each atom from the first to the Nth. This can be represented by an ordered sequence of quantum numbers n, ranging from $n = 0$ to $n \to \infty$, as illustrated by 3 samples in the tabulation below:

Atom No.		1	2	3	$\cdots$	$N-1$	N
samples of microstates	(a)	$n = 5$	2	1	$\cdots$	0	4
	(b)	$n = 2$	5	1	$\cdots$	0	4
	(c)	$n = 3$	0	2	$\cdots$	1	5

To make the analogy with the coin tossing case as strongly as possible, you may consider N numbered "dice," one for each atom, with the quantum numbers $n = 0, 1, 2, 3, \ldots$ written on the faces of the dice. Each throw of these N dice then produces an outcome which can be represented by a sequence of numbers, such as the ones written above. Each throw of these "dice" generates a "microstate" of the crystal.

The physical reason behind introducing microstates of the crystal in this way lies in the "principle of detailed balance" of quantum dynamics, which leads to the result that *each microstate is equally probable*. This equal probability applies either to the time–history of a single system, or, at any time, to microstates found in a large ensemble of identical systems.

We are now also ready to define a macrostate of the crystal in complete analogy with its definition in the coin-tossing problem: A macrostate of the Einstein crystal, specified by the set of numbers $N_0, N_1, N_2, \ldots, N_n, \ldots$, indicates *how many* atoms (*irrespective of their location*) are in quantum state $n = 0$ (N_0 of them), how many in quantum state $n = 1$ (N_1 of them), etc. The numbers $N_0, N_1, \ldots$ are thus the analogues of the numbers H and T of heads and tails in the coin-tossing problem.

To illustrate this point, we list below four sample microstates for a crystal of $N = 12$ atoms, belonging to two different macrostates:

		1	2	3	4	5	6	7	8	9	10	11	12
	(a)	$n = 4$	2	1	0	0	3	1	5	4	2	0	4
Samples of	(b)	$n = 2$	5	1	3	0	4	0	4	2	1	4	0
microstates	(c)	$n = 1$	2	1	1	0	0	3	2	3	1	0	1
	(d)	$n = 0$	1	0	3	2	0	1	1	3	2	1	1

Samples (a) and (b) belong to the same macrostate for which

$$N_0 = 3, \quad N_1 = 2, \quad N_2 = 2, \quad N_3 = 1, \quad N_4 = 3, \quad N_5 = 1, \quad N_{>5} = 0.$$

Samples (c) and (d) both belong to another macrostate, for which

$$N_0 = 3, \quad N_1 = 5, \quad N_2 = 2, \quad N_3 = 2, \quad N_{>3} = 0.$$

Given that every microstate has the same probability, we can now express the probability $P(N_0, N_1, \ldots, N_n, \ldots)$ of a macrostate as follows:

$$P(N_0, N_1, \ldots, N_n, \ldots) = \text{(probability of a microstate)}$$
$$\times \text{ number of all microstates with common values of }$$
$$N_0, N_1, \ldots, N_n, \ldots. \tag{4-20}$$

This second factor will again be called the weight $W(N_0, N_1, \ldots, N_n, \ldots)$ of the macrostate. This weight factor W contains all the decisive information, in particular, the relative probability of two macrostates. Indeed, the first factor in (4-20) may just be considered as a normalization constant for P. Indeed, where needed, we can express it in terms of the sum of all weights, using the fact that the sum of all probabilities $P(N_0, N_1, \ldots, N_n, \ldots)$ must add up to 1:

$$\sum P(N_0, N_1, \ldots, N_n, \ldots) = \text{(probability of a microstate)}$$
$$\times \sum W(N_0, N_1, \ldots, N_n, \ldots) = 1.$$

The probability of a single macrostate is therefore given by the reciprocal of the sum of all weights, and we can write, for (4-20):

$$P(N_0, N_1, \ldots, N_n, \ldots) = \frac{W(N_0, N_1, \ldots, N_n, \ldots)}{\sum W}. \tag{4-20a}$$

In the case of coin tossing, the factor W was the binomial coefficient. What is that factor in the present case? We will present, in an appendix to this chapter, a simple proof, which shows that the number W here is a generalization of the binomial coefficient, the so-called multinomial coefficient. This number is defined as $N!$, divided by the product of all $N_n!$ that are not zero

$$W(N_0, N_1, \ldots, N_n) = \frac{N!}{N_0!\, N_1! \cdots N_n! \cdots}. \tag{4-21}$$

Since this weight W gives the relative probability for the occurrence of various values of N_0, N_1, etc., it contains all the information needed to calculate the mean value $\overline{N}_0$ of N_0, for instance. This mean value is given, according to (4-20a), by

$$\overline{N}_0 = \frac{\sum N_0 W(N_0, N_1, \ldots, N_n, \ldots)}{\sum W(N_0, N_1, \ldots, N_n, \ldots)}. \tag{4-22}$$

The sum in this expression extends over all values of all N_n's, subject to the restrictions imposed by (4-17) and (4-18). To help us to get a "feel" for this procedure, we will illustrate it in the next subsection by means of a numerical example.

(iv)   Numerical Example for a Very Small Crystal

The number of microstates in a crystal of N atoms increases exponentially with the number of particles N and with energy E. We therefore choose an example in which N is quite small

$$N = 6.$$

We will also use the simple equidistant energy spectrum of a harmonic oscillator

$$\varepsilon_n = n\varepsilon_0, \qquad n = 0, 1, 2, \ldots.$$

We also assume that the total energy E of this six-atom crystal is $E = 8\varepsilon_0$, so that the mean energy per particle is $\frac{4}{3}\varepsilon_0$. Even in this case there are 1107 distinct microstates, as we shall see.

In Table 4.1 we list all the macrostates identified by the set of numbers

$$N_0, N_1, \ldots, N_8,$$

as well as the weight W defined by (4-21).

Table 4.1. Values of the Weight $W(N_0, N_1, \ldots, N_8)$ for the Various Macrostates of a Crystal with $N = 6$ Atoms and Total Energy $E = 8\varepsilon_0$.

N_0	N_1	N_2	N_3	N_4	N_5	N_6	N_7	N_8	W
5	0	0	0	0	0	0	0	1	6
4	1	0	0	0	0	0	1	0	30
4	0	1	0	0	0	1	0	0	30
4	0	0	1	0	1	0	0	0	30
4	0	0	0	2	0	0	0	0	15
3	2	0	0	0	0	1	0	0	60
3	1	1	0	0	1	0	0	0	120
3	1	0	1	1	0	0	0	0	120
3	0	2	0	1	0	0	0	0	60
3	0	1	2	0	0	0	0	0	60
2	3	0	0	0	1	0	0	0	60
2	2	1	0	1	0	0	0	0	180
2	2	0	2	0	0	0	0	0	90
2	0	4	0	0	0	0	0	0	15
1	4	0	0	1	0	0	0	0	30
1	3	1	1	0	0	0	0	0	120
1	2	3	0	0	0	0	0	0	60
0	5	0	1	0	0	0	0	0	6
0	4	2	0	0	0	0	0	0	15

For each macrostate, $(N_0, N_1, \ldots, N_8)$, the last entry lists the weight W, which is the number of distinct microstates, $W(N_0, N_1, N_2, \ldots)$, as given by (4-21). In order to illustrate how these weights W arise, let us look at the first and second row of this table. The first row lists the macrostate $N_0 = 5$, $N_8 = 1$, all other $N_n = 0$. This means one particle has all the energy $8\varepsilon_0$. This energy can be located at any of the six lattice points of our small crystal. Therefore, there are six microstates for this macrostate, and W is 6. In the second row we have a macrostate with $N_0 = 4$, $N_1 = 1$, $N_7 = 1$, all other $N_n = 0$. One particle carries energy $7\varepsilon_0$. It can be any of the six particles. One of the remaining five particles has an energy $1\varepsilon_0$. It can be any of the remaining five particles. This gives $6 \times 5 = 30$ microstates, hence $W = 30$.

Table 4.1 also shows that for each macrostate (row) the total number of particles is $N = \sum N_n = 6$, and the total energy $E = \sum \varepsilon_n N_n = \sum n\varepsilon_0 N_n = 8\varepsilon_0$, as required. We can use this table to calculate the mean values $\overline{N}_n$ for each n, using (4-22). For instance, $\overline{N}_0$, the mean number of atoms with energy zero, is obtained by taking each value of N_0, as listed in the first *column*, multiplying it by the weight W at the right end, adding them up, and finally dividing by the total weight $\sum W = 1107$:

$$\overline{N}_0 = (5 \times 6 + 4 \times 105 + 3 \times 420 + 2 \times 345 + 1 \times 210)/1107 = 2.36.$$

Dividing by $N = 6$ then gives the mean *fraction* of atoms that will be found in quantum state $n = 0$, with energy 0:

$$\frac{\overline{N}_0}{N} = \frac{2.36}{6} = 0.393$$

According to (4-16), this then is the *probability* $P(n = 0)$ *that an atom of the lattice will be in the lowest state* $n = 0$. In this manner we can assemble the table of probabilities $P(n)$ that an atom will be in the state n for the chosen value of the total energy. These probabilities are tabulated in Table 4.2.

Table 4.2. Probability $P(n)$ that an Atom Will Be in State n (Energy $\varepsilon_n = n\varepsilon_0$) for a Crystal of $N = 6$ Atoms, and Energy $E = 8\varepsilon$.

$n =$	0	1	2	3	4	5	6	7	8
$P(n) =$	0.393	0.279	0.135	0.087	0.063	0.032	0.013	0.004	0.001

In Figure 4.12 these probabilities $P(n)$ are plotted, along with the values of $P(n)$ obtained from the "Boltzmann" formula, valid for large values of N and N_n, to be derived later:

$$P(n) = \text{constant } e^{-n\beta\varepsilon_0} \qquad \text{[see also (4-39)]}, \qquad (4\text{-}23)$$

β is adjusted so that the mean energy per particle

$$\bar{\varepsilon} = \sum_n (n\varepsilon_0) P(n)$$

is equal to $E/N = \frac{4}{3}\varepsilon_0$. This gives $(\beta\varepsilon_0) = \ln\left(\frac{7}{4}\right) = 0.56$. In Figure 4.13 we plot the values of $P(n)$ for three different energies, $E = 4\varepsilon_0$, $E = 6\varepsilon_0$, and $E = 12\varepsilon_0$.

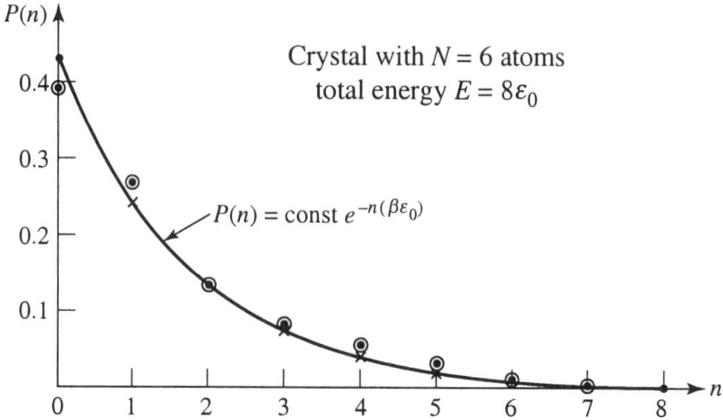

Figure 4.12. Plot of probability $P(n)$ for an atom of the Einstein crystal to be in energy state n. Circled points, $\odot$ = values of $P(n)$ as given in Table 4.1. The number of atoms in the crystal is $N = 6$, and the total energy is $E = 8\varepsilon_0$. Quantized energies are ε_n.

$$\varepsilon_n = n\varepsilon_0, \qquad n = 0, 1, 2, \ldots,$$

the continuous line = $P(n)$ as calculated from the Boltzmann formula [(4-39)] for the same value of total energy.

A number of very interesting features emerges from these two figures:

First, the probability $P(n)$ that an atom is in quantum state n with energy ε_n decreases monotonically with increasing ε_n. The lower the energy of the state, the larger the probability for the atom to be in it, irrespective of the total energy E of the crystal. For small E/N, the decrease of $P(n)$ with increasing ε_n is steeper than for large E/N, but all curves have the same general character.

Second, this decrease of $P(n)$ with increasing energy ε_n is fairly well approximated by an exponential. This is shown quantitatively in Figure 4.12, where a curve is drawn for $P(n)$ as expressed by the Boltzmann formula (4-39), which we will derive in Subsection 4.2.A(vi):

$$P(n) = \frac{e^{-\beta\varepsilon_n}}{Z},$$

with $\varepsilon_n = n\varepsilon_0$ and $Z = (1 - e^{-\beta\varepsilon_0})^{-1}$. This formula for $P(n)$ is not strictly valid for a small crystal ($N = 6$) with fixed energy E, but Figure 4.12 shows that it

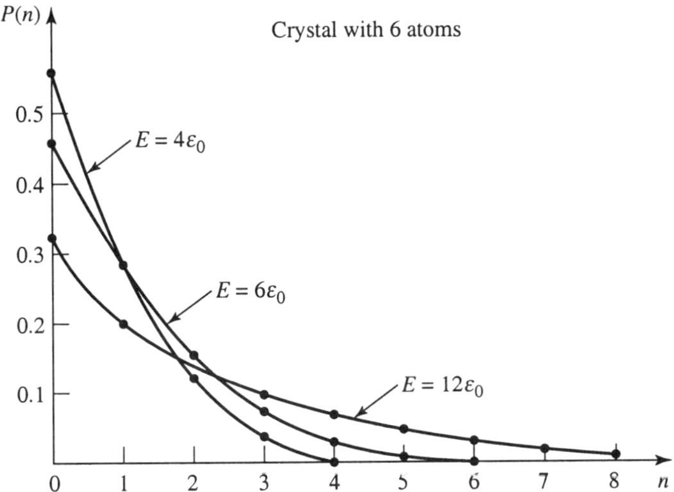

Figure 4.13. Comparison of probabilities $P(n)$ for various values of the total energy of the crystal with $N = 6$ atoms. These values are calculated from (4-16) and (4-22), and energies $\varepsilon_n = n\varepsilon_0$. The energies of the crystal are $E = 4\varepsilon_0$, $6\varepsilon_0$, and $12\varepsilon_0$, respectively. (See also the caption for Figure 4.12.)

gives a remarkably good representation of the values of $P(n)$, calculated by means of (4-22). The reason why the values of $P(n)$, calculated by this latter equation, are "bumpy" is that we have, for $N = 6$ and $E = 8\varepsilon_0$, only 19 macrostates, and the $\overline{N}_n$ are only small numbers.

This concludes our numerical example. Clearly the method used here is not feasible, as the total number of atoms and the available energy both become large. In the next subsection, we will therefore develop a more general technique suitable for large values of the numbers N and N_n.

(v)   Finding the Most Probable Macrostate

In this subsection we will show that the probability $P(n)$ for finding an atom of the crystal in energy state n (and energy ε_n) indeed decreases exponentially with ε_n, as suggested by Figure 4.12 and 4.13. This exponential behavior holds provided the number of atoms in the crystal is very large. Dealing with very large numbers allows us to use *two* simplifying procedures.

(a) We may use Stirling's approximation to represent numbers like $N!$ and $N_n!$:

$$N! = \sqrt{2\pi N}\, e^{N \ln N - N}$$

or

$$\ln N! = \tfrac{1}{2} \ln(2\pi) + \tfrac{1}{2} \ln N + N \ln N - N. \qquad (4\text{-}24)$$

We will see that what matters is to have a fair representation of $\ln W$, and for large values of N and N_n, even an approximation somewhat poorer than (4-24) will be good enough. Neglecting $\tfrac{1}{2} \ln(2\pi)$ and $\tfrac{1}{2} \ln N$ as small, compared to N and to $N \ln N$, we have

$$\ln(N!) \cong N \ln N - N. \qquad (4\text{-}25)$$

This will be our working approximation.

(b) For large values of N and N_n, the quantity $W(N_0, N_1, \ldots, N_n, \ldots)$ has a rather sharp maximum, located at the values $\overline{N}_0, \overline{N}_1, \ldots, \overline{N}_n$ of $N_0, N_1, \ldots, N_n$, and falls off like a Gaussian as the $\overline{N}_n$ deviate from the N_n. This is illustrated by the familiar behavior of the binomial coefficient occurring in the coin-tossing problem

$$W(H, T) = \frac{N!}{H!\,T!} \rightarrow \sqrt{\frac{2}{\pi N}}\, 2^N e^{-\frac{(H - (N/2))^2}{N}}\, e^{-\frac{(T - (N/2))^2}{N}}.$$

This shows that the *maximum* of W occurs at the position $H = T = N/2$, which are also the *mean* values of H and T for unbiased coins. This is a general property of the weight $W(N_0, N_1, \ldots, N_n, \ldots)$ if the numbers involved are large. It has the important consequence that we may determine the *mean* values $N_0, N_1, \ldots$ simply by locating the *maximum* of the weight function W.

We now proceed to locate this maximum. It will be easier to work with the logarithm $\ln W$, which is a simpler function of the numbers N_n than W itself. Also, W and $\ln W$ have their maximum for the same values N_n of the variables N_n, as can be seen from the relation

$$d(\ln W) = \frac{dW}{W}.$$

Since W is never zero, whenever dW is zero, $d(\ln W)$ is also zero. Using the approximation equation (4-25) for $N!$ and $N_n!$, we use the expression (4-21) for

W to write

$$\ln W \cong N \ln N - N - \sum_n N_n \ln N_n + \sum N_n$$

$$= \sum_n N_n (\ln N - \ln N_n). \tag{4-26}$$

To get the second line, we used $N = \sum N_n$. To locate the maximum of $\ln W$, we write each number N_n as the sum of its mean value $\overline{N}_n$ and a deviation δN_n from this mean value

$$N_n = \overline{N}_n + \delta N_n. \tag{4-27}$$

We will also write an analogous relation for the total number N of particles

$$N = \overline{N} + \delta N = \sum \overline{N}_n + \sum_n \delta N_n. \tag{4-27a}$$

In the two examples to be discussed in this subsection, N will be constant, and δN will be equal to zero. But at a later stage we will discuss the case of a solid in equilibrium with its vapor, and chemical equilibria, in which N is not constant. So we shall make allowance for this case right at the outset. Expressions (4-27) and (4-27a) may now be entered into (4-26). Since δN_n and δN are small increments, we will use the Taylor expansion

$$\ln N = \ln(\overline{N} + \delta N) \cong \ln \overline{N} + \frac{\delta N}{\overline{N}} - \frac{1}{2} \frac{\delta N^2}{\overline{N}^2} + \cdots$$

and similar expressions for $\ln N_n = \ln(\overline{N}_n + \delta N_n)$: This gives

$$\ln W \cong (\overline{N} + \delta N)\left(\ln \overline{N} + \frac{\delta N}{\overline{N}} - \frac{1}{2} \frac{\delta N^2}{\overline{N}^2} \right)$$

$$- \sum_n (\overline{N}_n + \delta N_n)\left(\ln \overline{N}_n + \frac{\delta N_n}{\overline{N}_n} - \frac{1}{2} \frac{\delta N_n^2}{\overline{N}_n^2} \right)$$

$$= \overline{N} \ln \overline{N} - \sum \overline{N}_n \ln \overline{N}_n - \sum_n \delta N_n \ln \left(\frac{\overline{N}_n}{\overline{N}} \right)$$

$$- \frac{1}{2}\left(\sum_n \frac{\delta N_n^2}{\overline{N}_n} - \frac{\delta N^2}{\overline{N}} \right). \tag{4-28}$$

This equation expresses $\ln W$ through its maximum value, and the first- and second-order deviations from the maximum

$$\ln W = \ln W_{\max} + \delta^{(1)} \ln W + \delta^{(2)} \ln W,$$

where

$$\ln W_{\max} = \overline{N} \ln \overline{N} - \sum \overline{N}_n \ln \overline{N}_n \qquad (4\text{-}29)$$

and $\delta^{(1)} \ln W$ is the term linear in the δN's, and $\delta^{(2)} \ln W$ is the term quadratic in the δN_n's. The identification of $\ln W_{\max}$ as the maximum of $\ln W$ implies that the difference

$$\ln W - \ln W_{\max}$$

should depend only *quadratically* on the deviations δN_n and δN. In other words, the term $\delta^{(1)} \ln W$ must be zero. The values of the $\overline{N}_n$ must therefore be chosen in such a way that

$$\delta^{(1)} \ln W = -\sum_n \delta N_n \ln \left(\frac{\overline{N}_n}{\overline{N}}\right) = 0 \qquad (4\text{-}30)$$

for all allowed values of the deviations δN_n.

If the deviations $\delta N_0, \delta N_1, \ldots, \delta N_n$ could be assigned values independent of each other, then the only way to satisfy (4-30) would be to require that all coefficients $\ln(\overline{N}_n/\overline{N})$ be zero. However, the premise is not correct. We are describing an isolated crystal, with a fixed number N of atoms, and a fixed energy E. Therefore, the deviations δN_n are *constrained* by the requirement that they leave the total number of particles and the total energy unchanged. These conditions of constraint are

$$\delta N = \sum_n \delta N_n = 0, \qquad (4\text{-}31a)$$

$$\delta E = \sum \varepsilon_n \, \delta N_n = 0. \qquad (4\text{-}31b)$$

How do these constraints on the possible values of δN_n affect the solution of (4-30)? This equation requires that the coefficients $\ln(\overline{N}_n/\overline{N})$ be chosen in such a way that $\delta^{(1)} \ln W = 0$ for all δN_n's which also satisfy equations (4-31). Of course, if we assume $\ln(\overline{N}_n/\overline{N}) = 0$ for all n, we have satisfied (4-30). But this solution is too restrictive and not a meaningful solution. Another solution is $\ln(\overline{N}_n/\overline{N}) = \alpha$ for all n, where α is some constant. We can verify this by inserting it into (4-30); this gives

$$\delta^{(1)} \ln W = -\alpha \sum_n \delta N_n$$

and this is indeed zero if the δN_n are constrained so as to obey (4-31a). We see now also that (4-31b) allows us to generalize the admissible values of $\ln(\overline{N}_n/\overline{N})$ even further. Let us try

$$\ln(\overline{N}_n/\overline{N}) = \alpha - \beta\varepsilon_n, \tag{4-32}$$

where we now have two unspecified constants α and β. Inserting into (4-30), we find

$$\delta^{(1)} \ln W = -\alpha \sum_n \delta N_n + \beta \sum_n \varepsilon_n \, \delta N_n$$

and this is indeed zero for all δN_n's that obey (4-31a ,b). We also see that in (4-32) we have exhausted the possibilities for the form of the coefficients $\ln(\overline{N}_n/\overline{N})$, and (4-32) is the *general* solution of (4-30) consistent with the *two* conditions of constraint. (From now on we will write N for $\overline{N}$, since it is actually, in this present case, a fixed constant, and not a mean value. See, however, Section 4.2.C.)

With the solution, (4-32), which we may also write as

$$\overline{N}_n = Ne^{\alpha}e^{-\beta\varepsilon_n}. \tag{4-33}$$

The most striking feature of this result is that the mean number of particles in each energy state decreases exponentially with the energy ε_n. We saw this already for $N = 6$ in our detailed example. This exponential factor $e^{-\beta\varepsilon_n}$ is the Boltzmann factor that we have previously mentioned. We will presently discuss its significance much more fully. At present, let us tidy things up a bit by examining the values of the parameters α and β.

It is not altogether astonishing that two such parameters should introduce themselves into the solution (4-33). The numbers N_n, as given by (4-33), must satisfy two conditions:

(a) They must add up to N:

$$\sum N_n = N. \tag{4-17a}$$

(b) They must yield a total energy equal to E:

$$\sum \varepsilon_n \overline{N}_n = E. \tag{4-18a}$$

Inserting the solution [(4-33)] into the first of these conditions, we find

$$\sum \overline{N}_n = \overline{N} e^\alpha \sum_n e^{-\beta \varepsilon_n} = N$$

or

$$e^\alpha = 1 \Big/ \Big(\sum_n e^{-\beta \varepsilon_n} \Big) \equiv 1/Z. \tag{4-34}$$

This shows that in fact e^α is just a normalization constant, and can be expressed in terms of β. In (4-34) we have also introduced the useful abbreviation

$$Z = \sum_n e^{-\beta \varepsilon_n}. \tag{4-35}$$

The second condition, (4-18a), gives

$$E = N e^\alpha \sum_n \varepsilon_n e^{-\beta \varepsilon_n}$$

$$= N \Big(\sum_n \varepsilon_n e^{-\beta \varepsilon_n} \Big) \Big/ Z.$$

We see that the parameter β is determined in an implicit manner by the value (E/N) of the energy per particle

$$\Big(\frac{E}{N} \Big) = \Big(\sum_n \varepsilon_n e^{-\beta \varepsilon_n} \Big) \Big/ Z. \tag{4-36}$$

We will take up the discussion of the physical interpretation of the parameter β in the next subsection. Let us here summarize our findings so far.

For fixed values of the number N of atoms and the total energy E of the crystal, the *mean numbers* $\overline{N}_n$ of atoms in quantum state n (of energy ε_n) are given by the expression

$$\overline{N}_n = N e^{-\beta \varepsilon_n}/Z, \tag{4-37}$$

where the constant Z is defined in (4-35). The parameter β is dependent on the energy per particle (E/N), and this dependence is expressed in (4-36).

With the help of the mean number $\overline{N}_n$, we can now also calculate the actual maximum value $W_{\max}$ of W, as given by (4-29):

$$\ln W_{\max} = N \ln N - \sum \frac{N}{Z} e^{-\beta \varepsilon_n} (\ln N - \ln Z - \beta \varepsilon_n).$$

Using the result that $\sum_n e^{-\beta \varepsilon_n} = Z$, and $\sum_n \varepsilon_n e^{-\beta \varepsilon_n} = (E/N)Z$, this reduces to

$$\ln W_{\max} = N \ln N - N \ln N + N \ln Z + \beta E$$

or

$$\ln W_{\max} = N \ln Z + \beta E. \tag{4-38}$$

We will come back, in Section 4.3, to this important expression, that is in fact one of the connecting links between statistical physics and thermodynamics.

Another available result should be mentioned here for the record. In (4-28) we have also an expression for the decrease of $\ln W$ as we move away from its maximum value

$$\ln W = \ln W_{\max} - \frac{1}{2} \sum_n \delta N_n^2 / \overline{N}_n.$$

(We omit the term $\sim \delta N^2$, since N is constant in our case.) Returning from the logarithm $\ln W$ to W itself, we can write this as

$$W = W_{\max} e^{-\frac{1}{2}\Sigma(\delta N_n^2/\overline{N}_n)}.$$

In words: the decrease of the weight W of a macrostate near its maximum is given by a product of Gaussians $e^{-(\delta N_n^2/2N_n)}$, in which δN_n represents the deviation of the number of atoms in state n from it most probable value $\overline{N}_n$.

(vi)   The Probability Distribution $P(n)$ for the Energies of Atoms in the Einstein Crystal. The Boltzmann Factor

Having found an expression for the mean number $\overline{N}_n$ of atoms in quantum state n, we can now construct the probability distribution function $P(n)$. $P(n)$ is the probability that a particular atom will be in quantum state n, and thus have an energy ε_n. The relation between $P(n)$ and the mean number N_n of particles in state n was given in (4-18) as

$$P(n) = \left(\frac{\overline{N_n}}{N}\right).$$

Using our expression [(4-37)] for $\overline{N}_n$, we find that

$$P(n) = e^{-\beta \varepsilon_n}/Z, \tag{4-39}$$

where the normalization constant Z is given by (4-35), which we repeat here

$$Z = \sum_n e^{-\beta \varepsilon_n}. \tag{4-35a}$$

Equation (4-39), the Boltzmann formula, is our most important result so far. It shows that indeed the probability $P(n)$ for an atom to be in quantum state n of energy ε_n decreases *exponentially* with this energy: The lower of any two energy states is always the more probable, irrespective of the total available energy for the crystal. This is not at all an obvious result. It is, however, supported by innumerable experimental data.

There remains the task of clearly identifying the significance of the parameter β occurring in the Boltzmann formula. We know already from (4-36) that β is determined by the mean energy per particle (E/N). We had, on a previous occasion, namely in Section 2.1.C and the Appendix to Chapter 2, connected the mean energy per particle with the temperature T. This connection is given by the Equipartition Theorem, of classical mechanics, according to which the mean energy of oscillation of a particle (oscillating in one direction only) should be equal to kT, k being the Boltzmann constant and T the Kelvin temperature.

In order to get an explicit connection between the energy per particle and β, and hence a connection between β and the temperature T, we will again assume that the energy levels are equidistant, as befits a harmonic oscillator. (Notice, however, that this assumption was *not* needed to get the *form*, (4-39), for the probability distribution $P(n)$.) So using again $\varepsilon_n = n\varepsilon_0, n = 0, 1, 2, \ldots,$ we find

$$\left(\frac{E}{N}\right) = \frac{\sum (n\varepsilon_0) e^{-\beta n \varepsilon_0}}{\sum e^{-\beta n \varepsilon_0}} = \varepsilon_0 \frac{\sum_{n=0}^{\infty} n (e^{-\beta \varepsilon_n})^n}{\sum_{n=0}^{\infty} (e^{-\beta \varepsilon_0})^n}$$

$$= \varepsilon_0 \frac{\sum_{n=0}^{\infty} n x^n}{\sum x^n} \qquad \text{where} \qquad x = e^{-\beta \varepsilon_0}.$$

We have to deal here with a geometric series, whose sum is well known

$$\sum_{n=0}^{\infty} x^n = \frac{1}{1-x} \qquad \text{for} \quad |x| < 1.$$

To calculate the sum $\sum n x^n$, we use a device learned in Chapter 1:

$$\sum_n n x^n = \sum_n x \frac{d}{dx}(x^n) = x \frac{d}{dx} \sum x^n$$

$$= x \frac{d}{dx} \left(\frac{1}{1-x}\right) = \frac{x}{(1-x)^2}.$$

Thus we have finally, for the energy per particle,

$$\left(\frac{E}{N}\right) = \varepsilon_0 \frac{x}{1-x} = \varepsilon_0 \frac{e^{-\beta \varepsilon_0}}{1 - e^{-\beta \varepsilon_0}} = \frac{\varepsilon_0}{e^{-\beta \varepsilon_0} - 1}. \tag{4-40}$$

In our discussion in Chapter 2, based on classical mechanics, we associated the energy per particle with the temperature T. Since E/N intuitively is a function of the temperature, so (4-40) implies that β must in some way represent temperature, but how? To help us to see this connection, we plot in Figure 4.14 the energy per particle (E/N) as a function of $(\beta \varepsilon_0)$.

There are clearly two regimes in this curve:

(a) Consider first the region of high energy $(E/N) \gg \varepsilon_0$. For this to occur, the denominator of (4-40) must be much smaller than unity. This requires

$$\beta \varepsilon_0 \ll 1.$$

Under these conditions, we find

$$\left(\frac{E}{N}\right) = \frac{\varepsilon_0}{e^{\beta \varepsilon_0} - 1} \cong \frac{\varepsilon_0}{(1 + \beta \varepsilon_0 + \cdots - 1)} = \frac{1}{\beta}. \tag{4-41}$$

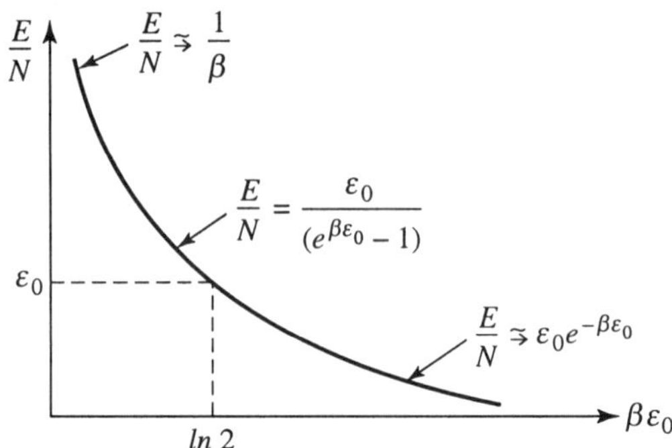

Figure 4.14. Energy per atom in a crystal, as a function of $\beta\varepsilon_0$, where ε_0 is the spacing of vibrational energy states.

What is the physical interpretation of this parameter β? We may take a hint given us by the data on specific heats, which we discussed in Section 2.1.C. These data indicate that at sufficiently high temperature T, the mean energy of a one-dimensional oscillator is equal to kT, k being the Boltzmann constant. This suggests the identification

$$\beta = \frac{1}{kT}, \tag{4-42}$$

at least for high temperatures. We will settle the issue definitely in Sections 4.2.B and 4.3 of this chapter, where we will show the general validity of (4-42). At present we ask the student to accept (4-42) on faith.

(b) For $\beta\varepsilon_0 \gg 1$, $e^{\beta\varepsilon_0}$ is very much larger than 1, and we can write

$$\frac{E}{N} = \frac{\varepsilon_0}{e^{\beta\varepsilon_0} - 1} \simeq \frac{\varepsilon_0}{e^{\beta\varepsilon_0}} = \varepsilon_0 e^{-\beta\varepsilon_0} \qquad (\beta\varepsilon_0 \gg 1). \tag{4-43}$$

If β is indeed $1/kT$, then this result indicates that in the case where kT is small compared to the spacing ε_0 of energy levels, the energy per particle is no longer kT, but *much smaller* and given by

$$\left(\frac{E}{N}\right) = \varepsilon_0 e^{-\varepsilon_0/kT} \qquad (\varepsilon_0 \gg kT). \tag{4-44}$$

We had indeed already commented on this very effect in Chapter 2 when we talked about the freezing of thermal motion due to quantum effects. In (4-44) we have the quantitative formulation of this phenomenon: If the "energy gap" between the lowest energy state and the excited energy states is much larger than kT, it is very unlikely to find an atom in any of these excited states, and the mean energy per atom will then be very much below the "expected" value of kT.

The fact that in the high-energy limit the value of (E/N) becomes independent of the energy level structure (here determined by the level spacing ε_0) is a general feature and not restricted to the quantized oscillation of a particle. The results of quantum physics merge into the results of classical mechanics in the limit that the available energy of a particle is large compared to the spacing of the energy levels. In this case, we may then calculate the mean energy per particle (be it bound or free), by using a classical description of particle motion based on Newton's second law. It is in this classical framework that one can prove the so-called equipartition theorem, which we have mentioned in Section 2.1.C. We shall give, in Appendix 4.A.3 to this chapter, a precise statement of this theorem, as well as a sketch of its proof.

Adopting, for the time being, the yet incompletely established statement that the parameter β is *generally* related to the Kelvin temperature T by

$$\beta = 1/kT, \tag{4-45}$$

we can *summarize* our main result by the statement: In an idealized crystal, consisting of a lattice of identical atoms, each of which has quantized energy states n of energies ε_n, representing oscillation about its equilibrium position, the probability that an individual atom will be found in a particular energy state n is given by

$$P(n) = \frac{1}{Z} e^{-\varepsilon_n/kT}. \tag{4-46a}$$

Z is a normalization constant given by

$$Z = \sum_n e^{-\varepsilon_n/kT}. \tag{4-46b}$$

The exponential factor $e^{-\varepsilon_n/kT}$ is the already mentioned *Boltzmann factor*.

(vii)   Physical Interpretation of the Boltzmann Factor

In the previous subsection we obtained the Boltzmann factor by means of a mathematical analysis of the conditions needed to maximize W. In the present section we will take that mathematical analysis apart and try to get at the fundamental physical reasons that produced the Boltzmann factor. This analysis is appropriate because the Boltzmann factor is a quite general result that applies for a very wide class of systems with large numbers of particles. By the present analysis we will be able to see how this general result emerges.

Let us begin by considering (4-30), viz:

$$\delta^{(1)} \ln W = \frac{\delta^{(1)} W}{W} = -\sum_n \delta N_n \ln\left(\frac{\overline{N_n}}{\overline{N}}\right) = 0, \qquad (4\text{-}30)$$

which is the condition that W be maximized. The δN_n's in this sum are, as we saw, not independent of one another. They must be coupled in such a way that the total number of particles N, and the total energy of the system, remains constant. These constraints play a vital role in the appearance of the Boltzmann factor. We can see this quite clearly by considering a very particular example of a possible variation in the δN_n's. This variation is illustrated in Figure 4.15.

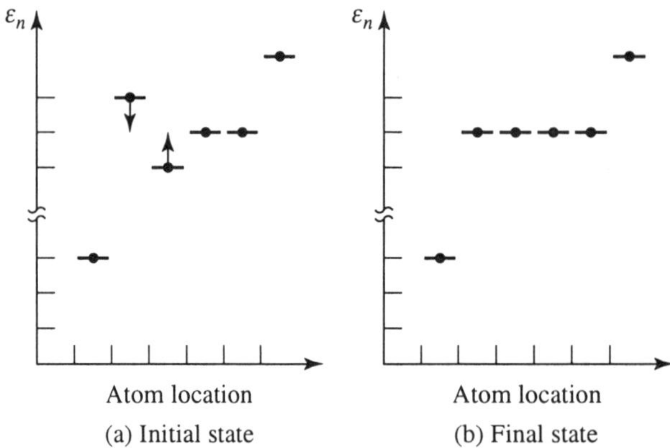

Figure 4.15. Illustration of a change in which two oscillator atoms exchange one unit of energy ε_0.

Here we show a particular change in the N_n's that occurs when two oscillating atoms initially in the states $(n' + 1)$ and $(n' - 1)$ exchange a single unit of energy ε_0, thereby elevating the atom in state $(n' - 1)$ to the state n' and depressing the atom in state $(n' + 1)$ to the state n'. Clearly the change δN_n in the numbers of atoms is given by

$$\delta N_{n'+1} = -1, \qquad \delta N_{n'-1} = -1, \qquad \delta N_{n'} = 2,$$

all other $\delta N_{n'} s = 0$. On inserting these values for the δN_n's into (4-30), we find the following condition on the $\overline{N}_n$'s:

$$0 = -\ln\left(\frac{\overline{N}_{n'+1}}{N}\right) + 2\ln\left(\frac{N_{n'}}{N}\right) - \ln\left(\frac{\overline{N}_{n'-1}}{N}\right).$$

This can be rewritten as

$$\ln\left(\frac{\overline{N}_{n'+1}}{\overline{N}_{n'}}\right) = \ln\left(\frac{\overline{N}_{n'}}{\overline{N}_{n'-1}}\right).$$

Since such a change can be conducted for pairs of atoms regardless of the state n', we must conclude that the ratio of the numbers $\overline{N}_n$ of atoms in two adjacent states is a constant regardless of n', i.e., independent of energy ε_n. Thus, we have

$$\ln\left(\frac{\overline{N}_{n'+1}}{\overline{N}_{n'}}\right) = (\text{constant}).$$

$\overline{N}_{n'+1}$ must be less than $\overline{N}_{n'}$, otherwise the numbers $N_{n'}$ would increase as one went to higher and higher energy states, and the total energy would become unbounded. The equation above shows the remarkable fact that the occupancy of successive states is a monotonic function of n' or the energy. The occupancy must monotonically *decrease* as the energy n' increases. We can see this quite explicitly by writing the constant in the form

$$\text{constant} = -\beta\varepsilon_0.$$

(This constant is negative because the ratio $\overline{N}_{n'+1}/\overline{N}_{n'}$ is less than 1. Hence, its logarithm is negative.) Putting this into the equation above gives

$$\overline{N}_{n'+1} = \overline{N}_{n'}e^{-\beta\varepsilon_0}.$$

From this result we see that the occupancy $N_{n'}$ of the state n' can be constructed step by step from the occupancy of the underlying levels as

$$\overline{N}_{n'} = \overline{N}_{n'-1}e^{-\beta\varepsilon_0} = \overline{N}_{n'-2}e^{-\beta\varepsilon_0}e^{-\beta\varepsilon_0}$$
$$= \overline{N}_{n'-3}e^{-3\beta\varepsilon_0} = \cdots = \overline{N}_0 e^{-\beta\varepsilon_0 n'}.$$

Thus, we see again that the occupancy of the various states is a monotonically decreasing function of the energy

$$\overline{N}_{n'} = \overline{N}_0 e^{-\beta\varepsilon_n'}. \tag{4-47}$$

In general, this decrease in occupancy is a direct result of two factors. The first is the appearance of the $\ln \overline{N}_n$ term in δW. The second is the crucial requirement that δN_n's must be adjusted so that the total energy of the system and the total number of particles is kept constant.

We may obtain even more insight into the origin of the Boltzmann factor. To see this, let us return to our basic formula for W [(4-21)]:

$$W = \left(\frac{N!}{N_0!\,N_1!\cdots N_n!\cdots}\right). \tag{4-21}$$

The most probable state is that for which W is as large as possible. To make W as large as possible with a total of N particles, it would appear obvious that we must make the various N_n's either 0 or 1. This is the situation in which no more than one atom has a particular energy ε_n. This is in fact the configuration of maximum disorder. Each atom has a different energy. However, such a situation leads to an enormous energy E for the system. If each atom had a different energy, then at the very least the most energetic atom would have energy $N\varepsilon_0$. The average energy per atom would be $\sim (N\varepsilon_0/2)$, and the total energy of the N atoms would be

$$E = \left(\frac{\varepsilon_0}{2}\right) N^2.$$

Accordingly, the energy of the system would increase as the *square* of the number of particles in the system. This is inconsistent with the physical requirement that the energy be bounded. By bounded we mean that each particle has some mean thermal energy $\overline{\varepsilon}$ determined by the temperature, and the resulting energy is $E = N\overline{\varepsilon}$. Physically, the total energy must be proportional to N, the number of particles, and not to N^2.

We see therefore that while the occupancies $\overline{N}_n = 0, 1$ leads to the largest possible value for W, they are inconsistent with the requirement of boundedness; i.e., that the total energy be linearly proportional to the number of particles. The fixed value of the total energy *forces* W to have a much smaller value than $N!$. To achieve such a smaller value, there must be many atoms in the same energy state. That is, the actual state must contain some *order* in the form of many atoms having the same energy. The equilibrium values of the N_n's tell us how many atoms are forced into the same energy state ε_n.

The actual values of $\overline{N}_n$ are partly determined by the fact that a change δN_n in one of the occupancies changes the value of $\ln W$ by $\delta N_n \ln(N/N_n)$. As we can see from (4-32) and (4-33), the fact that the derivative $(\partial \ln W/\partial N_n)$ is equal to the logarithm of $(N/\overline{N}_n)$ is at the origin of the Boltzmann factor for $\overline{N}_n$. It is, therefore, useful that we examine again how the value of this derivative arises. We will do this without invoking Stirling's approximation. Rather, we examine how W, as given by (4-21), changes if we change the number N_n of particles in a specific state n by

$$\delta N_n = k,$$

where k is a small integer. The numbers N_n, of particles in states n' $(n' \neq n)$, will be left unchanged. In this process, W, as given by (4-21), changes into W_k:

$$\underset{\substack{k \text{ particles in} \\ \text{state } n \text{ added}}}{W \to W_k} = \frac{(N+k)!}{N_0!\, N_1! \cdots (N_n+k)! \cdots}.$$

We can write this changed quantity W_k as W times a correction factor

$$W_k = W \frac{(N+k)!}{N!} \frac{N_n!}{(N_n+k)!}$$

$$= W \frac{(N+1)(N+2)\cdots(N+k)}{(N_n+1)(N_n+2)\cdots(N_n+k)}.$$

Now we make use of the fact that N and N_n are both very large numbers, whereas k is a small integer, 1, or 2, or 3, etc. We can, therefore, within an accuracy of order k^2/N_n, write the correction factor as N^k/N_n^k:

$$W_k \cong W \left(\frac{N}{N_n}\right)^k = W e^{k \ln(N/N_n)}. \tag{A}$$

We may take the logarithms of the quantities on either side, and then obtain the relation

$$\ln W_k = \ln W + k \ln(N/N_n). \tag{B}$$

Mathematically, it does not matter whether we use relation (A) or (B) to define the derivative provided that we may go smoothly to the limit $k = 0$. From (B) we may obtain this limit accurately for integer values of k, i.e.,

$$\left(\frac{\ln W_k - \ln W}{k}\right) \cong \left(\frac{\partial \ln W}{\partial N_n}\right) = \ln(N/N_n).$$

By contrast, the quantity N_k is such a rapidly varying function of k that it is not possible to define a derivative of W itself by using the expression

$$\frac{W_k - W}{k}$$

for integer values of k. Even for the smallest integer $k = 1$, the above ratio is still far from its limiting value for $k \to 0$, as can be verified by a careful examination of (A).

Thus the use of $\ln W$ rather than W arises from the need of having a function that varies sufficiently slowly with $k \,(= \delta N_n)$ so that the change associated with an integer value of k can be expressed accurately by its derivative.

The use of the function $\ln W$ is simply a convenient device to *scale down* the very rapid variation of W so that a derivative can be defined with integer changes of the number of particles. If we had chosen some other monotonic function $F(W)$ to scale down the rapid changes of W, the variation of $F(W)$ associated with a change δN_n would still be proportional to $\delta N_n \ln(N/N_n)$. The appearance of $\ln(N/N_n)$ is therefore *not* due to our choice of $F(W) = \ln W$ as a convenient scaling function.

4.2.B.   Energy Distribution for the Atoms of an Ideal Monoatomic Gas

In this section we will pursue the same objective as in the previous Section 4.2.A. Considering a gas of N atoms with a total energy E, we ask the question: How is this energy distributed over the various particles?

In detail the program followed in this section will be somewhat different than in the previous one. The reason for this lies in the fact that in the gas each particle has a *continuous* range of energies, and also has the total volume V of the container available for its travels. Its energy will be the sum of its kinetic energy $\vec{p}^2/2m$ ($\vec{p}$ being the momentum $m\vec{v}$ of the particle) and of a potential energy $u(\vec{r})$; an example of the latter is the gravitational potential energy

$$u = mgz$$

of a gas particle in the atmosphere. Since the energy depends in general on both momentum $\vec{p}$ and position $\vec{r}$ of the particle, and since both $\vec{p}$ and $\vec{r}$ have a continuous range, the proper question to ask is: What is the probability ΔP that an atom is to be found in a certain position range $\Delta\vec{r}$ and a certain momentum range $\Delta\vec{p}$? Just as in the case of the crystal lattice, we will answer this question by showing that ΔP is equal to the *mean* number of atoms to be found in the above-mentioned range of positions and momenta divided by the total number of atoms.

Let us then outline the different steps to be taken in the subsections that follow.

First, we introduce the notion of *phase space*, a six-dimensional space, each point of which represents both the instantaneous position and momentum of a particle. Each small volume element of this space then represents a joint position–momentum range, and will enable us to define the concept of a probability density or probability per unit phase-space volume.

Second, we will identify the state of thermal equilibrium as a statistically stationary distribution of points (atoms) in phase space. By dividing the phase space into cells labeled i, of size Ω_i, we can characterize the stationary population by the mean values of $\overline{N}_i$ of the numbers N_i of particles in each cell i. The numbers N_i fluctuate about their mean values. Due to their motion and collisions, particles constantly move in and out of a particular cell i. A particular set of cell populations N_i represents the analog of the *macrostate* considered in the previous example of the crystal. The set of mean values $\overline{N}_i$ represents the most probable macrostate.

Third, to determine the most probable macrostate, we face the problem of assigning a weight W to each macrostate which represents the probability that it occurs. It is at this point that we must make contact again with quantum physics. Without it we obtain only the relative but not the absolute weight W of a macrostate. These absolute weights are *not* strictly needed if all we want is the probability distribution for the energy of particles in the gas. It *is* needed, however, if we want to determine the vapor pressure of a liquid or solid, or to calculate the equilibrium constants of chemical reactions. In this Section 4.2.B, we will not enter into these latter questions that are reserved for Section 4.2.C and 4.3, but we will lay the foundation. Our last explicit objective of this section is to establish the well-known Maxwell–Boltzmann distribution function for the velocities of particles in a gas, and to discuss its properties.

(i) Phase Space and Phase-Space Trajectories of Particles

The notion of phase space is really a very simple one. We want a graphic representation not only of the position of a particle, but also of its state of motion, as expressed by either its velocity $\vec{v}$ or its momentum $\vec{p} = m\vec{v}$. For the latter, one has introduced the concept of *momentum space* (see Figure 4.16). In this space a point P represents a particular value of the momentum components p_x, p_y, p_z. The vector $\vec{p}$:

$$\vec{p} = p_x\hat{i} + p_y\hat{j} + p_z\hat{k}$$

is then represented by the arrow drawn from the coordinate origin O to the point P.

In creating the concept of phase space, we simply join "ordinary" space or position space with momentum space so as to create a six-dimensional space. In this space a point P has the six coordinates x, y, z, p_x, p_y, p_z, that specify both its position (through x, y, z) and its momentum p (through p_x, p_y, p_z). Unfortunately, we cannot represent this very easily on paper. But since it is so essential that this

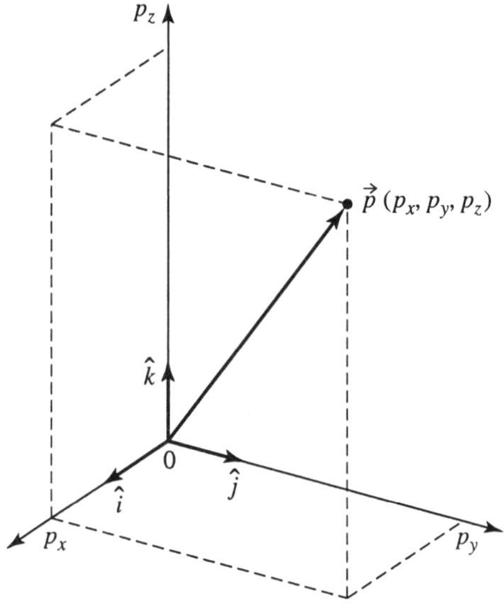

Figure 4.16. Graphical representation of momentum space. Three perpendicular coordinate axes are defined by the three unit vectors $\hat{i}$, $\hat{j}$, $\hat{k}$. They define the p_x-, p_y-, and p_z-axes, respectively. A point P in this space with coordinates p_x, p_y, p_z represents the vector $\vec{p} = p_x\hat{i} + p_y\hat{j} + p_z\hat{k}$.

concept of phase space be well understood, we will offer in Figure 4.17 a graphic representation of the phase space for the one-dimensional motion of a particle. In this case, there is a single position coordinate, say z, and a single momentum coordinate, $p_z = mv_z$, and the phase space is two-dimensional. We will assume that the moving particle is subject to a force, say the gravitational force $F_z = -mg$. The particle has then a total energy

$$\varepsilon = p_z^2/2m + mgz, \tag{4-48}$$

the sum of its kinetic energy $p_z^2/2m$, and its potential energy mgz. We see that (4-48) also attaches to each point in phase space a definite value of the particle *energy*. That is, giving the location of a particle in phase space implies assigning it a certain value of its energy. This is illustrated in Figure 4.17 where, in a graphical representation of two-dimensional phase space, we have drawn the curves of constant particle energy.

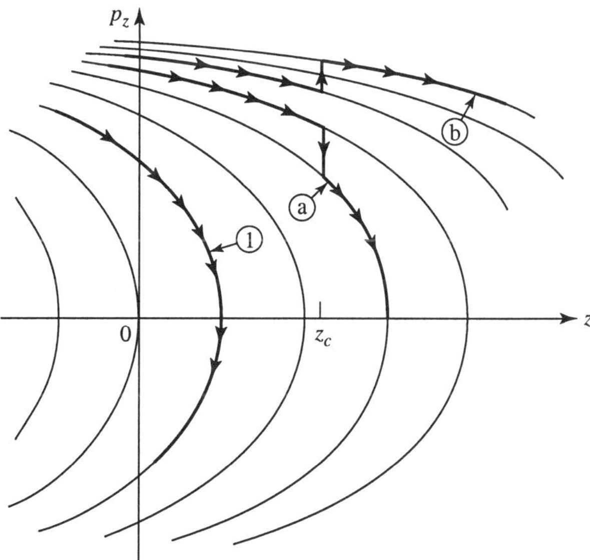

Figure 4.17. Phase space for the one-dimensional motion of a particle of mass m with energy $\varepsilon = p_z^2/2m + mgz$. Except for collisions, a particle moves on a curve of constant energy in phase space, such as trajectory (1). Also shown is the collision of two particles on trajectories (a) and (b), colliding at position $z = z_c$.

The actual motion of a particle in space–time is represented in phase space as a motion along a curve (or "trajectory") of constant energy. In a collision, the two colliding particles exchange both energy and momentum. They then both "jump" onto a different trajectory in phase space. This is also illustrated in Figure 4.17.

(ii)   Thermal Equilibrium as a Stationary Population in Phase Space

A completely detailed classical description of a gas could thus be given by representing each atom by a moving point in phase space. This distribution of these N points in phase space is constantly changing. Between collisions they move along their assigned trajectories; in a collision they jump onto a different one.

It is this *population* of moving points in phase-space points, each point representing an atom, that we attempt to describe *statistically*. This will give us the means to define thermal equilibrium. We are already very familiar with one aspect of thermal equilibrium of a gas enclosed in a volume V. We know that the particles arc distributed "uniformly" over the volume. What do we mean exactly by this statement? If we consider a small partial volume ΔV of the gas, we find that the actual number of particles in ΔV is constantly changing. However, in equilibrium this variation is in the nature of a fluctuation about a well-defined, stationary ($\equiv$ time independent) mean value. We call such a situation *statistically* stationary.

If we apply that same concept of a statistically stationary distribution to a *phase-space* population, it acquires the extended meaning that the number of particles, in a given spatial domain *and* a given momentum domain, fluctuates about a *stationary* mean value. In such a stationary situation, the mean number of particles in a given volume element does not change anymore, nor does the mean number of particles in a given momentum or energy range.

This then must be thermal equilibrium. We can summarize this by the statement that *thermal equilibrium is represented by a statistically stationary population of phase space.*

A snapshot of a phase-space population is represented in Figure 4.18.

In order to exploit this view of thermal equilibrium, and to establish the properties of a statistically stationary population in a quantitative fashion, we need a way to identify different *population patterns* and to assign to each of them a probability. This will be our next task.

(iii)   The Counting of Population Patterns. The Role of Planck's Constant h.   All Microstates Are Equally Probable

We will call a *population pattern* any particular distribution of points (atoms) in phase space, such as is represented in the "snapshot" in Figure 4.17. In the course

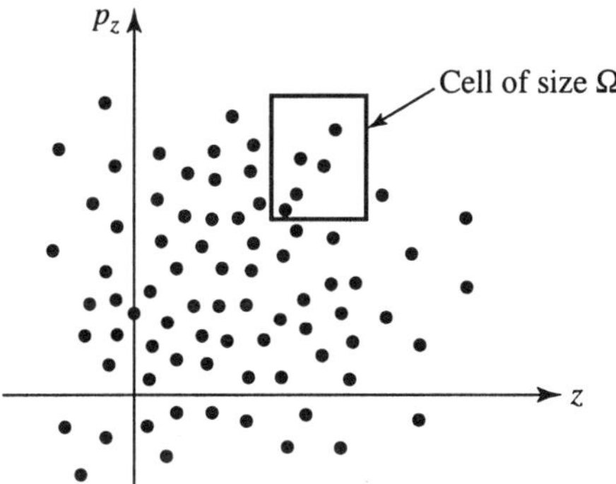

Figure 4.18. Snapshot of a phase-space population. For a *stationary* population the number of particles in cell Ω fluctuates about a stationary mean value.

of time, this pattern is constantly changing due to particle motion and interparticle collisions. In a statistical description of these patterns, we are called to make statements about the probability that a particular population pattern occurs. But to do this, we must be able to *count* patterns, and to recognize too, patterns as being either *distinct* or *identical*. How do we achieve this?

There is a very simple way, in fact, that we will illustrate for a two-dimensional phase space. (See Figure 4.19.) We cover all of phase space with a very fine *mesh* of squares, fine enough so that no two particles are likely to be found in the same square. For every population pattern we have then a set of occupied squares in the mesh and a large number of empty squares.

We see from Figure 4.19 that if a fine mesh is chosen, and the number N of atoms is large, there is a staggering number of distinct patterns. The *actual* number of distinct patterns of course must depend on the mesh size we use; the finer the mesh, the larger the number of distinct patterns.

It was one of the major problems of prequantum statistical physics that no "natural" mesh size could be established. The number of distinct patterns therefore remained ambiguous. The advent of quantum physics has eliminated that problem: According to quantum physics each quantum state "uses up" a domain of size h, Planck's constant ($h = 6.625 \times 10^{-27}$ erg/s) in two-dimensional phase space, or of size h^3 in six-dimensional phase space. This fact establishes a natural mesh size, where each square in the two-dimensional phase space has size h, and each "hypercube" in the six-dimensional phase space has a size h^3. In Appendix 4.A2 we will show that this is indeed an extremely fine mesh. This is illustrated by the fact that for a gas at room temperature and atmospheric pressure, only a very small

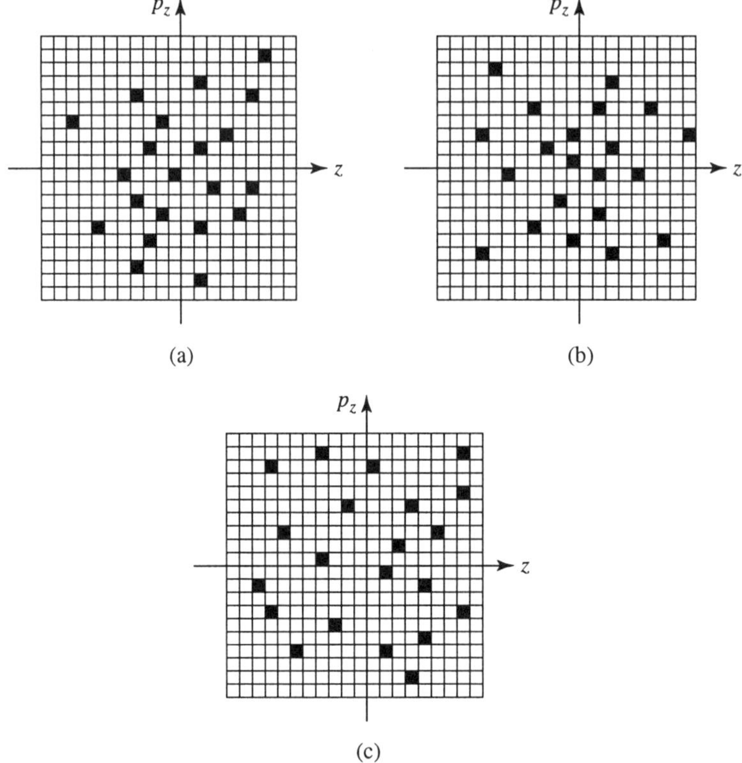

Figure 4.19. Three population patterns in phase space. Distinct patterns are identified by means of a square mesh. All three patterns shown are distinct. Patterns (a) and (b) have the same energy. Pattern (c) has a distinctly higher energy.

fraction of the mesh cubes are occupied by an atom, the overwhelming majority of them remain empty! These cubes of size h^3 will also be called *microcells*.

The dynamic evolution of the system of N particles forming the gas can now be described as a time sequence of patterns. However, it would be impossible and useless to attempt to establish the exact sequence of these patterns. We must rather seek a statistical statement relating to the frequency of occurrence of these patterns. Such a statement follows from the already previously invoked "principle of detailed balance" of quantum physics: In thermal equilibrium *all patterns having the same energy are equally likely to occur*; they have the same a priori probability.

Each detailed pattern, as illustrated in Figure 4.19, is the exact equivalent of the *microstate* that we introduced in the case of the crystal. We will from now on adopt this terminology and call each pattern a *microstate*. Their equiprobability gives us the means to determine the equilibrium population density ρ in phase space. This is our next task.

(iv)   The Equilibrium Population Density in Phase Space. Macrocells and Macrostates

For any population, be it the people of the United States, or mice in a corn field, one can introduce the concept of population *density*: The number of inhabitants per square mile, or the number of mice per acre in a field. It is a familiar concept. We also apply it to the population of atoms in phase space.

In order to define this density in phase space, we subdivide this space into cells that we will call *macrocells*. Their volume should be large enough so that they contain on average a large number of particles, yet small enough so that all particles in a single cell have essentially the same energy. This will help us to keep track of the total energy of all particles.

We label these macrocells by an index "i," and call their volume Ω_i. This volume must of necessity be very large compared to the size of a mesh-cube or microcell, whose volume is h^3. The ratio

$$\omega_i = \frac{\Omega_i}{h^3} \tag{4-49}$$

represents the number of mesh-cubes contained in the macrocell "i"; it is a very large number. For each population pattern or microstate of the gas, we have a certain number N_i of atoms falling inside the macrocell "i." This is illustrated in Figure 4.20.

A *set* of such numbers N_i, one for each macrocell "i," characterizes what we will call a *macrostate*. It is apparent at once that not all macrostates are equally probable. The macrocells in effect represent a coarse mesh in phase space, and the set of macrocell population numbers, N_i, is a *coarse-grained* way of describing the phase-space population pattern. For *each* of these coarse-grained patterns (macrostates), there is a *large number* of fine-grained patterns or microstates. These latter are each equally probable. Therefore, the probability of a macrostate is proportional to the *number of distinct microstates* that produce the same macrostates:

probability of macrostate $(N_1, N_2, \ldots, N_i)$
$\qquad$ = probability of a single microstate
$\qquad\qquad \times$ number of microstates with common values of $N_1, N_2, \ldots, N_i$.

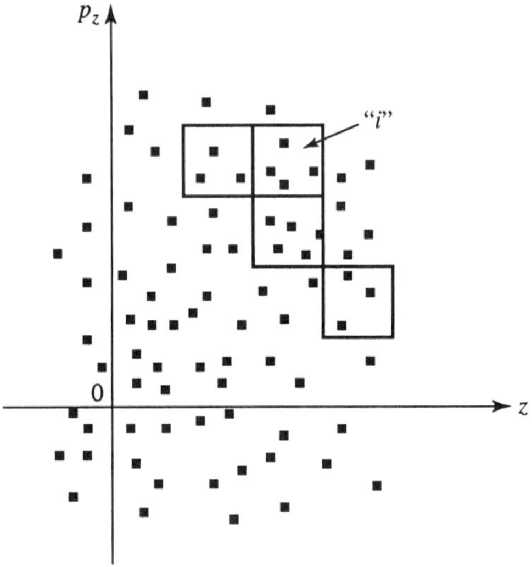

Figure 4.20. Portion of two-dimensional phase space. A particular population pattern or microstate is represented. Four macrocells are shown. Cell "i" contains $N_i = 4$ particles.

We will call that second factor again the *weight* of the macrostate, and denote it by the symbol

$$W(N_1, N_2, \ldots, N_i, \ldots).$$

The importance of establishing the values of these weights lies in the fact that for large numbers N_i, the N_i that produce the macrostate of largest weight can be identified with the average values $\overline{N}_i$ of the N_i. These $\overline{N}_i$ then give us a coarse-grained definition of the *average* density of particles in phase space. At the location of macrocell "i" this density is given by

$$\rho(i) = \frac{\overline{N}_i}{\Omega_i}, \tag{4-50}$$

the ratio of the mean number of points in the cell divided by the volume of the cell.

(v)   The Weight W of a Macrostate

Our program now calls for assigning a weight W to each macrostate. This weight is given by the *number* of population patterns or microstates for which we obtain the same occupation numbers N_i for each macrocell "i."

We can find this number W as follows. Let $W(i)$ be the number of distinct patterns for N_i particles in macrocell "i," irrespective of the rest of phase space. Then the number W is simply the *product* of all $W(i)$'s. That this is so can easily be seen from an analogy with dice. One die has six possible patterns (faces up); two dice jointly have $6 \times 6 = 36$ patterns (faces up); a number k of dice produces $(6)^k$ distinct patterns. Let this suffice as an argument.

Now what is W_i? Macrocell "i" of size Ω_i contains, according to (4-49), a number $\omega_i = \Omega_i / h^3$ of microcells. If we put one particle into the macrocell "i," it can settle in any of the ω_i different microcells. A second particle added can do the same. This appears to give rise to $(\omega_i)^2$ distinct patterns. However, these $(\omega_i)^2$ patterns are actually composed of $\frac{1}{2}(\omega_i)^2$ pairs. Such a pair is shown in Figure 4.21.

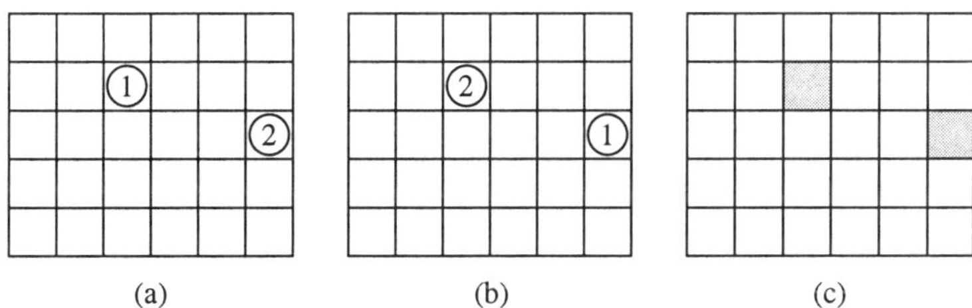

(a)   (b)   (c)

Figure 4.21. Two particles in a macrocell with $\omega = 30$ microcells. (1) represents the first, (2) the second particle placed into the cell. (a, b) represent a pair of patterns that are distinct only by a permutation of particle position. For a gas of *identical* particles, (a) and (b) do *not* represent distinct patterns. They represent a single pattern, as properly represented in (c).

In each pair, such as illustrated in Figure 4.21, the *same* microcells are occupied, and for identical ($=$ undistinguishable) particles, only a single population pattern is generated. The correct number of distinct patterns with two particles in the macrocell "i" is therefore

$$\tfrac{1}{2}(\omega_i)^2.$$

For three particles in the cell this number is $\frac{1}{6}(\omega_i)^3$. And, if N_i particles are placed into the cell "i," the number of distinct patterns is*

$$W_i = \frac{1}{N_i!}(\omega_i)^{N_i}. \tag{4-51}$$

The factor $(\omega_i)^{N_i}$ would represent the number of distinct patterns if all N_i atoms in the cell were distinguishable from each other. Since the atoms are identical, all $N_i!$ patterns that differ from each other only by a permutation of the positions of the N_i atoms must be counted as a *single* pattern. This accounts for the $1/N_i!$.

What is then the total weight W of a macrostate for which there are N_1 particles in cell "1," N_2 in cell "2," N_i in cell "i"? As we reasoned before, it is simply the product of the partial weights W_i for each macrocell

$$W(N_1, N_2, \ldots, N_i, \ldots) = W_1 \cdot W_2 \cdots W_i \cdots. \tag{4-52}$$

We will later work with the logarithm of W, which is a simpler expression to use. According to (4-52), we can indeed write it as

$$\ln W(N_1, N_2, \ldots, N_i, \ldots) = \ln W_1 + \ln W_2 + \cdots + \ln W_i + \cdots$$
$$= \sum_i \ln W_i = \sum_i \ln\left(\frac{(\omega_i)^{N_i}}{N_i!}\right). \tag{4-53}$$

---

*This result is correct only if $\omega_i \gg N_i$; that is, if the number of available microcells in "i" vastly exceeds the numbers N_i of particles to be placed into it. This case represents the dilute gas limit. Multiple occupancy of a microcell is then very improbable. For a *dense* gas for which N_i approaches ω_i, experience shows that two types of gases of identical particles exist:

(a) Bose particles for which multiple occupancy of a microcell is "permitted."

(b) Fermi particles in which multiple occupancy of a microcell is "forbidden."

An example of a Bose gas is helium, an example of a Fermi gas is the "electron gas" in metals. The exact expressions for the weight W_i for dense gases are:

(a) Bose gases: $W_i = \frac{1}{N_i!}\omega_i(\omega_i + 1)\ldots(\omega_i + N_i - 1)$;

(b) Fermi gases: $W_i = \frac{1}{N_i!}\omega_i(\omega_i - 1)\ldots(\omega_i - N_i + 1)$.

Both these expressions approach the value of W given by (4-51) in the limit where $\omega_i \gg N_i$. It may be observed that the recognition that the two patterns (a, b) of Figure 4.21 are *not distinct*, and must be counted as *one* only, is not a trivial point. This point was not properly understood until the advent of quantum physics, and generated considerable confusion in the early attempts at a statistical description of thermal equilibrium.

The numbers N_i are assumed to be large enough to allow us to use Stirling's approximation for $\ln N!$. We will thus introduce it here in the form

$$\ln N! \cong N \ln N - N. \tag{4-54}$$

This allows us to express the weight W of the macrostate in the simple form

$$\ln W = \sum_i (N_i \ln \omega_i - N_i \ln N_i + N_i). \tag{4-55}$$

We record again that the index i labels the diverse macrocells; that macrocell "i" has a volume Ω_i and contains $\omega_i = \Omega_i / h^3$ microcells. W is the weight of the macrostate characterized by the numbers N_i of particles in cell "i." This weight represents a relative (not normalized) probability of occurrence of a "coarse-grained" population pattern in phase space specified by the numbers N_i.

(vi)   Finding the Most Probable Macrostate

Each macrostate describes by means of the numbers N_i a "coarse-grained" population density in phase space. The weight W of the macrostate gives the relative probability that this particular density occurs. Two such coarse-grained phase-space densities are illustrated in Figure 4.22.

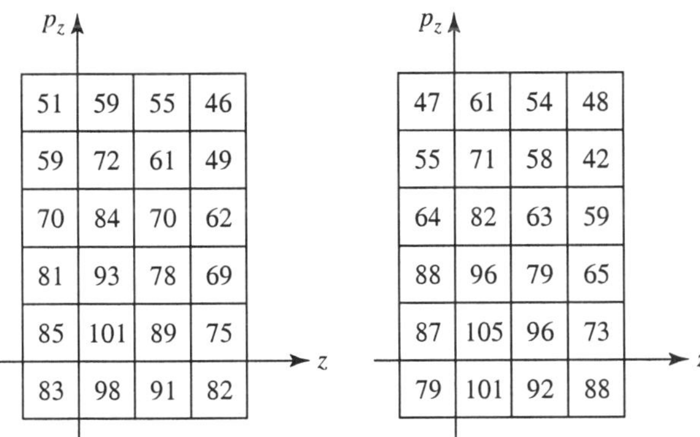

Figure 4.22. Two-dimensional phase space divided into *macrocells*. Two *macrostates* are represented by the numbers N_i of particles in each macrocell. These provide a coarse-grained representation of the *population density* in phase space.

As already mentioned in the introduction, we really want the mean numbers $\overline{N}_i$ in each macrocell. Again we exploit the fact that if the numbers N_i are generally large, then the weight has a *sharp* maximum essentially at the mean values $\overline{N}_i$ of the numbers N_i. This gives us the means to locate the values $\overline{N}_i$ by simply determining for what values of the N_i this maximum of W occurs. These values we take as the $\overline{N}_i$. Once the $\overline{N}_i$ are found, we will be able to define a mean density $\rho(i)$ of particles

$$\rho(i) = \left(\frac{\overline{N}_i}{\Omega_i} \right) \tag{4-56}$$

in phase space. Divided by the total number N of particles in the gas, this becomes the *probability density* $p(i)$ that an atom of the gas will be located at the site of the cell "i:"

$$p(i) = \frac{1}{N}\rho(i) = \frac{1}{N}\left(\frac{\overline{N}_i}{\Omega_i} \right). \tag{4-57}$$

So we must now find the values $\overline{N}_i$ for which the weight W, as given by (4-55), assumes its maximum value. Since the maximum of W is also the maximum for the logarithm $\ln W$, we propose to find the maximum of the latter. As was done in the case of the crystal, we expand the function $\ln W(N_1, N_2, \ldots, N_i, \ldots)$ about the presumed values $\overline{N}_i$ for which the maximum occurs by writing each N_i as the sum

$$N_i = \overline{N}_i + \delta N_i,$$

δN_i being the deviation of the value of N_i from the fixed value of $\overline{N}_i$. The familiar expansion of the logarithm

$$\ln N_i = \ln(\overline{N}_i + \delta N_i)$$
$$\cong \ln \overline{N}_i + \frac{\delta N_i}{\overline{N}_i} - \frac{1}{2}\frac{\delta N_i^2}{\overline{N}_i^2} + \cdots, \tag{4-58}$$

introduced into (4-55) then gives us

$$\ln W = \sum_i (\overline{N}_i + \delta N_i) \ln \omega_i - \sum_i (\overline{N}_i + \delta N_i) \left(\ln \overline{N}_i + \frac{\delta N_i}{\overline{N}_i} - \frac{1}{2} \frac{\delta N_i^2}{\overline{N}_i^2} \right)$$
$$+ \sum_i (N_i + \delta N_i).$$

We regroup this expression as follows:

$$\ln W = N - \sum_i \overline{N}_i \ln \left(\frac{\overline{N}_i}{\omega_i} \right)$$
$$- \sum_i \delta N_i \left(\ln \left(\frac{\overline{N}_i}{\omega_i} \right) \right)$$
$$- \frac{1}{2} \sum_i \frac{\delta N_i^2}{\overline{N}_i} + \cdots . \tag{4-59}$$

You will now recognize this equation as the analog of (4-28) for the case of the crystal.

The maximum of $\ln W$ is characterized by the fact that there should be no change of $\ln W$ that depends *linearly* on the δN_i. That is, we need

$$\sum_i \delta N_i \left(\ln \left(\frac{\overline{N}_i}{\omega_i} \right) \right) = 0 \tag{4-60}$$

for all allowed values of δN_i. We refer you here to the discussion following (4-28) that addresses the same issue: The deviations δN_i are not quite independent of each other since we may change the N_i only to the extent of preserving the total number $N = \sum N_i$ of particles, and the total energy $E = \sum \varepsilon_i N_i$. This subjects the deviations δN_i to the constraints

$$\sum_i \delta N_i = 0, \tag{4-61}$$

$$\sum_i \varepsilon_i \, \delta N_i = 0. \tag{4-62}$$

The consequence of this constraint condition is that (4-60) can be satisfied by the requirement

$$\ln\left(\frac{\overline{N}_i}{\omega_i}\right) = \alpha - \beta\varepsilon_i.$$

(4-63)

[If the δN_i were all independent, we would have to require $\ln(\overline{N}_i/\omega_i) = 0$ for all i!] Indeed, we can check that with (4-63):

$$\sum \delta N_i\left(\ln\left(\frac{\overline{N}_i}{\omega_i}\right)\right) = \alpha\sum \delta N_i - \beta\sum \varepsilon_i\,\delta N_i,$$

that is indeed zero if (4-61) and (4-62) are obeyed. The values N_i that give the maximum value of W consistent with a fixed number of particles and a fixed energy are thus given by the relation

$$\overline{N}_i = \omega_i e^{\alpha - \beta\varepsilon_i}.$$

(4-64)

The mean fraction of particles that are in macrocell "i" is therefore given by

$$\left(\frac{\overline{N}_i}{N}\right) = \frac{\omega_i e^{\alpha}}{N} e^{-\beta\varepsilon_i}.$$

(4-65)

This result is the analog of (4-33) for the crystal, giving the mean number of particles $\overline{N}_n$ in the quantum state n. Again, we observe the occurrence of the Boltzmann factor $e^{-\beta\varepsilon_i}$, indicating that the mean phase-space density decreases *exponentially* with the particle energy ε_i associated with the location "i" in phase space. The lowest energy regions in phase space are therefore the most densely populated, and the density decreases monotonically with increasing particle energy ε_i. It is again worthwhile to trace this remarkable energy dependence to its roots. These are somewhat hidden behind the elegant argument leading to (4-63). Let us start again with the expression for the *first-order* change of ln W due to a change of the numbers N_i by δN_i. According to (4-59), this is given by

$$\delta(\ln W) = -\sum_i \delta N_i \ln\left(\frac{N_i}{\omega_i}\right)$$

(4-66)

and should be *zero* for the values $\overline{N}_i$ of the N_i. This condition $\delta\ln W_{N_i=\overline{N}_i} = 0$ holds for *any* values of the variations δN_i compatible with the constraints (4-61) and (4-62).

Let us now consider a special variation of the simplest possible type. Let us take just three macrocells, call them 1, 2, 3, at three different energies (see Figure 4.23) $\varepsilon_3 > \varepsilon_2 > \varepsilon_1$:

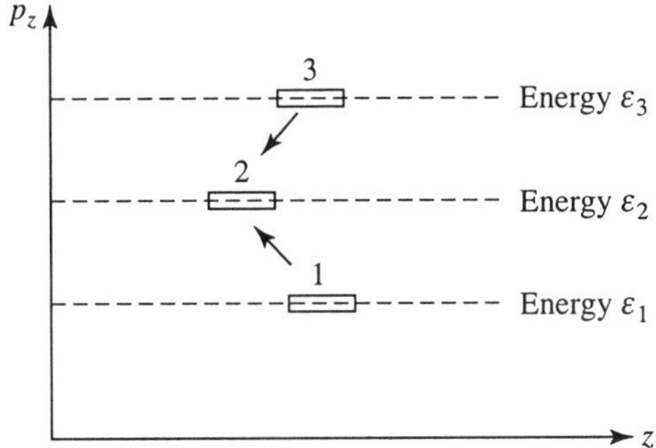

Figure 4.23. Three macrocells in two-dimensional phase space at three different values of the energy.

The variations δN_i that we consider now represent a transfer of particles from cells 1 and 3 into cell 2. We can express the variations that conserve the particle number and energy as follows:

$$\begin{aligned}
\delta N_3 &= (\varepsilon_1 - \varepsilon_2)\xi, \\
\delta N_2 &= (\varepsilon_3 - \varepsilon_1)\xi, \\
\delta N_1 &= (\varepsilon_2 - \varepsilon_3)\xi,
\end{aligned} \tag{4-67}$$

where the arbitrary constant ξ is the same in all three expressions. It is easily seen that the constraint conditions [(4-61), (4-62)] are satisfied for any value of ξ:

$$\delta N_1 + \delta N_2 + \delta N_3 = 0,$$
$$\varepsilon_1 \delta N_1 + \varepsilon_2 \delta N_2 + \varepsilon_3 \delta N_3 = 0.$$

Inserting the three δN's as given by (4-67) into (4-66), we find

$$\delta \ln W = \xi \left\{ (\varepsilon_1 - \varepsilon_2) \ln\left(\frac{N_3}{\omega_3}\right) + (\varepsilon_3 - \varepsilon_1) \ln\left(\frac{N_2}{\omega_2}\right) + (\varepsilon_2 - \varepsilon_3) \ln\left(\frac{N_1}{\omega_1}\right) \right\}.$$

This must be zero for all ξ when $N_3 = \overline{N}_3$, $N_2 = \overline{N}_2$, $N_1 = \overline{N}_1$. The braces must therefore be *zero*. Writing

$$\varepsilon_3 - \varepsilon_1 = (\varepsilon_3 - \varepsilon_2) + (\varepsilon_2 - \varepsilon_1),$$

we can express this result in the form

$$(\varepsilon_1 - \varepsilon_2)\left(\ln\left(\frac{\overline{N}_3}{\omega_3}\right) - \ln\left(\frac{\overline{N}_2}{\omega_2}\right) \right) + (\varepsilon_2 - \varepsilon_3)\left(\ln\left(\frac{\overline{N}_1}{\omega_1}\right) - \ln\left(\frac{\overline{N}_2}{\omega_2}\right) \right) = 0$$

or

$$\frac{1}{\varepsilon_2 - \varepsilon_3} \ln\left(\frac{\overline{N}_3/\omega_3}{\overline{N}_2/\omega_2}\right) = \frac{1}{\varepsilon_1 - \varepsilon_2} \ln\left(\frac{\overline{N}_2/\omega_2}{\overline{N}_1/\omega_1}\right).$$

To appreciate the meaning of this last equation, we must realize that it applies to *any* three macrocells and their corresponding energies. The result states therefore that the ratio

$$\frac{1}{\varepsilon_j - \varepsilon_i} \ln\left(\frac{\overline{N}_i/\omega_i}{\overline{N}_j/\omega_j}\right)$$

has the same value for any pair (i, j) of macrostates. Let us call this value β. This gives

$$\ln\left(\frac{\overline{N}_i}{\omega_i}\right) - \ln\left(\frac{\overline{N}_j}{\omega_j}\right) = \beta(\varepsilon_j - \varepsilon_i).$$

Since i and j are any two macrocells, we infer that

$$\ln\left(\frac{\overline{N}_i}{\omega_i}\right) + \beta\varepsilon_i = \ln\left(\frac{\overline{N}_j}{\omega_j}\right) + \beta\varepsilon_j = \text{same for any cell} = \alpha.$$

In this manner we have rederived (4-63):

$$\ln\left(\frac{\overline{N}_i}{\omega_i}\right) = \alpha - \beta\varepsilon_i \qquad \text{for all } i$$

in a way that demonstrates the crucial role of the constraint conditions.

As a final word, it should be clear that the appearance of the logarithm: $\ln(N_i/\omega_i)$ in expression 4-66) lies in the fact that the weight W depends on N_i through the combinatorial factor

$$W_i = \frac{1}{N_i!}(\omega_i)^{N_i}. \tag{4-51}$$

Using Stirling's approximation, we wrote

$$\frac{1}{N_i!} \cong e^{N_i - N_i \ln N_i}.$$

We can also write $(\omega_i)^{N_i}$ as $e^{N_i \ln \omega_i}$ so that

$$W_i \cong e^{N_i - N_i \ln(N_i/\omega_i)}. \tag{4-68}$$

It follows that the change δW_i produced by a change of N_i is

$$\delta W_i = \frac{\partial W_i}{\partial N_i} \delta N_i$$

$$= -W_i \ln\left(\frac{N_i}{\omega_i}\right) \delta N_i$$

or

$$\frac{\delta W_i}{W_i} = \delta(\ln \omega_i) = -\delta N_i \ln\left(\frac{N_i}{\omega_i}\right).$$

The appearance of the factor $\ln(N_i/\omega_i)$ results directly from the structure of (4-51). On computing δW_i, we obtain the $-\ln N_i$ term from the $N_i!$ in the denominator, and we obtain the term $\ln \omega_i$ from the numerator $(\omega_i^{N_i})$.

We terminate this subsection by constructing a more convenient notation for representing the mean density $\rho(i)$ of particles in phase space. This density was defined in (4-56) as the ratio $\overline{N}_i/\Omega_i$ between the mean number $\overline{N}_i$ of particles in cell "i" and the cell volume Ω_i. Using the relation $\Omega_i = \omega_i h^3$, we find

$$\rho(i) = \left(\frac{\overline{N}_i}{\Omega_i} \right) = \frac{1}{h^3} e^{\alpha} \cdot e^{-\beta \varepsilon_i}. \tag{4-69}$$

What we want to do now is to get rid of the subdivision of phase space into cells "i"; these cells have now served their purpose and are no longer needed. We observe that each cell has its center located at some point $(\vec{r}, \vec{p})$ in phase space. So we may write $\rho(\vec{r}, \vec{p})$ in place of $\rho(i)$. The other reference to "i" is in the "cell energy" ε_i, which is the sum of the kinetic and potential energy of a particle located in cells "i," that is, at a point $(\vec{r}, \vec{p})$ in phase space. We can get rid of the reference to the index "i" too, and simply write

$$\varepsilon_i \rightarrow \varepsilon(\vec{r}, \vec{p}) = \frac{p^2}{2m} + u(\vec{r}). \tag{4-70}$$

With these adjustments we may rewrite (4-69) for the mean density $\rho(\vec{r}, \vec{p})$ of particles at the location $(\vec{r}, \vec{p})$ in phase space

$$\rho(\vec{r}, \vec{p}) = \frac{e^{\alpha}}{h^3} e^{-\beta \varepsilon(\vec{r}, \vec{p})}. \tag{4-71}$$

This notational change also allows us to write the sum over all macrocells "i" in phase space by an integration over the phase-space volume. We will now write $d\Omega$ for a volume element in phase space. $d\Omega$ is the *product* of a position space volume element dV and a momentum space volume element dV_p. The relation

$$\overline{N}_i = \rho(i)\Omega_i$$

will therefore now be written as

$$dN = \rho(\vec{r}, \vec{p})\, d\Omega,$$

dN standing for the mean number of particles in the phase-space volume element $d\Omega$.

The total number of particles is then found by summing $\overline{N}_i$ over all cells "i," or, in our new notation, by integrating dN over all volume elements $d\Omega$:

$$N = \sum_i N_i = \sum_i \rho(i)\Omega_i$$

or

$$N = \int dN = \int \rho(\vec{r}, \vec{p})\, d\Omega. \tag{4-72}$$

The same change of notation can be used to rewrite the probability density $P(i)$, given by (4-57), for finding an atom at a particular region in phase space. We now write $P(\vec{r}, \vec{p})$ instead of $p(i)$ for this density

$$P(\vec{r}, \vec{p}) = \frac{1}{N}\rho(\vec{r}, \vec{p}) = \frac{e^{\alpha}}{Nh^3}e^{-\beta\varepsilon(\vec{r},\vec{p})}. \tag{4-73}$$

Let us reiterate that in this notation the probability that a particular atom is in phase-space volume element $d\Omega$ at location $(\vec{r}, \vec{p})$ is given by

$$P(\vec{r}, \vec{p})\, d\Omega = \frac{e^{\alpha}}{Nh^3}e^{-\beta\varepsilon(\vec{r},\vec{p})}\, d\Omega.$$

(vii)   The Boltzmann Factor and Temperature

Our point of departure here is (4-69) or (4-71) for the mean equilibrium density of particles in phase space. It contains the two, yet undetermined, constants α and β. Of these, α is essentially a normalization constant and can be expressed in terms of β by inserting the form (4-71) for the density $\rho(\vec{r}, \vec{p})$ into the normalization equation (4-72). Recall that the phase-space volume element $d\Omega$ is the product of dV, the volume element in position space, and dV_p, the volume element in momentum space. We may therefore write (4-72) as

$$N = \int dV \int dV_p \frac{e^{\alpha}}{h^3}e^{-\beta\varepsilon(\vec{r},\vec{p})},$$

where $\varepsilon(\vec{r}, \vec{p})$ is the particle energy

$$\varepsilon(r, p) = \frac{p^2}{2m} + u(\vec{r}).$$

The fact that ε is the *sum* of the kinetic and potential energies allows us to factor the two contributions in the Boltzmann factor

$$e^{-\beta\varepsilon} = e^{-\beta(p^2/2m)} e^{-\beta u(r)}.$$

The expression for the particle number N can therefore be written as

$$N = \frac{e^{\alpha}}{h^3} \int dV e^{-\beta u(r)} \int dV_p e^{-\beta(p^2/2m)}. \tag{4-74}$$

Let us first consider the case where *no potential energy* is present. (All gases are, of course, subject to the Earth's gravitational force; for a small amount of gas in a bottle, however, the potential energy is essentially constant, and we can put it equal to zero, or absorb it in the constant factor e^{α}.) We assume, in this case, that the gas is contained in a finite volume V. The absence of a variation of potential energy within V gives us a density $\rho(r, p)$ that docs not depend on $\vec{r}$. The spatial density of the gas is therefore *uniform*. The only integral to be calculated is then

$$\int dV_p e^{-\beta(p^2/2m)}.$$

Since $p^2 = p_x^2 + p_y^2 + p_z^2$ and $dV_p = dp_x\, dp_y\, dp_z$, this integral is equal to

$$\int dp_x e^{-\beta(p_x^2/2m)} \times \int dp_y e^{-\beta(p_y^2/2m)} \times \int dp_z e^{-\beta(p_z^2/2m)}.$$

Each single integral in this product has the same value (see Appendix 2.A3 to Chapter 2), that is $\sqrt{2\pi m/\beta}$. Hence

$$\int dV_p e^{-\beta(p^2/2m)} = \left(\frac{2\pi m}{\beta}\right)^{3/2}. \tag{4-75}$$

Inserted into (4-74), we find

$$N = \left(\frac{e^{\alpha}}{h^3}\right) V \left(\frac{2\pi m}{\beta}\right)^{3/2},$$

which gives

$$\frac{e^{\alpha}}{h^3} = \left(\frac{N}{V}\right)\left(\frac{\beta}{2\pi m}\right)^{3/2}.$$ (4-76)

With this we can write the normalized density distribution function $\rho(\vec{r}, \vec{p})$, as given by (4-71):

$$\rho(\vec{r}, \vec{p}) = \left(\frac{N}{V}\right)\left(\frac{\beta}{2\pi m}\right)^{3/2} e^{-\beta(\vec{p}^2/2m)}.$$ (4-77)

This is for the case of zero potential energy and a gas of N particles confined in a finite volume V. The function ρ depends on the momentum p only in this case. It has in fact the form of a product of the spatial density (N/V) times a function of the particle momentum $\vec{p}$ only

$$\rho(\vec{r}, \vec{p}) = \left(\frac{N}{V}\right) f(\vec{p}),$$ (4-78)

where

$$f(p) = \left(\frac{\beta}{2\pi m}\right)^{3/2} e^{-\beta(\vec{p}^2/2m)}.$$ (4-79)

This function $f(p)$ is called the momentum distribution function; we will look at it in more detail in Subsection 4.2.B(viii) below. As of now, our objective is to nail down β by calculating the value of the total energy. All particles in a volume element dV_p of momentum space have a kinetic energy $\varepsilon(\vec{p}) = \vec{p}^2/2m$. The total energy E is therefore

$$\begin{aligned}
E &= \int_V dV \int dV_p \left(\frac{\vec{p}^2}{2m}\right) \rho(\vec{r}, \vec{p}) \\
&= \frac{N}{V} \int dV \int dV_p \left(\frac{\vec{p}^2}{2m}\right) f(\vec{p}) \\
&= N \int dV_p \left(\frac{\vec{p}^2}{2m}\right) f(\vec{p}),
\end{aligned}$$ (4-80)

$f(\vec{p})$ is given in the preceding equation. The V_p integral is easily evaluated as follows. $(\vec{p}^2/2m)^{-\beta(\vec{p}/2m)}$ is expressible as a partial derivative with respect to β:

$$(\vec{p}^2/2m)e^{-\beta(\vec{p}^2/2m)} = -\frac{\partial}{\partial\beta}\left(e^{-\beta(\vec{p}^2/2m)}\right).$$

This β-derivative can be carried out *after* the V_p integration

$$\int dV_p \frac{\vec{p}^2}{2m} e^{-\beta(\vec{p}^2/2m)} = -\frac{\partial}{\partial\beta} \int dV_p e^{-\beta(\vec{p}^2/2m)}$$

$$= -\frac{\partial}{\partial\beta}\left(\frac{2\pi m}{\beta}\right)^{3/2} = \frac{3}{2\beta}\left(\frac{2\pi m}{\beta}\right)^{3/2}.$$

It follows that the energy E has the value

$$E = \frac{3}{2}\left(\frac{N}{\beta}\right). \tag{4-81}$$

We *know* that the energy of an ideal gas at temperature T is $E = \frac{3}{2}NkT$, for all temperatures. This identifies the parameter β as proportional to the inverse Kelvin temperature

$$\beta = \frac{1}{kT}. \tag{4-82}$$

We shall show in the next Section, 4.2.C, that thermal equilibrium *between* two systems is expressed by a common value of β. This property of β makes it possible, in hindsight, to reverse the argument and to *define* temperature by means of β and (4-82). This must be done, for instance, when one is dealing with very low temperatures, where *no* gas behaves like an ideal gas anymore, and therefore no "ideal gas standard" for T is available.

(viii)   The Barometric Formula

We have so far neglected the effect of potential energy. Now that our identification of β is achieved, let us briefly return to the case where there *is* a gravitational potential energy $u(z) = mgz$, so that

$$\varepsilon(\vec{r}, \vec{p}) = \vec{p}^2/2m + mgz.$$

In this case (4-71) states that the phase-space density is

$$\rho(z, \vec{p}) = \frac{e^{\alpha}}{h^3} e^{-\beta m g z} e^{-\beta(\vec{p}^2/2m)}.$$

The density therefore changes with z. If we are interested in the spatial density $\rho(z)$, irrespective of momentum of the particle, we must sum over all momentum space for fixed z:

$$\rho(z) = \frac{e^{\alpha}}{h^3} e^{-\beta m g z} \int dV_p e^{-\beta(p^2/2m)}$$

$$= \rho(z = 0) e^{-\beta m g z},$$

where we have absorbed all the constants and the value of the V_p integral into a constant that we recognize as the spatial density at $z = 0$. This is indeed a familiar result. Writing it in terms of the temperature, it reads as

$$\rho(z) = \rho(z = 0) e^{-(mg/kT)z}. \tag{4-83}$$

Alternatively, we may write it in terms of the molecular weight $M = N_0 m$, and the gas constant $R = N_0 k$ as

$$\rho(z) = \rho(z = 0) e^{-(Mg/RT)z}. \tag{4-83a}$$

This is the well-known "barometric formula" expressing the exponential decrease of the density of a gas (at constant temperature) with height. This relation, or rather the corresponding relation for the height dependence of pressure, was derived in Volume I, Section 3.5.B. Only the equation of state of the ideal gas

$$p(z) = \frac{RT}{M} \rho(z) \tag{4-84}$$

is, in fact, needed to obtain (4-83a). It follows at once from (4-81) of this volume, connecting the pressure gradient dp/dz to the body force per unit volume

$$\frac{dp}{dz} = \frac{\Delta F_z}{\Delta V}.$$

In the case of gravity, the body force ΔF_z acting on ΔV is

$$-\Delta V \rho g$$

so that we find $dp/dz = -\rho g$. For constant temperature we may use the equation of state (4-84) to express dp/dz in terms of $d\rho/dz$ and get

$$\frac{d\rho}{dz} = \frac{M}{RT}\frac{dp}{dz} = -\left(\frac{Mg}{RT}\right)\rho(z).$$

This is the familiar first-order differential equation for an exponential, and has the solution [(4-83a)]:

$$\rho(z) = \rho(z = 0)e^{-(Mg/RT)z}.$$

(ix)   The Maxwell–Boltzmann Distribution Function of Velocities

We have seen that in the absence of body forces on the gas particles (i.e., for constant potential energy), the phase-space density $\rho(\vec{r}, \vec{p})$ is the product of the uniform spatial density (N/V) of the particles and a momentum distribution function $f(p)$:

$$\rho(\vec{r}, p) = \left(\frac{N}{V}\right) f(p),$$

where

$$f(p) = (1/2\pi mkT)^{3/2}\, e^{-\vec{p}^2/2mkT}. \tag{4-85}$$

(We have now inserted $1/kT$ for β, and will continue to do so.) We can attach a simple interpretation to this function $f(\vec{p})$. The quantity

$$f(\vec{p})\, dV_p$$

represents the *fraction* of particles that have a momentum in the range dV_p. Alternatively, it may be viewed as the *probability* that a given particle, in the course of its history, has a momentum in that range. So $f(\vec{p})$ is a probability per unit momentum range, or a *probability distribution* for momentum. Starting with this distribution function f, we can construct several related distribution functions of interest. We will discuss three of them here: the distribution functions for velocities, for speeds, and for energies.

The Distribution Function for Velocities.   The velocity $\vec{v}$ of a particle is related to its momentum $\vec{p}$ by

$$\vec{p} = m\vec{v}.$$

Recall that velocity is a vector, like momentum. A given velocity range is therefore represented by the statement that the x-component of $\vec{v}$ is within a range Δv_x, and the y-component within a range Δv_y, the z-component within a range Δv_z. Geometrically, this means that the vector $\vec{v}$ has its end point within a parallelpiped of sizes Δv_x, Δv_y, Δv_z in "velocity space." (See Figure 4.24.)

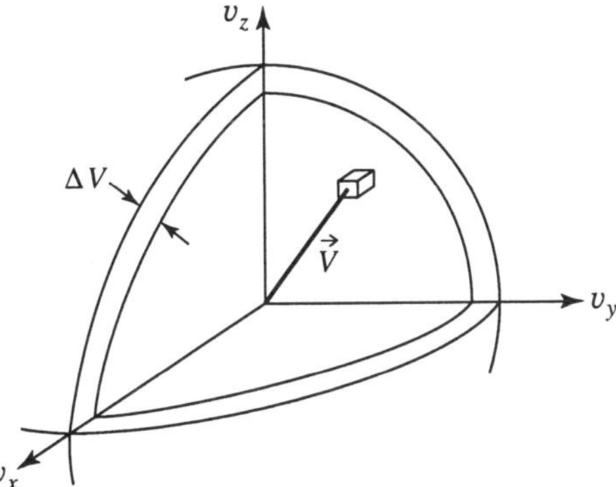

Figure 4.24. Velocity space. (a) A parallelpiped with sides Δv_x, Δv_y, Δv_z defines a velocity range. All velocity vectors in the range Δv_x, Δv_y, Δv_z have their end points inside the box. (b) Shown are two concentric spheres of radii v and $v + \Delta v$. They enclose a shell of volume $4\pi v^2 \Delta v$. All vectors with end points inside this shell belong to the speed range Δv.

The probability that a particle has a velocity in a range Δv_x, Δv_y, Δv_z can be found by expressing the corresponding momentum range in terms of the velocities

$$\Delta p_x = m\Delta v_x, \text{ etc.,}$$

and expressing the energy $p^2/2m$ as $(m/2)v^2$. This probability is

$$f(\vec{p})\Delta p_x \Delta p_y \Delta p_z = \left(\frac{1}{2\pi mkT}\right)^{3/2} e^{-p^2/2mkT} \Delta p_x \Delta p_y \Delta p_z$$

$$= \left(\frac{m}{2\pi kT}\right)^{3/2} e^{-mv^2/2kT} \, \Delta v_x \, \Delta v_y \, \Delta v_z$$

$$= \overline{f}(\vec{v}) \, \Delta v_x \, \Delta v_y \, \Delta v_z.$$

The last line of this equation *defines* the probability density distribution function, that is, the probability per unit velocity range. It has the form

$$\overline{f}(\vec{v}) = \left(\frac{m}{2\pi kT}\right)^{3/2} e^{-m\vec{v}^2/2kT}. \tag{4-86}$$

This function is known as the Maxwell–Boltzmann distribution function for velocities. Just as the corresponding momentum distribution function, it is the product of three Gaussians. Indeed, writing $\vec{v}^2 - v_x^2 + v_y^2 + v_z^2$, we see that $\overline{f}(\vec{v})$ is in fact a product

$$\overline{f}(\vec{v}) = f_x(v_x) f_y(v_y) f_z(v_z),$$

where

$$f_x(v_x) = \left(\frac{m}{2\pi kT}\right)^{3/2} e^{-mv_x^2/2kT} \tag{4-87}$$

and $f_y(v_y) f_z(v_z)$ have the same structure.

This product form of $\overline{f}(\vec{v})$ has the interesting consequence that the probability that a particle v_x lies in a certain range Δv_x is *independent* of the concurrent values of its v_y- and v_z-components of the velocity. A remarkable result!

We leave it as a problem assignment to calculate some mean values by means of this distribution function, in particular, to verify the statements

$$\overline{v_x} = \int dv_x \, v_x f_x(v_x) = 0, \tag{4-88a}$$

$$\overline{v_x^2} = \int dv_x \, v_x^2 f_x(v_x) = \left(\frac{kT}{m}\right). \tag{4-88b}$$

The mean square values of v_x, v_y, and v_z are all the same. Thus, from (4-88b) we reconfirm the result that

$$\frac{m\overline{v}^2}{2} = \frac{m}{2}(\overline{v_x^2} + \overline{v_y^2} + \overline{v_z^2}) = \frac{m}{2} 3 \left(\frac{kT}{m}\right) = \tfrac{3}{2}kT,$$

in words, the mean kinetic energy of a particle in the ideal gas at temperature T is $\frac{3}{2}kT$.

The Distribution Function for Speeds v.   This is a distribution derived from the previous one, and answers the question: What is the probability that a particle's speed

$$v = \sqrt{v_x{}^2 + v_y{}^2 + v_z{}^2}$$

lies in the range Δv?

Consulting Figure 4.23, we see that to answer this question we must find the probability that the velocity vector $\vec{v}$ lies inside the spherical shell of radius v and thickness Δv. Making Δv small, we can assume a common speed v to all velocity vectors ending in the shell. The volume of the shell is $4\pi v^2 \Delta v$. Thus, the desired probability that the speed is in range Δv is

$$\begin{aligned} \text{probability} &= \text{probability density} \times \text{volume of shell with speed range} \Delta v \\ &= \left(\frac{m}{2\pi kT}\right)^{3/2} e^{-mv^2/2kT} \times 4\pi v^2 \Delta v. \\ &\equiv F(v)\Delta v. \end{aligned} \tag{4-89}$$

This relation *defines* the distribution function of speeds. $F(v)$ is the probability density per unit speed range and is clearly expressed by

$$F(v) = 4\pi \left(\frac{m}{2\pi kT}\right)^{3/2} v^2 e^{-mv^2/2kT}. \tag{4-90}$$

This is the famous Maxwell–Boltzmann distribution function of *speeds*. We plot it in Figure 4.25.

Again we leave it as an exercise to calculate the following mean values:

- The mean speed $\bar{v} = \int_0^\infty dv\, v F(v)$. This turns out to be

$$\bar{v} = 2\sqrt{\frac{2}{\pi}} \left(\frac{kT}{m}\right)^{1/2} = 1.596 \times \left(\frac{kT}{m}\right)^{1/2}.$$

- The most probable speed v_M. This is the value of v for which the function $F(v)$ has its maximum. This occurs at the value

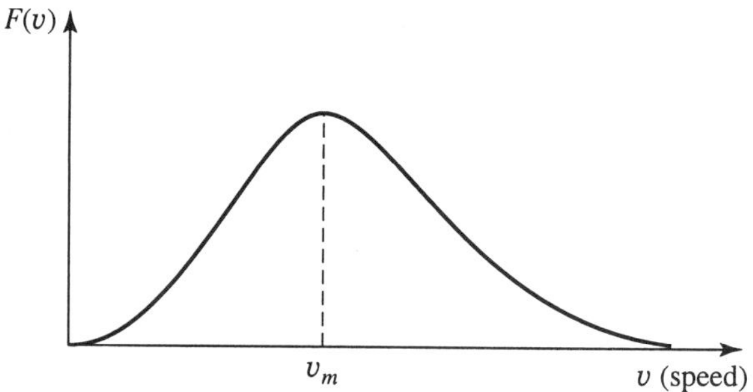

Figure 4.25. The Maxwell–Boltzmann distribution function of speeds.

$$v_M = \sqrt{2}\left(\frac{kT}{m}\right)^{1/2} = 1.414\left(\frac{kT}{m}\right)^{1/2}$$

of the speed.

- Finally, there is the root mean square (rms) speed $\sqrt{\overline{v^2}} \equiv v_{\mathrm{rms}}$. We know that

$$\overline{v^2} = 3\left(\frac{kT}{m}\right)$$

and have therefore

$$v_{\mathrm{rms}} = \sqrt{3}\left(\frac{kT}{m}\right)^{1/2} = 1.732\left(\frac{kT}{m}\right)^{1/2}.$$

We see that all three mean values are similar. The most probable speed has the smallest value, followed by the mean speed; the root mean square speed is the largest quantity. For numerical values of the root mean square speed for various particles, we refer back to Table 2.1 of Chapter 2.

The Distribution Function for Energy.   This function is used relatively little, but it serves as a good illustration of how a distribution changes if the nature of the independent variable changes. Here we address the question: What is the probability that the energy ε of a particle in the gas is in the range $\Delta\varepsilon$?

We can find this probability from the probability distribution of speeds by constructing the relation between a range Δv of speed and the corresponding energy range. From the relation

$$\varepsilon = \frac{m}{2} v^2$$

it follows that

$$\Delta\varepsilon = mv\,\Delta v$$

or

$$\Delta v = \frac{\Delta\varepsilon}{mv} = \frac{\Delta\varepsilon}{\sqrt{2m\varepsilon}}. \tag{4-91}$$

Inserting this into (4-90), and rewriting all speeds in terms of energy, we find

$$F(v)\,\Delta v = 4\pi \left(\frac{m}{2\pi kT}\right)^{3/2} \left(\frac{2\varepsilon}{m}\right) e^{-\varepsilon/kT}\,\frac{\Delta\varepsilon}{\sqrt{2m\varepsilon}}$$

$$= \overline{F}(\varepsilon)\,\Delta\varepsilon. \tag{4-92}$$

This relation *defines* the probability density function $\overline{F}(\varepsilon)$ for energy. It is

$$\overline{F}(\varepsilon) = \frac{2}{\sqrt{\pi}} \left(\frac{1}{kT}\right)^{3/2} \sqrt{\varepsilon}\, e^{-\varepsilon/kT}. \tag{4-93}$$

This function is plotted in Figure 4.26.

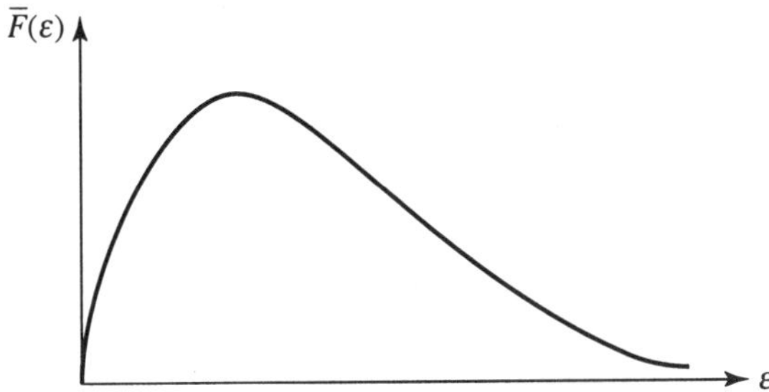

Figure 4.26. The probability distribution function for the energy ε of a gas particle. $\overline{F}(\varepsilon)\,\Delta\varepsilon$ represents the probability that the particle has an energy in the range $\Delta\varepsilon$.

4.2.C. Thermal Equilibrium Between Solid and Gas. Vapor Pressure of a Solid

In Section 4.2.B we saw that the probability distribution for the energy of an atom, both in the case of the crystal and in that of the gas, contained a parameter β. In the gas we could identify this parameter as being equal to $1/kT$, k being the Boltzmann constant and T the Kelvin temperature. In the case of the crystal this identification was possible only in the high-temperature limit. In this present section we show that the relation

$$\beta = \frac{1}{kT} \tag{4-94}$$

is *generally* valid; β is *always* related to the temperature by (4-94). In order to carry out this demonstration, we introduce one novel feature: two distinct systems in thermal contact. The "thermal contact" of two systems is a statement of their ability to exchange energy. We know from experience that the result of the thermal contact of two systems is that they exchange energy until their temperatures become equal. We will show that our statistical method of describing the thermal equilibrium between two systems leads to the result that the parameters β in the two probability distributions for energy must be the same. Thus, the equality of β becomes the correlate of the observational fact of equal temperatures. By choosing as our two systems in contact an ideal gas and a crystal, and using the validity of (4-94) for the ideal gas, we will have demonstrated it for the crystal too.

In a second part of this Section 4.2.C, we will go one step further and consider two special systems in contact, namely, the solid crystalline phase of a material and its gas phase. We have already shown in the introduction to Section 4.2 how every solid (and liquid) at finite temperature is in equilibrium with its own vapor. The equilibrium in this case results not only from the exchange of energy between the two systems (solid and vapor), but also from the exchange of particles. In investigating this equilibrium, we will show how the saturation vapor pressure depends on temperature mainly through a Boltzmann factor

$$p_{\text{vapor}} \sim e^{-E_B/kT},$$

that is responsible for the dramatic changes of this pressure with temperature.

(i)   Gas and Solid in Thermal Contact. Demonstration that $\beta = 1/kT$ at All Temperatures

In Section 4.2 we had shown that the thermal equilibrium for a crystal, or for an ideal gas, was characterized by a set of mean numbers $\overline{N}_n$ or $\overline{N}_i$ obtained by finding the "most probable macrostate." To find this most probable macrostate, we attached a *weight* W to each macrostate, and then determined the numbers $\overline{N}_n$ (or $\overline{N}_i$) for which the weight W, or its logarithm, assumed its maximum value. In finding this maximum, we had to keep in mind that the numbers N_n and N_i were not quite free, but restricted by the conditions [(4-31) and (4-61), (4-62)], expressing the fact that we dealt with a fixed number of particles and with a fixed total energy.

Let us now consider two systems in thermal contact. The first system is the Einstein crystal, a lattice of N_c atoms. (We disregard, at this point, the fact that this crystal has its own vapor phase by simply stating that the number N_c is a constant.) The second system is an ideal gas of N_g atoms in a volume V (see Figure 4.27).

We will assume that the solid and gas are two different materials like a copper block and helium gas. We exclude the possibility of particle exchange between these two systems.

We must now answer the question: What characterizes the thermal equilibrium of this composite system? Clearly we must have both thermal equilibrium within each component system (solid and gas), as well as equilibrium *between* the two components.

The key to the answer is the same as used in Sections 4.2.A, B. We introduce the macrostates of the joint system, and then look for the most probable macrostate. A macrostate of the joint system is represented by the two sets of numbers $\{N_n\}$ and $\{N_i\}$. N_n represents the number of atoms of the crystal that are in quantum state n, and N_i represents the number of gas atoms in phase-space macrocell "i." The total weight W of this macrostate is given by the total number of microstates it contains. Now, if for given sets $\{N_n\}$ and $\{N_i\}$ there are W_c microstates of the

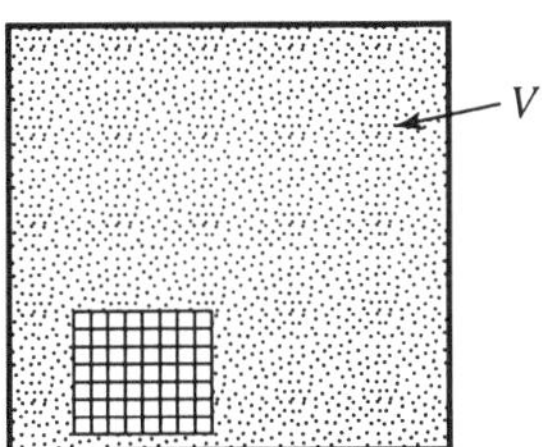

Figure 4.27. A solid (crystal) in thermal contact with an ideal gas enclosed in a volume V.

crystal and W_g microstates of the gas, there will be a total number

$$W = W_c \times W_g \qquad (4\text{-}95)$$

microstates for the joint system. (If you get hung up on this argument, remind yourself of the analogy: The number of outcomes for the throw of one die is 6, for the throw of two dice is 6×6.) Once this product rule for the weight of the macrostate of the joint system is established, we can proceed as before.

The most probable macrostate of the joint system is given by the maximum of W or, equivalently, by the maximum of $\ln W$:

$$\ln W = \ln W_c + \ln W_g. \qquad (4\text{-}96)$$

We now proceed to find the values of N_n and N_i for which $\ln W$ has its maximum value. For this we write again $N_n = \overline{N}_n + \delta N_n$, $N_i = \overline{N}_i + \delta N_i$, and use the previously derived expressions [(4-28) and (4-59)] for $\ln W_c$ and $\ln W_g$ in the neighborhood of the values $N_i = \overline{N}_i$ and $N_n = \overline{N}_n$:

$$\ln W = \sum_n \overline{N}_n \ln\left(\frac{\overline{N}_n}{N_c}\right) + N_g - \sum_i \overline{N}_i \ln\left(\frac{\overline{N}_i}{\omega_i}\right)$$

$$- \sum_n \delta N_n \ln\left(\frac{\overline{N}_n}{N_c}\right) - \sum_i \delta N_i \ln\left(\frac{\overline{N}_i}{\omega_i}\right)$$

$$+ \quad \text{quadratic terms in the } \delta N_i, \delta N_n. \qquad (4\text{-}97)$$

If the first line of this equation is to represent the maximum of $\ln W$, then the second line expressing the first-order change $\delta(\ln W)$ as a function of the δN_i and δN_n must vanish

$$\delta \ln W = -\sum_n \delta N_n \ln\left(\frac{\overline{N}_n}{N_c}\right) - \sum_i \delta N_i \ln\left(\frac{\overline{N}_i}{\omega_i}\right) = 0. \qquad (4\text{-}98)$$

This sum must be zero for all changes δN_n and δN_i compatible with the constraints imposed as the N_n and N_i. These constraints in our case are

$$\sum N_n = N_c = \text{constant: number of atoms in crystal fixed,}$$

$$\sum N_i = N_g = \text{constant: number of atoms in gas fixed,}$$

$$\sum \varepsilon_i N_n + \sum \varepsilon_i N_i = \text{constant: total energy fixed.}$$

The variations δN_i and δN_n therefore must be restricted so as to obey

$$\sum_n \delta N_n = 0, \tag{4-98a}$$

$$\sum_i \delta N_i = 0, \tag{4-98b}$$

$$\sum_n \varepsilon_n \, \delta N_n + \sum_i \varepsilon_i \, \delta N_i = 0. \tag{4-99}$$

We now have to ask again: What are the most general values of the coefficients $\ln(\overline{N}_n/N_c)$ and $\ln(\overline{N}_i/\omega_i)$ occurring in (4-98) which guarantee that $\delta(\ln W)$ is indeed zero?

We see at once that the constraints [(4-98)] permit a solution where all coefficients $\ln(\overline{N}_n/N_c)$ are equal to a single constant α_c, and again all coefficients $\ln(\overline{N}_i/\omega_i)$ are equal to another constant α_g. Indeed, inserting the values

$$\ln\left(\frac{\overline{N}_n}{N_c}\right) = \alpha_c \quad \text{for all } n,$$

$$\ln\left(\frac{\overline{N}_i}{\omega_i}\right) = \alpha_g \quad \text{for all } i,$$

into (4-98), we find

$$\delta(\ln W) = -\alpha_c \left(\sum_n \delta N_n\right) - \alpha_g \left(\sum_i \delta N_i\right),$$

and this is indeed zero because of the restrictions [(4-98)] imposed on the δN_n and δN_i.

But this is not the most general solution. The existence of the additional constraint [(4-99)] makes a slightly more general form possible, in which $\ln(\overline{N}_n/N_c)$ and $\ln(\overline{N}_i/\omega_i)$ also depend on energy

$$\ln\left(\frac{\overline{N}_n}{N_c}\right) = \alpha_c - \beta \varepsilon_n, \tag{4-100a}$$

$$\ln\left(\frac{\overline{N}_i}{\omega_i}\right) = \alpha_g - \beta\varepsilon_i. \tag{4-100b}$$

Inserting these tentative expressions into (4-98), we find

$$\delta(\ln W) = -\sum_n \delta N_n(\alpha_c - \beta\varepsilon_n) - \sum_i \delta N_i(\alpha_g - \beta\varepsilon_i)$$

$$= -\alpha_c \sum_n \delta N_n - \alpha_g \sum_i \delta N_i + \beta\left(\sum_n \varepsilon_n\,\delta N_n + \sum_i \varepsilon_i\,\delta N_i\right). \tag{4-101}$$

We see that this is indeed zero if the δN_i and δN_n are restricted so as to obey the constraints [(4-98) and (4-99)]. But *notice* that the constraint [(4-99)] requires that in (4-100) the parameter β must be the *same* for the crystal and the gas. This common value of β, instead of two separate values for the gas and the crystal, arises because we have only a single energy conservation constraint, (4-99), rather than two separate energy constraints for the gas and crystal. *We see that the common value of β in the two expressions [(4-100a, and b)] is a direct consequence of the ability of the two systems to exchange energy so that only their total energy, but not their separate energies, must be considered constant.* The appearance of this common parameter β leads to the equilibrium values

$$\overline{N}_n = N_c e^{\alpha_c} e^{-\beta\varepsilon_n},$$
$$\overline{N}_i = \omega_i e^{\alpha_g} e^{-\beta\varepsilon_i}, \tag{4-102}$$

and the common value β in the two Boltzmann factors is the expression of the common temperature that characterizes two systems in thermal equilibrium.

It follows that the identification of β with $1/kT$ is valid throughout, in particular also in the low-temperature regime of the crystal. In this regime, where kT is small compared to the level distance ε_0, the mean thermal energy per atom is indeed much smaller than kT. This assertion, that had to be made without proof, is therefore substantiated.

(ii)   The Vapor Pressure of a Crystalline Solid

In this subsection we investigate the equilibrium between two *phases* of the same material, the solid and gas phase. In all solids the constituent particles are bound with a finite binding energy E_B, which represents the work that has to be done to detach a particle from the solid. This energy may range from 0.1 eV or less for a "volatile" solid (such as solid carbon dioxide) to a few electron volts for tightly bound solids.

We have seen, in the example of the Einstein crystal (which will do further service as a model), that the fraction of atoms excited to an energy ε_n decreases exponentially by the Boltzmann factor

$$e^{-\varepsilon_n/kT}.$$

At any temperature there will be a few atoms which temporarily reach an energy that exceeds the binding energy E_B. If such an atom is located at the surface, it can "fly off." This is the origin of the phenomenon of *evaporation*.

Consider now a small piece of solid in an evacuated bottle of volume V. The evaporated atoms will constitute a gas, filling the volume V. The atoms in this gas will frequently collide with the solid surface and on such occasions may be reabsorbed; that is, they will be built into the crystal again, transferring their excess energy to their neighbors.

The interplay of the two processes of evaporation and reabsorption leads to the establishment of an equilibrium, at which the density of particles in the gas phase assumes a characteristic value that depends on temperature. We will treat this gas phase as an ideal gas, in which case we can express the equilibrium density

$$\rho_{eq} = \frac{\overline{N}_g}{V}$$

in terms of the pressure this gas generates

$$p_{eq} = \rho_{eq} kT = \frac{\overline{N}_g}{V} kT. \tag{4-103}$$

We now describe this equilibrium. We have two systems, a solid and a gas, with the *same* constituent atoms enclosed in a volume V and having total energy E. The two systems can now exchange *both energy* and *particles* between themselves. This latter possibility is the novel aspect in which this case differs from the previously discussed example of a solid and a gas; there we assumed two different materials, and excluded the possibility of particle exchange.

In all other respects the procedure that we must follow is closely analogous to the one followed in establishing the solid–gas equilibrium: We must find the macrostate of largest weight for the joint system, taking cognizance of the constraints that are imposed. That is, we can start directly with (4-97), where we expressed the logarithm of the weight W of a macrostate in terms of its presumed maximum and the deviation δW from that maximum value. This deviation δW must vanish to first order

$$\delta W = -\sum_n \delta N_n \ln \left(\frac{\overline{N}_n}{N_c} \right) - \sum_i \delta N_i \ln \left(\frac{\overline{N}_i}{\omega_i} \right) = 0, \tag{4-104}$$

for all values of δN_n and δN_i subject to the constraints imposed on these variations by the constancy of the total number of particles

$$N = N_c + N_g = \sum_n N_n + \sum_i N_i = \text{constant} \tag{4-105}$$

and the constancy of the total energy

$$\begin{aligned} E &= E_c + E_g \\ &= \sum_i (\varepsilon_n - E_B) N_n + \sum_i \varepsilon_i N_i. \end{aligned} \tag{4-106}$$

In this equation we have put the zero of energy at zero kinetic energy of a gas particle. The lowest energy of a particle in the crystal is then $-E_B$ (where E_B is the binding energy), and in state n an atom has energy $\varepsilon_n - E_B$. This is illustrated in Figure 4.28.

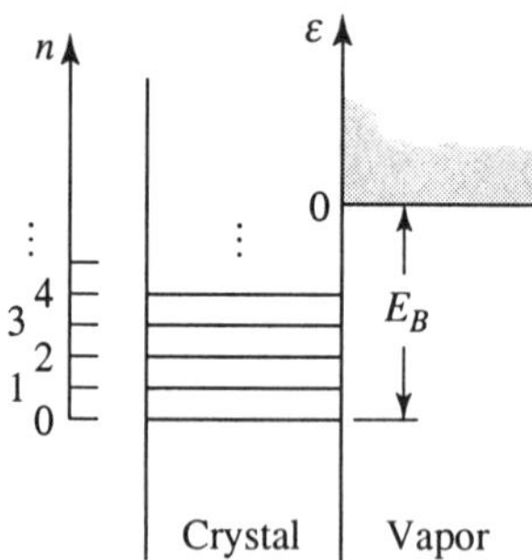

Figure 4.28. Energies of atoms in the crystal and vapor phase. The state $n = 0$ of the crystal has an energy $-E_B$ relative to a zero energy particle in the gas phase.

The two conditions of constraint now limit the variations δN_n and δN_i in a manner to assure that $\delta N = 0$ and $\delta E = 0$:

$$\sum_n \delta N_n + \sum_i \delta N_i = 0, \tag{4-107}$$

$$\sum_n (\varepsilon_n - E_B)\, \delta N_n + \sum_i \varepsilon_i\, \delta N_i = 0. \tag{4-108}$$

The solution for (4-104), that is consistent with these *two* equations of constraint, is

$$\ln\left(\frac{\overline{N}_n}{N_c}\right) = \alpha - \beta(\varepsilon_n - E_B), \tag{4-109}$$

$$\ln\left(\frac{\overline{N}_i}{\omega_i}\right) = \alpha - \beta\varepsilon_i. \tag{4-110}$$

Notice that we now have both the same α and the same β. This is needed since we have only two constraints. Let us verify that (4-109) indeed is a solution. Inserting it into (4-104), we find

$$\delta W = -\alpha \left(\sum \delta N_n + \sum \delta N_i\right)$$
$$+ \beta \left(\sum \delta N_n(\varepsilon_n - E_B) + \sum \delta N_i \varepsilon_i\right)$$

and this is indeed zero if (4-107), (4-108) are obeyed by the variations δN_i and δN_n.

Equations (4-109), (4-110) are our result. They give the desired value of $\overline{N}_n$ and $\overline{N}_i$ for the most probable macrostate of the joint system. The common value of β indicates the common temperature of the solid phase and vapor phase. The fact that we now also have a *common* value of α (as distinct from the case discussed in Section 4.2.C(i)) indicates that the numbers of particles in the gas phase and solid phase are not independently fixed. Let us see what these numbers are. From (4-109) we find, for the crystal

$$\overline{N}_n = N_c e^{\alpha + \beta E_B} e^{-\beta \varepsilon_n}$$

and summing over all energy states n, we find

$$\sum_n \overline{N}_n = N_c = N_c e^{\alpha + \beta E_B} \sum_n e^{-\beta \varepsilon_n}$$

or

$$e^{\alpha} = \frac{e^{-\beta E_B}}{Z_c}, \tag{4-111}$$

where Z_c is the previously introduced sum [see (4-35)]:

$$Z_c = \sum_n e^{-\beta \varepsilon_n}. \tag{4-112}$$

We see that α is independent of the number N_c of atoms in the crystal, but depends on the binding energy E_B and on the sum Z_c.

Looking now at the number of atoms in the vapor phase, we find from (4-110):

$$\begin{aligned}
N_{\text{gas}} &= \sum_i \overline{N}_i = \sum_i \omega_i e^{\alpha - \beta \varepsilon_i} \\
&= \frac{e^{-\beta E_B}}{Z_c} \left(\sum_i \omega_i e^{-\beta \varepsilon_i} \right) \\
&= e^{-\beta E_B} \frac{Z_{\text{gas}}}{Z_c}.
\end{aligned} \tag{4-113}$$

Here we have introduced the notation Z_{gas} for the sum

$$\sum_i \omega_i e^{-\beta \varepsilon_i} \equiv Z_{\text{gas}}. \tag{4-114}$$

We see at once from (4-113) that the number of particles in the vapor phase depends exponentially on the binding energy E_B of the atoms in the crystal. The term

$$e^{-\beta E_B} = e^{-E_B / kT}$$

is, in fact, the main factor responsible for the dramatic variation of the vapor pressure of a material with temperature.

Let us now put (4-113) into a more definite form by evaluating the two sums Z_{gas} and Z_c. Z_c can be found if we assume that the energy levels of the oscillations of atoms in the crystal are equidistant

$$\varepsilon_n = n\varepsilon_0.$$

In this case, we showed in (4-40) that

$$Z_c = \frac{1}{1 - e^{-\beta\varepsilon_0}}. \tag{4-115}$$

To evaluate the sum [(4-114)] for Z_{gas}, we convert this sum over macrocells in phase space into an integration over phase-space volume. To this end, recall that

$$\omega_i = \frac{\Omega_i}{h^3}, \qquad \Omega_i \text{ being the volume of the macrocell ``}i\text{''},$$

and

$$\varepsilon_i = (p^2/2m)_i = \text{ kinetic energy of an atom located in cell ``}i\text{.''}$$

Thus

$$Z_{\text{gas}} = \frac{1}{h^3} \sum \Omega_i e^{-\beta(p^2/2m)_i}$$

$$= \frac{1}{h^3} \int d\Omega\, e^{-\beta(p^2/2m)},$$

$$Z_{\text{gas}} = \frac{1}{h^3} \int_V dV \int dV_p\, e^{-\beta(p^2/2m)}.$$

Here we used the fact that the phase-space volume element $d\Omega$ is the product $dV\, dV_p$ of position space element dV and momentum space element dV_p. The integral $\int_v dV_p$ just gives the volume of the container holding the vapor. The integral over momentum space was calculated previously with the result [see (4-75)]:

$$\int dV_p\, e^{-\beta(p^2/2m)} = \left(\frac{2\pi m}{\beta}\right)^{3/2}.$$

The sum Z_{gas} therefore has the value

$$Z_{\text{gas}} = \frac{V}{h^3} \left(\frac{2\pi m}{\beta}\right)^{3/2}.$$

Since Z_{gas} is a pure number, and is proportional to the vapor volume V, it must be the ratio of V to another volume. We call this second volume the "Fermi volume" $V_F(T)$. It depends on temperature

$$Z_{\text{gas}} = \frac{V}{V_F(T)}, \qquad V_F(T) = \left(\frac{h^2}{2\pi m k T}\right)^{3/2}. \tag{4-116}$$

We will see shortly that except for T very close to zero, V_F is a very small volume of the order $(10^{-8} \text{ cm})^3$. For a dilute gas the volume per particle, (V/N), is large compared to V_F.

Returning now to (4-113), into which we insert the expressions, (4-115) for Z_c and (4-116) for Z_{gas}, we find

$$\left(\frac{N_{\text{gas}}}{V}\right) = e^{-E_B/kT}(1 - e^{\varepsilon_0/kT})\frac{1}{V_F(T)}. \tag{4-117}$$

We see here the interesting fact that the thermal equilibrium between the two phases determines the density (N_{gas}/V) of the particles in the vapor phase, as a function of temperature. We may consider this vapor as an ideal gas, as long as it is a dilute gas, and write the density (N_{gas}/V) in terms of the equilibrium vapor pressure p_{eq} and the temperature

$$p_{\text{eq}} = \left(\frac{N_{\text{gas}}}{V}\right) kT. \tag{4-118}$$

With this we can write our final result for the vapor pressure of an Einstein solid

$$p_{\text{eq}} = e^{-E_B/kT}(1 - e^{\varepsilon_0/kT})\frac{kT}{V_F(T)}, \tag{4-119}$$

with the "Fermi volume" $V_F(T)$ given by (4-116). This formula for p, the equilibrium vapor pressure of a solid, is interesting in many respects. Let us identify its principal features.

The *main* temperature dependence of the vapor pressure is given by the *Boltzmann factor* $e^{-E_B/kT}$ in which E_B is the binding of the particles in the crystal. This feature is dramatically demonstrated in Figure 4.29. In this figure we plot the logarithms of the vapor pressure of four solids:

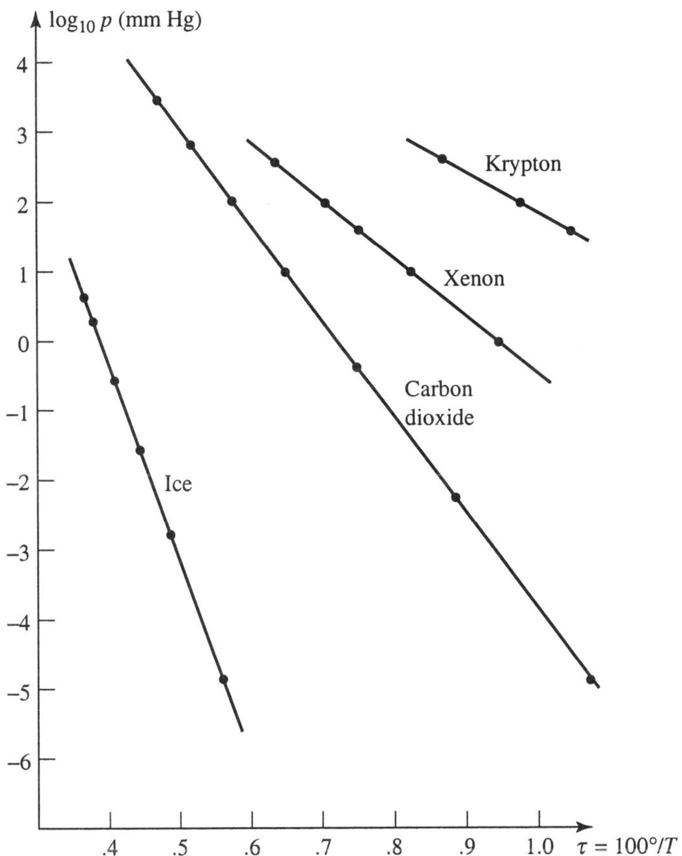

Figure 4.29. Vapor pressure of various solids. This graph plots the logarithm of the vapor pressure (as measured in mmHg) as a function of the inverse temperature as expressed by the parameter $\tau = 100°\text{K}/T$. Data are shown for ice, solid carbon dioxide, solid krypton, and solid xenon. (Data from *Handbook of Chemistry and Physics* [4]).

ice (solid H_2O);

solid carbon dioxide;

solid xenon and solid krypton;

as functions of the inverse temperature. For convenience we have introduced the dimensionless parameter $\tau = T_0/T$ with $T_0 = 100°\text{K}$. We then write

$$\frac{E_B}{kT} = \left(\frac{E_B}{kT_0}\right) \cdot \left(\frac{T_0}{T}\right) = \left(\frac{E_B}{kT_0}\right) \tau = \gamma \tau$$

so that the Boltzmann factor can be written as

$$e^{-E_B/kT} = e^{-\gamma\tau} : \gamma = E_B/kT_0.$$

In a plot of $\log p$ this factor appears as a term depending linearly on τ:

$$\log_{10} p = -0.434\gamma\tau + \log\left(\left(1 - e^{-\varepsilon_0/kT}\right)\frac{kT}{V_F(T)}\right),$$

when the factor 0.434 is $\log_{10} e$. We will show that the second term in this expression for $\log p$ depends very weakly only on temperature so that we have very nearly the relation

$$\log_{10} p = \text{constant} - 0.434\gamma\tau, \tag{4-120}$$

which is a straight line in a plot of $\log p$ versus τ.

The plots of $\log p$ versus τ are indeed straight lines within the accuracy of the data. Their slope can be measured and used to determine the value of γ, and therefore the value of the binding energy E_B of the particles in the crystal.

In Table 4.3 below we tabulate the values of γ extracted from Figure 4.28, and the values of the binding energy E_B deduced from them.

Table 4.3. Binding Energies E_B from Vapor Pressure Plots.

Material	Slope γ (from Figure 4.29)	E_B (eV)	$N_0 E_B$ (kcal/mol)
Ice	65.8	0.56	13.1
Carbon dioxide	31.9	0.27	6.38
Xenon	18.9	0.16	3.78
Krypton	12.8	0.11	2.56

These binding energies are quite interesting in themselves. We notice that they are much smaller than typical homopolar bond energies that are in the range of 50–100 kcal/mol. In the solid noble gases, xenon and krypton, the atoms bind together only by the weak so-called van der Waals forces. These forces also bind carbon dioxide, but are stronger there because the bound "particle" is a triatomic molecule. In ice, finally, we see the effect of hydrogen bonds. They may contribute ~ 5 kcal/mol per bond, and the value of E_B for ice is a composite of the effect of weak van der Waals forces and hydrogen bonds.

It should not pass unnoticed that the lines in Figure 4.28 represent an extraordinarily strong dependence of vapor pressure on temperature. For example, in the range

$$0.5 \leq \tau \leq 1.0,$$

that corresponds to a temperature range from $T = 100\,°K$ to $T = 200\,°K$ (a factor of 2 change only), the pressure of carbon dioxide changes by a factor 10^7!

Finally, the Boltzmann factor explains another feature of the plot in Figure 4.28. We see that the substance with the *lower* vapor pressure also has a vapor pressure curve with the *steeper* slope. Ice has the steepest slope of all four materials presented, and the lowest vapor pressure at any temperature in the range reported. Krypton has the highest pressure and the flattest slope. This feature is to be expected if the pressure p is indeed approximately of the form

$$p(\tau) = \text{``constant''} \times e^{-\gamma\tau},$$

and the values of the "constant" are not too different from case to case. In this case a comparison of any two substances will show that the one with the larger γ (i.e., the steeper slope) will also have the smaller value of p since $e^{-\gamma\tau}$ will be smaller for the material with the larger γ.

This brings us to the last and most intriguing question. Do the lines, as plotted in Figure 4.29, converge to a common point? And, if yes, what is the significance of this point?

In order to answer this question we must have a harder look at (4-119), and in particular add a few improvements. After all, this equation derived from an Einstein model of a crystal with oscillations restricted to one dimension. For a quantitative comparison with data, the result we obtained on this basis is not quite good enough. There are two simple modifications that occur in this equation if we go to a more realistic model of the solid. We will simply state these modifications without any attempt at a derivation. The first one arises from letting the atoms of the Einstein solid vibrate in all three space directions. The effect of this is to replace the factor $Z_c^{-1} = (1 - e^{\varepsilon_0/kT})$ in (4-119) by $(1 - e^{\varepsilon_0/kT})^3$. We mentioned this modification in a footnote previously. The second arises from replacing the

Einstein model by the more realistic Debye model of the crystal in which the oscillations of the atoms are strongly coupled. This introduces an extra factor in the expression for the pressure p of the form

$$e^{-D(\varepsilon_0/kT)},$$

where $D(x)$ is the so-called Debye function that has the value

$$D(x) \simeq 1 - \tfrac{3}{8}x \qquad \text{for} \qquad 0 \le x \le 1.$$

Here $x = (\varepsilon_0/kT)$. These two corrections give a reasonably realistic expression for the vapor pressure p:

$$p(\tau) = e^{-E_B/kT}(1 - e^{-\varepsilon_0/kT})^3 e^{-D(\varepsilon_0/kT)}\frac{kT}{V_F(T)}. \tag{4-119a}$$

Recall also that $V_F(T)$ is the "Fermi" volume [(4-116)]:

$$V_F(T) = \frac{h^3}{(2\pi mkT)^{3/2}}.$$

With the help of this expression we will show that the pressure has indeed the approximate form

$$p \simeq \text{"constant"}\, e^{-E_B/kT}$$

in the limited temperature range considered in Figure 4.29. We will now determine the value of this constant, compare it with the data and point out its significance. To do this, we rewrite (4-119a) again in terms of the parameter $\tau = T_0/T$ (using $T_0 = 100\,°\mathrm{K}$); we also isolate the complete dependence of p on T and on the mass m of the atoms: This mass dependence we express by means of the ratio of m to the mass m_H of hydrogen using $m = (m/m_\mathrm{H})m_\mathrm{H} = Mm_\mathrm{H}$; M being the "atomic weight" of the vapor in grams. This gives, from (4-119a):

$$p(\tau) = e^{-\gamma\tau} M^{3/2} \left(\frac{(1 - e^{-\eta\tau})^3\, e^{-D(\eta\tau)}}{\tau^{5/2}}\right) p_0(T_0), \tag{4-121}$$

where $\gamma = E_B/kT_0$, $\eta = \varepsilon_0/kT_0$, and

$$p_0(T_0) = kT_0(2\pi m_\mathrm{H}kT_0)^{3/2}/h^3. \tag{4-122}$$

The last factor, $p_0(T_0)$ in (4-121), is a "reference" pressure. It has the same value for all materials, and we now determine its value for $T_0 = 100\,°K$. Using the data

$$k = 1.38 \times 10^{-16} \text{ erg/°K (Boltzmann's constant)},$$
$$m_H = 1.67 \times 10^{-24} \text{ g} \qquad \text{(mass of hydrogen atom)},$$
$$h = 6.625 \times 10^{-27} \text{ erg s (Planck's constant)},$$

we find

$$\left(\frac{2\pi m_H k T_0}{h^2}\right) = 3.3 \times 10^{15} \text{ cm}^{-2}$$

and

$$p_0(T_0) = 1.38 \times 10^{-14} \text{ erg} \times (3.3 \times 10^{15} \text{ cm}^{-2})^{3/2}$$
$$= 1.38 \times 10^{-14} \text{ erg} \times 1.9 \times 10^{23} \text{ cm}^{-3}$$
$$= 2.6 \times 10^9 \text{ erg/cm}^3.$$

Atmospheric pressure has the value 10^6 erg/cm^3. Therefore

$$p_0(T_0) = 2.6 \times 10^3 \text{ atm}$$
$$= 2 \times 10^6 \text{ mmHg}.$$

We see that this is an enormous pressure. The fact then that the actual vapor pressures of solids are so small lies entirely in the circumstance that the Boltzmann factor $e^{-E_B/kT} = e^{-\gamma\tau}$ is exceedingly small. If we measure pressure in mmHg, then $\log_{10} p_0$ has the value

$$\log_{10} p_0(T_0) = 6.3$$

and we can write, for $\log_{10} p(\tau)$:

$$\log_{10} p(\tau) = 6.3 + \tfrac{3}{2}\log M - 0.434\gamma\tau$$
$$- \{\tfrac{5}{2}\log\tau + 0.434 D(\eta\tau) - 3\log_{10}(1 - e^{-\eta\tau})\}. \qquad (4\text{-}123)$$

We will apply this relation for a comparison of the behavior of the vapor pressure of four very similar monoatomic gases in equilibrium with their solid phase, namely, the noble gases—neon, argon, krypton, and xenon, with atomic weights $M = 20$, 40, 84, and 131, respectively. The value of the constant $\eta = \varepsilon_0/kT_0$ can be established from specific heat measurements at low temperature, which give $\eta \simeq 0.65$ for all of them. A calculation of the quantity in the braces in (4-123) shows then that it has a rather constant value in the range $0.5 \leq \tau \leq 3$:

$$\{\tfrac{5}{2} \log \tau + 0.434 D(\eta\tau) - 3 \log(1 - e^{-\eta\tau})\} \simeq 1.3 \qquad (0.5 < \tau < 3),$$

and we will add it in as a constant in (4-123). In this range of τ-values we have therefore, in good approximation, the relation

$$\log_{10} p(\tau) - \tfrac{3}{2} \log_{10} M = (6.3 - 1.3) - \gamma\tau = 5.0 - \gamma\tau.$$

This means that in a plot of $\log_{10} p(\tau) - \tfrac{3}{2} \log_{10} M$ for any of the four noble gases, we predict four straight lines that extrapolate to a *common point* at $\tau = 0$ at a value 5.0 of the ordinate. In Figure 4.30 we present the data taken from the tables of Landolt–Börnstein [5]. We see that indeed the plots of $\log p - \tfrac{3}{2} \log M$ are straight lines that extrapolate to a common point at $\tau = 0$ with ordinate $\sim 4.9 \pm 0.1$. This is to be compared with the predicted value of 5.0. We have indeed substantial agreement.

The decisive importance of this result lies in the fact that it confirms experimentally the quantum mechanical argument that the volume of a microcell in phase space should be given by the cube of Planck's constant. Indeed, we see from (4-121) and (4-122) *that the vapor pressure of a solid is inversely proportional to the size h^3 of a microcell in phase space.* The fact that we obtain the correct numerical value for $\log p_0(T_0)$ demonstrates that the "h" in (4-122) is indeed numerically equal to Planck's constant $h = 6.62 \times 10^{-27}$ erg s. We may, in fact, say that with the data of Figure 4.30 we have extracted a value for h with an accuracy of about 10%.

In conclusion, in this section, let us restate the main results obtained:

(a) In considering the equilibrium between two systems capable of exchanging energy, we have shown that the equality of temperature, that is an experimental attribute of two systems in thermal contact at equilibrium, is expressed by a common value of the parameter β. We concluded that, in general, β is equal to $1/kT$.

(b) Considering the equilibrium of a solid with its own gas phase, we found that in equilibrium the gas phase has a density or pressure that does not depend on

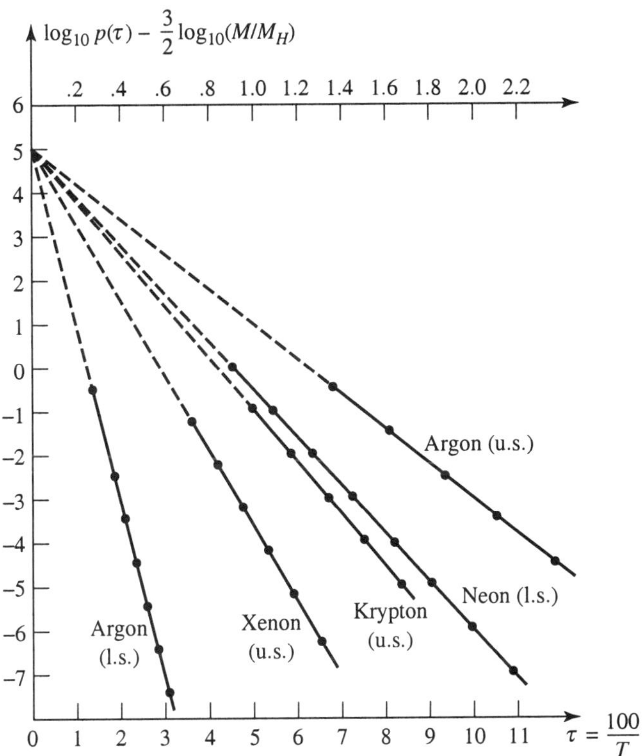

Figure 4.30. Plot of the vapor pressure $p(\tau)$ of solid noble gases as a function of temperature. Lines give $\log_{10} p(\tau) - \frac{3}{2}(M/M_H)$ as a function of $\tau = 100°/T$. Two τ-scales are used, upper scale (u.s.) and lower scale (l.s.).

the relative amounts of solid or gas. The gas pressure depends only on such quantities as the temperature, the binding energy, the mass, and the vibrational energy of the atoms in the crystal. The dominance of the Boltzmann factor in the pressure–temperature relation was experimentally confirmed. We also showed that the absolute value of the pressure is proportional to the inverse volume of a microcell in phase space, and that this volume is indeed given by h^3, h being Planck's constant.

4.3   Macroscopic Statement of Equilibrium Conditions. Entropy and the Second Law of Thermodynamics, Minimum Principle for Free Energy. Chemical Potentials

In this section we will demonstrate that the conditions for thermal equilibrium between components of a composite system can be stated in terms of a purely *macroscopic* (or "phenomenological") description of the component systems. A *macroscopic* description is one that makes no appeal to the underlying atomic–molecular structure of matter, nor to the principles of quantum physics. Instead, it uses only measurable "bulk parameters," such as total mass, mole numbers, volume, and density, to describe a material and phenomenological concepts such as pressure, temperature, electric fields, etc., to specify the conditions under which it is observed.

4.3.A.   Introduction. Simple and Composite Systems. Equations of State. Equilibrium in Composite Systems

In order to appreciate the subject matter of this section, it is well to distinguish clearly between two quite distinct aspects of the equilibrium problem: One aspect pertains to the internal equilibrium of simple systems, the other to equilibrium *between* the simple components of a composite system.

A *simple* system consists of a certain amount (specified by the mole number n) of material existing in a single phase, solid or liquid or gaseous, and being made of a single chemical species or a nonreactive mixture of chemical species of a fixed composition.

A simple system, when it is in internal thermal equilibrium, has a uniform temperature, and is under uniform pressure (in the absence of external forces applied to the system). It also has certain mechanical, thermal, electrical, etc., bulk properties, that are described by means of so-called "equations of state." Thus, the *mechanical* properties of a simple system may be expressed by an equation relating the volume V of the system to mole number, pressure, and temperature

$$V = nv(T, p). \tag{4-124}$$

In this equation $v(T, p)$ is the molar volume of the material, and is a function of temperature T and pressure p only. For fixed T and p, V is proportional to the number of moles n, that is, to the amount of material present.

Another equation of state describes the *thermal* properties of the material. Let $C_V(n, T, p)$ represent the heat capacity of the material at constant volume. Its dependence on n, T, p has the form

$$C_V = nc_V(T, p),$$

c_V being the molar heat capacity. For each material, c_V is a certain function of T and p. The quantities $v(T, p)$ and $c_V(T, p)$ can, in principle, be determined from the statistical description of equilibrium in a simple system. The information one needs for that is contained in the parameters $\overline{N}_i$ or $\overline{N}_n$ that characterize the most probable macrostate of the system. Indeed, it is one of the goals of the microscopic approach of statistical physics to establish the functional form of the equations of state.

On the other hand, the functions $v(T, p)$ and $c_V(T, p)$ can be determined by measurements, and may be used as purely *empirical* relations, describing the properties of a simple system in thermal equilibrium. It is rather obvious that the information contained in these relations is much less detailed than that which a microscopic description of thermal equilibrium provides. This latter description gives us the equilibrium density in phase space (for a gas), or the mean numbers $\overline{N}_n$ of lattice sites in quantum state n (for an Einstein crystal), a description that is much richer in information. This additional information in the statistical approach gives us a means of finding the probability for *deviations* of the above-mentioned distributions from their mean, enabling us to predict fluctuations about the mean values.

But let us return to the macroscopic description of a simple system. Since all its bulk properties are described by appropriate equations of state, that can be found empirically, there seems to be no equilibrium problem at all. Only if we want to use statistical physics to establish these equations of state, then we must solve the equilibrium problem of maximizing $\ln W$ in order to identify the most probable macrostate. The value of $\ln W_{\mathrm{max}}$ found in this process seems to have no further significance, and certainly no obvious macroscopic interpretation.

Let us now pass to composite systems. We have already seen two examples of a composite system: The first consisted of two different simple systems, a solid and a gas, in thermal contact. The second was a solid in contact with its own vapor. These two examples illustrate a common feature of composite systems: The components can exchange energy, or particles, or volume (e.g., two gases separated by a moveable piston). If the components of a system are assembled, such exchanges may indeed take place. They represent the adjustments needed to establish thermal equilibrium *between* the components of a system. Let us illustrate this by the example of the solid in equilibrium with its own vapor. We will choose carbon dioxide as an example. We consider an initial state, where the two components are separate (see Figure 4.31). A thermos, or vacuum bottle, of volume V, is filled with carbon dioxide gas; it also contains a piece of dry ice (solid carbon dioxide) wrapped in a tight, insulating cover.

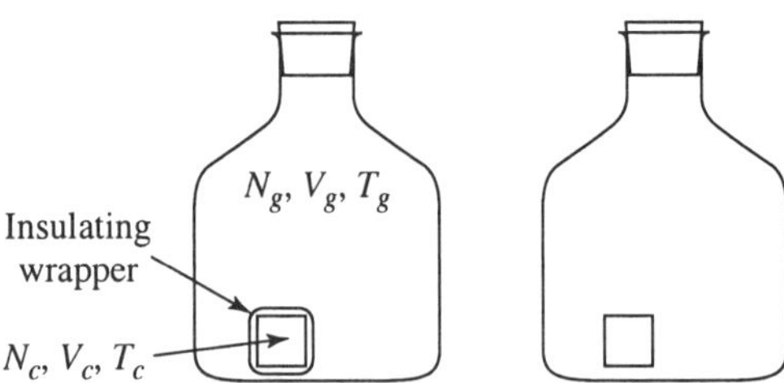

Figure 4.31. Gas–vapor equilibrium. A thermos filled with gaseous carbon dioxide and a piece of solid carbon dioxide wrapped. (a) Solid and gas separated: We have two separate, simple systems. Each is in equilibrium but at different temperatures. (b) The wrapper has been removed: A new equilibrium is formed after the exchange of energy and particles between the systems.

Let us now review what characterizes the equilibrium in the initial state, Figure 4.31(a), in the language of statistical physics. The two components are disconnected; each is a simple system with a fixed number N of particles, and fixed energy E (thermos, or vacuum bottle, and wrapper!). Each component has had time to establish internal thermal equilibrium. That is, each of the two systems can be assumed to be in (or very near) a macrostate of maximum weight W. The numerical value of this maximum weight is a function of the particle number, available energy and, possibly, the volume of the system

$$W_{\max} = f(E, N, V).$$

The same holds for $\ln W_{\max}$, of course. Recall that $W_{\max}$ represents the maximum value of the expression for W, consistent with a *fixed* value of E, N, V.

If we now remove the wrapper around the solid carbon dioxide, we join the two simple systems into one composite system: We open up the possibility of energy and particle exchange. Maybe some of the gaseous carbon dioxide will condense on the solid, or some additional carbon dioxide will evaporate from it, as the case may be. Also, heat (that is, energy) will be exchanged until the temperatures of the two components are equal. The final equilibrium of the total system is characterized by the maximum value for the total weight

$$W = W \text{ (gas)} \times W \text{ (solid)}.$$

of the system. This may also be expressed as the maximum of the logarithm of the total weight

$$\ln W \text{ (system)} = \ln W \text{ (gas)} + \ln W \text{ (solid)}.$$

Now $\ln W$ (gas) had already attained its maximum value consistent with the initial values E_g of its energy, and N_g of the number of gas particles. The same can be said for the solid. But this does not exclude the possibility that if the two components trade (exchange) energy and particles, the value of the combined weight W (gas) $\times$ W (solid) might still be increased. Such an exchange could be described by the changes

$$E_g \to E_g + \delta E,$$
$$E_s \to E_s - \delta E,$$
$$N_g \to N_g + \delta N,$$
$$N_s \to N_s - \delta N.$$

In such an exchange the value of $\ln W$ (system) goes over from

$$\ln W_{\max}(E_g, N_g) + \ln W_{\max}(E_s, N_s)$$

to a new value

$$\ln W_{\max}(E_g + \delta E, N_g + \delta N) + \ln W_{\max}(E_s - \delta E, N_s - \delta N).$$

If that new value is larger than the original value, the components were not in equilibrium with each other; the described shift will then actually occur, and continue until $\ln W$ (system) has reached its largest possible value consistent with the values $E = E_g + E_s$ and $N = N_g + N_s$ of total energy and particle number.

At this maximum point of $\ln W$ (system) a further small particle and energy transfer, as described above, will leave $\ln W$ (system) unchanged. Since this change can be expressed as

$$\delta \ln W \text{ (system)} = \frac{\partial \ln W_{\max} \text{ (gas)}}{\partial E_g} \delta E + \frac{\partial \ln W_{\max} \text{ (gas)}}{\partial N_g} \delta N$$

$$+ \frac{\partial \ln W_{\max} \text{ (solid)}}{\partial E_s}(-\delta E) + \frac{\partial \ln W_{\max} \text{ (solid)}}{\partial N_s}(-\delta N).$$

we see that at the equilibrium point the right-hand side of this expression must vanish. This will be the case if the relations

$$\frac{\partial \ln W_{\max}(E_g, N_g)}{\partial E_g} = \frac{\partial \ln W_{\max}(E_s, N_s)}{\partial E_s} \tag{4-124a}$$

and

$$\frac{\partial \ln W_{\max}(E_g, N_g)}{\partial N_g} = \frac{\partial \ln W_{\max}(E_s, N_s)}{\partial N_s} \tag{4-124b}$$

hold. These equations then represent the conditions of equilibrium *between* the components of the vapor–solid system.

There is a third equilibrium condition that is somewhat less obvious in the context of the example given here. In the process of exchanging energy and particles, the two components also trade volume. If the solid evaporates, its volume will *decrease* by δV, and that of the gas will *increase* by the same amount since the total volume is constant. Since $\ln W_{\max}$ of both vapor and solid do depend on volume (although this volume dependence was absent in our oversimplified crystal model), these volumes will also adjust in equilibrium so as to make $\ln W$ (system) as large as possible. We have therefore the third condition

$$\frac{\partial \ln W_{\max}(E_g, N_g, V_g)}{\partial V_g} = \frac{\partial \ln W_{\max}(E_s, N_s, V_s)}{\partial V_s}. \tag{4-124c}$$

What is the significance of these results? They state that in order to formulate the conditions for equilibrium between components of a composite system, we need to know the numerical values of $\ln W_{\max}$ for each component, as a function of the energy, total particle number (and in a more general case, volume), of the component. What we *don't* need is the general form of $\ln W$ for all possible macrostates. All that is important in the present context is how, for each component, the value $\ln W_{\max}$ for the most probable macrostate depends on the *macroscopic parameters* energy, E, particle number, N (or mole number $n = N/N_0$) and volume V of that component.

We thus see that in order to deal with composite equilibria, we need additional information about the simple systems that are the components of any composite system. We need the value of $\ln W_{\mathrm{max}}$ as a function of the energy, volume, and particle number of the system.

A statistical approach to the equilibrium in a simple system gives us the value of $\ln W_{\mathrm{max}}$ as one of the results of having solved the equilibrium problem. But, if we want to operate in a purely phenomenological (thermodynamic) framework, the intriguing question arises: How do we get hold of this $\ln W_{\mathrm{max}}(E, N, V)$ *without* passing through the microscopic statistical consideration, as we did?

The answer to this is as follows: Since $\ln W_{\mathrm{max}}$ depends only on the macroscopic parameters of the system, there should be some recognizable physical meaning attached to it. Such an identification would make it possible to get a hold of $\ln W_{\mathrm{max}}$ empirically by experimental observation. We will show that this is indeed the case.

The following Section 4.3.B of this part of Chapter 4 will be devoted to a demonstration of the physical interpretation of $\ln W_{\mathrm{max}}$. We will show that $\ln W_{\mathrm{max}}$ is a quantity proportional to what is historically called the *entropy* of the system. The entropy has an operational definition that permits an experimental determination of $\ln W_{\mathrm{max}}$ (up to a constant) as a function of E, V, and N.

This makes it possible to state the equilibrium conditions between the parts of a composite system purely in terms of experimentally determinable quantities. thus, the macroscopic, thermodynamic method is sufficient to establish the equilibrium condition in composite systems without any need for a microscopic, statistical description. In the later Section 4.3.C, we will introduce alternative, *equivalent* statements of the general equilibrium conditions, in terms of the so-called Helmholtz and Gibbs free energies. These alternative statements have some practical advantages, which we shall explain.

4.3.B.   Entropy of a Simple System and its Properties. The Second Law of Thermodynamics

In the introduction to this Section 4.3, we demonstrated that the equilibrium conditions between the components of a composite system could be formulated in terms of the derivatives of the functions $\ln W_{\mathrm{max}}$ for the components. $\ln W_{\mathrm{max}}$ for each component is a function of the energy, particle number, and volume of that component. At equilibrium, corresponding derivatives of $\ln W_{\mathrm{max}}$ with respect to E, N, and V have to be the same for all components.

In this section we establish the general form of $\ln W_{\mathrm{max}}$ as a function of E, N, and V, and identify the physical significance of this function. We will show that it is proportional to the quantity which in thermodynamics is known as the *entropy* S of a system. The numerical relation between $\ln W_{\max}$ and S is

$$S = k \ln W_{\max}, \qquad (4\text{-}125)$$

k being the Boltzmann constant. We will henceforth use both the symbol S and the name of entropy when talking about $\ln W_{\max}$, being aware that by giving it a name, we have not yet identified the physical meaning of this quantity.

In order to get a first insight into the functional form of S viewed as a function of E, N, and V, we take up the cases of the Einstein crystal and the ideal gas. After these examples we will make some more general statements about the properties of the entropy. In particular, we will relate the entropy change in a system to heat transfer. This latter relation represents the original context in which the concept of entropy arose independent of any microscopic interpretation.

(i)   The Entropy of the Einstein Crystal

The value of the entropy S in this case follows from (4-29) (we write N for $\overline{N}$):

$$\frac{1}{k} S = \ln W_{\max} \equiv N \ln N - \sum \overline{N}_n \ln \overline{N}_n$$

$$= -\sum_n \overline{N}_n \ln \left(\frac{\overline{N}_n}{N} \right). \qquad (4\text{-}126)$$

Here $\overline{N}_n$ represent the equilibrium numbers of atoms in quantum state n, and are given by (4-32):

$$\ln \left(\frac{\overline{N}_n}{N} \right) = \alpha - \beta(\varepsilon_n - E_B). \qquad (4\text{-}127)$$

(We have written here the energy of an atom in state n as $\varepsilon_n - E_B$, where E_B is the binding energy; this will allow us to compare the energy of the crystal and its vapor.) The total number of particles determines α:

$$N = \sum \overline{N}_n = N e^\alpha \sum_n e^{-\beta(\varepsilon_n - E_B)} \qquad (4\text{-}128)$$

or

$$e^{-\alpha} = \sum_n e^{-\beta(\varepsilon_n - E_B)} \equiv Z(\beta). \tag{4-129}$$

β in turn is related to the energy per particle

$$\frac{E}{N} = \sum_n (\varepsilon_n - E_B) \left(\frac{\overline{N}_n}{N} \right)$$

$$= e^{\alpha} \sum_n (\varepsilon_n - E_B) e^{-\beta(\varepsilon_n - E_B)}$$

$$= -\frac{1}{Z} \frac{\partial Z}{\partial \beta} = -\frac{\partial}{\partial \beta} \ln Z(\beta). \tag{4-130}$$

Inserting (4-127) into (4-126), we may now obtain an expression for S in terms of E and N, viz:

$$S = -k \sum_n \overline{N}_n (\alpha - \beta(\varepsilon_n - E_B))$$

$$= -k\alpha N + k\beta E. \tag{4-131}$$

With this expression we can now establish how the entropy S changes if we change the energy E by a small increment ΔE, and the particle number N by a small increment ΔN. We must keep in mind, however, that α is a function of β by (4-129), and β a function of E/N by (4-130). We then have for the small change ΔS of S:

$$\Delta S = k\{-\alpha \Delta N + \beta \Delta E - N \Delta \alpha + E \Delta \beta\}.$$

To calculate $\Delta \alpha$, we use (4-129) for α:

$$\alpha = -\ln Z(\beta),$$

hence

$$\Delta \alpha = \left(\frac{d\alpha}{d\beta} \right) \Delta \beta = -\frac{\partial}{\partial \beta} (\ln Z) \Delta \beta = \left(\frac{E}{N} \right) \Delta \beta$$

using (4-130) in the final step. We see therefore that $-N\Delta\alpha + E\Delta\beta = 0$ so that our expression for ΔS simply becomes

$$\Delta S = -k\alpha\,\Delta N + k\beta\,\Delta E. \tag{4-132}$$

This relation identifies the terms $-k\alpha$ and $k\beta$ as the partial derivatives of S with respect to N and E, respectively. Indeed, for any function $S(E, N)$ of E and N, the change of S due to a small change of E and N is expressed as

$$\Delta S(E, N) = \left(\frac{\partial S}{\partial N}\right)_E \Delta N + \left(\frac{\partial S}{\partial E}\right)_N \Delta E. \tag{4-133}$$

The subscript E on $(\partial S/\partial N)$ indicates that this derivative is calculated while keeping E constant; similarly $(\partial S/\partial E)_N$ is an E-derivative, calculated while keeping N constant. In the kind of procedures we are engaged in here, we have to deal with functions of many variables, some of them interrelated, and it is important always to spell out clearly what is meant by a derivative. As a safeguard, it is therefore customary to append a subscript (or subscripts) to any partial derivatives so as to clearly indicate what quantities are being kept constant. It should then be clear from (4-133) also what the meaning of the partial derivative actually is: $(\partial S/\partial N)_E$ means the limit, for small ΔN, of the ratio $(\Delta S/\Delta N)$, for a constant value of E:

$$\left(\frac{\partial S}{\partial N}\right)_E = \lim_{\Delta N \to 0}\left\{\frac{1}{\Delta N}(S(E, N + \Delta N) - S(E, N))\right\}.$$

The other derivative, $(\partial S/\partial E)_N$, is similarly defined. Equation (4-133) shows that, given an expansion as (4-132), one can simply read off the partial derivatives by comparing it with the general equation (4-133). In our case

$$\left(\frac{\partial S}{\partial N}\right)_E = -k\alpha, \qquad \left(\frac{\partial S}{\partial E}\right)_N = k\beta.$$

Since these partial derivatives play such a crucial role in the statement of the equilibrium condition, we should, whenever possible, identify their physical meaning. We know already, for instance, that $k\beta$ is the inverse temperature so that

$$\left(\frac{\partial S}{\partial E}\right)_N = \frac{1}{T}.$$

By contrast, the quantity $-(k\alpha)$ has a much less transparent meaning, but because of its importance the N-derivative of S is written in terms of a named quantity, the so-called chemical potential μ; this latter is generally defined as

$$\mu \equiv \frac{\alpha}{\beta} = -kT\alpha = -T\left(\frac{\partial S}{\partial N}\right)_{E,V}. \tag{4-134}$$

We can therefore write

$$\left(\frac{\partial S}{\partial E}\right)_N = \frac{1}{T}, \qquad \left(\frac{\partial S}{\partial N}\right)_E = -\frac{\mu}{T}. \tag{4-135}$$

Thus, we can reexpress (4-133) as

$$T\Delta S(E, N) = \Delta E - \mu \Delta N.$$

For the crystal the chemical potential (μ) is given by the expression

$$\mu = kT\alpha = kT \ln Z(\beta)$$
$$= -kT \ln\left(\sum_n e^{-\beta(\varepsilon_n - E_B)}\right). \tag{4-136}$$

For equidistant energy levels, $\varepsilon_n = n\varepsilon_0$, this reduces to

$$\mu = kT \ln(1 - e^{-\beta\varepsilon_0}) - E_B. \tag{4-136a}$$

In the Einstein crystal we have no dependence of S on volume. This is a deficiency of this "model system"; real crystalline solids do show a dependence of the entropy on volume.

We turn now to the example of the ideal gas, and here we will pay special attention to this volume dependence of its entropy.

(ii)   The Entropy of the Ideal Gas

We refer here to (4-59) and to (4-63) and (4-64) which show that the entropy is given by

$$S = k\left\{N - \sum_i \overline{N}_i \ln\left(\frac{\overline{N}_i}{\omega_i}\right)\right\}, \tag{4-137}$$

with $(\overline{N}_i/\omega_i)$, the mean density in phase space, given by $\ln(N_i/\omega_i) = \alpha - \beta\varepsilon_i$. It follows that the value of S is

$$S = k\left\{N - \sum_i N_i(\alpha - \beta\varepsilon_i)\right\}$$
$$= k\{N(1 - \alpha) + \beta E\}. \tag{4-138}$$

We have previously determined an expression for α. Using (4-64), we see that α is fixed in terms of the total number of particles N by the expression

$$N = \sum_i \overline{N}_i = \sum \omega_i e^{\alpha - \beta\varepsilon_i} = e^\alpha \sum \omega_i e^{-\beta\varepsilon_i}$$

or

$$\alpha = \ln(N/Z(\beta, V)). \tag{4-139}$$

Here the sum $Z(\beta, V)$ is defined as

$$Z = \sum \omega_i e^{-\beta\varepsilon_i}.$$

If we now use the expression for Z obtained, for example, in (4-116), we find

$$Z = \frac{V}{V_F(\beta)} = V\left/\left(\frac{\beta h^2}{2\pi m}\right)^{3/2}\right. . \tag{4-140}$$

The constant β is determined by the energy per particle, E/N:

$$\left(\frac{E}{N}\right) = \sum_i \varepsilon_i \omega_i e^{-\beta\varepsilon_i}\left/\sum \omega_i e^{-\beta\varepsilon_i}\right.$$
$$= -\frac{\partial Z}{\partial \beta}\left/Z = -\frac{\partial}{\partial \beta}(\ln Z)\right. \tag{4-141}$$

We recall also that using expression (4-140) for the value of z, we found that (E/N) has the value $(3/2)/\beta$, a result that we used to identify the physical significance of β as $1/kT$. At this point we have in hand the form of the dependence of all the terms in S on E and N.

Again, in the statement of the equilibrium condition, we will need the derivatives of the entropy S with respect to E, N and V. To find these, it is best to start with an expression for the change ΔS of S associated with a change in the values of E, N and V, taking into account that α and β, too, depend on these parameters. From (4-138) we find that

$$\Delta S = k\{(1 - \alpha)\Delta N + \beta\Delta E - N\Delta\alpha + E\Delta\beta\}.$$

The parameter α depends on N and Z, and Z in turn on β and V. From (4-139):

$$\alpha = \ln N - \ln Z(\beta, V).$$

Hence

$$\Delta\alpha = \frac{\Delta N}{N} - \left(\frac{\partial \ln Z}{\partial \beta}\right)_V \Delta\beta - \left(\frac{\partial \ln Z}{\partial V}\right)_\beta \Delta V.$$

With the help of (4-140) and (4-141), this can be reduced to

$$\Delta\alpha = \frac{\Delta N}{N} + \frac{E}{N}\Delta\beta - \frac{\Delta V}{V}.$$

Inserting this into the expression for the change ΔS of S, we find

$$\Delta S = k\left\{-\alpha\Delta N + \beta\Delta E + \frac{N}{V}\Delta V\right\}.$$

From this expression we can now read off the partial derivatives of S with respect to E, N, V:

$$\left(\frac{\partial S}{\partial E}\right)_{N,V} = k\beta = \frac{1}{T}, \tag{4-142a}$$

$$\left(\frac{\partial S}{\partial N}\right)_{E,V} = -k\alpha \equiv -\frac{\mu}{T}, \tag{4-142b}$$

$$\left(\frac{\partial S}{\partial V}\right)_{E,N} = \frac{Nk}{V} = \frac{p}{T}. \tag{4-142c}$$

In (4-142a) we have made use of the identification

$$\beta = \frac{1}{kT}.$$

In (4-142b) we have introduced the chemical potential μ, defined as

$$\mu \equiv \left(\frac{\alpha}{\beta}\right) = -T\left(\frac{\partial S}{\partial N}\right)_{E,N}.$$

For the ideal gas we have an explicit form for μ from (4-139) and (4-140):

$$\mu = kT \ln\left(\frac{N V_F(\beta)}{V}\right) = kT \ln\left(\frac{p V_F(\beta)}{kT}\right). \tag{4-143}$$

Finally, in (4-142c) we have used the ideal gas equation of state $pV = NkT$ to express the volume derivative of S in terms of the pressure

$$\left(\frac{\partial S}{\partial V}\right)_{E/N} = \frac{p}{T}.$$

We must mention here an important point. We know from experience that if two materials in contact are in equilibrium, they are under the same pressure. For instance, an elastic solid, surrounded by a gas at pressure p, is itself under pressure p.

The fact then that we could identify $(\partial S/\partial V)_{E,N}$ as p/T in the ideal gas case, combined with the condition [(4-124c)] that the values of $(\partial S/\partial V)_{E,N}$ of any two materials in equilibrium must be the same, identifies $(\partial S/\partial V)_{E,N}$ as p/T for any material, not just the ideal gas.

We can now use the results [(4-142a, b, c)] for the derivatives of S to make the decisive step of establishing the physical significance of the entropy. These equations tell us that upon a change of E by a small increment ΔE, a change of N by a small increment ΔN, and a change of V by a small increment ΔV, the change of the entropy S:

$$\Delta S = \left(\frac{\partial S}{\partial E}\right)_{N,V}\Delta E + \left(\frac{\partial S}{\partial N}\right)_{E,V}\Delta N + \left(\frac{\partial S}{\partial V}\right)_{E,N}\Delta V$$

can be written for the *ideal gas* in the form

$$\Delta S = \frac{1}{T}\Delta E + \frac{p}{T}\Delta V - \frac{\mu}{T}\Delta N. \tag{4-144}$$

We see that this equation tells us how the changes of S, E, N, V are interrelated as one goes from an equilibrium state (E, N, V) to a neighboring equilibrium state $(E + \Delta E, N + \Delta N, V + \Delta V)$.

We can also turn this equation (4-144) around in many ways; we can, for instance, make it into a statement of *how the energy E changes* if we change S by ΔS, N by ΔN, and V by ΔV. Of course, we wouldn't know yet what it means "to change S by ΔS," operationally, but (4-144) gives us a means to find out. So let us multiply (4-144) by T and solve for the energy change ΔE; we find

$$\Delta E = T\Delta S - p\Delta V + \mu\Delta N. \tag{4-145}$$

This is a most remarkable equation. It describes the change in energy of a simple, i.e., one-component, system in thermal equilibrium if we change its entropy, or its volume, or the number of particles it contains by small amounts. Here at last we are again on more familiar ground. We know something about how the energy of the system can be changed by the physical processes of letting the system do mechanical work, or letting it exchange heat with its environment.

In the next Subsection 4.3.B(iii), we will relate these physical processes to the items in the mathematical statement [(4-145)].

(iii)   Physical Interpretation of the Entropy. Ideal Gas Case

Let us first consider an ideal gas with a fixed number of particles N so that $\Delta N = 0$. We will assume that this gas is enclosed in a volume V at temperature T. Its pressure is then determined by the ideal gas equation of state $p = NkT/V$.

There are two ways in which we can change the energy of the gas:

(a) We can let the gas do mechanical work by changing its volume. If the volume increases by ΔV, then we know that the work done by the gas is

$$\Delta W = p\Delta V.$$

This then represents energy *lost* by the gas

$$(\Delta E)_{\text{gas}} = -p\Delta V.$$

We recognize this item of course in (4-145).

(b) We can let the gas pick up an amount of heat ΔQ from its surroundings (or give up heat to its surroundings). Counting ΔQ as positive, if it is fed into the gas, we have then

$$\Delta E = \Delta Q.$$

In the general case, both energy transfers may occur simultaneously; an example is isothermal expansion of the ideal gas. Since the energy of the ideal gas depends only on temperature, there is *no* energy change in isothermal expansion. Yet work is being done. Hence, in the expansion process heat must be drawn in by the gas from the environment; this heat inflow must exactly match the mechanical work done.

The fact that the two items, mechanical work and heat transfer, represent the ways in which the energy of a system may change, is expressed in the *first law* of thermodynamics (see Volume I, Chapter 5)

$$\Delta E = \Delta Q - p\Delta V. \tag{4-146}$$

Upon confronting this relation with (4-145), we recognize that (since $\Delta N = 0$ in the present discussion) the term ΔQ in (4-146) must be identified with $T\Delta S$ of (4-145):

$$\Delta Q = T\Delta S. \tag{4-147}$$

Here at last we have the looked-for relation, that relates the change in entropy ΔS to the heat transfer ΔQ to the system. ΔQ is a quantity that can be measured, and we thus have here a connection between the entropy change and an experimentally measurable quantity. This relation, as it stands, does give us the means to determine the *entropy difference* between any two equilibrium states of the system: Starting from an equilibrium state (1), with pressure p_1, temperature T_1, and volume $V_1 = nRT_1/p_1$, we change p and T in small increments Δp, ΔT, and determine the heat ΔQ picked up during this change. The sum of all changes Δp, ΔT leads us eventually to a new equilibrium state (2), with values p_2, T_2, and $V_2 = nRT_2/p_2$ of pressure, temperature, and volume. The entropy difference between the states (2) and (1) is then the sum of all increments ΔQ, divided by the temperature T at which ΔQ was added

$$S_2 - S_1 = \sum \frac{\Delta Q}{T} = \int_1^2 \frac{\Delta Q}{T}. \tag{4-148}$$

This formula is the basic *operational* definition of the entropy, that allows the value of the entropy of a system, relative to a reference state, to be determined for every equilibrium state of the system. Because of the importance of this relation, and to ensure that it is clearly understood, we will first illustrate its proper interpretation. In the next Subsection 4.3.B(iv), we will then show for the ideal gas how (4-148) may be used to calculate the entropy difference between any two states.

We start our discussion of (4-148) with the construction of a state diagram, as illustrated in Figure 4.32. In this diagram a "p–T plane" is introduced. Each pair of values (p, T) is represented by a point in this plane. This point represents an equilibrium state of the gas having the values of p and T associated with the point, and a value of V given by the equation of state, $V = nRT/p$. This equation shows that the set of all equilibrium states having a common value of V is a straight line in the p–T plane, passing through the origin.

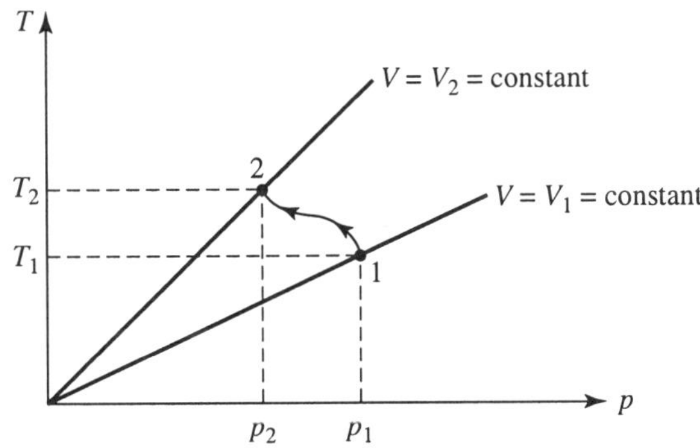

Figure 4.32. The state diagram for an ideal gas. Each point in the p–T plane represents a possible equilibrium state of the gas. Constant volume is represented by straight lines passing through the origin. A path drawn from (1) to (2) represents a sequence of reversible changes, by means of which the system can be pushed from state (1) into state (2).

Our next objective is now to show how to use the fundamental relation [(4-148)] to determine the entropy difference $S_2 - S_1$ between the two equilibrium states (2) and (1), represented in Figure 4.32.

The entropy S is a function of the values of E, N, V representing the gas at equilibrium. Since E and V (for a fixed mole number $n = N/N_0$) are determined by T and p, we may, of course, also view the entropy for fixed n as a function of T and p. This means that to each point in the T–p plane of Figure 4.32 belongs a well-defined value of S. Since each point in this plane represents a possible equilibrium state of the system, we call S a *state function*. The total energy of the system is also a state function for the same reason: To every point in the p–T plane, we can attach a well-defined value of the energy E, in this case (ideal gas) $E = \frac{3}{2}nRT$. The fundamental equation [(4-148)] $S_2 - S_1 = \int_1^2 dQ/T$ was based on the argument that the difference between the entropies at states (2) and (1) in Figure 4.32 could be represented as the sum (or integral) of a sequence of small changes $S = \Delta Q/T$, that move the system from state (1) through a set of intermediate states to state (2):

$$S_2 - S_1 = \sum_1^2 \Delta S = \int_1^2 dS. \tag{4-149}$$

It does *not matter* through what particular sequence of intermediate states we bring the system from (1) to (2). What does matter, however, is that every intermediate state must also be an *equilibrium* state of the system; otherwise, the entropy difference would not be defined. This means, in more physical terms, that if we have to heat up or compress the gas to go from (1) to (2), we must do it *slowly*. How slowly? Assume we compress the gas by moving a piston inward. This motion of the piston imparts additional energy to the gas molecules colliding with it. This extra energy must be given time to redistribute itself over the system before the piston moves by a significant amount. Practically, this means that the piston speed must be very small compared with the root mean square speed $\sqrt{3kT/m}$ of the gas molecules. This is achieved, in the case of compression or expansion, by keeping the force exerted by the piston per unit area, F/A, very close to the ambient gas pressure p. If F/A is slightly larger than p, it will compress the gas; if F/A, on the other hand, is very slightly less than p, the gas will be allowed to expand. A process carried out in this manner is called *reversible*. In a reversible process, one never departs *significantly* from an equilibrium situation. A similar argument can be made for heating and cooling. To be reversible, it must be done by thermal contact with a "heat bath" whose temperature is only slightly above or below the ambient temperature of the gas.

In a reversible change the system goes through a sequence of equilibrium states, and this sequence can be represented by a definite *path* in a state diagram such as in Figure 4.32. The proper interpretation of (4-148) for the entropy difference between two equilibrium states (2) and (1) of the gas is then:

(a) Choose a *path* in the p–T plane connecting state (2) to state (1). It does not matter what the path is, as long as it connects the two points.

(b) Break up the path into small segments, representing a small change Δp of p and ΔT of T. Determine the heat ΔQ absorbed while the system changes from p to $p + \Delta p$, and from T to $T + \Delta T$.

(c) Add up the increments $\Delta Q / T$ for each segment of the path.

To carry out this program, one naturally exploits the fact that any path leading from (1) to (2) must give the same result for the entropy difference. It is then convenient to choose the actual path in such a way that it consists of a piece in which only the temperature changes (where the gas is heated at constant pressure), and of a piece in which only the pressure changes (i.e., the gas is expanded at constant temperature). This procedure will be illustrated in the next subsection.

(iv)   Illustrative Calculation of the Entropy Difference for Two States of the Ideal Monoatomic Gas

The purpose of this illustration is to show how such a "path integral" as given in (4-149) is actually calculated. We will choose two different, simple paths and show the identity of the results obtained for the entropy difference. These paths are illustrated in Figure 4.33.

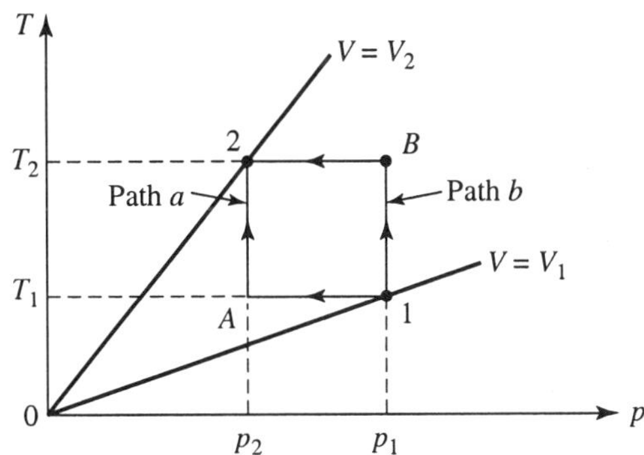

Figure 4.33. State diagram for an ideal gas, showing two different paths a and b, used to calculate the entropy difference between states (2) and (1). Notice that $V_2 > V_1$.

The input information used to calculate the entropy change is the following:

(a) The known equation of state

$$V = nR\frac{T}{p},\qquad(4\text{-}150)$$

n being the mole number.

(b) The known dependence of energy on temperature alone

$$E = \tfrac{3}{2}nRT.\qquad(4\text{-}151)$$

From these inputs we can calculate the heat input ΔQ associated with a small change of pressure and temperature, Δp and ΔT:

$$\Delta Q = \Delta E + p\Delta V.$$

Using (4-150), (4-151), and assuming a constant value of n, ΔQ can be related to ΔT and Δp as follows:

$$\Delta Q = nR\left(\tfrac{3}{2}\Delta T + p\left(\frac{\Delta T}{p} - \frac{T}{p^2}\Delta p\right)\right)$$

or

$$\Delta Q = nR\left(\tfrac{5}{2}\Delta T - \frac{T}{p}\Delta p\right).\qquad(4\text{-}152)$$

Consider now path (a). Along the part $(1) \to (A)$, T is constant and equal to T_1; along the part $(A) \to (2)$, p is constant and equal to p_2. So, we have

$$S_2 - S_1 = \int_1^A \frac{dQ}{T} + \int_A^2 \frac{dQ}{T}$$

$$= nR\left\{\int_{p_1}^{p_2} \frac{-T_1\,dp/p}{T_1} + \int_{T_1}^{T_2} \frac{\tfrac{5}{2}\,dT}{T}\right\}$$

$$= nR\left\{-\ln\left(\frac{p_2}{p_1}\right) + \tfrac{5}{2}\ln\left(\frac{T_2}{T_1}\right)\right\}.\qquad(4\text{-}153a)$$

If, alternatively, we choose path (b), then from $(1) \rightarrow (B)$ we have $p = \text{constant} = p_2$, and from $(B) \rightarrow (2)$ we have $T = \text{constant} = T_2$. This gives

$$
\begin{aligned}
S_2 - S_1 &= nR \left\{ \int_{T_1}^{T_2} \frac{\frac{5}{2} dT}{T} + \int_{p_1}^{p_2} \frac{-T_2 \, dp/p}{T_2} \right\} \\
&= nR \left\{ \tfrac{5}{2} \ln \left(\frac{T_2}{T_1} \right) - \ln \left(\frac{p_2}{p_1} \right) \right\},
\end{aligned}
\tag{4-153b}
$$

the same result.

We leave it as a problem assignment to show that the total amount of heat $\int_1^2 dQ$ picked up by the system is *not* the same for paths (a) and (b). This result is important in that it demonstrates that there is no state function $Q(n, T, p)$. The amount of heat one has put into a system does not depend on the *state* of the system, but on the *history* of the system.

In a later section (4.3.C(ii)), we will show how in a more realistic case one would actually calculate the entropy difference. For a real gas, for instance, we do not have the information on the energy E of the system as a function of T and p. We will show, however, that the knowledge of the two equations of state for the molar volume v as a function of p and T, and the molar heat capacity c_p as a function of p and T:

$$
v(p, T) \qquad \text{and} \qquad c_p(p, T)
$$

provides all the input needed to calculate the entropy difference between any two equilibrium states of the real gas, or indeed for any simple system.

(v)   A Note on the Chemical Potential μ

The chemical potential is defined in (4-142b) as

$$
\mu = -T \left(\frac{\partial S}{\partial N} \right)_{E,V}.
$$

In this form we can see that the chemical potential is important in characterizing the equilibrium state of a composite system. For example, in the case of the composite gas–solid system, we saw in (4-124b) that the condition for equilibrium, insofar as exchange of particles between the gas and solid, is

$$\left(\frac{\partial \ln W_{\max}(E_g, N_g)}{\partial N_g}\right) \equiv \left(\frac{\partial \ln W_{\max}(E_s, N_s)}{\partial N_s}\right).$$

Since $S = k \ln W_{\max}$, we see that when the temperature of the gas and solid are the same, that the remaining condition for equilibrium is that

$$\left(\frac{\partial S_g}{\partial N}\right)_{E,V} = \left(\frac{\partial S_s}{\partial N}\right)_{E,V}.$$

or

$$\mu_{\text{gas}} = \mu_{\text{solid}}.$$

Because of its importance in the statement of the equilibrium conditions, it is worthwhile to examine the meaning of the chemical potential for a gas carefully.

With this objective in mind, let us first study (4-145), viz:

$$\Delta E = T\Delta S - p\Delta V + \mu\Delta N.$$

From this equation we see that the chemical potential can also be expressed quite rigorously as

$$\mu = \left(\frac{\partial E}{\partial N}\right)_{S,V}.$$

This equation looks quite different from $\mu = -T(\partial S/\partial N)_{E,V}$, but in fact it is perfectly equivalent to it as we will demonstrate.

If one looks at the equation $\mu = (\partial E/\partial N)_{S,V}$, superficially one can go astray in identifying the meaning of μ. If we calculate μ from this equation without much thought, we might argue like this. Clearly the energy of the gas is

$$E = \tfrac{1}{2}NkT.$$

Since this is the case, should not we expect that μ is given by

$$\left(\frac{\partial E}{\partial N}\right) = \tfrac{3}{2}kT \ ?$$

We have, however, already obtained an expression for μ for an ideal gas using the basic definition $\mu = T(\partial S/\partial N)_{E,V}$. This gives [see (4-143)]:

$$\mu = kT \ln \left(\frac{N V_F}{V} \right). \tag{4-154}$$

This expression for μ is very different from $(3kT/2)$. What is the source of the discrepancy? The problem arises because in computing $(\partial E/\partial N)$, we have not carefully held constant the entropy and volume. Physically, the chemical potential is equal to the change in energy associated with the addition of a particle *provided* that the entropy and volume are held constant on the addition. In deriving (4-153) from $E = \frac{3}{2}NkT$, we did not hold S and V constant. Instead we held the temperature constant. When particles are added to a gas while the temperature is held constant, the entropy changes. We can see this quite clearly by expressing the change in entropy of a gas, when the temperature is held constant as

$$\Delta S(T, V, N) = \left(\frac{\partial S}{\partial V} \right)_{N,T} \Delta V + \left(\frac{\partial S}{\partial N} \right)_{V,T} \Delta N. \tag{4-155}$$

Thus the change in energy associated with a change in the number of particles and a change in the volume at constant temperature can be found from (4-145), using (4-155) for ΔS. This gives

$$\Delta E = T \left(\frac{\partial S}{\partial V} \right)_{N,T} \Delta V + T \left(\frac{\partial S}{\partial N} \right)_{V,T} \Delta N - p\Delta V + \mu\Delta N$$

for T constant. On rearranging this, we find

$$\Delta E = \left(T \left(\frac{\partial S}{\partial V} \right)_{N,T} - p \right) \Delta V + \left(\mu + T \left(\frac{\partial S}{\partial N} \right)_{V,T} \right) \Delta N. \tag{4-156}$$

Now we know that for an ideal gas, $E = 3NkT/2$. From this equation we see that $(\partial E/\partial V)_{T,N} = 0$. We therefore expect that the coefficient of the ΔV term in (4-156) must equal zero. This is in fact the case as we now verify: To do this we first find the entropy of the gas. Using (4-138) and

$$\alpha = \ln(N V_F / V)$$

obtained from (4-139) and (4-140), we see that

$$S(T, V, N) = Nk \left\{ \tfrac{5}{2} - \ln N + \ln \left(\frac{V}{V_F} \right) \right\}. \qquad (4\text{-}157)$$

From this we see that

$$T \left(\frac{\partial S}{\partial V} \right)_{N,T} = + \frac{NkT}{V}.$$

But in an ideal gas we know that

$$p = \frac{NkT}{V}.$$

So that the coefficient of ΔV in (4-157) is indeed zero, and we have left the expression

$$\Delta E = \left(\mu + T \left(\frac{\partial S}{\partial N} \right)_{V,T} \right) \Delta N \qquad \text{(for } T \text{ constant)}. \qquad (4\text{-}156a)$$

Thus we see quite clearly that $(\partial E / \partial N)_T$ is not equal to μ, but differs from it by an amount given by

$$\left(\frac{\partial E}{\partial N} \right)_T = \mu + T \left(\frac{\partial S}{\partial N} \right)_{V,T}.$$

The left-hand side of this equation is $\tfrac{3}{2}kT$. This equation can be written as

$$\mu = \left(\frac{\partial E}{\partial N} \right)_T - T \left(\frac{\partial S}{\partial N} \right)_{V,T}. \qquad (4\text{-}158)$$

It states that the chemical potential is equal to the energy change produced on adding a particle at constant temperature *minus* T times the entropy change associated with adding the particle at constant temperature and volume. This second term then *corrects* for the entropy change associated with adding another particle at constant temperature. Let us now show quite explicitly that the subtraction of the term $T(\partial S/\partial N)_{T,V}$ from $\tfrac{3}{2}kT = (\partial E/\partial N)_{T,V}$ does in fact produce the correct expression [(4-154)] for μ. From (4-157) we find that

$$T\left(\frac{\partial S}{\partial N}\right)_{T,V} = \tfrac{3}{2}kT - kT \ln\left(\frac{NV_F}{V}\right).$$

On using this in (4-158), we find

$$\mu = \tfrac{3}{2}kT - \tfrac{3}{2}kT + kT \ln\left(\frac{NV_F}{V}\right)$$

or

$$\mu = kT \ln\left(\frac{NV_F}{V}\right),$$

which is identical with the result of (4-154).

We see therefore that one can look upon μ as being given either by

$$\mu = \left(\frac{\partial E}{\partial N}\right)_{S,V}$$

or by

$$\mu = -T\left(\frac{\partial S}{\partial N}\right)_{E,V}.$$

Both formulas give the same result for μ provided that one keeps in mind the crucial importance of computing the changes in E or S while keeping constant those quantities specified as subscripts in the partial derivatives.

(vi) General Statement of the Equilibrium Conditions. The Second Law of Thermodynamics

We have already stated the equilibrium conditions in some detail. In this subsection we will summarize them in a concise way and present them clothed in their phenomenological, macroscopic garb, that is, in the language of thermodynamics.

Let us recall the definition of a simple system. It consists of a homogeneous piece of material in a single phase (liquid, solid, or gaseous). The material may be of a single chemical species or else a nonreactive and nonseparable mixture of such species.

A composite system is a set of simple systems (the components), separated from each other by internal boundaries. These boundaries may be material boundaries (walls), or just interfaces, such as the surface separating a liquid from its vapor, or the surface of contact of two nonmiscible fluids. It is across these boundaries that heat, or particles, may be exchanged between the components of the system. By moving a boundary, work may be done by one component against another. On the other hand, boundaries may also act as *constraints*. Walls or membranes may selectively prohibit or permit the passage of heat, or particles, or, by being rigid, prevent work from being done. In short, internal boundaries may represent internal constraints on the system. Such internal constraints also affect chemical equilibria. A mixture of hydrogen and chloride represents a constrained equilibrium; the addition of a catalyst is needed to open up the path for the formation of HCl.

Every equilibrium in a composite system is therefore dependent on the nature of the constraints operating. Any release of an internal constraint produces a *shift* in the equilibrium. The question we must answer is which way, and how far, does this shift go?

In formulating the equilibrium conditions for a system, we must define the nature of its external boundaries. If these boundaries are rigid, and permit neither the transfer of energy nor of matter, we speak of a *closed* system.

On the other hand, if the system has boundaries across which energy or matter or volume may be transferred, we speak of an *open* system. After this preamble we are now in a position to state the general equilibrium condition as follows. *The equilibrium in a closed composite system is characterized by the maximum value of the total entropy of the system consistent with the internal constraints. This statement is one of a number of equivalent forms of the second law of thermodynamics.* Associated with this statement is the premise that the entropy is a state function. The difference in entropy between the two states (1) and (2) of the system can be found by using the equation

$$S_2 - S_1 = \int_1^2 \left(\frac{dQ}{T} \right)$$

taken along a reversible path.

The statement just given is not the form in which the second law of thermodynamics is generally presented. The second law of thermodynamics was formulated by Rudolf Clausius and Lord Kelvin, in 1850. Clausius' statement of the second law reads: "There exists no process whatever whose *sole* effect is to extract heat from a cooler reservoir and to deliver it to a hotter reservoir." It can be shown that on the basis of this deceptively simple statement, one can determine the minimal amount of *work* needed to transfer an amount ΔQ_1 of heat from a reservoir at an ideal gas temperature T_1 to a reservoir at higher temperature T_2. This work, ΔW (that is delivered in the form of additional heat to the higher temperature reservoir), is such that

$$\frac{\Delta Q_1}{T_1} = \frac{\Delta Q_1 + \Delta W}{T_2} = \frac{\Delta Q_2}{T_2}.$$

A detailed study of this process then reveals that the integral $\int_a^b dQ/T$, taken along a reversible path from an equilibrium state a to an equilibrium state b, is independent of the path and defines a state function S:

$$\int_a^b \frac{dQ}{T} = S_b - S_a.$$

In 1865, Clausius introduced the term *entropy* as a name for this function.

There remained to be shown that any irreversible process in a closed system is accompanied by an increase of the entropy of the system. By careful analysis of the work and heat transfer in reversible and irreversible processes, it can indeed be shown that in an irreversible process taking place in a closed system the entropy always increases. This purely phenomenological or thermodynamic approach to the equilibrium conditions is perfectly equivalent to the microscopic statistical formulation that we have presented. In the following subsection we will play through some simple illustrative applications of the entropy maximum condition for *closed* systems.

On the other hand, most systems one deals with experimentally are open systems. It will therefore be necessary to reformulate the equilibrium conditions so as to apply to open systems. This will be done in Section 4.3.C. Following this section we will provide a variety of applications of the equilibrium conditions for such open systems.

(vii)   Simple Illustrative Applications of the Entropy Maximum Principle

The two first examples to be presented here will illustrate the well-known fact that in equilibrium two simple systems with a heat-conducting interface will have the same temperature, and also the same pressure, if that interface is movable.

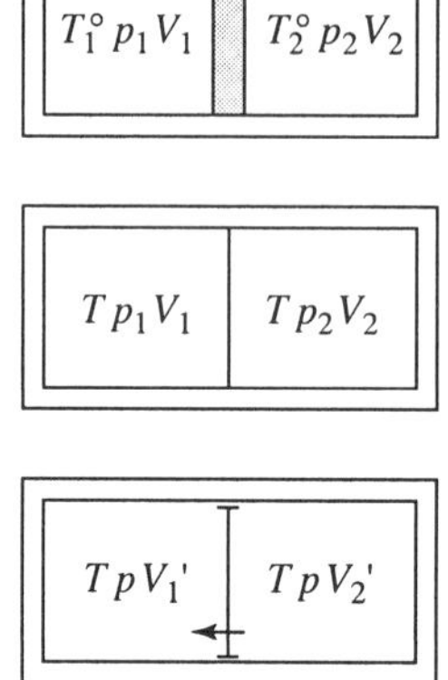

(0) Two simple systems separated by a rigid, insulating wall. Neither p nor T need be the same for equilibrium.

(a) Two simple systems separated by heat-conducting wall. Temperatures are equal in equilibrium. Pressures may be different.

(b) Two simple systems separated by mobile, heat-conducting wall. Temperature and pressure are the same at equilibrium.

Figure 4.34. Types of separation for spatially adjacent simple systems: (0) Two simple systems separated by a rigid, insulating wall. Neither p nor T need be the same for equilibrium. (a) Two simple systems separated by a heat-conducting wall. Temperatures are equal in equilibrium. Pressures may be different. (b) Two simple systems separated by a mobile, heat-conducting wall. Temperature and pressure are the same at equilibrium.

These systems are illustrated in Figure 4.34. Notice that since we apply the entropy maximum principle, the total system must be closed.

In Figure 4.34(0) the two systems are separated by a rigid, insulating (nonheat conducting) wall. We have in fact two closed separate systems. Each is in internal equilibrium with a value of p and T corresponding to its energy contents.

Equalization of Temperature.   In Figure 4.34(a), the insulating wall has been replaced by a heat-conducting (permeable) wall. A constraint has been released. There is now the possibility of energy exchange by means of heat flow across the separating wall. The system will move to a new equilibrium state.

Let us look at this process in some detail. We will assume that the initial temperature T_1^0 is larger than T_2^0. We will also assume that the wall is a poor conductor, so that the heat transfer occurs slowly enough to have essentially uniform temperatures in the two-component systems. Also, the heat capacity of the wall will be assumed negligibly small, so that we need not be concerned with the wall when looking at the energy budget. In Figure 4.35 we plot the temperature profile across the wall separating the two components (enlarged).

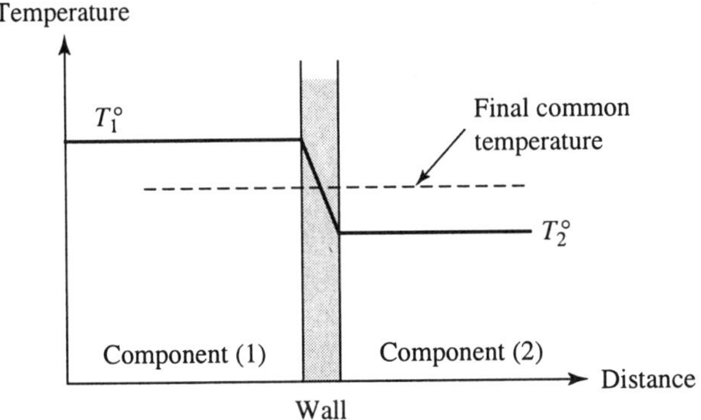

Figure 4.35. Temperature profile across a heat-conducting wall separating the two components of a closed system.

We now calculate the change in entropy of the system when heat flows across the wall under the conditions specified above.

Let us assume that a small amount ΔQ of heat has been transferred from compartment (1) to compartment (2). This entails a temperature change in both compartments. In compartment (1) it is lowered by

$$\Delta T_1 = \Delta Q / C_{V_1}$$

and in compartment (2) it is raised by

$$\Delta T_2 = +\Delta Q / C_{V_2}.$$

To keep the example simple, we now assume that the two heat capacities C_{V_1} and C_{V_2} are equal and independent of temperature: $C_{V_1} = C_{V_2} = C_V = \text{constant}$. In this case, the temperature changes of the system are equal and opposite

$$\Delta T_1 = -\Delta T_2 \equiv -\Delta T$$

for any amount of heat transferred across the wall. The entropy change in compartment (1) associated with a reduction of its temperature by ΔT is

$$\Delta S_1 = C_V \int_{T_1^0}^{T_1^0 - \Delta T} \frac{dT_1}{T_1} = C_V \ln \left(\frac{T_1^0 - \Delta T}{T_1^0} \right)$$

and the corresponding change in compartment (2) is

$$\Delta S_2 = C_V \int_{T_2^0}^{T_2^0 + \Delta T} \frac{dT_2}{T_2} = C_V \ln \left(\frac{T_2^0 + \Delta T}{T_2^0} \right).$$

The total entropy change in this transaction is therefore

$$\Delta S = \Delta S_1 + \Delta S_2 = C_V \ln \left(\frac{(T_1^0 - \Delta T)(T_2^0 + \Delta T)}{T_1^0 T_2^0} \right). \qquad (4\text{-}159)$$

In Figure 4.36 we plot this entropy change ΔS as a function of ΔT. We see at once that $\Delta S = 0$ for $\Delta T = 0$ and $\Delta T = T_1^0 - T_2^0$, and has a maximum in between these two values.

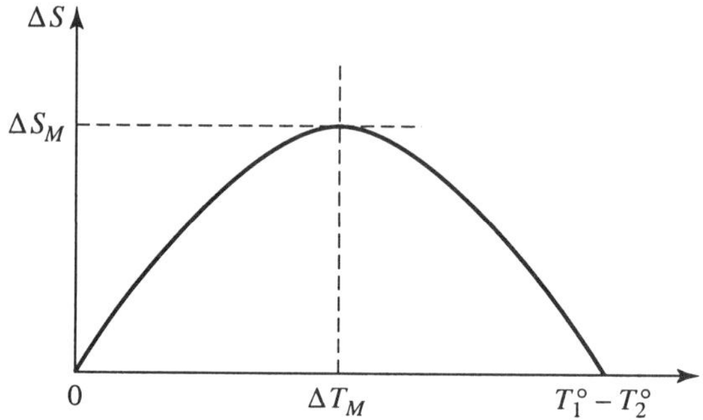

Figure 4.36. Entropy change in a system with two compartments of equal heat capacities, and initial temperatures T_1^0, T_2^0, as a function of ΔT, the temperature change in a compartment.

The maximum of ΔS occurs at the maximum of the function $(T_1^0 - \Delta T)(T_2^0 + \Delta T) = T_1^0 T_2^0 + (T_1^0 - T_2^0)T - \Delta T^2$. We know that the maximum of a function $a + bx + cx^2$ occurs at

$$x = -\frac{b}{2c}.$$

Applied to the case at hand, we find that $\Delta S_{\max}$ occurs at

$$\Delta T = \Delta T_M = \tfrac{1}{2}(T_1^0 - T_2^0). \qquad (4\text{-}160)$$

It is now easily seen that for this value of T, corresponding to the maximum entropy gain, the temperatures of the two compartments have become equal

$$T_1 = T_1^0 - \Delta T_M, \qquad T_2 = T_2^0 + \Delta T_M,$$

or

$$T_1 = \left(\frac{T_1^0 + T_2^0}{2}\right), \qquad T_2 = \left(\frac{T_2^0 + T_1^0}{2}\right).$$

Thus, $T_1 = T_2 = (T_2^0 + T_1^0)/2$ in the state of equilibrium.

We leave it as a problem assignment to demonstrate the analogous result for the case that the two compartments have unequal heat capacities.

On a more formal level of reasoning, we may simply ask ourselves: What characterizes the ultimate equilibrium? Since the entropy must be maximum, it must be stationary (unchanged) if, at the equilibrium point, a small amount of energy ΔE is transferred between systems (1) and (2):

$$\Delta S_{eq} = \left(\frac{\partial S_1}{\partial E_1}\right)_{V_1, N_1} \Delta E + \left(\frac{\partial S_2}{\partial E_2}\right)_{V_2, N_2} (-\Delta E) = 0. \qquad (4\text{-}161)$$

In (4-135) and (4-142a) we saw that these derivatives of S were the inverse temperatures

$$\left(\frac{\partial S_1}{\partial E_1}\right)_{V_1, N_1} = \frac{1}{T_1}, \qquad \left(\frac{\partial S_2}{\partial E_2}\right)_{V_2, N_2} = \frac{1}{T_2},$$

the condition $\Delta S_{eq} = 0$ therefore requires that

$$\left(\frac{1}{T_1} - \frac{1}{T_2}\right) \Delta E = 0 \qquad (4\text{-}162)$$

or $T_1 = T_2$.

Equalization of Pressure and Temperature.   We now pass to part (b) of Figure 4.34. The easiest way to visualize what happens is to assume that we have already equalized temperatures. We also assume that initially the pressure p_1^0 at the left is larger than the pressure p_2^0 at the right. What happens if we simply allow the separating wall to move? It will accelerate under the initial pressure difference $\Delta p^0 = p_1^0 - p_2^0$; a damped oscillation will occur that will gradually transfer the available kinetic and potential energy into heat.

Clearly, this is not a sequence of equilibrium states, and we cannot use such a situation to calculate the entropy change ΔS associated with the equalization of pressure. What we can do to control the speed of the piston movement is to introduce a strong, viscous damping system, as illustrated in Figure 4.37.

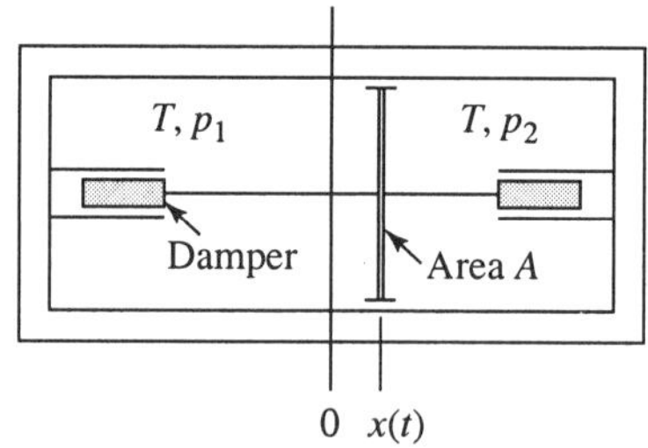

Figure 4.37. Movable piston separating two systems at different pressures p_1 and p_2. Temperatures are equal. Damping assumes very slow motion of the piston.

By attaching suitable dampers to the wall, its velocity can be kept so low that the temperatures of the two components can at all times be kept equal. (Notice that a rapid decompression of compartment (1) would lower its temperature, whereas the corresponding compression of (2) would raise the temperature there. Time must therefore be allowed for heat to flow from (2) to (1) in the process of equalizing the pressure.) The work done by the moving piston (see Figure 4.37 for symbols)

$$\Delta W = A \int (p_1 - p_2)\, dx$$

is equal to the frictional work done on the dampers, and returns to the system as a heat input ΔQ. Provided this process occurs slowly enough, the two compartments have equal temperatures T (which may change during the process) and the frictional heat is fed into the system under equilibrium conditions. The entropy change ΔS is therefore (assuming $p_1 \geq p_2$)

$$\Delta S = \int \frac{dQ}{T} = A \int \frac{(p_1 - p_2)}{T}\, dx > 0. \tag{4-163}$$

We thus see that every step toward pressure equalization in a closed system represents an increase in entropy. In each successive displacement step Δx the pressure difference $p_1 - p_2$ is reduced, until a value of x is reached at which $p_1 = p_2$. At this point mechanical equilibrium is reached, and no further displacement occurs. The entropy has increased to the maximum value consistent with mechanical equilibrium. In fact, mechanical equilibrium in this case can be viewed as a consequence of the entropy maximization principle. For if the piston is displaced to the right of the equilibrium position, $p_1 - p_2$ becomes negative, and the entropy will begin to decrease below its value at $p_1 = p_2$.

Again, we could have bypassed this detailed argument and simply asked: What characterizes the equilibrium between two components of a closed system having a common temperature T, and being able to trade off volume against each other?

In answer to this question we can invoke the conditions that the total entropy $S = S_1 + S_2$, at *equilibrium*, must remain stationary (unchanged) under a small trade-off ΔV of volume between components (1) and (2):

$$\Delta S = \left(\frac{\partial S^{(1)}}{\partial V_1}\right)_{E_1, N_1} \Delta V + \left(\frac{\partial S^{(2)}}{\partial V_2}\right)_{E_2, N_2} (-\Delta V) = 0.$$

We know from (4-142b) that the V-derivatives of S have the value p/T; we have thus the condition

$$\frac{p_1}{T}\Delta V - \frac{p_2}{T}\Delta V = 0 \tag{4-164}$$

or $p_1 = p_2$.

The Entropy of Mixing.   This example deals with the increase of entropy in the process of mixing two gases (or miscible fluids for that matter) that previously were in separate compartments.

We notice that even without stirring the two components into each other, this mixing will take place by the process of *diffusion*, that we had studied extensively in Chapter 2. In that chapter we had described the diffusion process, starting from a random walk picture of the process. In the present context we observe that if two separate miscible components (all gases are miscible, but not all fluids) are allowed to mix, they will do so. Mixing, therefore, leads to a new equilibrium state, whose entropy will be larger than that of the initial state. The change in entropy is called the entropy of mixing.

This is a good example to illustrate the provision made in the definition [(4-148)] of the entropy difference between two states of the system, the final state

(f) and the initial state (i):

$$S_f - S_i = \int_i^f \frac{dQ}{T},$$

namely, that the intermediate states of the system on the path from (i) to (f) must always be equilibrium states of the system. (They may depart from equilibrium "infinitesimally," so as to make the system undergo a slow change.)

In Figure 4.38 we illustrate the mixing of two gases following the simple removal of the wall separating the two compartments. We assume that pressure and temperature are equal, initially. Observe that the system is a closed one. There is no heat exchange with the outside. For mixing of ideal gases there is no temperature change in the process either. In this special case we see that

$$\int_i^f \frac{dQ}{T} = \frac{1}{T} \int_i^f dQ = \frac{1}{T} Q_{fi} = 0.$$

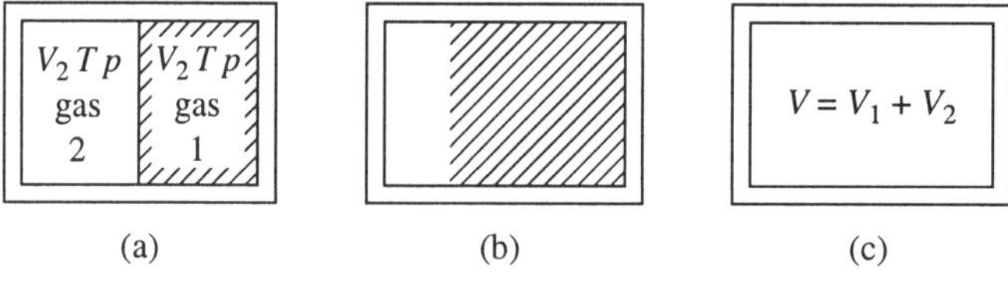

Figure 4.38. Mixing of two gases. (a) Gases originally separated by a wall but at the same pressure and temperature. (b) The wall has been removed; transition period with diffusional flow. This is *not* an equilibrium state. (c) The final equilibrium state, the uniform density of both gases over the whole volume $V = V_1 + V_2$.

But we must be aware that this integral does *not* represent the entropy change between states (f) and (i) of the system because the intermediate states are *not* at all equilibrium states.

To calculate the entropy change in the mixing process, we must let this process proceed in such a way as to go through a sequence of equilibrium states. This can be achieved with the aid of semipermeable walls, as illustrated in Figure 4.39. In this setup, wall M_1 is subject to the partial pressure p_1 of gas (2), M_2 to the partial pressure p_2 of gas (2). By attaching viscous frictional dampers to these walls, this

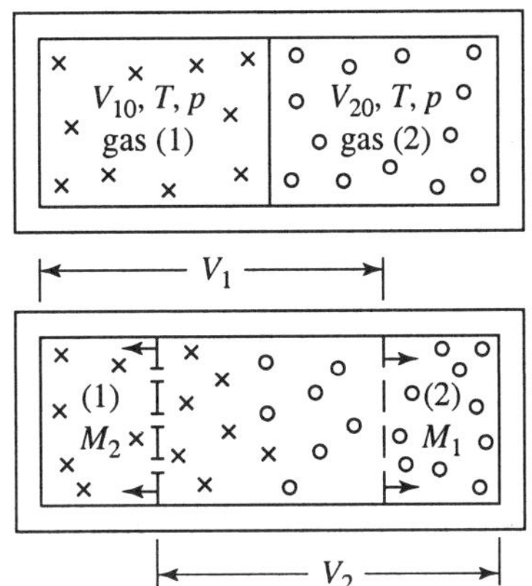

Figure 4.39. Mixing of two different gases (1) and (2) by means of the semi-permeable walls M_1 and M_2. M_1 permits the passage of gas (2), but not of (1); M_2 permits the passage of gas (1), but not of (2).

motion can be reduced to a creeping advance, and the work of expansion is fed back into the system in the form of heat. This heat is fed into the system under conditions as close to equilibrium as we choose. This heat then, equal to the work done by the two slowly moving walls, is responsible for the increase of entropy in the mixing process. If we consider the case of two ideal gases, we are in a position to be more explicit about the amount of entropy increase. For ideal gases the *partial pressures* p_1 and p_2 are given by the expressions (see Figure 4.39 for V_1 and V_2):

$$p_1 = n_1 \frac{RT}{V_1}, \qquad p_2 = n_2 \frac{RT}{V_2}.$$

Also, since the energy of the ideal gas does not depend on volume but only on temperature, the heat fed into the system by the frictional devices, being equal to the work done by the moving walls, will just be such as to keep the temperature of the two gases constant, and equal to the initial temperature T. Neither of the gases changes its energy in the process of mixing. The total heat input ΔQ is equal to the total *work* ΔW done by the two semipermeable walls as they move from their original position to their final position. This gives us, for the entropy change,

$$\Delta S_{fi} = S_f - S_i = \frac{\Delta Q_{fi}}{T} = \frac{\Delta W_{fi}}{T}.$$

Now this work done by the walls can be written as

$$\Delta W_{fi} = \int p_1 \, dV_1 + \int p_2 \, dV_2$$

$$= n_1 RT \int_{V_{10}}^{V} \frac{dV_1}{V_1} + n_2 RT \int_{V_{20}}^{V} \frac{dV_2}{V_2},$$

where V_{10} and V_{20} are the original volumes occupied by gases (1) and (2), and V is the common final volume. The volume integrals give $\ln(V/V_{10})$ and $\ln(V/V_{20})$, respectively, so that the total entropy change upon mixing is

$$\Delta S_{fi} = S_f - S_i = n_1 R \ln\left(\frac{V}{V_{10}}\right) + n_2 R \ln\left(\frac{V}{V_{20}}\right). \qquad (4\text{-}165)$$

We now want to show that this same result also follows from the microscopic interpretation of the entropy as $S = k \ln W_{\max}$, where $W_{\max}$ is the weight of the most probable macrostate, and represents the number of distinct microstates belonging to that single macrostate. This definition of S gave us the following expression [see (4-166)], valid for an ideal monoatomic gas of $n = N/N_0$ moles in volume V at temperature T:

$$S = Nk\left\{\tfrac{5}{2} + \ln\left(\frac{V}{N V_F(T)}\right)\right\}$$

$$= nR\left\{\tfrac{5}{2} + \ln\left(\frac{V}{N V_F(T)}\right)\right\}. \qquad (4\text{-}166)$$

Before actually going into investigating the statistical origin of the entropy of mixing, we want to point out that it is implied by (4-166) that there is *no* entropy of mixing if the two gases, that are allowed to intermingle after removal of a separating wall, are of the *same kind*, say helium. If we have two compartments, one of volume V_1, containing n_1 moles, the other of volume V_2, containing n_2 moles, and if then the temperatures and pressures are equal, their joint entropy will be

$$S = S_1 + S_2 = n_1 R\left(\tfrac{5}{2} + \ln\left(\frac{V_1}{N_1 V_F(T)}\right)\right)$$

$$+ n_2 R\left(\tfrac{5}{2} + \ln\left(\frac{V_2}{N_2 V_F(T)}\right)\right). \qquad (4\text{-}167)$$

If we remove the wall separating the two boxes, we create a system of $n = (n_1+n_2)$ moles of a single gas in volume $V = V_1 + V_2$. Its entropy is given by (4-166). Because of the assumption of equal temperatures and pressures in the two separate boxes, the unified system will have the same temperature as the previously separate components, and also have the same density

$$\frac{N}{V} = \frac{N_1}{V_1} = \frac{N_2}{V_2} \quad \text{since if} \quad \frac{a}{b} = \frac{c}{d} \quad \text{then} \quad \frac{a}{b} = \frac{a+c}{b+d}.$$

Its entropy can therefore be written as

$$(n_1 + n_2) \left\{ \frac{5}{2} + \ln\left(\frac{V}{NV_F}\right) \right\}$$

$$= n_1 \left\{ \frac{5}{2} + \ln\left(\frac{V_1}{N_1 V_F}\right) \right\} + n_2 \left\{ \frac{5}{2} + \ln\left(\frac{V_2}{N_2 V_F}\right) \right\} = S_1 + S_2.$$

In summary, we see that removal of a wall separating two compartments containing the *same* gas at the same pressure and temperature does not lead to a change in entropy. Let us now look at the case where the two separate compartments contained *different* gases, say helium and argon. As long as a wall separates the two compartments, the entropy of this system is given by (4-166). After the separating wall has been removed, we have n moles of a mixture composed of n_1 moles of helium and n_2 moles of argon ($n_1 + n_2 = n$) in a common volume $V = V_1 + V_2$. The entropy of this mixture is *not* given by (4-166), however.

To understand the reason for this, we must look at the *microstates* of the mixture. Recall that the weight W represents the *number* of microstates for a given macrostate. Each microstate represents a population pattern in phase space. Such a pattern is illustrated in Figure 4.40. Figure 4.40(A) shows such a pattern for four identical atoms. Figures 4.40 (B_1)–(B_6) illustrate the fact that for *each* pattern made with four identical atoms, we have six distinct patterns for a gas mixture of two species with two atoms each.

In general, if we replace a gas of N identical atoms with a gas mixture of N_1 atoms of one kind, N_2 atoms of a second kind ($N_1 + N_2 = N$), then for each microstate of the one species gas, we have

$$W_{\text{mix}} = \frac{N!}{N_1! \, N_2!}$$

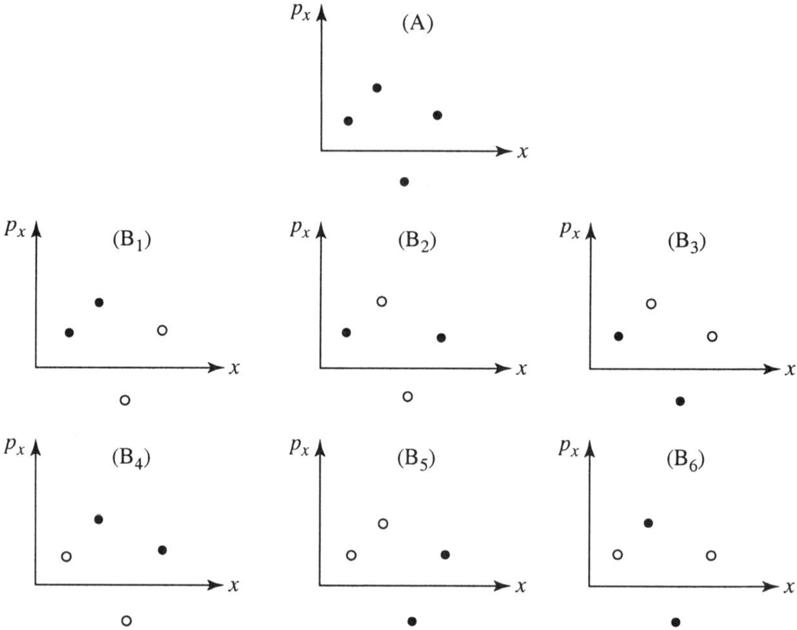

Figure 4.40. Microstates for a single gas and gas mixtures. (A) shows a microstate for $N = 4$ identical particles. B_1 to B_6 show the six microstates that, for a gas mixture of four particles with $N_1 = N_2 = 2$, correspond to *each* microstate of the single gas.

distinct microstates of the mixture. To get the correct expression for W for a gas mixture, we must multiply the W appropriate for a pure gas with the above factor. The logarithm of this extra factor can be expressed using Stirling's approximation as

$$\ln W_{\text{mix}} = \ln \left(\frac{N!}{N_1!\, N_2!} \right) \cong N_1 \ln \left(\frac{N}{N_1} \right) + N_2 \ln \left(\frac{N}{N_2} \right).$$

The entropy S for a *mixture* of $N = N_1 + N_2$ particles in a volume V is therefore *larger* than the entropy of N identical particles in the same volume V by

$$\Delta S_{\text{mix}} = k \ln W_{\text{mix}} = N_1 k \ln \left(\frac{N}{N_1} \right) + N_2 k \ln \left(\frac{N}{N_2} \right). \tag{4-168a}$$

We can write this alternatively in terms of mole numbers

$$n = N/N_0, \qquad n_1 = N_1/N_0, \qquad n_2 = N_2/N_0,$$

or in terms of the ratios V_{10}/V and V_{20}/V between the original volumes of the components and the total volume $V = V_{10} + V_{20}$:

$$\Delta S_{\text{mix}} = n_1 R \ln \left(\frac{n}{n_1} \right) + n_2 R \ln \left(\frac{n}{n_2} \right)$$

$$= n_1 R \ln \left(\frac{V}{V_{10}} \right) + n_2 R \ln \left(\frac{V}{V_{20}} \right). \qquad (4\text{-}168\text{b})$$

Thus we see that there is complete agreement between the microscopic result for the entropy of mixing [(4-168)], and the thermodynamic result [(4-165)].

As a final remark, we want to point out the operational significance of this mixing entropy. If we let this mixing take place by means of the displacement of semi-permeable pistons, as illustrated in Figure 4.38, we can extract energy in the form of work from the system. (The moving pistons could lift a weight, etc.) If the system is enclosed in heat-conducting external walls, as in this case, and immersed in a heat bath of temperature T, then $T \Delta S_{\text{mix}}$ represents both the heat drawn in by the system from the heat bath, and also the mechanical work done by the moving walls.

Conversely, we can separate the two components of the mixture by pushing the semipermeable pistons back. In this case, the quantity $T \Delta S_{\text{mix}}$ represents the isothermal work needed for separating the system into its components.

We will, in Section 4.3.C, show how the entropy of mixing is responsible for lowering the chemical potential of the solvent in a solution, and is thus directly at the source of such phenomena as the rise of the boiling point and the depression of the freezing point of a fluid due to the addition of solute.

4.3.C.   The Free Energy Minimum Principles

In this section we show that the entropy maximum principle for the equilibrium in a *closed* system can be used to establish alternative, equivalent equilibrium conditions for *open* systems.

The motivation for formulating alternative equilibrium conditions lies in the fact that the notion of a closed system, although conceptually useful, is somewhat artificial. Almost none of the systems the physicist or chemist works with are closed in the sense defined in Section 4.3.B. Most systems one wants to investigate are kept in thermal contact with the "outside" so that their temperatures, and most often also their pressures, are kept constant and equal to the values of T and p of the immediate environment. In such systems an internal rearrangement will in general be associated with an exchange of energy with the environment in the form of heat and/or work. We will show that for such an open system the equilibrium is characterized by the *minimum* value of the so-called *free energy*.

(i)   A System at Constant Temperature and Volume.
The Helmholtz Free Energy

A composite system may be kept at constant temperature T by surrounding it with a "heat bath." This bath is essentially a large heat reservoir. We shall consider it large enough so that it may donate or pick up finite amounts of heat from the system without changing its temperature. This is illustrated in Figure 4.41.

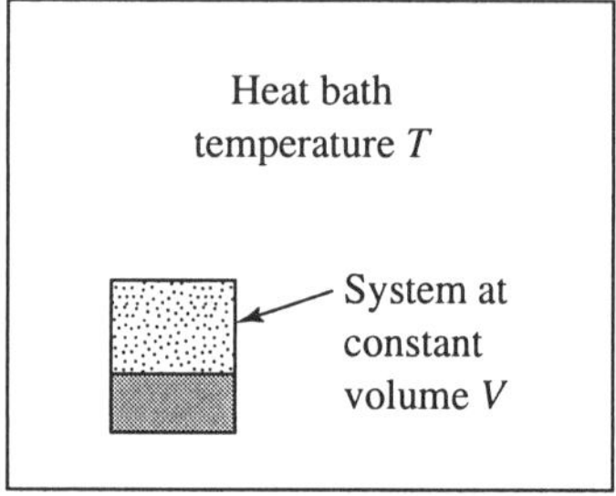

Figure 4.41. A composite system surrounded by a heat bath at temperature T. The system has a fixed volume V; it is in thermal contact with the bath, and has temperature T. There is no exchange of matter between the system and bath.

We assume explicitly that the system has a *fixed total volume* V so that no energy exchange between the system and bath in the form of mechanical work is possible. Also, no exchange of matter is allowed. Such a situation may be realized by enclosing the system with a rigid but heat-conducting wall.

To formulate the equilibrium condition for the system in this case, we make use of the possibility of viewing the system *plus* the bath as a single *closed* composite supersystem. To that supersystem the entropy maximum principle does apply.

We see from this right away that, in equilibrium, the bath and system have the same temperature. They need not have the same pressure. The rigid walls surrounding the system maintain a *constrained* equilibrium in which a pressure difference can persist.

Assume now that in the system an internal constraint is released; mixing of components may occur, or work may be done by one component against another, etc. We cannot assert that the entropy of the system increases since the system is not closed, and heat may be exchanged with the heat bath. But the entropy of the *closed* supersystem, system and bath, must increase

$$\Delta S \text{ (system)} + \Delta S \text{ (bath)} > 0. \tag{4-169}$$

The change ΔS (bath) can only result from a heat transfer to or from the system, occurring at constant temperature T:

$$\Delta S \text{ (bath)} = \frac{\Delta Q \text{ (bath)}}{T}. \tag{4-170}$$

This heat transfer ΔQ (bath) is equal and opposite to the energy change ΔE (system) of the system

$$\Delta Q \text{ (bath)} = -\Delta E \text{ (system)}. \tag{4-171}$$

We write ΔE (system) rather than ΔQ (system) for the simple reason that although this energy is transferred to the system as a heat flow across the walls, it is not necessarily transferred to all components of the system in the form of heat flow. All we can say is that $-\Delta Q$ (bath) represents *the sum of the energy changes* in the various components of the system.

Combining (4-170) with (4-169), we reach the result that in the process of adjustment of the system to a less constrained equilibrium situation at *constant* temperature, we have

$$\Delta S \text{ (system)} - \frac{1}{T}\Delta E \text{ (system)} > 0. \tag{4-172}$$

We can rewrite this in the form

$$\Delta E \text{ (system)} - T\Delta S \text{ (system)} < 0,$$

or even, since these changes occur at constant T:

$$\Delta(E - TS) \underset{T = \text{constant.}}{\text{(system)};} \quad < 0 \qquad (4\text{-}173)$$

This is the desired new result. It brings on the scene a new state function F:

$$F = E - TS, \qquad (4\text{-}174)$$

the so-called *Helmholtz free energy*. It is the quantity, in terms of which the equilibrium conditions in an isothermal, isochorous (constant volume) system can be stated: *The equilibrium in a composite system, held at constant temperature T confined in a constant volume V, is characterized by the minimum value of the Helmholtz free energy E.*

To make use of this statement, we must express the free energy of the system in terms of the free energy of its components. We will do this in the next section on applications. Here, in preparation for this, we will simply look at the free energy of a *simple* system.

Since this system is characterized by specified values of temperature T, volume V, and particle number N, these are the independent variables that determine the value of the free energy F. We will therefore regard F as a function of the variables T, V, and N. The manner in which F depends on these variables can be established by constructing the expression that describes the change ΔF of F which occurs if T, V, and N are changed by ΔT, ΔV, and ΔN, respectively. We can obtain this expression by combining the definition [(4-174)] of F with (4-145) for ΔE. Indeed, combining (4-145):

$$\Delta E = T\Delta S - p\Delta V + \mu\Delta N,$$

with $F = E - TS$, we find for the change in F:

$$\Delta F = \Delta E - T\Delta S - S\Delta T,$$

which gives us

$$\Delta F = -S\Delta T - p\Delta V + \mu\Delta N. \qquad (4\text{-}175)$$

Since the change ΔF of any function $F(T, V, N)$ can be written in terms of its partial derivatives, viz:

$$\Delta F = \left(\frac{\partial F}{\partial T}\right)_{N,V} \Delta T + \left(\frac{\partial F}{\partial V}\right)_{T,N} \Delta V + \left(\frac{\partial F}{\partial N}\right)_{T,V} \Delta N, \qquad (4\text{-}176)$$

we see that the quantities S, p, and μ are the partial derivatives of the free energy F; that is,

$$\left(\frac{\partial F}{\partial T}\right)_{V,N} = -S, \tag{4-177a}$$

$$\left(\frac{\partial F}{\partial V}\right)_{T,N} = -p, \tag{4-177b}$$

$$\left(\frac{\partial F}{\partial N}\right)_{T,V} = \mu. \tag{4-177c}$$

The first of these equations, (4-177a), has no intuitive interpretation. We will show, however, in the next subsection that it has its use as a means to determine the entropy differences between the states of a simple system. The second equation (4-177b) expresses the relation

$$(\Delta F)_{\substack{T=\text{constant} \\ N=\text{constant}}} = -p\Delta V = -\Delta W,$$

and shows that the work ΔW done by the system at constant temperature and mole number is equal to its loss of free energy. It is this relation that is responsible for the name given to this quantity.

The last equation [(4-177c)] finally shows that the chemical potential can be expressed as the rate of change of free energy with particle number at constant temperature and volume: Breaking F up into E and $-TS$, we see that

$$\mu = \left(\frac{\partial E}{\partial N}\right)_{T,V} - T\left(\frac{\partial S}{\partial N}\right)_{T,V},$$

which is again (4-156), which we had derived in subsection (v) of Section 4.3.B on the chemical potential.

We leave it as a problem assignment to verify the three equations [(4-177a) to (4-177c)] when the free energy is that appropriate for the ideal monoatomic gas.

(ii)   Mathematical Interlude: Maxwell Relations and Their Use

There is a simple but important property of derivatives of a function of several variables that we want to introduce here.

Let us recall that for a continuous function $F(x)$ of a single variable x, the difference between the value of F at x and at $x + \Delta x$, Δx being a small increment of x, can be written as a series of increasing powers of Δx:

$$F(x + \Delta x) = F(x) + A(x)x + \tfrac{1}{2}\alpha(x)(\Delta x)^2 + O(\Delta x^3). \tag{4-178}$$

One then shows that the function $A(x)$ is the first derivative of $F(x)$:

$$A(x) = \lim_{\Delta x \to 0} \left(\frac{F(x + \Delta x) - F(x)}{\Delta x} \right) \equiv \left(\frac{dF}{dx} \right). \tag{4-179}$$

Similarly, one may show that $\alpha(x)$ is the first derivative of $A(x)$, or the second derivative of F:

$$\alpha(x) = \lim_{\Delta x + x} \left(\frac{A(X + \Delta x) - A(x)}{\Delta x} \right) = \frac{dA}{dx} = \frac{d^2 F}{dx^2}. \tag{4-180}$$

Completely analogous relations hold for a function $F(x, y)$ of two independent variables x and y. In place of (4-178) one has the double series

$$\begin{aligned}
F(x + \Delta x, y + \Delta y) = {} & F(x, y) + A(x, y)\Delta x + B(x, y)\Delta y \\
& + \tfrac{1}{2}\alpha(x, y)(\Delta x)^2 + \tfrac{1}{2}\beta(x, y)(\Delta y)^2 + \gamma(x, y)\Delta x\,\Delta y \\
& + \text{third or higher powers of } \Delta x, \Delta y.
\end{aligned} \tag{4-181}$$

Again one shows that

$$A(x, y) = \lim_{\Delta x \to 0} \left(\frac{F(x + \Delta x, y) - F(x, y)}{\Delta x} \right) \equiv \left(\frac{\partial F}{\partial x} \right)_y$$

and

$$B(x, y) = \lim_{\Delta y \to 0} \left(\frac{F(x, y + \Delta y) - F(x, y)}{\Delta y} \right) \equiv \left(\frac{\partial F}{\partial y} \right)_x.$$

As in the case of a function $F(x)$ of x only, the functions $\alpha(x, y)$, $\beta(x, y)$, and $\gamma(x, y)$ are derivatives of A and B:

$$\alpha(x, y) = \lim_{\Delta x \to 0} \left(\frac{A(x + \Delta x, y) - A(x, y)}{\Delta x} \right) = \left(\frac{\partial A}{\partial x} \right)_y = \left(\frac{\partial^2 F}{\partial x^2} \right)_y, \tag{4-182a}$$

$$\beta(x, y) = \lim_{\Delta y \to 0} \left(\frac{B(x, y + \Delta y) - B(x, y)}{\Delta y} \right) \equiv \left(\frac{\partial B}{\partial y} \right)_x = \left(\frac{\partial^2 F}{\partial y^2} \right)_x. \quad \text{(4-182b)}$$

An interesting relation is found for the function $\gamma(x, y)$. It can be expressed in two ways

$$\gamma(x, y) = \left(\frac{\partial A(x, y)}{\partial y} \right)_x = \left(\frac{\partial B(x, y)}{\partial x} \right)_y \quad \text{(4-183)}$$

or

$$\gamma(x, y) = \left(\frac{\partial}{\partial y} \left(\frac{\partial F}{\partial x} \right)_y \right)_x = \left(\frac{\partial}{\partial x} \left(\frac{\partial F}{\partial y} \right)_x \right)_y.$$

We will leave it as a problem assignment to establish the proofs of the relations (4-182) and (4-183). The important relation [(4-183)] can be simply illustrated with the example of the function

$$F(x, y) = \tfrac{1}{2}px^2 + \tfrac{1}{2}qy^2 + rxy,$$

p, q, r being constants. We then have

$$F(x + \Delta x, y + \Delta y) = F(x, y) + (px + ry)\Delta x + (qy + rx)\Delta y$$
$$+ \tfrac{1}{2}p(\Delta x)^2 + \tfrac{1}{2}q(\Delta y)^2 + r\Delta x \Delta y.$$

In this case

$$A(x, y) = px + ry,$$
$$B(x, y) = rx + qy,$$

and we see indeed that

$$\left(\frac{\partial A}{\partial y} \right)_x = r = \left(\frac{\partial B}{\partial x} \right)_y.$$

Relations such as (4-183) are widely and profitably used in thermodynamics. They are called *Maxwell relations*. We will now illustrate to what use they can be put by an example taken from the previous subsection on the Helmholtz free energy.

We had obtained the following expression for the change ΔF of the free energy F associated with a change ΔT of temperature T and ΔV of volume V (we keep N constant in the example):

$$\Delta F(T, V) = -S(T, V)\Delta T - p(T, V)\Delta V. \qquad (4\text{-}184)$$

We see that $-S(T, V)$ and $-p(T, V)$ are the analogies of $A(x, y)$ and $B(x, y)$ in the expansion of the function $F(x, y)$. Equation (4-183) then tells us that the V-derivative of S is equal to the T-derivative of p:

$$\left(\frac{\partial S}{\partial V}\right)_{T,N} = \left(\frac{\partial p}{\partial T}\right)_{V,N}. \qquad (4\text{-}185)$$

This equation has physical contents as we will now show.

Assume we want to determine the entropy difference $S_2 - S_1$ between two states (2) and (1) of a simple system by making use of *empirical* data. What data do we need? Recall that

$$S_2 - S_1 = \int_1^2 \frac{dQ}{T} = \sum_1^2 \frac{\Delta Q}{T},$$

where the "path" from (1) to (2) may be chosen in any way we like as long as it is reversible. We may choose it in particular to consist of a piece along which T changes and V is constant, and a second piece along which T is constant and V changes. Such a path is illustrated in Figure 4.42.

Along the path from (1) to (A), we need to know the heat inputs ΔQ accompanying the expansion of the gas at *constant temperature*. These heat inputs may be written as

$$\Delta Q_{T=\text{constant}} = \Delta V \left(\frac{\Delta Q}{\Delta V}\right)_{T=\text{constant}} = T\,\Delta V \left(\frac{\Delta S}{\Delta V}\right)_{T=\text{constant}}.$$

But since $S(T, V)$ is a state function, $(\Delta S/\Delta V)_T$ is the partial derivative

$$\left(\frac{\partial S}{\partial V}\right)_T.$$

It is almost impossible to measure the rate of heat input with volume expansion at constant temperature. The Maxwell relation, (4-183) or (4-185), gives us the

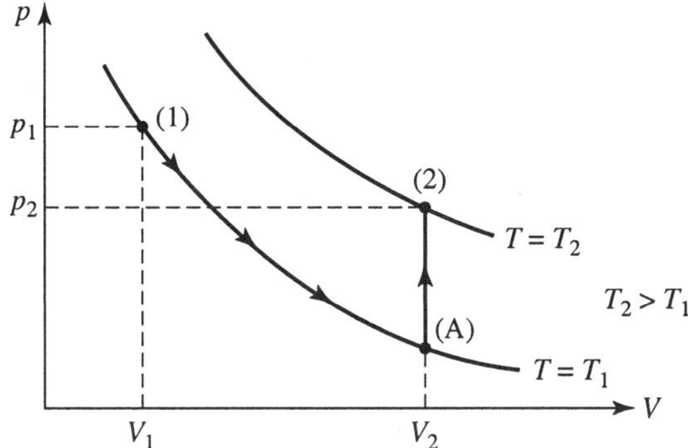

Figure 4.42. State diagram for a simple system showing a reversible path from state (1) to state (2). The path consists of a piece along which $T = $ constant (isothermal expansion), and a piece along which V is constant (isochorous heating).

means to equate it with a different, much more easily established quantity, namely, the rate of pressure change with temperature at constant volume, i.e., $(\partial p/\partial T)_V$. We can write

$$S_2 - S_1 = (S_A - S_1) + (S_2 - S_A)$$
$$= \int_{V_1}^{V_2} \left(\frac{\partial p}{\partial T}\right)_V dV + \int_{T_1}^{T_2} \frac{C_V\, dT}{T}. \tag{4-186}$$

The first integral represents the change of entropy associated with the path between (1) and (A), and this change is expressed in terms of the quantity $(\partial p/\partial T)_V$ evaluated for each volume V along the isothermal path. The second integral represents the change in entropy associated with the path from (A) to (2). Along this path the volume is constant, and the entropy change dS is equal to $(C_V/T)\, dT$. Both the quantities $(\partial p/\partial T)_V$ and (C_V) are determinable directly from the equations of state of the system.

(iii)   Equilibrium at Constant Pressure and Temperature. The Gibbs Free Energy

Most equilibria in a composite system studied in the laboratory are observed under conditions of constant temperature and constant pressure rather than constant volume.

The system studied is therefore *open* with the ability to exchange heat with a heat bath, and to do work. We show in this subsection that under these conditions the equilibrium is characterized by the minimum possible value of the so-called Gibbs free energy G, defined as

$$G = E - TS + pV. \tag{4-187}$$

The demonstration of this result again makes use of the device of joining system and heat bath into one, *closed* supersystem, and to apply the entropy maximum principle to it. This supersystem is illustrated in Figure 4.33.

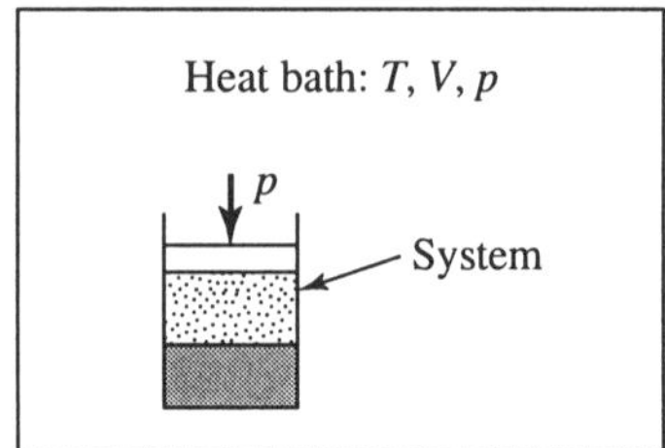

Figure 4.43. A composite system enclosed by a heat bath that represents an environment of constant pressure and temperature. A movable piston covers the system and keeps it at the same (constant) pressure as the heat bath.

The heat bath is a large enough system so that any finite heat transfer ΔQ to or from the system does not change the temperature perceptibly, nor does any volume change of the system change the pressure exerted by the heat bath. The bath therefore represents an environment of constant pressure p and temperature T.

One of the equilibrium conditions arising from the entropy maximum principle for the closed supersystem is that the system has the same pressure p and temperature T as the bath. These values are therefore constant whatever changes may occur inside the system.

Let us now see what the effect of a release of an internal constraint in the system is. We know that the total entropy will increase

$$\Delta S \text{ (bath)} + \Delta S \text{ (system)} > 0. \tag{4-188}$$

The entropy change of the bath can be related to its energy and volume change by the relation

$$\Delta E \text{ (bath)} = T \Delta S \text{ (bath)} - p \Delta V \text{ (bath)} ,$$

that is,

$$\Delta S \text{ (bath)} = \frac{\Delta E \text{ (bath)} + p \Delta V \text{ (bath)}}{T}. \tag{4-189}$$

Since the supersystem is closed, its total energy and volume are constants, we have therefore

$$\Delta E \text{ (bath)} = -\Delta E \text{ (system)},$$
$$\Delta V \text{ (bath)} = -\Delta V \text{ (system)}.$$

Combining (4-189) and (4-188), and incorporating the above relations, we find then

$$\Delta S \text{ (system)} - \frac{1}{T} (\Delta E \text{ (system)} + p \Delta V \text{ (system)}) > 0. \tag{4-190}$$

Since T and p are constants, and T is not zero, we can write

$$p \Delta V = \Delta (pV),$$
$$T \Delta S = \Delta (TS),$$

and restate (4-190), after multiplying it with T, in the form

$$\{\Delta (TS) - \Delta E - \Delta (pV)\}_{\text{system}} > 0$$

or

$$\Delta \{E + pV - TS\}_{\text{system}} \equiv \Delta G < 0. \tag{4-191}$$

This relation introduces a new state function

$$G = E + pV - TS, \tag{4-192}$$

the so-called Gibbs free energy. We therefore have the statement: *In an open composite system held at constant temperature and pressure, the release of an internal constraint shifts the equilibrium in such a way as to reduce the value of the Gibbs free energy of the system. Such an open system has therefore, in equilibrium, the smallest value of the Gibbs free energy consistent with existing internal constraints.*

Again we prepare the ground for the application of this principle by looking at the properties of the Gibbs free energy G for a simple system. In particular, we will see that G is most naturally viewed as a function of the variables T, p, and the mole number n (or particle number N). The reason for this, again, is the fact that these quantities specify the state of this open system, just as T, V, and N did in the case of the Helmholtz free energy. We may express the dependence of G on T, p, and N as follows.

From the definition of G, we have

$$\Delta G = \Delta E + \Delta(pV) - \Delta(TS),$$

and using (4-145) for ΔE, we can write this as

$$\Delta G = T\Delta S - p\Delta V + \mu\Delta N + (p\Delta V + V\Delta p) - (T\Delta S + S\Delta T)$$

or

$$\Delta G = -S\Delta T + V\Delta p + \mu\Delta N. \tag{4-193}$$

We see that G, viewed as a function of N, T, and p, has the following partial derivatives

$$\left(\frac{\partial G}{\partial T}\right)_{p,N} = -S, \tag{4-194a}$$

$$\left(\frac{\partial G}{\partial p}\right)_{T,N} = +V, \tag{4-194b}$$

$$\left(\frac{\partial G}{\partial N}\right)_{T,p} = \mu. \tag{4-194c}$$

These equations are the analogies of the equations (4-177a, b, c) for the Helmholtz free energy.

Before going into a discussion of applications, we want to conclude this subsection by drawing attention to the importance of the almost trivial fact that quantities as energy, entropy, volume, and the free energies F and G are what we call *extensive* quantities. This means, for fixed values of pressure, temperature, density, their value is *proportional* to the amount of material in the system, as expressed by the mole number n or the particle number $N = N_0 n$.

Using the Gibbs free energy as an illustration, this means that $G(n, T, p)$ must be proportional to n:

$$G(n, T, p) = ng(T, p),\qquad(4\text{-}195)$$

that is, G is n times a function of T and p alone. The meaning of $g(T, p)$ is then that of the *molar free* energy. Similarly, we have

$$\text{for the entropy:}\quad S(n, T, p) = ns(T, p),\qquad(4\text{-}196)$$
$$\text{for the volume:}\quad V(n, T, p) = nv(T, p).\qquad(4\text{-}197)$$

(This latter relation was already introduced in Section 4.3.A.) $s(T, p)$ is the molar entropy, and v is the molar volume.

Applying these relations to (4-194a, b, c), we find

$$\left(\frac{\partial g(T, p)}{\partial T}\right)_p = -s(T, p),\qquad(4\text{-}198a)$$

$$\left(\frac{\partial g(T, p)}{\partial p}\right)_T = +v(T, p),\qquad(4\text{-}198b)$$

and

$$g(T, p) = N_0\mu(T, p).\qquad(4\text{-}198c)$$

This last equation shows that the chemical potential of a simple system is just the molar Gibbs free energy, divided by Avogadro's number N_0. We may say it is the Gibbs free energy per molecule.

We may observe that chemists generally would call the Gibbs free energy per mole $(N_0\mu)$ the chemical potential, whereas μ, the Gibbs free energy per molecule, is used by physicists. Chemists think in terms of mole numbers, physicists in terms of particle numbers. We will use in *what follows* the term chemical potential of the species "i," either the physicist's definition

$$\mu_i = T\left(\frac{\partial S}{\partial N_i}\right)_{E,V,N_j} = \left(\frac{\partial G}{\partial N_i}\right)_{T,p,N_j},\qquad(4\text{-}199)$$

or the chemist's definition in which S and G are viewed as functions of the mole numbers $n_i = N_i/N_0$:

$$\mu_i = T\left(\frac{\partial S}{\partial n_i}\right)_{E,V,n_j} = \left(\frac{\partial G}{\partial n_i}\right)_{T,p,n_j}. \tag{4-200}$$

It will always be clear from the context what definitions we are actually using.

4.4 Applications of the Equilibrium Conditions to Problems in Physics, Chemistry, and Biology

In this section we will discuss a variety of equilibrium problems. Some of these examples will illustrate the power of the purely macroscopic approach to equilibrium. Others will demonstrate how a combination of the microscopic and macroscopic views can be very fruitful.

4.4.A. Equilibrium Between Phases. The Clausius–Clapeyron Equation

Most pure substances, provided they are chemically stable under a sufficient range of temperatures and pressures, may occur in either of the three phases—solid, liquid, or gaseous. Each phase represents the stable state in a certain domain of the pressure temperature plane. (See Figure 4.44.)

These domains are bounded by the coexistence curves. For values of p and T lying on such a curve, the two adjacent phases can coexist in equilibrium. The curve separating the liquid phase from the gas phase thus represents the vapor pressure curve of the liquid, and the curve separating the solid phase from the gas phase represents the vapor pressure curve of the solid. Finally, the curve separating solid from liquid represents the relation between melting temperatures and pressures.

We now propose to describe these curves. In particular, we will show that their slope (dp/dT) can be obtained from readily available, apparently independent data, namely, the latent heat and volume change associated with the change of phase.

For each value of p and T (i.e., for each point in the p–T plane), the equilibrium of the system is characterized by the minimum value of the Gibbs free energy G. Let n_s, n_l, and n_v stand for the mole numbers of material in the solid, liquid, and gas phases, respectively, and let $g_s(p, T)$, $g_l(p, T)$, and $g_v(p, T)$ stand for the

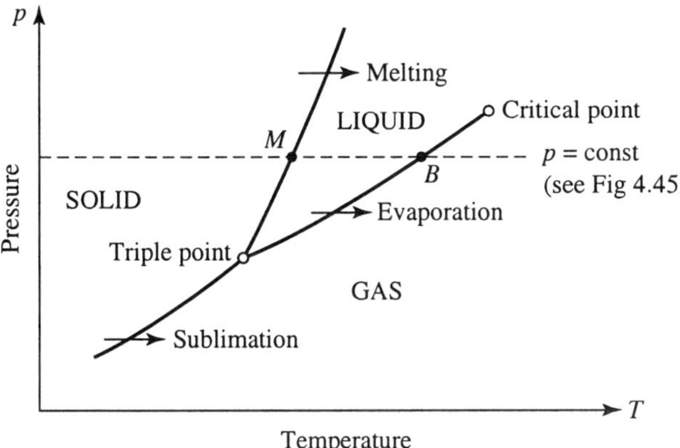

Figure 4.44. A p–T plane showing domains of stability of the solid, liquid, and gas phases. The boundaries of these domains are the coexistence curves for the two phases on each side.

corresponding *molar* free energies. Then the total free energy G is given by the expression

$$G(n_s, n_l, n_v, p, T) = n_s g_s(p, T) + n_l g_l(p, T) + n_v g_v(p, t), \qquad (4\text{-}201)$$

that is, it is the sum of the molar free energies of each phase, each multiplied with the appropriate mole number. The total amount of material is constant

$$n_s + n_l + n_v = n = \text{constant}.$$

In general, the three molar free energies g have different values. In Figure 4.45, we plot schematically the values of g_s, g_l, g_v for the points on the dashed line in Figure 4.44 (constant pressure, variable temperature).

We see that for $T < T_M(p)$, $g_s < g_l < g_v$. In this domain, the minimum of G is obtained when *all* the material is in the solid phase

$$G_{\min} = n g_s, \qquad \Delta T < T_M(p).$$

Similarly, for T in the range

$$T_M(p) < T < T_B(p),$$

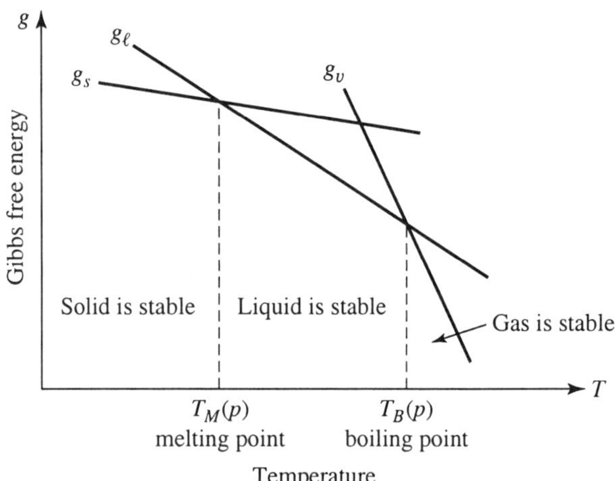

Figure 4.45. A schematic plot of the molar free energies g (or chemical potentials μ) for the solid, liquid, and gas phases at a fixed value of p. The phase with the lowest value of g represents stable equilibrium.

that is, for temperatures between melting and boiling points, the liquid phase is stable, and

$$G_{\min} = ng_l, \qquad T_M(p) < T < T_B(p),$$

and finally, for $T > T_B(p)$ the gas phase is stable. The melting point $T_M(p)$ and the boiling point $T_B(p)$ are marked in Figure 4.44 as the points M and B, respectively.

If we were to construct a series of graphs like that shown in Figure 4.45 for different pressures, then the values of T_M or T_B as a function of p are the coexistence curves shown in Figure 4.44. This implies, for instance, that along the vapor pressure curve of the liquid, $p = p_v(T)$, we have equality of the molar free energies g_l and g_v for liquid and vapor

$$g_l(T, p_v(T)) = g_v(T, p_v(T)). \qquad (4\text{-}202)$$

This fundamental equation determines the equilibrium vapor pressure curve

$$p = p_v(T).$$

The manner in which we use it depends very much on the information we possess on the value of the molar free energies g_l and g_v. In a purely empirical, macroscopic approach to the equilibrium problem, based on the equations of state of the materials involved, we can obtain *at best* the *difference* between the values of $g_l(T, p)$ at two different points in the T–p plane; the same holds for g_v. We can therefore only make statements based on that limited knowledge. Figure 4.46 illustrates how this can be done.

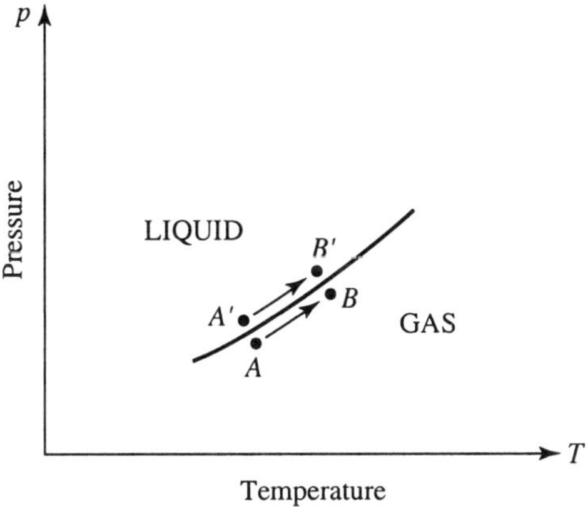

Figure 4.46. Liquid–gas coexistence curve. The liquid–gas equilibrium condition, (4-202), requires that $g_l(B') - g_l(A') = g_v(B) - g_v(A)$.

The condition that $g_l = g_v$ along the coexistence curve implies that if we move along the curve from the pair of adjacent points (A, A') to the pair of adjacent points (B, B'), both g_l and g_v must remain equal to each other, and hence individually change by the same amount

$$\Delta g_l = g_l(B') - g_l(A')$$
$$= \Delta g_v = g_v(B) - g_v(A). \tag{4-203}$$

At this point we can use the equations [(4-198a, b)] to express the changes Δg_l and Δg_v of the free energies

$$\Delta g_l = -s_l \Delta T + v_l \Delta p, \tag{4-204a}$$
$$\Delta g_v = -s_v \Delta T + v_v \Delta p, \tag{4-204b}$$

where s_l, s_v and v_l, v_v are the molar entropies and molar volumes of the liquid and vapor, respectively. According to (4-203), these two changes must be the same if we move along the coexistence curve $p = p_v(T)$:

$$(s_v - s_l)\Delta T - (v_v - v_l)\Delta p = 0,$$

which we can write as

$$\left(\frac{\Delta p}{\Delta T}\right)_{\text{along curve}} = \frac{(s_v - s_l)}{(v_v - v_l)}. \tag{4-205}$$

The left-hand side of this equation, in the limit of very small Δp, ΔT, is just the derivative of the function $p_v(T)$:

$$\left(\frac{\Delta p}{\Delta T}\right)_{\text{along curve}} \rightarrow \left(\frac{dp_v(T)}{dT}\right),$$

so that our result can be written as a differential equation for the vapor pressure curve $p = p_v(T)$:

$$\frac{dp_v(T)}{dT} = \left(\frac{s_v - s_l}{v_v - v_l}\right). \tag{4-206}$$

This fundamental equation is known as the Clausius–Clapeyron equation. It states that the rate of change of the equilibrium vapor pressure of a liquid with temperature is given by the ratio of the molar entropy difference between liquid and vapor, and the corresponding difference between molar volumes. Both these quantities are measurable. The entropy difference $(s_v - s_l)$ is simply $1/T$ times the heat of vaporization ΔQ_{vap} of 1 mole of liquid into 1 mole of gas at the same pressure and temperature

$$(s_v - s_l) = \frac{1}{T}\Delta Q_{\text{vap}}. \tag{4-207}$$

That this is so follows from the definition of the entropy difference between points A and A' of Figure 4.46:

$$S_A - S_{A'} = \int_{A'}^{A} \frac{dQ}{T} = \frac{\Delta Q_{\text{vap}}}{T}.$$

Since, in general, the molar volume v_v of the gas is very much larger than the molar volume of the liquid, we can write v_v for $v_v - v_l$. In a further approximation we may view the vapor as an ideal gas so that

$$v_v \cong RT/p_v(T). \tag{4-208}$$

Inserting (4-207) and (4-208) into the Clausius–Clapeyron equation, we obtain the approximate relation

$$\frac{dp_v}{dT} \cong p_v \cdot \frac{\Delta Q_{\text{vap}}}{RT^2}. \tag{4-209}$$

We can solve this equation by using the further approximation that ΔQ_{vap}, the molar heat of vaporization, is a constant, independent of temperature. This is not exactly true, as the data on the water–steam equilibrium presented in Tables 4.4 and 4.5 show.

The first column lists temperatures, in degrees Centigrade, from room temperature to 229 °C; the second column lists the corresponding Kelvin temperature. The third column gives the equilibrium vapor pressures in atmospheres. The most striking feature of Table 4.4 is to be found in the third column. The range of steam pressures is a spectacular 640:1, even though the absolute temperature changes only by a factor of 1.6. This dramatic effect is produced by the Boltzmann factor. We can show this explicitly by first approximating ΔQ_{vap} by a constant. This allows us to simply integrate (4-209). First, we write it as

$$\frac{dp_v}{p_v} = d(\ln p_v) = \frac{\Delta Q_{\text{vap}}}{R}\frac{dT}{T^2} = -\frac{\Delta Q_{\text{vap}}}{R}d\left(\frac{1}{T}\right) = -d\left(\frac{\Delta Q_{\text{vap}}}{RT}\right),$$

Table 4.4. Water–Steam Equilibrium at High Temperatures.

t (°C)	T (°K)	$p_v(T)$ (atm.)	$\Delta Q_{\text{vap}}/M$ (cal/g)	ΔQ_{vap} (cal/mol)
27	300	.036	581.2	10,462
77	350	.427	552.9	9,952
127	400	2.51	520.7	9,372
177	450	9.53	483.1	8,696
220	493	23.04	446.0	8,028

Table 4.5. Equilibrium Vapor Pressure of Water in the Physiological Temperature Range.

T (°C)	T (°F)	T (°K)	$p_v(T)$ (mmHg)	$\Delta Q_{vap}/M$ (cal/g)	$dp_v/p_v\,dT$ (°K)$^{-1}$	$\Delta Q_{vap}/RT^2$ (°K)$^{-1}$
(1)	(2)	(3)	(4)	(5)	(6)	(7)
10	50	283	9.205	590.2		
12	53.6	285	10.513	589.1	0.0660	0.0657
14	57.2	287	11.98	588.1	0.0649	0.0647
16	60.8	289	13.624	587.0	0.0638	0.0637
18	64.4	291	15.46	585.9	0.0628	0.0627
20	68	293	17.51	584.9	0.0618	0.0617
22	71.6	295	19.79	583.9	0.0608	0.0608
24	75.2	297	22.32	582.8	0.0598	0.0599
26	78.8	299	25.13	581.8	0.0590	0.0590
28	82.4	801	28.25	580.7	0.0582	0.0581
30	86	303	31.71	579.6	0.0574	0.0572
32	89.6	305	35.53	578.6	0.0566	0.0564
34	93.2	307	39.75	577.4	0.0558	0.0555
36	96.8	309	44.40	576.4	0.0550	0.0547
38	100.4	311	49.51	575.3		

from which we infer that

$$\ln p_v(T) \cong \text{constant} - \frac{\Delta Q_{vap}}{RT}$$

or

$$p_v(T) \cong \text{constant}\, e^{-\Delta Q_{vap}/RT}. \tag{4-210}$$

This approximate result displays the characteristic exponential temperature dependence that we had found in the discussion of the vapor pressure of the crystalline solid in Section 4.2.C. In (4-210), the Boltzmann factor appears in a macroscopic guise; the exponent is the ratio of a molar energy difference ΔQ_{vap} to the thermal energy per mole $RT = N_0 kT$. Notice that the macroscopic approach, based on the Clausius–Clapeyron equation, cannot predict the absolute value of the vapor pressure; the constant in (4-210) must be taken from the empirical data. By contrast, the statistical, microscopic approach does in principle permit a prediction of the absolute value of vapor pressure. However, it can do this successfully only for crystalline solids. In the case of a liquid one does not yet know how to determine all the microstates and their corresponding energies.

But let us see whether the *data* on the water–steam equilibrium even support the validity of the thermodynamic *prediction* as expressed by the Clausius–Clapeyron equation (4-209). To check this, we present, in Table 4.5, data on the equilibrium vapor pressure of water, with a finer mesh of data points, so as to be able to determine derivatives dp_v/dT from the data. For added interest we choose a physiological range of temperatures. (The corresponding vapor pressures play a role in the physiology of mammalian respiration!)

In Table 4.5 the first three columns on the left list the temperatures, in degrees Centigrade, Fahrenheit, and Kelvin; the fourth column lists the equilibrium pressure $p_v(T)$, and the fifth column the heat of vaporization per gram in cal/g. Multiplying this entry with the molecular weight, 18 g/mol of water, we find ΔQ_{vap} in cal/mol.

The two last entries on the right (columns (6) and (7)) contain the data designed to *test* the validity of the thermodynamic formula (4-209). The first entry lists the values $dp_v/p_v\,dT$ compiled from the data in columns (3) and (4) by means of the approximate relation

$$\frac{dp_v}{p_v\,dT} \simeq \frac{p_v(T+2) - p_v(T-2)}{4p_v(T)}.$$

According to formula (4-209), this should be equal to $\Delta Q_{\text{vap}}/RT^2$. This latter quantity is presented in the last column (7) of Table 4.5. It is calculated from the data given in columns (3) and (5). We see that there is indeed an excellent agreement between the data and the theory, the latter represented here by the approximate Clausius–Clapeyron formula (4-209). Thermodynamics works!

While we cannot predict $p_v(T)$ in the absence of a microscopic picture of the substance, we can relate changes in p_v to other macroscopically measurable quantities. Thermodynamics thus provides the *relationships* between separate and apparently disconnected macroscopic data. The usefulness of the thermodynamic approach lies in the fact that it provides such relationships.

4.4.B.   Dilute Solutions of Nonelectrolytes

A solution is a mixture of two or more components. The "main" component, also called the *solvent*, is usually assumed to be liquid. The other components are called *solutes*. They occur in the solution as free molecules. In their pure form, they may be either a solid, or liquid, or gas. The distinction between solute and solvent is conceptually somewhat arbitrary, but practically useful.

Solutions are composite systems that lend themselves to the study of many important and interesting equilibrium problems. Let us list a few of them:

(a) The equilibrium between a solid (crystalline) solute and the solvent liquid. This equilibrium represents the *saturated* solution. The saturation *concentration* of solute, as a function of pressure and temperature, defines the *solubility* of the solute.

(b) If the solvent is a volatile liquid (like water, alcohol, etc.), its vapor pressure depends on the solute concentration. We can relate the change of vapor pressure to the solute concentration quantitatively by using the equilibrium conditions.

(c) A problem of physiological importance is the solubility of gases in a fluid. The equilibrium in this case is between the pure gas in the gas phase and the gas dissolved in the solution. We will investigate how the saturation concentration of gas in the solution depends on pressure and temperature.

We will study these equilibria by means of the free energy minimum principle. Our first task is therefore to find an expression for the free energy of the solution in terms of pressure, temperature, and the mole numbers of the components. Once this is done, the free energy minimum condition provides the means to determine the equilibria between solution and solvent vapor (vapor pressure), between solution and pure solvent (osmotic pressure), and between solution and solid solute (solubility).

(i)   The Gibbs Free Energy of a Dilute Solution.
The Concept of the Ideal Solution

The Gibbs free energy for any system, occupying a volume V with all components at a common temperature T and pressure p, is defined by the relation [see (4-192)]:

$$G(T, p, n_1, n_2, \ldots) = E + pV - TS. \tag{4-211}$$

To specify the solution, which is our system in this case, we must introduce the mole numbers n_i of the various components. We will adhere to the convention that n_1 *always* represents the mole number of *solvent*, and $n_2, n_3, \ldots$ the mole number of various *solutes*. The symbol n:

$$n = n_1 + n_2 + n_3 + \cdots, \tag{4-212}$$

stands for the total mole number of particles of the solution, and the quantities

$$\left(\frac{n_1}{n}\right), \quad \left(\frac{n_2}{n}\right), \text{ etc.,}$$

represent the *mole fractions* of solvent and solute particles, respectively.

In what follows, we will restrict the discussion to binary solutions with only one solute species. For this case, we will now construct the expressions for the various items in (4-211) for G.

We start with the volume V. At first, one might think that V is just the sum of the volumes of the pure constituents of the solution, at the same pressure and temperature. This is not at all true. 1 l of water, at $t = 15\,°C$ and $p = 760$ mmHg, can dissolve 1 l of carbon dioxide; yet the volume of the solution is not 2 l but remains very close to 1 l: The volumes are not additive. Mixing 0.5 l of water with 0.5 l of alcohol does not produce 1 l of dilute alcohol, but considerably less. Volumes are *not* additive. Nevertheless, we can make use of the *extensive* property of the volume to obtain a functional relationship between the volume and the mole numbers of the constituent. The extensive property of V implies that for a fixed ratio of the mole numbers of solute and solvent, as expressed by the ratio

$$x = \left(\frac{n_2}{n_1}\right), \tag{4-212a}$$

the volume of the solution is proportional to the amount of solvent present, as measured by the mole number n_1. The mathematical expression of this statement is

$$V(T, p, n_1, n_2) = n_1 v(T, p, x). \tag{4-213}$$

(As a check, let us double n_1 and n_2, keeping p and T the same; this should double the volume. Indeed, by (4-213), and keeping in mind that $2n_2/2n_1 = n_2/n_1 = x$:

$$V(T, p, 2n_1, 2n_2) = 2n_1 v(T, p, x) = 2V(T, p, n_1, n_2).)$$

We now make the assumption that we are dealing with a *dilute* solution only. In this case, the parameter x, (4-212), is small: $x \ll 1$. We can therefore write in (4-213):

$$v(T, p; x) = v(T, p; 0) + x \left(\frac{\partial v}{\partial x}\right)_{x=0} + \text{ terms of order } x^2$$

$$\equiv v_1(T, p) + x\bar{v}_2(T, p) + O(x^2). \tag{4-214}$$

The quantity

$$v_1(T, p) = v(T, p, x = 0) \tag{4-215a}$$

represents the molar volume of the pure solvent; the quantity

$$\bar{v}_2(T, p) = \left(\frac{\partial v(T, p, x)}{\partial x}\right)_{x=0} \tag{4-215b}$$

is defined as the *partial molar volume* of the solute at zero concentration. Its physical meaning is that of the *rate of change* of the volume of the solution with increasing mole fraction x of solute for 1 mol of solvent. With (4-215a, b), and using $x = n_2/n_1$, we can now write for the volume of the dilute solution

$$V(T, p, n_1, n_2) \cong n_1 v_1(T, p) + n_2\bar{v}_2(T, p), \tag{4-216}$$
$$(n_2 \ll n_1).$$

An analogous argument can be made for the energy. From the extensive property of the energy E, it follows that we can express $E(T, p, n_1, n_2)$ as

$$n_1 e(T, p, x),$$

and for $x \ll 1$, this can be approximated by

$$E(T, p, n_1, n_2) \cong n_1 e_1(T, p) + n_2\bar{e}_2(T, p), \tag{4-217}$$

where $e_1(T, p) = e(T, p, x = 0)$ is the molar energy of the pure solvent; and

$$\bar{e}_2(T, p) = \left(\frac{\partial e(t, p, x)}{\partial x}\right)_{x=0}$$

is the partial molar energy of the solute.

For the entropy S, one might at first glance again expect an expression analogous to those just given for the volume and energy

$$S(T, p, n_1, n_2) = n_1 s(T, p, x) = n_1 s_1(T, p) + n_2 \bar{s}_2(T, p), \qquad (4\text{-}218a)$$

where

$$s_1(T, p) = s(T, p, x = 0)$$

and

$$\bar{s}_2(T, p) = \left(\frac{\partial s(T, p, x)}{\partial x} \right)_{x=0}$$

However, this expression for S is incomplete. The reason for this lies in the entropy of mixing that we considered in Section 4.3.B(iii). Now, many equilibrium phenomena involving solutions indicate that the entropy of mixing the solvent and solute has approximately the same structure as that for mixing two gases, and is given by the expression [see (4-168b)]:

$$\Delta S_{\mathrm{mix}} = +n_1 R \ln \left(\frac{n_1 + n_2}{n_1} \right) + n_2 R \ln \left(\frac{n_1 + n_2}{n_2} \right). \qquad (4\text{-}218b)$$

The correct expression for the entropy of the solution is obtained by adding this expression for ΔS_{mix} to the right-hand side of (4-218a). This gives

$$S = n_1 s_1(T, p) + n_2 \bar{s}_2(T, p) + \Delta S_{\mathrm{mix}}. \qquad (4\text{-}218)$$

Solutions for which (4-218) holds are called *ideal solutions*. Many dilute solutions of nonelectrolytes approximate ideal behavior. On the other hand, even dilute solutions of electrolytes generally deviate significantly from ideal behavior. All our subsequent discussions will be based on the approximation of an ideal, dilute, binary solution.

Since we deal with dilute solutions, for which $n_2 \ll n_1$, we can simplify (4-218b) by writing

$$\ln \left(\frac{n_1 + n_2}{n_2} \right) \cong \ln \left(\frac{n_1}{n_2} \right) = \ln \left(\frac{1}{x} \right) = - \ln x$$

and

$$\ln\left(\frac{n_1 + n_2}{n_1}\right) = \ln\left(1 + \frac{n_2}{n_1}\right) = \ln(1 + x) \cong x.$$

Thus

$$\Delta S_{\text{mix}} \cong n_1 R x - n_2 R \ln x. \tag{4-219}$$

Thus, the entropy of the solution may be written in the form

$$S = n_1 s_1(T, p) + n_2 \bar{s}_2(T, p) + n_1 R x - n_2 R \ln x. \tag{4-220}$$

If we now use the forms for E, V, and S from (4-216), (4-217), (4-219), and (4-220) in (4-211) for the Gibbs free energy, we find the following result for an ideal, dilute ($n_2 \ll n_1$) solution

$$G(T, p, n_1, n_2) = n_1\{g_1(T, p) - RTx\} + n_2\{\bar{g}_2(T, p) + RT \ln x\}. \tag{4-221a}$$

In this equation $g_1(T, p)$ and $\bar{g}_2(T, p)$ are given by

$$g_1 = (e_1 + pv_1 - Ts_1) \tag{4-221b}$$

$$g_2 = \left(\frac{\partial e(T, p, x)}{\partial x}\right)_{x=0} + p\left(\frac{\partial v(T, p, x)}{\partial x}\right)_{x=0} - T\left(\frac{\partial s(T, p, x)}{\partial x}\right)_{x=0}. \tag{4-221c}$$

It may be noted that the principal result [(4-221a)] given above for $G(T, p, n_1, n_2)$ represents the Gibbs free energy in terms of the number of moles n_1 and n_2 of solvent and solute, respectively. Alternatively, one can regard G as being expressed in terms of the total number $N_1 = N_0 n_1$ of solvent molecules, and the total number $N_2 = N_0 n_2$ of solute molecules. Here N_0 is Avogadro's number. In this case, we can denote G as $G(T, p, N_1, N_2)$.

(ii) Connection Between the Gibbs Free Energy and the Chemical Potentials of Each Species in a Multicomponent Solution

In the case of a single component system, we previously introduced [(4-194c)] the concept of the chemical potential $\hat{\mu}$ per particle as the change in G associated with the addition of one more particle into the system, viz:

$$\left(\frac{\partial G(T, p, N)}{\partial N}\right)_{T,p} \equiv \hat{\mu}.$$

In our present discussion, we prefer to regard G as a function of the mole number $n = (N/N_0)$. Thus, the corresponding chemical potential μ:

$$\mu = \left(\frac{\partial G(T, p, n)}{\partial n}\right) = N_0\hat{\mu},$$

provides the molar chemical potential that is N_0 times larger than that of the single particle chemical potential $\hat{\mu}$.

We may extend the definition of the molar chemical potential μ to the case of a multicomponent solution. Indeed, let the solution contain r distinct chemical components each designated by the symbol i. Let the mole number of each component be designated as n_i, $i = 1, \ldots, r$. Let $i = 1$ correspond to the solvent species. The Gibbs free energy of such a multicomponent solution has the form

$$G = G(T, p, n_1, \ldots, n_i, \ldots, n_r).$$

We now define the corresponding chemical potential for a mole of each of the components as

$$\left(\frac{\partial G(T, p, n_1, \ldots, n_i, \ldots)}{\partial n_i}\right)_{T,p,n_{j\neq i}} = \mu_i. \tag{4-222a}$$

With this definition we can show quite generally that the Gibbs free energy may be written in the following very simple form

$$G(T, p, n_1, \ldots, n_i, \ldots) = \sum_{i=1}^{r} n_i \mu_i. \tag{4-222b}$$

This result follows directly from the extensive property of G. This property can be expressed by the condition

$$G(T, p, \lambda n_1, \ldots, \lambda n_i, \ldots) = \lambda G(T, p, n_1, \ldots, n_i, \ldots), \tag{4-223}$$

where λ is an arbitrary multiplier. If we now choose $\lambda = 1 + \varepsilon$, we have

$$G(T, p, \lambda n_1, \ldots, \lambda n_i, \ldots) = G(T, p, (1 + \varepsilon)n_1, \ldots, (1 + \varepsilon)n_i, \ldots).$$

Now let us choose ε as a small number ($\varepsilon \ll 1$). If we then expand the right-hand side of this equation, we have

$$G(T, p, \lambda n_1, \ldots, \lambda n_i, \ldots) = G(T, p, n_1, \ldots, n_i, \ldots) + \sum_{j=1}^{r}(\varepsilon n_j)\left(\frac{\partial G}{\partial n_j}\right)_{T,p,n_k}.$$

Note now that according to the extensive property of G as expressed in (4-223) the left-hand side of this equation is equal to $(1 + \varepsilon)G(T, p, n_1, \ldots, n_i, \ldots)$. If we also note that $(\partial G/\partial n_j)_{T,p,n_k} = \mu_j$, we see that

$$(1 + \varepsilon)G(T, p, n_1, \ldots, n_i, \ldots) = G(T, p, n_1, \ldots, n_i) + \varepsilon \sum_{j=1}^{r}n_j\mu_j. \quad (4\text{-}224)$$

Thus, (4-222b) is proven. Note that if we used, instead of the mole numbers $n_1 \ldots n_r$, the actual number of particles of each species $N_1 \ldots N_r$, we would have obtained

$$G(T, p, N_1, \ldots, N_i, \ldots, N_r) = \sum_{j=1}^{r} N_i\hat{\mu}_i, \quad (4\text{-}225)$$

where $\hat{\mu}_i$ is the chemical potential per particle of species i. Note, furthermore, that since the chemical potentials are intensive quantities, they can depend only on the mole fractions

$$x_i = n_i \bigg/ \sum_{i=1}^{r} n_i. \quad (4\text{-}226)$$

Thus,

$$\mu_i = \mu_i(T, p, x_1, \ldots, x_i, \ldots). \quad (4\text{-}227)$$

(iii)   Expressions for the Chemical Potentials of Solvent, and Solutes in a Dilute, Ideal Solution

We may return to (4-221a), which is the form of the Gibbs free energy in the case of an ideal dilute solution. We see that for a solvent (1) and a single solute (2) the Gibbs free energy $G(T, p, n_1, n_2)$ has the form

$$\mu_1 = \mu_1^0 - RTx \qquad \text{(solvent)}, \qquad\qquad \text{(4-228a)}$$

$$\mu_2 = \mu_2^0 + RT \ln x \qquad \text{(solute)}, \qquad\qquad \text{(4-228b)}$$

where $\mu_1^0 = g_1$ and $\mu_2^0 = \overline{g}_2$, as defined in (4-221b). In (4-228a) the term RTx represents the lowering of the chemical potential of the solvent by the presence of solute molecules. This is at the basis of the phenomenon of osmotic pressure, that we will revisit shortly in Section 4.4.B(v) below.

In the case that the solution contains r different solute species ($i = 2, 3, \ldots, r$), each present with mole fraction x_i, it is easily seen that the analysis given above gives the following forms for the chemical potentials of solvent (1) and solute $(2, 3, \ldots, r)$ molecules in the case of dilute, ideal solutions

$$\mu_1 = \mu_1^0 - RT \sum_{i=2}^{r} x_i \qquad \text{(solvent)}, \qquad\qquad \text{(4-229a)}$$

$$\mu_i = \mu_i^0 + RT \ln x_i \qquad \text{(solute)}. \qquad\qquad \text{(4-229b)}$$

Here, the so-called standard part of the chemical potentials μ_1^0 and μ_i^0 are given by (4-221b, c), viz:

$$\mu_1^0 = \varepsilon_1 + pv_1 - Ts_1 = g_1,$$

and

$$\mu_i^0 = \left(\frac{\partial e}{\partial x_i} \right)_{x_j, x_i=0} + p \left(\frac{\partial v}{\partial x_i} \right)_{x_j, x_i=0} - T \left(\frac{\partial s}{\partial x_i} \right)_{x_j, x_i=0} \equiv \overline{g}_i.$$

Of course, if the Gibbs free energy is expressed in terms of the actual number of solvent and solute molecules, the corresponding chemical potentials are those for a single particle. $R = N_0 k$ becomes simply k, Boltzmann's constant, in Eqns. (4-229a) and (4-229b).

(iv) Raoult's Law; the Lowering of the Solvent Vapor Pressure by the Presence of Solute. Elevation of Boiling Point

We consider here the case of a solution in which the solvent is "volatile"; that is, has a nonnegligible vapor pressure. Water, alcohol, and organic solvents, such as ether, carbon tetrachloride, etc., are good examples. If a nonvolatile (by comparison) solute is dissolved in the solvent, experience shows that the equilibrium vapor pressure of the solvent is lowered. Another way of stating the same effect is to observe that, at a fixed pressure, the boiling point temperature of the solution is higher than that of the pure solvent. These phenomena can be understood as a simple extension of the previous analysis of the liquid–vapor equilibrium (Section 4.4.A). In the present case we must, however, employ the expression (4-221) for the Gibbs free energy of the solution since the fluid is here a mixture of two components: The solution contains n_1 moles of solvent and n_2 moles of solute ($n_2 \ll n_1$). The system under consideration consists of this solution and the solvent vapor above it. We will denote the number of moles of solvent vapor by n_v. The free energy of this total system is given by the sum of free energies of the vapor ($n_v g_v(p, T)$), and of the solution, viz:

$$G = n_v g_v(p, T) + n_1 g_1(T, p) + n_2 \bar{g}_2(p, T)$$

$$+ n_2 RT \left\{ \ln\left(\frac{n_2}{n_1}\right) - 1 \right\}. \tag{4-230}$$

At the equilibrium point between vapor and solution, G must be stationary (unchanged) with respect to an infinitesimal transfer Δn of solvent from the liquid to the vapor phase, or vice versa:

$$\Delta G = G(T, p, n_v + \Delta n, n_1 - \Delta n, n_2) - G(T, p, n_v, n_1, n_2)$$

$$= \Delta n \left\{ \left(\frac{\partial G}{\partial n_v}\right)_{T,p,n_1,n_2} - \left(\frac{\partial G}{\partial n_1}\right)_{T,p,n_v,n_2} \right\} = 0.$$

This requires that the two derivatives be equal. Now

$$\left(\frac{\partial G}{\partial n_v}\right)_{T,p,n,n} = g_v(T, p) \tag{4-231a}$$

is the molar free energy (chemical potential) of the solute vapor. The n_1-derivative is found from (4-230) as

$$\left(\frac{\partial G}{\partial n_1}\right)_{T,p,n_v,n_2} = g_1(T, p) - RT\left(\frac{n_2}{n_1}\right), \tag{4-231b}$$

since $\partial/\partial n_1 \ln(n_2/n_1) = -1/n_1$. Equation (4-231b) represents the chemical potential of the *solvent in the solution*. We see that it is smaller, by $RT(n_2/n_1) = RTx$, than the chemical potential $g_1(T, p)$ of the pure solvent. (The physical reason for this is that the *entropy* of the solution is increased by the entropy of mixing!)

Equating the chemical potentials (4-231a, b), we get an equation between p and T, which defines the *coexistence* curve between solution and vapor, $p = p_v(T)$:

$$g_v(T, p_v(T)) = g_1(T, p_v(T)) - RT\left(\frac{n_2}{n_1}\right). \tag{4-231c}$$

This equation is thus the generalization of the previously considered equation [(4-202)] that describes the condition of equilibrium between a pure liquid and its vapor. The quantity $g_1(T, p)$ of (4-231c) is the same as $g_l(T, p)$ of (4-202), namely, the molar free energy of pure solvent. We now proceed as follows. Let the vapor pressure curve for the pure solvent be represented by $p = p_v^0(T)$; that is,

$$g_v(T, p_v^0(T)) = g_1(T, p_v^0(T)). \tag{4-232}$$

Equation (4-232) leads to the differential equation for $p_v^0(T)$ as we discussed previously, and we need not go into that again. We want the difference $\delta p = p_v(T) - p_v^0(T)$ of the equilibrium pressures for solution and pure solvent at a given temperature T. To find this, we write $p_v(T) = p_v^0(T) + \delta p$ in (4-231c):

$$g_v(T, p_v^0(T) + \delta p) = g_1(T, p_v^0(T) + \delta p) - RT\left(\frac{n_2}{n_1}\right).$$

Anticipating δp to be small, we expand

$$g_v(T, p_v^0 + \delta p) = g_v(T, p_v^0) + v_v \delta p$$

and

$$g_1(T, p_v^0 + \delta p) = g_1(T, p_v^0) + v_1 \delta p,$$

where we used the familiar derivative of g with respect to pressure [i.e., (4-198b)]: $(\partial g/\partial p)_T = +v$. Now, since (4-232) holds, inserting the above expansions into (4-224) leaves us with

$$(v_v - v_1)\,\delta p = RT\left(\frac{n_2}{n_1}\right). \qquad (4\text{-}233)$$

Again we may, within the accuracy of the approximation of the solution by an ideal solution, neglect v_1, the molar volume of liquid solute compared to the molar volume of solute vapor (v_v). If, in addition, we treat this vapor as an ideal gas, then

$$\frac{RT}{v_v} = p_v^0(T).$$

With these assumptions, (4-233) reduces to the extremely simple statement

$$\frac{\delta p(T)}{p_v^0(T)} = -\left(\frac{n_2}{n_1}\right), \qquad \left(\frac{n_2}{n_1}\right) \ll 1. \qquad (4\text{-}234)$$

In words: the fractional reduction of the equilibrium vapor pressure of a fluid due to the addition of solute is equal to the mole fraction of solute in the solution.

This is *Raoult's* law for dilute, ideal solutions.

We can easily derive from this result an expression for the elevation δT_B of the boiling point $T_B(p)$ of a fluid due to the presence of a small amount ($n_2/n_1 \ll 1$) of solute. Figure 4.47 will help to illustrate the reasoning involved. The word "boiling point" denotes that temperature at which the *partial* pressure of the vapor reaches a value equal to the total imposed pressure p. At this temperature the vapor begins to drive out all the foreign gas above the liquid, and creates an atmosphere of pure vapor.

The dotted curve in Figure 4.47 shows the vapor pressure curve for the solution, while the solid curve shows the vapor pressure for the pure solvent. The difference δp between the two curves at a given value of T is the depression of the vapor pressure described in Raoult's law. Associated with this, however, we see clearly that at fixed pressure the boiling temperature increases by δT_B. Since the two curves are parallel to one another to within a small error of order (n_2/n_1), we see that the ratio ($\delta p/\delta T_B$), the indicated slope of the dotted curve, is nearly equal to the indicated slope of the solid curve at point B^0; that is,

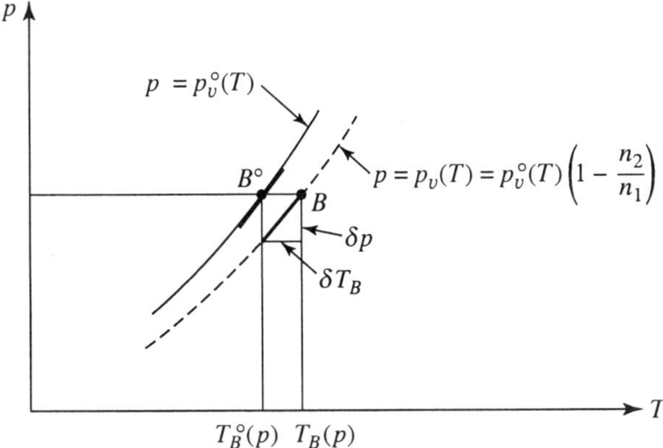

Figure 4.47. Vapor pressure curves $p = p_v^0(T)$ for the pure solvent, and $p = p_v(T)$ for the solution. δp is the reduction of the vapor pressure due to solute at temperature $T_B^0(p)$. δT_B is the elevation of the boiling point at fixed pressure p.

$$\left(\frac{\delta p}{\delta T_B}\right) = \left(\frac{dp_v^0(T)}{dT}\right)_{B_0}. \tag{4-235}$$

In the right-hand side we can use (4-209) for the slope of the vapor pressure curve (this formula applies to a pure solvent; the p_v of (4-209) is, therefore, the p_v^0 of the present discussion)

$$\left(\frac{dp_v^0}{dT}\right)_{B_0} = p_v^0 \frac{\Delta Q_{\text{vap}}}{R(T_B^0)^2}. \tag{4-236}$$

Combining this with (4-235), we have

$$\delta T_B \cong \left(\frac{\delta p}{p_v^0}\right) \frac{R(T_B^0)^2}{\Delta Q_{\text{vap}}}$$

and by using Raoult's law for $(\delta p / p_v^0)$, we find

$$\delta T_B \cong + \left(\frac{n_2}{n_1}\right) \frac{R(T_B^0)^2}{\Delta Q_{\text{vap}}}. \tag{4-237}$$

Let us now insert some numbers into (4-237) for the case of water as a solvent. We assume a total pressure of $p = 760$ mmHg, in which case the boiling point is $T_B = 100\,°\text{C} = 373\,°\text{K}$. The heat of vaporization at this temperature and pressure is $\Delta Q_{\text{vap}} = 539$ cal/g $= 9702$ cal/mol. Since $R = 2$ cal/mol $°\text{K}$, we have $RT_B^2 = 2.78 \times 10^5$ cal $°\text{K}$, and

$$\left(\frac{RT_B^2}{\Delta Q_{\text{vap}}}\right)_{\text{water}} = 28.6\,°\text{K}.$$

This means that the boiling point elevation due to a mole fraction (n_2/n_1) of solute is

$$\delta T_B \,(\text{water}) = \left(\frac{n_2}{n_1}\right) \times 28.6\,°\text{K}. \tag{4-238}$$

It is evident that this phenomenon in principle offers a means to determine molecular weights of solutes. (See problem assignments.) In this context, we may point out that the prediction of (4-238) for δT_B is borne out by the experimental data with remarkable accuracy. The data are generally presented in the form

$$\delta T_B = K_b m, \tag{4-239}$$

where m is the *molality* of the solution, defined as the number of gram-moles of solute per *unit mass of solvent*; that is, m is related to (n_2/n_1) by

$$m = \frac{n_2}{n_1 M_1}, \tag{4-240}$$

M_1 being the molecular weight of the solvent. The relation between boiling point elevation and molality m is then

$$\delta T_B = \frac{R(T_B^0)^2}{\Delta Q_{\text{vap}}} M_1 \cdot m = K_B m. \tag{4-241}$$

Molality m is generally listed in units of mol/kg, and hence the boiling point constant K_B in ($°$K g/mol). In Table 4.6 we list some solvents, with their values of T_B^0, ΔQ_{vap}, M_1, and the calculated values of the constant K_B, based on (4-241).

These calculated values are so accurate that physico–chemical tables generally present calculated values of K_B rather than experimental ones. In Table 4.7 we

Table 4.6. Boiling Point Constants K_B Calculated on the Basis of (4-241).

Solvent	Formula	Molecular weight M_1 (g)	$T_B^0(°C)$ at 760 mmHg	$\Delta Q_{vap}/M$ cal/g	K_B (calculated) $°K$ kg/mol
Water	H_2O	18	100	539.5	0.513
Benzene	C_6H_6	78.11	80.2	94.3	2.59
Carbon-tetrachloride	CCl_4	153.84	77.0	46.4	5.25
Ethylalcohol	$C_2H_5–OH$	46.07	78.3	204	1.22

Table 4.7. The ratio $\delta T_B/m$ of Boiling Point Elevation to Molality of a Solution of Sucrose in Water (from Landolt–Bornstein [5]).

m (mol/kg)	δT_B ($°K$)	$\delta T_B/m$ (kg deg/mol)
0.00162	0.000842	0.519
0.01298	0.00674	0.520
0.0318	0.0169	0.519
0.0658	0.0335	0.509
0.132	0.0676	0.513
0.322	0.164	0.509
0.634	0.317	0.50

give experimental values of the ratio $\delta T_B/m$ for a solution of sucrose ($C_{12}H_{22}O_{11}$) in water of molality m. This ratio, according to (4-241), should be independent of m (for dilute solutions) and equal to the constant K_B.

We see that the values of $\delta T_B/m$ for small concentrations are indeed very close to the predicted value of $K_B = 0.512$.

(v)   Osmotic Pressure Revisited. Van t'Hoff's Law

We gave a molecular description of osmotic pressure in Chapter 2, Section 2.6. In this subsection we now present an alternative description based on the macroscopic equilibrium conditions. This method offers less "insight" into the actual molecular mechanism responsible for osmotic pressure. On the other hand, it ties in the osmotic pressure effect with other phenomena, such as the raising of the boiling point of a solution, or lowering its freezing point. All three have their origin in the fact that the chemical potential of the solvent in the solution is lowered by the presence of the solute.

To investigate the conditions of osmotic equilibrium, consider a system with two compartments, separated by a semipermeable membrane. This membrane permits passage of the solvent (water, generally) but not the passage of solute molecules. (See Figure 4.48.)

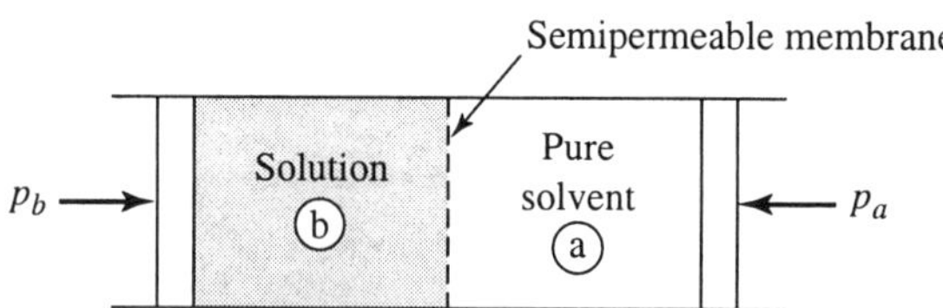

Figure 4.48. Osmotic pressure equilibrium. Compartments (a) and (b) are separated by a semipermeable membrane. Equilibrium requires $p_b - p_a = \Pi > 0$. Π is the osmotic pressure of the solution.

In this arrangement the two compartments can be kept at different pressures p_a and p_b. To establish the equilibrium conditions, we write the expression for the Gibbs free energy G of the system: G is the sum of the free energies of the two compartments. Let n_a, n_b stand for the mole numbers of *solvent* in the two compartments, and n_2 for the mole number of *solute* in compartment (b). From (4-230) we see that the Gibbs free energy of the system is

$$G = n_a g_1(T, p_a) + n_b g_1(T, p_b)$$
$$+ n_2 \bar{g}_2(T, p_b) + n_2 RT \left\{ \ln\left(\frac{n_2}{n_b}\right) - 1 \right\}. \qquad (4\text{-}242)$$

At equilibrium, the free energy is at its minimum and must be stationary with respect to a small transfer Δn of solvent from compartment (a) to (b) or vice versa:

$$\Delta G = \left(\frac{\partial G}{\partial n_a}\right)_{T,p_a} \Delta n + \left(\frac{\partial G}{\partial n_b}\right)_{T,p_b} (-\Delta n) = 0. \qquad (4\text{-}243)$$

In other words, the chemical potentials

$$\mu_a = \left(\frac{\partial G}{\partial n_a}\right)_{T,p_a} \qquad \text{and} \qquad \mu_b = \left(\frac{\partial G}{\partial n_b}\right)_{T,p_b} \tag{4-244}$$

of the *solvent* in the two compartments, must be equal. From (4-242) we find the expressions for these chemical potentials

$$\mu_a = g_1(T, p_a), \tag{4-245a}$$

$$\mu_b = g_1(T, p_b) - \left(\frac{n_2}{n_b}\right) RT. \tag{4-245b}$$

We see at once that no equilibrium is possible if $p_a = p_b$. To see how big the difference $\Delta p = p_b - p_a$ in pressure must be, we write

$$
\begin{aligned}
g_1(T, p_b) &= g_1(T, p_a + \Delta p) \\
&\cong g_1(T, p_a) + \left(\frac{\partial g_1}{\partial p}\right)_T \Delta p.
\end{aligned}
\tag{4-246}
$$

The p-derivative of g_1 is the molar volume of the solvent, v_1. Inserting this result into the equilibrium condition $\mu_b = \mu_a$, we find

$$v_1 \Delta p_{\text{equil}} = \frac{n_2}{n_b} RT. \tag{4-247}$$

This pressure difference Δp *at* equilibrium defines the value of the osmotic pressure π:

$$\Pi = (p_b - p_a)_{\text{equil}} = \Delta p_{\text{equil}}.$$

Dividing (4-247) by v_1, and observing that for $n_2 \ll n_1$ the product $n_b v_1$ represents simply the total volume v_b of compartment (b) containing the solution, we have

$$\Pi = n_2 \frac{RT}{v_b}. \tag{4-248}$$

This is Van t'Hoff's law for the osmotic pressure. It states that the osmotic pressure Π of a dilute solution containing n_2 moles of solvent in a volume v_b is equal to the pressure exerted by an ideal gas of the same mole number in the same volume.

(vi)   The Solubility of Gases. Henry's Law

We now consider a system consisting of a gas in equilibrium with a solution in which the gas molecules are dissolved. The gas molecules thereby play the role of solute in the solution. Such a system is, for example, atmospheric oxygen in equilibrium with oxygen dissolved in blood plasma. Thus, in general, the equilibrium we consider is that of pure solute in the gas phase, at pressure p, in contact with a dilute solution of that gas in a solvent. We neglect the vapor pressure of the solvent, but observe that if the gas whose equilibrium we study is mixed with other gases, then the formulas given below will remain valid provided p is understood to be the *partial pressure* of the gas in question.

In this subsection, we will be able to shorten our analysis since the argument leading to the equilibrium condition is essentially the same as we have discussed previously. The free energy is the sum of the free energy of the gas phase

$$G_{\text{gas}} = n_g \mu_g(T, p_g)$$
$$= n_g RT \ln\left(\frac{p_g V_F(T)}{kT}\right) \tag{4-249}$$

and the free energy of the solution is given by (4-221):

$$G_{\text{solution}} = n_1 g_1(T, p) + n_2 \bar{g}_2(T, p) + n_2 RT \left\{\ln \frac{n_2}{n_1} - 1\right\}. \tag{4-250}$$

Thus, the total free energy is

$$G = G_{\text{gas}} + G_{\text{solution}}.$$

In these equations p_g is the partial pressure of the gas, whose solubility is being investigated; p is the total pressure of all the gases above the solution; n_1 is the number of moles of solvent, and n_2 is the number of moles of gas molecules dissolved in the solvent. In this problem the free energy at equilibrium must be stationary with respect to a transfer of gas molecules between the gas phase and the solution, viz:

$$n_g \to n_g + \Delta n,$$
$$n_2 \to n_2 - \Delta n.$$

As we have seen previously, the equilibrium conditions require that the chemical potentials of the molecules under investigation in the gas phase and the solution must be equal; that is,

$$\mu_g \equiv \left(\frac{\partial G}{\partial n_g}\right)_{T,p,n_1,n_2} = \left(\frac{\partial G}{\partial n_2}\right)_{T,p,n_g,n_1} \equiv \mu_2.$$

The chemical potential of the gas in the gas phase follows from (4-249) as

$$\mu_g = RT \ln\left(\frac{p_g V_F(T)}{RT}\right). \tag{4-251a}$$

Calculating $(\partial G/\partial n_2)_{p,T,n,n_g}$ from G using (4-221) gives, after a little algebra,

$$\mu_2 = \bar{g}_2(T, p) + RT \ln\left(\frac{n_2}{n_1}\right). \tag{4-251b}$$

The equality $\mu_g = \mu_2$ leads to

$$\ln\left(\frac{n_2}{n_1}\right)_{eq} = \ln\left(\frac{p_g V_F(T)}{kT}\right) - \frac{\bar{g}_2(T, p)}{RT}$$

or

$$\left(\frac{n_2}{n_1}\right)_{eq} = \frac{p_g V_F(T)}{kT} e^{-\bar{g}_2/RT}. \tag{4-252}$$

The structure of this relation is

$$\left(\frac{n_2}{n_1}\right)_{equil} = K(T)p_g, \tag{4-253}$$

with $K(T) = (e^{-\bar{g}_2/RT})(V_F/kT)$. In words: At fixed temperature the mole fraction of gas in solution is proportional to the (partial) pressure p_g of the solute in the gas phase. This is Henry's law.

This law is experimentally well confirmed in the sense that (n_2/n_1) is indeed linear in p_g, provided that (n_2/n_1) remains small. In the process of verification of Henry's law, one can determine experimentally the magnitude of the quantity $K(T)$. Thermodynamic analysis does not provide the means to calculate K from the properties of the individual molecules. In Table 4.8, we give some values for the solubility of various gases in water at 25 °C.

Table 4.8. Solubility of Various Gases in Water at 25°C (after Daniels and Alberty [21]).

Gas	K in $(\text{mmHg})^{-1}$
Hydrogen, H_2	1.87×10^{-8}
Nitrogen, N_2	1.54×10^{-8}
Oxygen, O_2	3.03×10^{-8}
Carbon dioxide, CO_2	0.80×10^{-6}

We see that carbon dioxide is much more soluble in water than nitrogen and oxygen.

4.4.C.   The Binding of Ligands to Multi-subunit Proteins

(i)   Thermodynamic Equilibrium and the Binding of Ligands to Distinct Sites on Multi-subunit Proteins

a) Introduction. Central Role of Ligand Binding for Protein Function. Specification of Thermodynamic Variables and Definition of the Molecular Parameters.   The "work-horses" of living cells are the proteins. Frequently these proteins are oligomers made up of nearly identical subunits. Proteins perform an astonishing variety of functions. They serve as catalysts in all the metabolic pathways. They store and transport essential metabolites such as oxygen and carbon dioxide. They act as receptors and transducers, pumps and motors, and regulators of the expression of genetic information. In the performance of all these functions, proteins are activated and controlled by the binding of specific ligands. Consequently, it is of prime biological importance to be able to understand quantitatively the factors that control the probability of occupancy of the various binding sites on proteins. In the present section we will develop the theoretical apparatus needed for such understanding, and will apply it to the particular problem of the binding of oxygen to the storage and transport proteins myoglobin and hemoglobin.

We envision a solution in which is dissolved a single species of ligand molecules and a single type of protein molecule. Let each protein have n specific sites at which the ligand molecules can reside. We label each site by the symbol i where $i = 1, \ldots, n$, and will designate by the symbol α_i the number 1 or 0 depending upon whether the ith site is or is not occupied by a ligand molecule. We denote by the symbol $\{\alpha\}$ a particular pattern of occupancy of the protein. The symbol $\{\alpha\}$ can be envisioned as an n-component row vector, consisting of an array of zeros and ones. The n-component array $\{\alpha\}$ specifies a particular pattern of occupancy of the protein subunits. The total number $\nu(\{\alpha\})$ of ligands bound in $\{\alpha\}$ is

$$\nu(\{\alpha\}) = \sum_{i=1}^{n} \alpha_i. \tag{4-254}$$

It will be useful to realize that there are in general a multiplicity of patterns $\{\alpha\}$ having the same value of ν. If we designate this multiplicity by $M(\nu)$, it can be seen that

$$M(\nu) = \frac{n!}{(n-\nu)!\,\nu} \equiv \binom{n}{\nu}. \tag{4-255}$$

Here the symbol on the right-hand side is the familiar binomial coefficient.

In the solution under consideration we assume that a multiple chemical equilibrium takes place between the ligands free in solution, and the ligands bound to each configuration $\{\alpha\}$ on the protein. Let us denote as $\mu_{\{\alpha\}}$, the chemical potential of a protein in configuration $\{\alpha\}$, and as μ_1 the chemical potential of a ligand molecule free in solution. As we have seen previously, $\mu_{\{\alpha\}}$ depends upon the protein mole fraction $(X_{\{\alpha\}})$, and μ_1 depends upon the free ligand mole fraction (X_1). These two mole fractions in turn are related to $N_{\{\alpha\}}$, N_W, and N_1, that are, respectively, the number of proteins in solution, the number of water molecules, and the number of ligand molecules in the solution, by

$$X_{\{\alpha\}} = N_{\{\alpha\}}/(N_W + N_1 + \sum N_{\{\alpha\}}) \tag{4-256a}$$

or, approximately,

$$X_{\{\alpha\}} \cong N_{\{\alpha\}}/N_W, \tag{4-256b}$$

correspondingly,

$$X_1 \cong N_1/N_W. \tag{4-256c}$$

We seek to obtain the relation between the chemical potentials $\mu_{\{\alpha\}}$ and μ_1 at equilibrium. Alternatively, we seek the relationship between each $X_{\{\alpha\}}$ and the mole fraction X_1, of free ligand at equilibrium.

We must, of course, also include the effect of any external constraint upon the system. For example, one such constraint could be that the ligand molecules free in solution are not only in equilibrium with the protein molecules, but may also be in equilibrium with ligand molecules in an overlying gas phase. Such an equilibrium in effect fixes the value of X_1 at a value determined by the partial pressure of that molecule in the overlying gas phase. Another possible constraint, applicable in the case of nonvolatile ligand molecules, is that the value of X_1 is fixed at a chosen value by the experimenter by a suitable addition of ligand.

b) Free Energy Minimization and the Relationship Between the Chemical Potentials $\mu_{\{\alpha\}}$ and μ_1 at Thermal Equilibrium. According to the theory of multicomponent solutions discussed previously, the Gibbs free energy $(G - G_0)$, associated with the proteins and ligands in the aqueous solution, can be written as

$$(G - G_0) = \sum_{\{\alpha\}} N_{\{\alpha\}}\mu_{\{\alpha\}} + N_1\mu_1. \tag{4-257}$$

For equilibrium $(G - G_0)$ must be at a minimum with respect to the transfer of ligands between bound configurations on the protein and ligands free in solution. Consider δN proteins, each holding ligands in the configuration $\{\alpha'\}$. We now envision a transfer in which ligands are removed from each protein so as to produce in each a binding configuration $\{\alpha\}$. This transfer releases into the free ligand pool a number δN_1 of ligand molecules, where $\delta N_1 = (\nu' - \nu)\,\delta N$. Here $\nu'(\{\alpha'\}) = \sum_1^n \alpha'$ and $\nu(\{\alpha\}) = \sum_1^n \alpha$. Thus the transfer described involves the following changes:

$$\delta N_{\{\alpha'\}} = -\delta N, \tag{4-257a}$$

$$\delta N_{\{\alpha\}} = +\delta N, \tag{4-257b}$$

$$\delta N_{\{\alpha''\}} = 0 \quad \text{for} \quad \{\alpha''\} \neq \{\alpha'\} \quad \text{or} \quad \{\alpha\}, \tag{4-257c}$$

$$\delta N_1 = (\nu' - \nu)\,\delta N. \tag{4-257d}$$

If we use these variations in the condition that the free energy G must be minimized, we find

$$\delta(G - G_0) = 0 = (-\mu_{\{\alpha'\}} + \mu_{\{\alpha\}}) + (\nu' - \nu)\mu_1. \tag{4-258a}$$

This result can be expressed as

$$\frac{(\mu_{\{\alpha'\}} - \mu_{\{\alpha\}})}{(v' - v)} = \mu_1. \tag{4-258b}$$

This is the statement that the change in chemical potential of the protein, per ligand removed, is equal to the chemical potential of a single ligand free in solution. Alternatively, since (4-258a) must hold for all possible $\{\alpha\}$ and $\{\alpha'\}$, it follows that

$$(\mu_{\{\alpha\}} - v\mu_1) = \text{constant}, \tag{4-258c}$$

where the constant is independent of α. As will be seen, the value of the constant can be fixed by the total mole fraction of proteins in the solution.

c) The Protein Mole Fractions $X_{\{\alpha\}}$, the Relative Occupancies $R_{\{\alpha\}}$ of Each Configuration. The Central Role of the Free Ligand Chemical Potential. The Binding Free Energy Level Diagram.   The results obtained above for the relationship between $\mu_{\{\alpha\}}$ and μ_1 can be used to connect the mole fractions $X_{\{\alpha\}}$ of protein in configuration $\{\alpha\}$ with the chemical potential μ_1. This connection can be made if we use the following relationship connecting $\mu_{\{\alpha\}}$ and $X_{\{\alpha\}}$, as obtained in the previous Section 4.4.B(iii), viz:

$$\mu_{\{\alpha\}} = \mu^0_{\{\alpha\}} + kT \ln X_{\{\alpha\}}. \tag{4-259}$$

Here $\mu^0_{\{\alpha\}}$ is the standard part of the chemical potential of the protein in configuration $\{\alpha\}$. If this is placed into (4-258c), we find

$$X_{\{\alpha\}} = (\text{constant})' e^{-\mu^0_{\{\alpha\}}/kT} z^{v(\{\alpha\})}, \tag{4-260a}$$

where

$$z = e^{\mu_1/kT} \equiv \text{fugacity of free ligand molecule}, \tag{4-260b}$$

$$(\text{constant})' = \exp(\text{constant}/kT). \tag{4-260c}$$

Equations (4-260) show that the mole fraction of protein in configuration $\{\alpha\}$ is proportional to z^v where z is the fugacity of the free ligand in solution. Thus, this fugacity factor strongly diminishes the fraction of protein to be found in highly liganded configurations. The (constant)$'$ is fixed by the total mole fraction of proteins

X_0 placed in solution. Indeed, since

$$X_0 = \sum_{\alpha} X_{\{\alpha\}}, \tag{4-260d}$$

it follows that

$$(\text{constant})' = X_0 \bigg/ \sum_{\{\alpha\}} z^{\nu} \exp - \left[\mu^0_{\{\alpha\}}/kT \right]. \tag{4-260e}$$

Equations (4-260a) and (4-260e) fully determine the actual mole fractions $X_{\{\alpha\}}$ in terms of the fundamental free energies $\mu^0_{\{\alpha\}}$, and the fugacity z.

From an operational point of view it is simpler to consider not $X_{\{\alpha\}}$, but the quantity $R_{\{\alpha\}}$ which is defined as

$$R_{\{\alpha\}} = (X_{\{\alpha\}}/X_{\{\alpha-0\}}). \tag{4-261a}$$

The relative occupancy $R_{\{\alpha\}}$ is the ratio of the number of proteins to be found in configuration $\{\alpha\}$ with the number of proteins to be found in an unliganded configuration. From the definition in (4-261a) and (4-260a) we see that $R_{\{\alpha\}}$ has the simple form

$$R_{\{\alpha\}} = z^{\nu} \exp\left[-\Delta\mu^0_{\{\alpha\}}\right]/kT. \tag{4-261b}$$

Here

$$\Delta\mu^0_{\{\alpha\}} = \left(\mu^0_{\{\alpha\}} - \mu^0_{\{\alpha=0\}}\right). \tag{4-261c}$$

Equation (4-261b) shows quite clearly that the relative occupancy of each configuration $\{\alpha\}$ is determined by two factors. The first is the fugacity (z). According to the definition in (4-260b), z is related to the chemical potential of the free ligand by

$$z = \exp(\mu_1/kT). \tag{4-262a}$$

This fugacity can in fact be controlled experimentally.

If the ligand is a nonvolatile species, its chemical potential μ_1 is related to its mole fraction X_1 free in solution at equilibrium by

$$\mu_1 = \mu^0_1 + kT \ln X_1. \tag{4-262b}$$

Thus

$$z = X_1 e^{\mu_1^0/kT}. \tag{4-262c}$$

This equation shows that by changing experimentally the concentration of the free ligand, one can vary z, and hence the relative occupancy of each configuration.

If the ligand is a volatile species, and equilibrium is established between the dissolved ligand and ligand molecules in the overlying gas phase, then as we have seen in the section on Henry's law, the chemical potential μ_1 is equal to that of the ligand in the gas phase μ_g, i.e.,

$$\mu_1 = \mu_g. \tag{4-263a}$$

However, the chemical potential in the gas phase is well known if the gas is sufficiently dilute. In this case, if p is the partial pressure of the ligand in the gas phase

$$\mu_g = \mu_g^0 + kT \ln(p/P_F). \tag{4-263b}$$

Here

$$\mu_g^0 = -kT \ln Z_{\text{int}}. \tag{4-263c}$$

Z_{int} is the partition function for the internal degrees of freedom of the ligand molecule. For diatonic gases, the internal degrees of freedom are essentially only rotational. μ_g^0 is a well-determined quantity for diatomic molecules. The quantity P_F, known as the Fermi pressure, is given by

$$P_V = \left(\frac{kT}{V_F}\right), \tag{4-263d}$$

where

$$V_F = (h^2/2\pi mkT)^{3/2}. \tag{4-263e}$$

Here m is the mass of the ligand molecule. Thus in the case of volatile ligand molecules in equilibrium with an overlying gas phase, the fugacity can be varied experimentally by changing the pressure p in the gas phase. Indeed, here

$$z = \left(\frac{p}{p_F}\right) e^{(\mu_g^0/kT)}. \tag{4-264}$$

The basic result of (4-261b) shows that $R_{\{\alpha\}}$ is also determined by the magnitude of the free-energy differences $\Delta\mu_{\{\alpha\}}^0$. For each occupancy pattern $\{\alpha\}$ one can designate the value of $\Delta\mu_{\{\alpha\}}^0$ on a free-energy level diagram. This diagram has the form of a descending sequence of levels corresponding to larger negative values of binding free energy as the number of bound ligands increases.

d) The Species Fractions, the Binding of Polynomial and the Saturation Function.   If one imagines sampling randomly individual proteins in the solution and determining its occupancy pattern, then repeated trials will establish the probability $P_{\{\alpha\}}$ for a particular configuration. From the definition of the $R_{\{\alpha\}}$, it follows that

$$P_{\{\alpha\}} = R_{\{\alpha\}} \bigg/ \sum_{\{\alpha\}} R_{\{\alpha\}}. \tag{4-265}$$

It is convenient to define the normalizing factor in the denominator of (4-265) as the binding polynomial

$$B(z) = \sum_{\{\alpha\}} R_{\{\alpha\}}. \tag{4-266}$$

Since, according to (4-261b), $R_{\{\alpha\}}(z)$ is proportional to z^ν, it follows that $B(z)$ is a polynomial of order n in the fugacity z. Here n is simply the total number of binding sites on the protein. The coefficients of each power of z in the polynomial are determined by the values of the various $\Delta\mu_{\{\alpha\}}^0$'s corresponding to each ν. If the sites are all equivalent, then for each ν there will be a multiplicity $M(\nu)$ [see (4-255)] of configurations $\{\alpha\}$, each having the same value of $\Delta\mu_{\{\alpha\}}^0$.

The biological function of the protein is often determined by the average number of ligands $\bar{\nu}$ bound to the proteins. Clearly,

$$\bar{\nu} = \sum_{\{\alpha\}} \nu(\{\alpha\}) P_{\{\alpha\}}. \tag{4-267}$$

If n is the maximum number of ligands that can be bound to the protein, then we may define the fractional saturation function or, more simply, the saturation function $Y(z)$ as

$$Y(z) = \left(\frac{\overline{v}}{n}\right). \tag{4-268a}$$

Here $0 \leq Y \leq 1$. It is useful to observe that the saturation function $Y(z)$ can be obtained readily from the binding polynomial $B(z)$. Indeed, since

$$Y(z) = \left(\frac{1}{n}\right) \frac{\sum v(\{\alpha\}) R_{\{\alpha\}}(z)}{\sum R_{\{\alpha\}}(z)} \tag{4-268b}$$

and since, from (4-261b), we see that

$$v(\{\alpha\}) R_{\{\alpha\}}(z) = z \frac{\partial}{\partial z} R_{\{\alpha\}}, \tag{4-268c}$$

it follows at once, using the definition (4-266) of the binding polynomial, that

$$Y(z) = \left(\frac{1}{n}\right)\left(\frac{z}{B(z)}\right)\left(\frac{\partial B(z)}{\partial z}\right)$$

or

$$Y(z) = \left(\frac{1}{n}\right)\left(\frac{\partial \ln B(z)}{\partial \ln z}\right). \tag{4-269}$$

Thus we see that the binding polynomial determines completely the form of the saturation function.

Summarizing then we observe that the relative occupancies $R_{\{\alpha\}}(z)$ are fixed by the fugacity z of the free ligand, and the sequence of free energy levels $\Delta\mu^0_{\{\alpha\}}$. The species fractions $P_{\{\alpha\}}(z)$ are determined by the ratio $R_{\{\alpha\}}(z)/B(z)$ where $B(z)$ is the binding polynomial. Finally, the saturation function $Y(z)$ is determined according to (4-268a) from the logarithmic derivative of the binding polynomial $B(z)$.

At this point we have developed sufficiently the theory of ligand binding that we can apply it to the particular cases of the binding of oxygen to the proteins myoglobin and hemoglobin.

(ii)   The Structure and Function of the Oxygen Binding Proteins: Hemoglobin and Myoglobin

Hemoglobin and myoglobin are proteins whose main physiological role is the transport and storage of oxygen.

Hemoglobin, which has already been described in Chapter 1, serves the function of carrying molecular oxygen from the lungs to the body tissues. Each red blood cell contains about 2.8×10^8 hemoglobin molecules. With about 5×10^{12} red blood cells in one liter of blood, this amounts to about 1.4×10^{21} molecules of hemoglobin per liter of blood.

As we showed in Chapter 1, each hemoglobin molecule is composed of four folded protein chains. Each chain carries a so-called heme group with a ferrous iron ion (Fe^{++}) in its center. This heme group is the site of oxygen binding. Each hemoglobin molecule can therefore bind a maximum of four oxygen molecules. Fully oxygenated hemoglobin in blood is therefore capable of carrying about 5.7×10^{21} molecules of oxygen per liter of blood, or 0.93×10^{-2} mol of O_2/l of blood. This vastly exceeds the oxygen-carrying capacity of blood plasma in which oxygen is dissolved as a gas. From Henry's law, Table 4.6, we infer that at a partial pressure of oxygen of 105 mmHg (this is the O_2 partial pressure in the alveoli of the lung), the amount of oxygen carried by $1 \; l \sim (10^9/18)$ mol of plasma is

$$n_2 = \frac{10^3}{18} \times 3.03 \times 10^{-8} \times 105$$
$$= 1.76 \times 10^{-4} \text{ mol.}$$

We see that this is about 2% of the oxygen carried, at maximum saturation by the hemoglobin as bound oxygen. Hemoglobin is thus responsible for the transport of 98% of the oxygen reaching the body tissues.

The number $\bar{\nu}$ of oxygen molecules carried, on average, by each hemoglobin molecule is the result of a thermal (chemical) equilibrium between the oxygen dissolved in the blood plasma and that bound to hemoglobin. Let us describe briefly how this equilibrium becomes established. Suppose that deoxygenated blood enters the alveolar capillaries. Both the hemoglobin and the blood plasma contain a concentration of oxygen below the equilibrium value associated with the partial pressure of oxygen surrounding the capillaries in the alveoli. Because of the low concentration of oxygen in the plasma and the hemoglobin, oxygen gas from the alveoli will diffuse into the blood. The influx of oxygen will be snatched up by the hemoglobin almost completely until a compound equilibrium is reached. The compound equilibrium is one in which the oxygen concentration in the plasma is in equilibrium with *both* the external oxygen gas pressure and with the oxygen held on the hemoglobin. We thus expect a definite functional relationship to exist between the oxygen saturation of the hemoglobin and the partial pressure of oxygen in the gas above the blood. In fact, this relationship between p_{O_2} and the fractional saturation Y of the hemoglobin molecules is called the oxygen saturation curve or the oxygen disassociation curve of hemoglobin. The quantity Y is defined by (4-267) and (4-268a) as

$$Y = \frac{\text{number of oxygen molecules bound}}{4 \times \text{number of Hb molecules}}, \qquad (4\text{-}270)$$

since at saturation each Hb molecule may bind four oxygen molecules. In Figure 4.49 we plot the oxygen saturation curve for Hb for blood at body temperature (37 °C) and the normal value of the hydrogen ion concentration (pH) of the blood plasma.

As we have previously discussed in Volume I, Chapter 3, Section 3.5.D(iii), this curve shows some remarkable features:

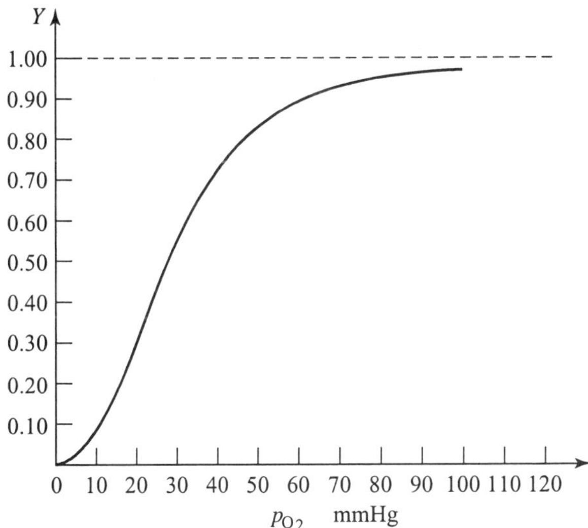

Figure 4.49. Oxygen saturation curve for hemoglobin for $T = 37\,°\text{C}$ and pH $= 7.2$. Ordinate: fractional saturation defined in (4-267) and (4-268a). Abscissa: partial pressure of oxygen.

(a) First, we observe at that oxygen pressure which prevails in the alveoli of the lung, the Hb molecule practically saturates itself with oxygen ($Y = 97.5\%$).

(b) In the muscular tissues when oxygen is consumed, the partial pressure of oxygen is of the order of 40 mmHg under resting conditions. This is evidenced by the residual oxygen pressure in venous blood. At this pressure we see that Y is still of the order of 75%. Blood retains, therefore, a large oxygen reserve to be utilized in case of an increased demand. Sudden muscular activity will result in an additional drop of the oxygen pressure p_{O_2} in the tissue, below 40 mmHg, whereupon the Hb molecules are capable of releasing a substantial oxygen reserve. Indeed, as can be seen from Figure 4.49, between $p_{O_2} = 40$ mmHg and $p_{O_2} = 20$ mmHg, hemoglobin can unload an additional 35% of its original oxygen content.

It is the sigmoidal, or S, shape of the oxygen saturation function that enables hemoglobin's biological function. In effect, the oxygen binding curve Y acts as a switch. For low values of p_{O_2}, the hemoglobin in effect binds very little oxygen, but as (p_{O_2}) approaches $(p_{1/2})$ the half-saturation pressure, the saturation curve rises sharply, and for $(p_{O_2}) > p_{1/2}$, the curve flattens and becomes essentially independent of p_{O_2}. This switching characteristic is repeated in the case of the binding of ligands to enzymes. In effect, multi-subunit enzymes are inactive below a certain ligand concentration X_s and "switch on" only above X_s. As we shall see, this sigmoidal characteristic is intimately related to energetically favorable interactions *between* bound ligands. The explanation of the molecular basis for this so-called "cooperative homotropic interaction" is a topic of great biophysical importance as it is the basis for successful protein function.

Let us now turn to a brief discussion of the second oxygen carrier, myoglobin. Myoglobin is found in muscular tissue. It is a protein consisting of a single folded chain of molecular weight $M = 16,000$ g. It contains one heme group, and is capable of binding a single molecule of oxygen. Its oxygen saturation curve is significantly different from that of hemoglobin. It does not show the characteristic "sigmoid" (S-shaped) form of the Hb curve, but is a simple hyperbola (see Figure 4.50).

In Figure 4.50 below we show the form of the oxygen saturation function $Y(p_{O_2})$ for human myoglobin at 35 °C. It is observed that $Y(p_{O_2})$ has the form of a hyperbola. This hyperbola has a half-saturation pressure $p_{1/2}$. The hyperbola has a horizontal asymptote at $Y = 1$ and a vertical asymptote at $p_{O_2} = -p_{1/2}$. Experimentally, it is observed that $p_{1/2}$ is a very strong function of the temperature. We observe from this figure that myoglobin is able to unload significant amounts of oxygen only when the partial pressure of oxygen has fallen below ~ 20 mmHg,

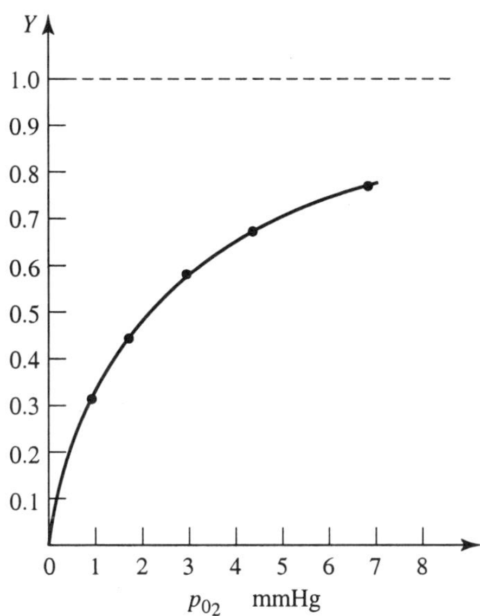

Figure 4.50. The oxygen saturation curve of human myoglobin at 35 °C. Tris buffer 0.05 M, pH 7.45. Adapted from A. R. Fanelli and E. Antonini, *Arch. Biochem. Biophys.* **77**, 478 (1958).

a pressure at which hemoglobin has discharged the main part of the oxygen it carries. Clearly, myoglobin may be viewed as an "emergency" store of oxygen to be drawn upon when the normal supply through hemoglobin has been used up. Not surprisingly, myoglobin is found most abundantly in the muscles of diving mammals such as whales, seals, and porpoises. The myoglobin serves as a reservoir of oxygen to be utilized while the mammal is unable to breathe atmospheric oxygen.

In the following sections we will provide an understanding of the microscopic basis for the dramatically different shapes of the hyperbolic myoglobin saturation curve and the sigmoidal saturation curve for hemoglobin. We will pay particular attention to the molecular implications of the hemoglobin saturation curve that is so ingeniously tailored to the physiological needs of the organism.

(iii)   The Theory of Oxygen Binding to Myoglobin

a) The Relative Occupancies R_0, R_1, the Binding Polynomial B, and the Oxygen Saturation Function Y.   The binding of oxygen to a single subunit containing a heme group provides the simplest example of the use of the theory given in

Section 4.4.C(i) above. In the present case myoglobin can be found either in configuration $\{0\}$ or $\{1\}$. $\{\alpha\}$ has only one component that is zero for the unliganded state and 1 for the liganded state. The relative occupancy $R_{\{\alpha\}}$ of these two states is given, according to (4-261b), by

$$R_0 = 1 \tag{4-271a}$$

$$R_1 = z\exp(-\beta\Delta\mu^0) \equiv x. \tag{4-271b}$$

Here

$$\beta = (1/kT), \tag{4-271c}$$

$$\Delta\mu^0 = (\mu^0_{\{1\}} - \mu^0_{\{0\}}). \tag{4-271d}$$

$\Delta\mu^0$ is the difference in standard chemical potential between a liganded and unliganded myoglobin molecule. We expect $\Delta\mu^0$ to be a negative number, corresponding to a favorable (negative) free energy advantage of the ligand bound protein compared to the unbound protein. The fugacity z of the dissolved oxygen molecule can be related to the partial pressure p_{O_2} of oxygen in the overlying gas phase using (4-264). Thus

$$z = \left(\frac{p_{O_2}}{P_F}\right)\exp\left(\mu^0_{O_2}/kT\right). \tag{4-272}$$

Here $\mu^0_{O_2}$ is the standard part of the chemical potential of the molecule in the gas phase and arises from the rotational degrees of freedom of the diatomic molecule. $\mu^0_{O_2}$ is well known from the statistical mechanics of diatomic molecules. From (4-272) and (4-271) we see that

$$R_1 = x = (p_{O_2}/P_F)\exp\left[-(\delta\mu^0/kT)\right], \tag{4-273a}$$

where

$$\delta\mu^0 \equiv (\Delta\mu^0 - \mu^0_{O_2}). \tag{4-273b}$$

We observe that R_1 is directly proportional to the fugacity z according to (4-271b). Indeed the variable x is simply a constant times z.

The binding polynomial has the form

$$B(x) = 1 + x. \tag{4-274}$$

The saturation function Y is given by

$$Y = \frac{\partial \ln B(z)}{\partial \ln z} = \left(\frac{\partial \ln B(x)}{\partial \ln x} \right). \tag{4-275a}$$

Thus, we find

$$Y = x/(1+x). \tag{4-275b}$$

If we use (4-273a) in this expression for Y, we find

$$Y(p_{O_2}) = \frac{p_{O_2}}{(p_{O_2} + p_{1/2})}, \tag{4-275c}$$

where

$$p_{1/2} = P_F(T) \exp(\beta \delta \mu^0). \tag{4-275d}$$

Thus, we see that the saturation function is indeed a hyperbola with vertical asymptote at $p_{O_2} = -p_{1/2}$, and a horizontal asymptote at $Y = 1$. The half-saturation pressure depends exponentially on $\delta \mu_0$ which, according to (4-273b), is the free-energy advantage of removing an oxygen molecule from the gas phase and binding it to the protein heme group.

b) Determination of the Free Energy Parameter $\delta \mu^0$. $\delta \mu^0$ is the basic molecular energy parameter that determines the biological function of the myoglobin molecule. Its magnitude and sign can be found from (4-275d) using the experimental values of $p_{1/2} \cong 3$ mmHg at $37\,^{\circ}$C and pH 7.4 for human myoglobin. The Fermi pressure for oxygen molecules at $37\,^{\circ}$C is $\sim 6 \times 10^9$ mmHg. Using these in (4-275d) we find, that at $37\,^{\circ}$C:

$$\left(\frac{\delta \mu^0}{kT} \right) = -21.4. \tag{4-276a}$$

Thus the free-energy advantage of transferring oxygen from the gas phase to binding states on the myoglobin is $\sim -21kT$. This is a very large exponential factor. Since $RT = 0.616$ kcal/mol, the energy advantage per mole is

$$\delta \mu^0 = -13.2 \text{ kcal/mol}. \tag{4-276b}$$

c) The Temperature Dependence of $p_{1/2}$.   The very large value of $\delta\mu^0$ in the exponent of the equation for $p_{1/2}$ suggests that the half-saturation pressure can be expected to increase strongly as the temperature increases. Indeed, we find, according to the experimental data of Rossi Fanelli and Antonini, that near $35\,°C$:

$$\left(\frac{d\ln p_{1/2}}{dT}\right)_{\exp t} = 0.076 \pm 0.01\ (°K)^{-1}. \tag{4-277a}$$

We may calculate theoretically the temperature dependence of $p_{1/2}$ under the assumption that $\delta\mu^0$ is essentially independent of the temperature. Then remembering that $P_F \sim T^{5/2}$ we see from (4-275d) that

$$\frac{d\ln p_{1/2}}{dT} = \left(\frac{5}{2} - \frac{d\delta\mu^0}{dT}\right)\frac{1}{T}. \tag{4-277b}$$

If we use $T = 35\,°C$, $(\delta\mu^0/kT) = -21.4$, we find theoretically that

$$\left(\frac{d\ln p_{1/2}}{dT}\right)_{\text{theory}} = 0.078\ (°K)^{-1}. \tag{4-277c}$$

Thus the theory described above is capable of explaining the strong increase of $p_{1/2}$ with temperature. This amounts to about 1 mmHg (i.e., a 30% increase in $p_{1/2}$) for a $5°$ rise in temperature.

d) The Species Fractions P_0 and P_1 as a Function of the Mean Number of Ligands Bound.   We conclude this subsection by describing the species fraction P_0 and P_1 as a function of the oxygen partial pressure, or alternatively the mean number $\bar{\nu}$ of ligands bound. The species fraction P_0 and P_1 are

$$P_0(x) = 1/B(x) = 1/(1+x) \tag{4-278a}$$

and

$$P_1(x) = x/B(x) = x/(1+x). \tag{4-278b}$$

Instead of the variable x that varies for zero to infinity, it is convenient to express P_0 in terms of the mean number $\bar{\nu} = Y(x)$ of oxygen molecules bound to the myoglobin. $\bar{v}$ is confined to the domain $0 \leq \bar{v} \leq 1$. We see from (4-275b) that

$$\bar{v} = x/(1+x).$$

If we eliminate x in favor of $\bar{v}$ and use this in (4-278a,b), we find

$$P_0(\bar{v}) = (1 - \bar{v}), \tag{4-279a}$$
$$P_1(\bar{v}) = \bar{v}. \tag{4-279b}$$

Thus the species fractions vary quite simply with $\bar{v}$. P_0 decreases linearly with $\bar{v}$ from 1 to 0 as $\bar{v}$ varies from 0 to 1. P_1 increases linearly with $\bar{v}$.

(iv)   The Theory of Oxygen Binding to Hemoglobin

a) The Saturation Function and Species Fractions for Equivalent, Distinguishable, Noninteracting Oxygen Binding Sites on its Hemoglobin Tetramer.
In the simplest model for oxygen binding to the hemoglobin tetramer, one assumes that the free-energy advantage $\Delta\mu^0 \equiv \mu^0_{\{\alpha+1\}} - \mu^0_{\{\alpha\}}$ for the protein upon binding an oxygen is the same regardless of the subunit position, and is independent of the pattern of occupancy of the other sites. Under these conditions, the state of occupancy can be specified by the single quantity v, the number of ligands bound. The sequence of protein free-energy levels corresponding to $v = 0, 1, 2, 3, 4$ is a descending ladder of five levels. The spacing between the rungs on this ladder is $\Delta\mu^0$. The free-energy level associated with v oxygens bound contains $\binom{4}{v}$ states corresponding to the total number of distinguishable patterns $\{\alpha\}$ of oxygen occupancy all of which have the same occupancy number v. For each pattern the free-energy level difference $\mu^0_{\{\alpha\}} - \mu^0_{\{\alpha=0\}}$ is equal to $v\Delta\mu^0$.

Thus $R_v = \binom{4}{v} R_{\{\alpha\}}$ and if we use our fundamental formula Eqn. 4-261 for $R_{\{\alpha\}}$, we find

$$R_v = \binom{4}{v} \left(z \exp -\beta\Delta\mu^0 \right)^v. \tag{4-280}$$

We may express the fugacity of the free ligand in terms of the partial pressure of oxygen on the overlying gas phase (4-264). Then, as in (4-271b), we can define a quantity x, that is directly proportional to the fugacity z, as

$$x = z \exp\left(-\beta\Delta\mu^0\right) = \left(\frac{p_{O_2}}{P_F}\right) \exp\left(-\beta\delta\mu^0\right), \tag{4-281a}$$

where

$$\delta\mu^0 = \left(\Delta\mu^0 - \Delta\mu^0_{O_2}\right). \tag{4-281b}$$

Here $\delta\mu^0$ is the free-energy advantage (a negative number) associated with removal of oxygen from the gas phase and its binding onto the protein subunit. In terms of x, R_ν simply becomes

$$R_\nu = \binom{4}{\nu}x^\nu. \tag{4-282}$$

It is now quite simple to calculate the binding polynomial B, viz:

$$B(x) = \sum_{\nu=0}^{4}\binom{4}{\nu}x^\nu = (1+x)^4. \tag{4-283}$$

From this we can calculate the saturation function Y:

$$Y = \frac{1}{4}\frac{\partial \ln \beta(z)}{\partial \ln z} = \frac{1}{4}\frac{\partial \ln \beta(x)}{\partial \ln x}.$$

Thus

$$Y = \frac{x(1+x)^3}{(1+x)^4} = \left(\frac{x}{1+x}\right). \tag{4-284}$$

Since x is directly proportional to the oxygen partial pressure p_{O_2}, we can write $Y(p_{O_2})$ using (4-281a) as

$$Y(p_{O_2}) = \frac{p_{O_2}}{(p_{O_2} + p_{1/2})}, \tag{4-285a}$$

where

$$(p_{1/2}) = P_F \exp(\beta\delta\mu^0). \tag{4-285b}$$

We see that under these assumptions the oxygen saturation function $Y(p_{O_2})$ remains simply a hyperbola with a half-saturation pressure $(p_{1/2})$ having the same structure as was found for the myoglobin case. The independent, identical binding energy model is quite incapable of producing the characteristic sigmoidal feature so essential to the function of the hemoglobin molecule. Clearly, one must consider the possible effect of interaction between bound ligands if we are to understand the sigmoidicity or "cooperativity" exhibited in oxygen binding to hemoglobin. We will in fact include such interactions in the subsequent subsection. However, we wish to complete the treatment of the present simple model by calculating the species fractions that emerge.

The probability P_ν that, on sampling the solution, one will find a hemoglobin to which ν oxygens are bound is given by

$$P_\nu = \frac{R_\nu}{B} = \frac{\binom{4}{\nu} x^\nu}{(1+x)^4}. \tag{4-286}$$

Here x is proportional to p_{O_2}, hence x varies from zero to infinity. It is more convenient to express P_ν as a function of $\bar{\nu}$, the mean number of oxygen molecules bound. $\bar{\nu}$ and x are related according to the hyperbolic character of the saturation function by

$$\bar{\nu} = 4Y(x) = 4x/(1+x). \tag{4-287a}$$

Thus

$$x = \left(\frac{\bar{\nu}}{4}\right) \frac{1}{\left(1 - \dfrac{\bar{\nu}}{4}\right)} \qquad (0 < \bar{\nu} < 4), \tag{4-287b}$$

and

$$P_\nu(\bar{\nu}) = \binom{4}{\nu} \left(\frac{\bar{\nu}}{4}\right)^\nu \left(1 - \frac{\bar{\nu}}{4}\right)^{4-\nu}. \tag{4-288}$$

In Figure 4.51 below we provide a graph of (4-288) that shows how each species fraction varies with the mean occupancy $\bar{\nu}$. We see from this graph that, while the mean occupancy $\bar{\nu}$ rises monotonically, the probabilities P_0 fall rapidly to small values at $\bar{\nu} = 2$, while above $\bar{\nu} = 2$, P_4 rises rapidly. Furthermore, the probabilities P_ν of the intermediate species ($\nu = 1, 2, 3$) are quite significant. Each P_ν rises to maximum value at $\bar{\nu} = \nu$ and declines for $\nu > \bar{\nu}$. Indeed Figure 4.51 shows clearly that as $\bar{\nu}$ ranges from 0 to 4, there is a very significant presence of intermediate species. This is the underlying reason that the saturation function is hyperbolic. In the case of more sigmoidal saturation functions, we will see that in the passage

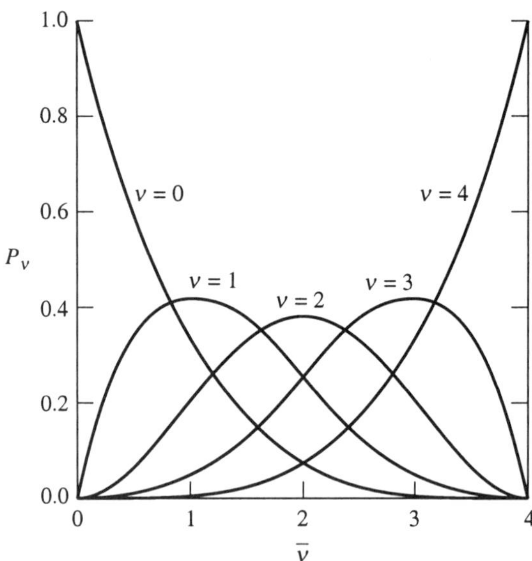

Figure 4.51. Each of the species fractions $P_0 \cdots P_4$ as a function of the mean occupancy $\overline{\nu}$ for the hemoglobin tetramers assuming no interaction between bound oxygens, and the same free-energy advantage for binding to each subunit.

from $0 < \overline{\nu} < 4$ the intermediate $\nu = 1, 2, 3$ species are suppressed in comparison with the $\nu = 0$ and $\nu = 4$ species.

The hyperbolic saturation function we have found is consistent with the view that the hemoglobin tetramer binds oxygen as would four separate subunits each free in solution. Each such subunit has a probability $p_1(\overline{\nu}) = (\overline{\nu}/4)$ of having an oxygen bound and a probability $p_0(\overline{\nu}) = (1 - \overline{\nu}/4)$ of being free of oxygen. Here $\overline{\nu}$ is the mean number of oxygen molecules bound to the tetramer. Both p_0 and p_1 are monotonic functions of $\overline{\nu}$. In view of this it is interesting to inquire how the species fractions P_ν for the parent tetramer has the strongly nonmonotonic character shown in Figure 4.51 and expressed in (4-288).

Indeed, we now show that $P_\nu(\overline{\nu})$ as given by (4-288) is precisely what one expects for the binding of oxygen to four separate subunits in solution. Consider then a solution that initially consists of hemoglobin tetramers which on average carry $\overline{\nu}$ oxygens. Let these oxygen bearing tetramers now be divided into their subunits, each of which contains either one or zero bound oxygen molecules with probability $p_1 = (\overline{\nu}/4)$ and $p_0 = (1 - \overline{\nu}/4)$, respectively. We now consider a trial in which four such subunits are sampled randomly and the total number of ν of the oxygen bound is determined. We seek the probability $P_\nu(\bar{\nu})$ that in repeated trials ν bound oxygen molecules will be found. This probability is immediately seen to be

$$P_\nu(\bar{\nu}) = \binom{4}{\nu} p_1^\nu p_0^{(4-\nu)} = \binom{4}{\nu} \left(\frac{\bar{\nu}}{4}\right)^\nu \left(1 - \frac{\bar{\nu}}{4}\right)^{4-\nu}. \tag{4-289}$$

Here, the factor $p_1^\nu p_0^{4-\nu}$ is the probability that, of the four subunits, ν will contain oxygen molecules and the remaining $(4 - \nu)$ subunits will not carry oxygen. The binomial coefficient $\binom{4}{\nu}$ represents the total number of distinct patterns of ν ones, and $(4-\nu)$ zeros, each of which pattern represents a successful outcome of the trial. Thus we see that the nonmonotonic dependence of the tetrameric species fraction on $\bar{\nu}$ is entirely consistent with the independent binding of oxygen to disconnected subunits dispersed throughout the solution.

b) The Pauling Tetrahedral Model for Homotropic Interaction Between Ligands.   We have seen above that, if the ligand binds with equal protein free-energy change to each of the binding sites, the resulting saturation function is a hyperbola. Equivalently, we note that if the descending free-energy ladder, that describes the value of the free-energy associated with protein occupancy by ν ligands, has equal rung spacings, then the saturation function Y does not exhibit the sigmoidal shape which is so important physiologically.

In order to obtain a sigmoidal saturation function one must consider a more general form for the sequences of free-energy levels. One such more general expression has the form

$$\Delta\mu_{\{\alpha\}}^0 \equiv (\mu_{\{\alpha\}}^0 - \mu_{\{\alpha=0\}}^0) = \sum_\alpha \alpha_i \Delta\mu_i^0 + \frac{1}{2} \sum_i \sum_{\substack{j \\ (i \neq j)}} \alpha_j \alpha_j \mu_{ij}^0. \tag{4-290}$$

This form for $\delta\mu_{\{\alpha\}}^0$ has the following features. First, the protein free-energy change $(\Delta\mu_i^0)$ associated with the binding of a single ligand may in general be different for each binding site i. Second, (4-290) allows for pairwise coupling between ligands on different sites. If both site i and site j are occupied, μ_{ij}^0 represents the free energy of interaction between these two bound ligands. If μ_{ij}^0 is a negative number corresponding to an attractive pairwise interaction, one uses the term "positive homotropic interaction." If μ_{ij}^0 is a positive number corresponding to a repulsive interaction between ligands, the term "negative homotropic interaction" is used. Clearly, $\mu_{ij}^0 = \mu_{ji}^0$ hence the factor $\frac{1}{2}$ is used before the double sum in (4-290). The value of μ_{ij}^0 is expected to vary strongly depending on the distance between the i and j sites. Also, the $\Delta\mu_i^0$ can be expected to depend strongly on the structure of the subunit in the vicinity of the binding site.

We may simply extend the earlier picture of noninteracting identical binding sites as follows. Since the α and β subunits of the hemoglobin tetramer show great similarity in the local structure in the vicinity of the heme binding site, we can assume that all the $\Delta\mu_i^0$ are the same and equal to $\Delta\mu^0$. Furthermore, if all the heme groups were equally spaced from one another as would be the case for the corners of a tetrahedral, then we expect that all the μ_{ij}^0's would be the same independent of i and j and could be designated by $\delta\overline{\mu}^0$. In fact this choice for the form of $\Delta\mu_{\{\alpha\}}^0$ was first made by Linus Pauling in 1935 (see [6]) to explain the sigmoidal shape of the oxygen saturation curve.

With such a choice for the position of the free-energy levels $\Delta\mu_{\{\alpha\}}^0$ we now have two parameters $\Delta\mu^0$, and the free energy $\delta\overline{\mu}^0$ rather than the single parameter used in Section 4.4.C.(ia). In this case Eqn. (4-290) becomes

$$\Delta\mu_{\{\alpha\}}^0 = \Delta\mu^0 \left(\sum_i \alpha_i\right) + \delta\overline{\mu}^0 \frac{1}{2} \sum_i \sum_{\substack{j \\ i \neq j}} (\alpha_i \alpha_j). \qquad (4\text{-}291)$$

Thus $\Delta\mu^0\{\alpha\}$ depends only upon the total number ν of ligands bound in configuration $\{\alpha\}$. For each such ν there are $\binom{4}{\nu}$ configurations, each of which has the free energy advantage

$$\Delta\mu_{\{\alpha\}}^0 = \nu\Delta\mu^0 + \frac{\nu(\nu-1)}{2} \delta\overline{\mu}^0. \qquad (4\text{-}292)$$

Thus we see that the descending ladder ($\Delta\mu^0$ is a negative number) of free-energy levels does not have equally spaced rungs. In fact, for an attractive interaction energy ($\delta\overline{\mu}^0$ negative) the rungs become more widely spaced as one descends the ladder. As we will see quantitatively below, the increase in rung spacing strongly enhances the occupancy of configurations having multiple bound ligands. This is the result of the strong effect of the Boltzmann factors $\exp(-\beta\Delta\mu_{\{\alpha\}}^0)$ in "pulling" ligands into advantageous low free-energy states. With this simple model for the free-energy levels ($\Delta\mu^0(\nu)$) we may readily obtain the relative occupancies R_ν of all configurations having the same number ν of bound ligands. Indeed, from

(4-261b) for $R_{\{\alpha\}}$ we find, that as a function of the monomer ligand fugacity z, that

$$R_\nu(z) = \binom{4}{\nu}\left(z\exp\left(-\frac{\Delta\mu^0}{kT}\right)\right)^\nu f^{\nu(\nu-1)}, \qquad (4\text{-}293a)$$

where

$$f \equiv \exp\left[-(\delta\overline{\mu}^0/2kT)\right]. \qquad (4\text{-}293b)$$

If we observe that $f^{\nu(\nu-1)} = f^{3\nu}f^{-\nu(4-\nu)}$, we can express (4-293a) as

$$R_\nu(z) = \binom{4}{\nu}y^\nu f^{-\nu(4-\nu)}, \qquad (4\text{-}294)$$

where

$$y = z\exp\left[-\frac{(\Delta\mu^0 + \tfrac{3}{2}\delta\overline{\mu}^0)}{kT}\right]. \qquad (4\text{-}294a)$$

The variable y is directly proportional to the fugacity z, thus $d\ln y = d\ln z$. Furthermore, if we wish to express y in terms of the partial pressure p_{O_2} of oxygen in the gas phase, we have

$$y = \left(\frac{p_{O_2}}{P_F}\right)\exp\left\{-\left[\frac{(\Delta\mu^0 - \mu^0_{O_2}) + \tfrac{3}{2}\delta\overline{\mu}^0}{kT}\right]\right\}. \qquad (4\text{-}295)$$

Here, the quantity $(\Delta\mu^0 - \mu^0_{O_2})$ is the free-energy advantage of transferring an oxygen molecule from the gas phase and binding it to the protein. Note that $\mu^0_{O_2} = \mu^0_g$ as defined in Eqn. (4-263c). The variable y has been defined so as to provide a manifestly symmetric structure to the relative occupancy function R_ν. Here the quantities $\nu(4-\nu)$ and $\binom{4}{\nu}$ have values which are the same symmetrically about $\nu = 2$. These symmetry properties will exhibit themselves in the binding polynomial, and will prove useful in analyzing the properties of the saturation function $Y(y)$ and the species fractions $P_\nu(y)$.

The binding polynomial can be obtained from the relative occupancies $R_\nu(y)$ giving

$$B(y) = \sum_{\nu=0}^{4} R_\nu(y) = (1 + 4yf^{-3} + 6y^2 f^{-4} + 4y^3 f^{-3} + y^4). \qquad \text{(4-296)}$$

From this we can deduce the saturation function since

$$Y = \frac{1}{4}\left(\frac{\partial \ln B(z)}{\partial \ln z}\right) = \frac{1}{4}\left(\frac{\partial \ln B}{\partial \ln y}\right). \qquad \text{(4-297a)}$$

Thus we find

$$Y(y) = \frac{y(f^{-3} + 3yf^{-4} + 3y^2 f^{-3} + y^3)}{(1 + 4yf^{-3} + 6y^2 f^{-4} + 4y^3 f^{-3} + y^4)}. \qquad \text{(4-297b)}$$

Having obtained the form of the oxygen saturation function we examine briefly some of its properties, and show how the fundamental energy parameters $\delta\overline{\mu}^0$ and $(\Delta\mu^0 - \mu_{O_2}^0)$ can be determined experimentally from observations of the dependence of the oxygen saturation versus the oxygen partial pressure.

We first observe that the half-saturation condition $Y = \frac{1}{2}$ occurs when $y = 1$. Thus, if we designate the experimentally observable oxygen pressure for half-saturation as $p_{1/2}(T)$, we see from (4-295) that we obtain one relation for the two parameters $\delta\overline{\mu}^0$ and $(\Delta\mu^0 - \mu_{O_2}^0)$, viz:

$$p_{1/2}(T) = P_F(T)\exp\left[\frac{(\Delta\mu^0 - \mu_{O_2}^0) + \frac{3}{2}\delta\overline{\mu}^0}{kT}\right]. \qquad \text{(4-298)}$$

A second measurement that can be made experimentally is the slope $(dY(p)/dp)_{p_{1/2}}$ of the saturation function at the half-saturation pressure $p_{1/2}$. Since y is directly proportional to p_{O_2} and $y = 1$ when $p_{O_2} = p_{1/2}$, it follows that

$$p_{1/2}\left(\frac{dY}{dp}\right)_{p_{1/2}} = \left(\frac{dY(y)}{dy}\right)_{y=1}. \qquad \text{(4-299a)}$$

The left-hand side of this equation can be found experimentally. The right-hand side can be calculated from (4-297b) and is given by

$$\left(\frac{dY}{dy}\right)_{y=1} = \frac{(1 + 5f^{-3} + 3f^{-4} + 8f^{-6} + 3f^{-7})}{(1 + 4f^{-3} + 3f^{-4})^2}. \qquad \text{(4-299b)}$$

Since f depends solely on $(\delta\mu^0)$ an experimental measurement of (dY/dy) using (4-299a) permits a determination of $\delta\mu^0$ using (4-299b). Using this value one can then determine $(\Delta\mu^0 - \mu_{O_2}^0)$ using the experimental value for $p_{1/2}(t)$ in (4-298). Indeed upon fitting the theory to data one may deduce that at $37\,^\circ$C and pH $= 7.4$:

$$\frac{(\Delta\mu^0 - \mu_{O_2}^0)}{kT} = -16.7 \tag{4-300a}$$

and

$$\left(\frac{\delta\overline{\mu}^0}{kT}\right) = -1.7. \tag{4-300b}$$

In Figure 4.52 we show how the theoretical fit to the data on sheep blood at $37\,^\circ$C and pH $= 7.4$ enables a determination that the interaction parameter f has the value $f^2 \cong 5.5$. The value of $p_{1/2}$ for human blood at pH $= 7.4$ and $T = 37\,^\circ$C is ~ 2.65 mmHg.

In Figure 4.53 we use the value of $f^2 = 5.5$ in (4-297b) for Y and plot the theoretical prediction for the fractional saturation versus y. This figure shows the characteristic sigmoidal shape of the fractional saturation curve. The effect of the attractive interaction between occupied heme sites has the effect of producing a sharp transition between states for which no oxygen is bound ($Y \cong 0$) to states for which four oxygens are bound ($Y \cong 1$). In effect $Y(y)$ now resembles a switch between zero oxygens bound to four oxygens bound over a relative narrow range of oxygen pressure.

Let us now return to (4-300a, b) and examine more closely the significance of the numerical results exhibited there. Using the fact that kT represents an energy of 0.616 kcal/mol at $37\,^\circ$C, we see that

$$(\Delta\mu^0 - \mu_{O_2}^0) = -10.2 \text{ kcal/mol} \tag{4-301a}$$

and

$$\delta\overline{\mu}^0 = -1.05 \text{ kcal/mol}. \tag{4-301b}$$

We may compare $(\Delta\mu^0 - \mu_{O_2}^0)$ as found for hemoglobin with the comparable quantity $\delta\overline{\mu}^0$ for myoglobin. The former being -10.2 kcal/mol, and the latter being -13.2 kcal/mol. We observe that oxygen is more tightly bound to the heme group in myoglobin than to the heme group in hemoglobin. This difference is likely due

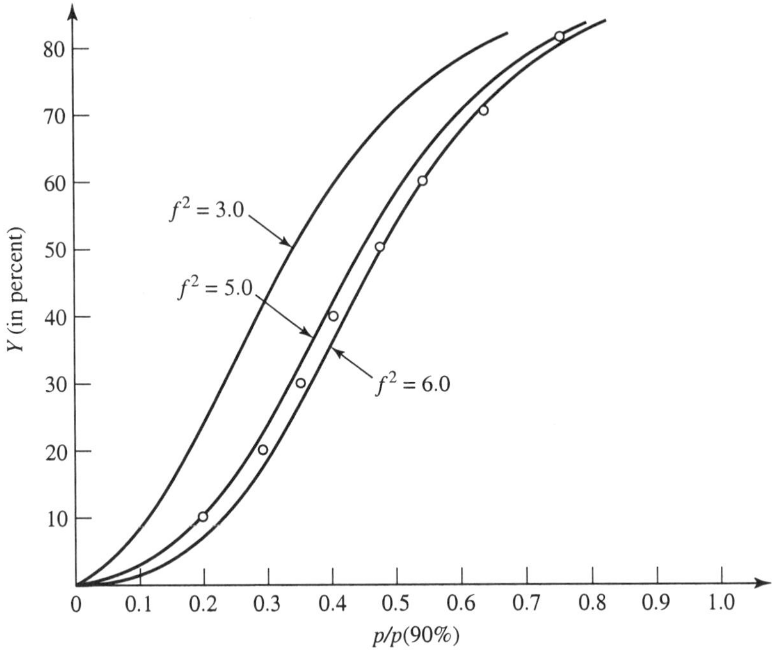

Figure 4.52. Theory and experimental data on the oxygen saturation curve. Circles represent data for sheep blood at 37 °C, pH = 7.4. (From Bartels and Harms, *Arch. Ges. Physiol.* **268**, 335 (1959)). The abscissa p is scaled so that all curves coincide at the value of p for which Y is equal to 0.90 or 90%.

to the fact that in the case of the hemoglobin tetramer there are ligands such as 2–3 diphosphoglycerate that bind to their own specific sites and interact repulsively with oxygen bound to the heme sites. These so-called "negative heterotropic linkages" are very important in modulating the value of the half-saturation pressure.

Turn now to the value of the attractive energy of interaction $\delta\bar{\mu}^0 = -1.7kT$. We must be impressed that the magnitude of this interaction energy is so small, indeed it is comparable to the thermal energy kT. Nevertheless, this small, subtle interaction is sufficiently strong so that it can convert the dissociation curve from a nonfunctional hyperbola to the physiologically essential sigmoidal form.

From a biological point of view the function of hemoglobin is determined by the two free energies $(\Delta\mu_0 - \bar{\mu}^0_{O_2})$ and $(\delta\bar{\mu}^0)$. The true connection between biological structures and function lies in a molecular explanation for the observed numerical values of these two energies.

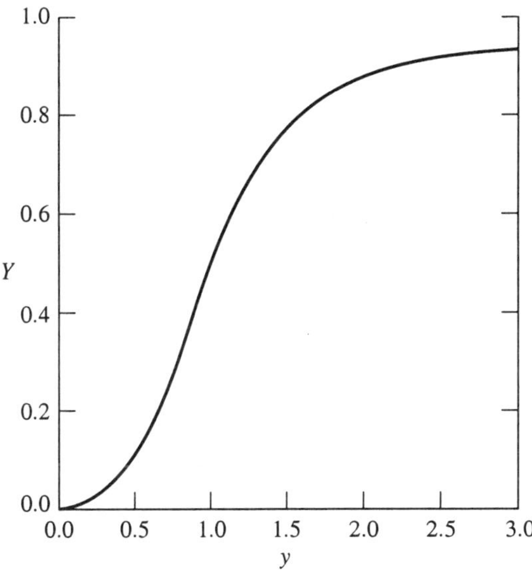

Figure 4.53. The saturation function $Y(y)$ as predicted using the Pauling tetrahedral model (4-297b) and the experimentally deduced value of $f^2 = 5.5$. This saturation function clearly exhibits the sigmoidal shape produced by the homotropic interaction between oxygen molecules.

We may conclude our examination of the Pauling tetrahedral model by considering the dependence of the species fraction P_ν on the mean occupancy $\bar{\nu}$. From the definition of the species fraction we can immediately write P_ν as a function of the variable y using (4-294) and (4-296), viz:

$$P_\nu(y) = \frac{R_\nu(y)}{B(y)} = \frac{\binom{4}{\nu} y^\nu f^{-\nu(4-\nu)}}{(1 + 4yf^{-3} + 6y^2 f^{-4} + 4y^3 f^{-3} + y^4)}. \tag{4-302}$$

On the other hand, the variable y is connected to the mean occupancy $\bar{\nu}$ by $\bar{\nu} = 4Y(y)$. Thus for each value of ν we may obtain the corresponding $\bar{\nu}$, and thereby determine the probability $P_\nu(\bar{\nu})$. The result of this determination, using (4-302), and $Y(y)$ as shown graphically in Figure 4.53 using $f^2 = 5.5$, is shown in Figure 4.54.

These species fractions exhibit the following interesting features. First, for each ν, P_ν has a maximum value $\bar{\nu} = \nu$. Second, the species fractions exhibit

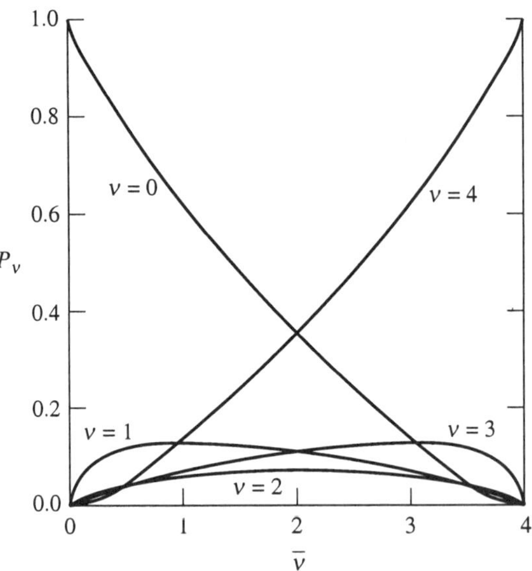

Figure 4.54. The species fractions $P_\nu(\overline{\nu})$. The probability of finding ν ligands bound to a hemoglobin molecule as a function of the mean number $\overline{\nu}$ bound. Prediction of the Pauling tetrahedral model using the interaction parameter $f^2 = 5.5$.

mirror reflection symmetry about the value $\overline{\nu} = 2$. That is,

$$P_\nu(\overline{\nu}) = P_{4-\nu}(4 - \overline{\nu}).$$

A proof of both these properties for $P_\nu(\nu)$, regardless of the values of the interaction parameters, will be left as a homework problem.

We may compare $P_\nu(\overline{\nu})$, as obtained using the Pauling model, and $f^2 = 5.5$ with that obtained previously and shown in Figure 4.51 for the case of identical free energies for each subunit and no interaction between subunits. We observe that in the Pauling model, even though the coupling free energy $(\delta\overline{\mu}^0)$ is very small $-1.7\,kT$, the intermediate species fractions for $\nu = 1, 2, 3$ are dramatically reduced in comparison with the dominant $\nu = 0$ and $\nu = 4$ species. The sigmoidal or switch-like character of the saturation function is closely connected with enhancement of the occupancies of the $\nu = 0$ and $\nu = 4$ species and the suppression of the intermediate $\nu = 1, 2, 3$ species. As we have seen above, the basic mechanism underlying the sigmoidicity is an increase in the spacing of the rungs of the free-energy ladder as the number of ligands ν increases. In the Pauling model this increase in rung spacing is the result of attractive interactions between pairs of oxygen molecules occupying the heme groups.

c) The Monod–Wyman–Changeaux (MWC) Allosteric Model for Oxygen Binding to Hemoglobin.

In 1965, Monod, Wyman, and Changeaux introduced an ingenious model [7] for oxygen binding to hemoglobin. This model also possesses the essential feature that the rung spacing of the ladder of free-energy levels increases as more ligands are bound. However, the basic mechanism for this increase in the MWC model is strikingly different from that envisioned in the Pauling model.

The basic ideas of the MWC model can be summarized qualitatively as follows. First, the hemoglobin tetramer is assumed to be able to take on two distinct structural (allosteric) forms. These two forms are designated by the symbols T (tense) and R (relaxed). Second, the oxygen molecules free in solution can bind to each heme site on the T form with the same free-energy advantage $\Delta\mu_T^0$ regardless of the occupancy of other sites. Similarly the ligand molecules can bind to each heme site on the R form with the fixed free-energy advantage $\Delta\mu_R^0$ regardless of the occupancy of other sites. Both $\Delta\mu_T^0$ and $\Delta\mu_R^0$ are negative numbers, however, $\Delta\mu_R^0$ is taken to be a larger negative number than $\Delta\mu_T^0$. Thus the T form of the hemoglobin tetramer has lower "affinity" for binding oxygen as compared to the high "affinity" R form. Third, the relative proportion of the unliganded T_0 and R_0 forms is determined by a reaction equilibrium of the form $T_0 \leftrightarrow R_0$. That is, the difference $\Delta\mu_{RT}^0$ between the standard free energy of the R form and the T form, $\Delta\mu_{RT}^0$, is assumed to be positive, i.e., the unliganded R form is energetically disadvantageous compared to the unliganded T form. Thus in the absence of ligands the hemoglobin tetramers are overwhelmingly likely to be found in the T form.

On the basis of these assumptions we can readily envision the evolution of the various ligand bound configurations as a function of the free ligand concentration. At low free ligand concentration the ligand binds primarily to the predominant low affinity T form with free-energy advantage $\Delta\mu_T^0$. However, as the ligand concentration increases, the free-energy advantage $\Delta\mu_R^0$ of binding to the higher affinity R forms results in an increase in the relative occupancy of these forms. This increase in R forms continues until the R forms actually begin to predominate over the T forms. Under these conditions further binding of oxygen takes place primarily on R forms, where the free energy is $\Delta\mu_R^0$. Thus, in this MWC model the effective spacing of the ladder of free-energy levels shows an increase as more ligands are bound. From the point of view of chemical equilibrium, we see that the liganding process has the effect of "pulling" the $T \leftrightarrow R$ reaction equilibrium from the low affinity T forms into the higher affinity R forms. Equivalently, the form of hemoglobin tetramers "switches" from the T form at low-ligand binding ($\bar{\nu} < 2$) to the R form for high-ligand binding ($\bar{\nu} > 2$).

A direct consequence of this picture is the conclusion that the crystal structure of fully deoxygenated hemoglobin should be different from that for fully oxygenated hemoglobin. Indeed such a finding had early been made by Felix Haurowitz [8] in 1938. He found that crystals of oxygenated hemoglobin had the form of scarlet needles, while deoxygenated hemoglobin had the form of purple plates. He concluded that this difference in crystal habit signified that the hemoglobin molecule had a different structure depending on whether it was oxygenated or deoxygenated. Perutz [9] has described vividly the dramatic events connected with the publication of this important but neglected discovery. Indeed, it was Haurowitz's discovery which initiated the brilliant series of X-ray diffraction studies that Perutz has conducted, throughout his life, on the detailed structure of the hemoglobin molecule [9].

Having outlined the conceptual foundation of the MWC model we are now in a position to determine the form of the free-energy level diagram and the relative occupancy of each possible species of hemoglobin tetramer. From this we will deduce the form of the binding polynomial, and hence the oxygen saturation function Y and the species fractions. According to the MWC model, for each value ν of the number of ligands bound there are two distinct tetrameric forms that we will designate as $T(\nu)$ and $R(\nu)$. The number of occupation patterns associated with each form $T(\nu)$ and $R(\nu)$ is the multiplicity or degeneracy $M(\nu) = \binom{4}{\nu}$. We chose as a reference state the unliganded $T(0)$ state. Relative to this state, the unliganded $R(0)$ state lies higher in free energy by the amount $\Delta\mu_{RT}^0$. The protein free-energy change on binding ν ligands to the T form is $\nu\Delta\mu_T^0$ where $\Delta\mu_T^0$ is a negative number. The protein free-energy change relative to the position of the $R(0)$ level upon binding ν ligands to the R form is $\nu\Delta\mu_R^0$ where $\Delta\mu_R^0$ is also a negative number but larger in magnitude than $\Delta\mu_T^0$. The two protein free-energy level ladders are shown side by side in Figure 4.55. We now consider the relative occupancy R of each state choosing the $T(0)$ state as the reference state for which $R_T(0) = 1$. Since the unliganded forms are in reversible equilibrium with one another, i.e., $\mu_T = \mu_R$, it follows that

$$R_T(0) = 1, \tag{4-303a}$$

$$R_R(0) = \exp\left(-\beta\Delta\mu_{RT}\right) \equiv \left(\frac{1}{L}\right). \tag{4-303b}$$

State	Free energy level		Relative occupancy	
	T	R	T	R
$v = 0$ $T(0); R(0)$	0 ———	——— $\Delta\mu^0_{RT}$	1	$(1/L)$
$v = 1$ $T(1); R(1)$	$\Delta\mu^0_T$ ———	——— $\Delta\mu^0_{RT} + \Delta\mu^0_R$	$4(cy)$	$4y/L$
$v = 2$ $T(2); R(2)$	$2\Delta\mu^0_T$ ———	——— $\Delta\mu^0_{RT} + 2\Delta\mu^0_R$	$6(cy)^2$	$6y^2/L$
$v = 3$ $T(3); R(3)$	$3\Delta\mu^0_T$ ———	——— $\Delta\mu^0_{RT} + 3\Delta\mu^0_R$	$4(cy)^3$	$4y^3/L$
$v = 4$ $T(4); R(4)$	$4\Delta\mu^0_T$ ———	——— $\Delta\mu^0_{RT} + 4\Delta\mu^0_R$	$(cy)^4$	y^4/L

Figure 4.55. Free-energy level diagram and relative occupancies associated with each state $T(v)$ and $R(v)$ in the MWC model.

Here $\beta = 1/kT$ and L is the so-called allosteric constant; it is very large compared to unity, since $\beta\Delta\mu^0_{RT} \gg 1$.

The binding of oxygen to each form of the protein is governed by (4-261b). Thus we may express the relative occupancies $R_T(v)$ and $R_R(v)$ as

$$R_T(v) = \binom{4}{v} \left(z\exp(-\beta\Delta\mu^0_T)\right)^v , \tag{4-304a}$$

$$R_R(v) = \binom{4}{v} \left(\frac{1}{L}\right) \left(z\exp(-\beta\Delta\mu^0_R)\right)^v . \tag{4-304b}$$

It is now convenient to define the variable y which is directly proportional to the fugacity z as

$$y \equiv z\exp(-\beta\Delta\mu^0_R). \tag{4-305a}$$

It is also convenient to define the quantity c as

$$c = \exp\left[\beta\left(\Delta\mu^0_R - \Delta\mu^0_T\right)\right]. \tag{4-305b}$$

Since we expect $\Delta\mu_R^0$ to be a larger negative number than $\Delta\mu_T^0$, we also expect that c, which can be regarded as the relative oxygen binding affinity of T and R forms, will be a number small compared to unity. With these definitions we find that for $0 \leq \nu \leq 4$ the relative occupancies $R_T(\nu)$ and $R_R(\nu)$ can be written as

$$R_T(\nu) = \binom{4}{\nu}(cy)^{\nu}, \tag{4-306a}$$

$$R_R(\nu) = \binom{4}{\nu}\left(\frac{1}{L}\right)y^{\nu} \tag{4-306b}$$

for all ν.

It is important furthermore to note that, if the fugacity z is expressed in terms of the partial pressure of oxygen in the overlying gas phase, then the variable y can be expressed as

$$y = \left(\frac{p_{O_2}}{P_F}\right)\exp(-\beta\delta\mu_R^0), \tag{4-307a}$$

where

$$\delta\mu_R^0 = (\Delta\mu_R^0 - \mu_{O_2}^0). \tag{4-307b}$$

Here $\delta\mu_R^0$ is the free-energy advantage associated with removing an oxygen molecule from the gas phase and its attachment to the protein in the R form. We exhibit the relative occupancies $R_T(\nu)$ and $R_R(\nu)$ alongside the free-energy ladders in Figure 4.55.

Having now established the relative occupancies of each of the possible oxygen binding states, we may now obtain the binding polynomial $B(y)$. Indeed, since

$$B(y) = \sum_{\nu=0}^{4}(R_T(\nu) + R_R(\nu)), \tag{4-308a}$$

it follows that

$$B(y) = (1 + cy)^4 + \frac{1}{L}(1 + y)^4. \tag{4-308b}$$

From this we may deduce the oxygen fractional saturation Y as a function of y on the oxygen partial pressure p_{O_2}, since according to (4-269):

$$Y = \left(\frac{1}{4}\right)\frac{\partial \ln B(z)}{\partial \ln z} = \frac{1}{4}\left(\frac{\partial \ln B(y)}{\partial \ln y}\right), \qquad (4\text{-}309)$$

since y is directly proportional to z from (4-305a). In carrying out the indicated operations in (4-309) we find that the oxygen saturation function predicted by the MWC model is given by

$$Y(y) = \frac{y(1+y)^3 + Lcy(1+cy)^3}{(1+y)^4 + L(1+cy)^4}. \qquad (4\text{-}310)$$

The MWC model contains three free-energy parameters: $\Delta\mu^0_{RT}$, $\Delta\mu^0_R$, $\Delta\mu^0_T$. These are obtained by fitting (4-310) to the experimental data on oxygen saturation versus the oxygen pressure. Such a fit will provide the values of L, c, and the numerical relationship between the scaled variable y and p_{O_2}. Using these parameters the three free-energy parameters can be obtained. The fitting procedure can, of course, be carried out by computer. However, it is useful to examine briefly the mathematical features of the MWC model to see how such a fit can be achieved conceptually.

Consider first the value of y above which the switchover from T forms to R forms occurs. If we call this the midpoint value y_m, we see from (4-308b) that the occupancy of all the T forms becomes equal to the occupancy of all the R forms when $y = y_m$ where

$$y_m = (L^{1/4} - 1)/(1 - cL^{1/4}). \qquad (4\text{-}311)$$

Though y_m is not identical to the half-saturation value $y_{1/2}$ (for which $Y = \frac{1}{2}$), it is nevertheless quite comparable to it. Thus, we see that the scale for y is set, when $cL^{1/4} \ll 1$, by the magnitude of the allosteric constant $L = \exp(\Delta\mu^0_{RT}/kT)$. The half-saturation of the oxygen association curve occurs for values of the pressure for which $y \sim (L^{1/4} - 1) \sim y_m$. The experimentally observed values of L range from $\sim 10^4$ to $\sim 10^8$ depending upon the solution conditions.

Furthermore, we see from (4-310) that at the foot of the oxygen dissociation curve ($y \ll y_m$) that Y rises linearly with y according to

$$Y(y) \cong \left(\frac{y}{L}\right), \qquad y \ll 1.$$

On the other hand, in the domain $y \gg y_m$, the oxygen saturation function becomes hyperbolic, i.e.,

$$Y(y) \cong \left(\frac{y}{1+y}\right), \qquad y \gg y_m.$$

The half-saturation pressure in this asymptotic domain is that for which $y = 1$.

Careful fitting of the MWC theoretical curve to the slope, and magnitude of the observed oxygen saturation function enables a determination of the parameters L and c. Monod, Wyman, and Changeaux reported [7] the following parameters to fit data obtained by Lyster on horse hemoglobin measured at pH $= 7.0$, $T = 19\,°\text{C}$. Here $(p_{O_2})_{1/2} = 10$ Torr and $L = 9.054$; $c = 0.014$. Using these parameters, we can deduce the following values of the fundamental free-energy differences

$$\left(\frac{\Delta\mu_{RT}^0}{kT}\right) = +9.11, \tag{4-312a}$$

$$\left(\frac{\Delta\mu_R^0}{kT}\right) = -22.3, \tag{4-312b}$$

$$\left(\frac{\Delta\mu_T^0}{kT}\right) = -18.0. \tag{4-312c}$$

Using $T = 19\,°\text{C}$, we find

$$\Delta\mu_{RT}^0 = 5.3 \text{ kcal/mol}, \tag{4-313a}$$

$$\Delta\mu_R^0 = -12.9 \text{ kcal/mol}, \tag{4-313b}$$

$$\Delta\mu_T^0 = -10.4 \text{ kcal/mol}. \tag{4-313c}$$

In Figure 4.56, we show the position of each of the rungs of the free energy ladder corresponding to each occupancy of the T and R forms. Also, in Figure 4.57, we plot the theoretical form of the oxygen saturation function versus y on the lower abscissa and the corresponding pressure p_{O_2} on the upper abscissa. We use the parameters reported in [7]. Using this curve, one may obtain for each value of y the corresponding mean number $\bar{\nu}$ of oxygen molecules bound. The figure shows clearly the sigmoidal characteristic that is produced in this model by the switchover

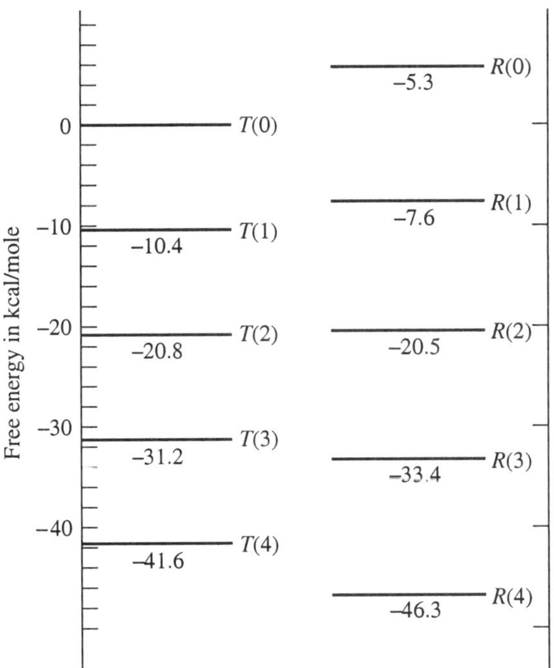

Figure 4.56. Free-energy differences in the T and R ladders for horse hemoglobin at $T = 19\,°\text{C}$ and pH $= 7.0$.

from occupancy of the T forms to occupancy of the R forms as y increases into the region around y_m.

We now examine the species fractions $P_T(\nu, \bar{\nu})$ and $P_R(\nu, \bar{\nu})$ for each of the possible ligand binding states ν as a function of the mean number of ligands ($\bar{\nu}$) bound. According to the definition of the species fractions and the relative occupancies $R_T(\nu, y)$ and $R_R(\nu, y)$, obtained previously in (4-306a, b), we see that

$$P_T(\nu, y) = \binom{4}{\nu}(cy)^2 \Big/ B(y),\qquad (4\text{-}314\text{a})$$

$$P_R(\nu, y) = \binom{4}{\nu}\frac{y^2}{L} \Big/ B(y),\qquad (4\text{-}314\text{b})$$

where the binding polynomial $B(y)$ is given by (4-308b). Instead of the variable y, we may express the species fractions in terms of the mean occupancy ($\bar{\nu}$) since

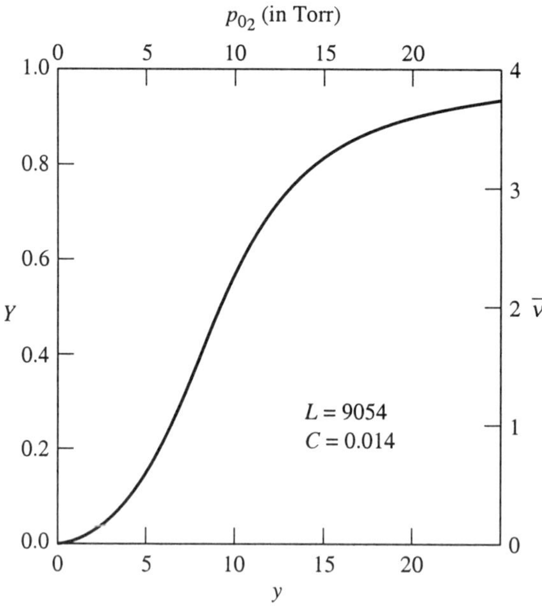

Figure 4.57. Theoretically calculated form of the saturation function $Y(y)$ using the parameters obtained by Monod, Wyman, and Changeaux [7] from data on horse hemoglobin.

$\bar{\nu} = 4B(y)$, which is shown in Figure 4.57. Using the parameters used above, we have plotted the various species fractions $P_T(\nu, \bar{\nu})$ and $P_R(\nu, \bar{\nu})$ in Figure 4.58. With these parameters, the MWC model predicts that for the T form only the $\nu = 0$ and $\nu = 1$ states are significantly occupied. The fractions $P_T(\nu = 2, 3, 4)$ are too small to be seen on the figure. Correspondingly, for the R form only the $\nu = 3$ and $\nu = 4$ forms are significantly occupied. The $\nu = 0, 1, 2$ R forms have negligible probability.

Figure 4.58 shows that as $\bar{\nu}$ increases from $\bar{\nu} = 0$, the species present are $T(0)$ and $T(1)$. The former declines steadily with increasing $\bar{\nu}$, the latter rises to a weak maximum at $\bar{\nu} = 1$ and then monotonically declines with increasing $\bar{\nu}$. Correspondingly, the $R(4)$ and $R(3)$ species grow increasingly important as $\bar{\nu}$ increases beyond $\bar{\nu} = 3$, while the latter rises to a weak minimum at $\bar{\nu} = 3$, after which it declines rapidly. Figure 4.58 shows clearly that the switchover mechanism from the T to R ladder occupancy has the effect of strongly suppressing intermediate species. The sigmoidal character of the dissociation curve, again in this model, is

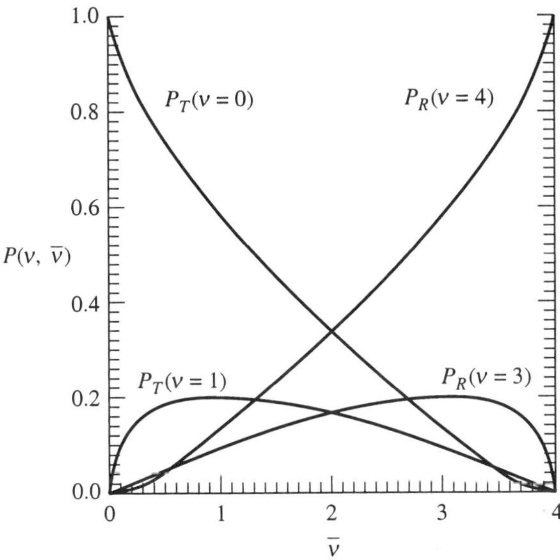

Figure 4.58. The species fractions $P_T(v, \bar{v})$ for $v = 0, 1$ and $P_R(v, \bar{v})$ for $v = 3, 4$ versus $\bar{v}$, according to the MWC model using $L = 9.054$ and $c = 0.014$.

connected with the predominance of the $\bar{v} = 0$ and $\bar{v} = 4$ species. The molecule acts more like an ideal switch, i.e., one which makes the transition from no ligand binding to fully saturated binding, over a relatively narrow range in the variable y, or the oxygen pressure p_{O_2}.

(v)   Further Models and Applications of the Theory of Ligand Binding

The Pauling model and MWC models presented above have been central to the further development of theories of ligand binding to proteins. It is useful at this point to draw attention to other models and other applications of the theory of ligand binding.

There is a class of models called "induced fit models" that represent an extension of the Pauling model. These models were introduced by Koshland [10] and Koshland, Nemethy, and Filmer [11]. In these models the overall quaternary structure of the protein is regarded as being fixed in a single form. However, the free energy with which a ligand binds is determined according to definite geometrical rules by the occupancy of binding sites on other subunits. Here, the occupancy of a given site is envisioned as altering the local environment of nearby ligand binding sites.

More recently, Di Cera, Robert, and Gill [16] have introduced a modification of the MWC model in order to explain very accurate measurements of the shape of the oxygen dissociation curve. In their model the basic postulate that the protein can exist in two conformationally distinct overall structures is retained. However, the free-energy ladder associated with each configuration of the T form is not equally spaced as was assumed in the original form of the MWC theory. Indeed, in the case of hemoglobin, the DRG model assumes that, in the T form of hemoglobin, oxygen can bind only to the two alpha subunits. No binding is allowed to the beta subunits of the T form. Furthermore, the free-energy advantage of binding of the first oxygen to the T form is assumed less than that for binding the second oxygen. This modification resembles the induced fit model. On the other hand, DRG follow the original MWC model by assuming that in the R form, the oxygen binding ladder rungs are all equally spaced. The ligand binds with equal affinity to each alpha and each beta subunit. The authors justify this hybrid model on the basis of sterochemical information obtained from high resolution X-ray crystallography and nuclear magnetic resonance. Furthermore, this model provides a much more accurate fit to high-resolution measurements of the dissociation curve.

It is important to realize that the theory of ligand binding is essential to understand the role of proteins as catalysts, motors, receptors and transducers, stores and transporters, genetic regulators, and more. For the student interested in learning more of the variety and depth of the biological physics of ligand binding, we draw attention to the following books and collections, listed in the references to this chapter, which contain a quantitative analysis of these functions. First, there is a collection of selected original classic papers describing the role of allosteric regulation in the function of enzymes (Tokushige [15]). Also, the more recent book by Herve [17] contains many further examples of the allosteric control of enzyme functions. Finally, we mention three books published in the period 1990–1998. The first of these is coauthored by Wyman, a pioneering figure in the field of allosteric regulation, viz, Wyman and Gill [18]. This book contains many biological examples and discussions of heterotropic linkage from a basic thermodynamic point of view. Second, a thorough mathematical representation of the theory of ligand binding is presented in Di Cera [19]. Third, we draw the reader's attention to Volume 15 in the series: *Advances in Protein Chemistry*. This volume is edited by Di Cera [20] and contains articles written by leading researchers on a variety of protein systems whose properties can be understood using the theory of ligand binding.

Appendices

4.A1   Multinomial Coefficients: Weight of a Macrostate for the Einstein Crystal

The question to be answered is the following: Consider a crystal made of N identical atoms. Each atom is in one of the possible quantum states n, where n assumes the values

$$n = 0, 1, 2, \ldots, n_M.$$

Let us call a "pattern" or "microstate" a specific arrangement of the N atoms, where each atom is in a specified quantum state n. Consider now the *set* of all patterns which have in common the property that N_0 of the N atoms are in quantum state $n = 0$, N_1 of them in quantum state $n = 1$, etc., irrespective of their location in space.

This set of all such patterns defines what we call the *macrostate* $\{N_0, N_1, \ldots, N_{n_M}\}$. The question is: How many distinct patterns do exist for a given macrostate? We call this number W, the weight of the macrostate.

We will now prove the proposition that this weight W is given by the multinomial coefficient

$$W = \frac{N!}{N_0!\, N_1! \cdots N_{n_M}!}.$$

To prove this, we proceed as follows: We consider first the macrostate specified by

$$N_0 \text{ atoms in state } n = 0$$

and

$$\overline{N}_1 = N - N_0 \text{ atoms in state } n = 1.$$

The weight W of this macrostate is the number of ways in which N_0 of the N atoms can be put into state $n = 0$, and the others into state $n = 1$. This number is identical to the number of distinct outcomes in N tosses of a coin in which a total of N_0 heads and $\overline{N}_1$ tails occur. It is given by the binomial coefficient

$$W(N_0, \overline{N}_1, 0, 0, \ldots, 0) = \frac{N!}{N_0! \, \overline{N}_1!}.$$

Consider now all $\overline{N}_1$ atoms in state $n = 1$. Let us keep N_1 of them in state 1, and move $\overline{N}_2 = (\overline{N}_1 - N_1)$ of them into state $n = 2$. For *each* pattern containing $\overline{N}_1$ atoms in state $n = 1$, there are

$$\frac{\overline{N}_1!}{N_1! \, \overline{N}_2!}$$

ways of keeping N_1 of them in state $n = 1$ and moving $\overline{N}_2$ of them into state $n = 2$. So, the the total number of patterns with N_0 particles in state $n = 0$, N_1 in state $n = 1$, and $\overline{N}_2 = N - N_0 - N_1$ in state $n = 2$ is

$$W(N_0, N_1, \overline{N}_2, 0, \ldots, 0) = \frac{N!}{N_0! \, \overline{N}_1!} \frac{\overline{N}_1!}{N_1! \, \overline{N}_2!} = \frac{N!}{N_0! \, N_1! \, \overline{N}_2!}.$$

We now repeat the procedure. Of the $\overline{N}_2$ particles in $n = 1$, we keep N_2 of them and move $\overline{N}_3 = (\overline{N}_2 - N_2)$ into state $n = 3$. This can be done in

$$\frac{\overline{N}_2!}{N_2! \, \overline{N}_3!}$$

ways, giving

$$W(N_0, N_1, N_2, \overline{N}_3, 0, \ldots, 0) = \frac{N!}{N_0! \, N_1! \, N_2! \, \overline{N}_3!}.$$

We repeat this until the final macrostate $(N_0, N_1, \ldots, N_{n_M})$ is reached. This establishes our formula for W.

4.A2   Occupancy of Microcells by Atoms of an Ideal Gas

It was stated without proof in Section 4.2.B(iii) that at room temperature and atmospheric pressure, most microcells (of size h^3) in phase space are empty. The truth of this proposition is a prerequisite for the validity of (4-51) for the number of distinct population patterns in the ith macrocell.

To prove the proposition, we start from expression, (4-77), for the equilibrium density of particles in phase space at temperature T. This formula shows that the highest density of phase-space population occurs at $\vec{p} = 0$, and is given thereby

$$\rho_0 = \frac{N}{V}\left(\frac{1}{2\pi mkT}\right)^{3/2}.$$

In this expression N is the number of gas atoms, V the volume in which they are contained, and m the mass of the gas atom.

The volume of a microcell in phase space is h^3, h being Planck's constant

$$h = 6.63 \times 10^{-27} \text{ erg s.}$$

The proposition that we have to prove is that the quantity

$$\rho_0 h^3,$$

that represents the mean number of particles per microcell or, equivalently, the probability of finding a gas atom in a particular microcell, is very small compared to 1:

$$\rho_0 h^3 \ll 1.$$

We will do this for the case of helium gas at atmospheric pressure p_0 and room temperature ($T = 300\,°\text{K}$). Under these conditions one mole (N_0 particles) of gas occupies a volume $V_0 = RT/p_0$. We may then write

$$\begin{aligned}
\rho_0 h^3 &= \frac{N_0}{V_0}\left(\frac{h^2}{2\pi mkT}\right)^{3/2} \\
&= \frac{N_0 V_F(T)}{V_0},
\end{aligned}$$

where the parentheses raised to the $\frac{3}{2}$-power represents the "Fermi volume" introduced in (4-116):

$$V_F(T) = \left(\frac{h^2}{2\pi mkT}\right)^{3/2}.$$

Let us start by calculating $V_F(T)$. With the numerical data

$$h = 6.63 \times 10^{-27} \text{ g cm}^2 \text{ s}^{-1},$$
$$m = 6.68 \times 10^{-24} \text{ g},$$
$$k = 1.38 \times 10^{-16} \text{ g cm}^2 \text{ s}^{-2}/^\circ\text{K},$$
$$T = 300\,^\circ\text{K},$$

we find

$$\left(\frac{h^2}{2\pi mkT}\right) = 2.53 \times 10^{-17} \text{ cm}^2,$$

and hence

$$V_F(T = 300\,^\circ\text{K}) = 1.27 \times 10^{-25} \text{ cm}^3.$$

Now the value V_0 of one mole of ideal gas at atmospheric pressure p_0 and temperature $T = 300\,^\circ\text{K}$ is

$$V_0 = 2.46 \times 10^4 \text{ cm}^3.$$

This gives us

$$\rho_0 h^3 = \frac{N_0 V_F(T)}{V_0}$$
$$= \frac{6 \times 10^{23} \times 1.27 \times 10^{-25}}{2.46 \times 10^4} \cong 3 \times 10^{-6}.$$

This demonstrates that under the state conditions for an ideal gas (helium), the probability that a microcell is occupied is of the order of 3×10^{-6}.

4.A3 The Equipartition Theorem of Classical Statistical Mechanics

In Section 4.2 we considered thermal equilibrium in two systems, the Einstein crystal and the ideal monoatomic gas. The equilibrium of the first system was described quantum mechanically, and we found

$$P(n) = \frac{1}{Z_c} e^{-\varepsilon_n/kT} \tag{A3-1}$$

for the probability that an oscillating atom of the crystal is in the nth quantum state of energy ε_n. (Z_c is the normalization constant.)

For the gas, where we applied a classical treatment, we described thermal equilibrium through a probability dP for finding a particular atom in a volume element $d\Omega$ of phase space

$$dP = \frac{d\Omega}{Z_G} e^{-\varepsilon(r,p)/kT}, \tag{A3-2}$$

where Z_G is a normalization constant, and $\varepsilon(r, p)$ is the energy of the particle located in $d\Omega$ at location $(\vec{r}, \vec{p})$ in phase space.

Every *classical* statistical description of motion defines thermal equilibrium by means of a probability distribution in phase space. What this phase space is depends on the system we describe. The elements of the system may be particles, as in a gas, and in the Einstein crystal, or standing waves, as in the Debye crystal. Classically, the motion of every such "element" or "quasi-particle" can be described as a moving point in a $2f$-dimensional phase space. The number f is the number of "degrees of freedom" of the element. For an atom in a gas, $f = 3$, since we need three coordinates xyz, and three momentum components p_x, p_y, p_z to describe the location and state of motion of an atom. For a diatomic molecule, $f = 6$. The simplest "elements" or "quasi-particles" have $f = 1$, such as the atoms in the crystal model introduced in Section 4.2.A, that could oscillate in just one direction.

In a classical statistical description of a system of such elements, the key problem is to find the probability that such a quasi-particle is located inside a volume element $d\Omega$ of phase space at a location corresponding to an energy ε of the element. Using methods entirely analogous to the ones applied in Section 4.2.B for the ideal gas, one finds that this probability dP is given by the product of the phase-space volume $d\Omega$ and a Boltzmann factor

$$dP = d\Omega e^{-\varepsilon/kT}/Z, \tag{A3-3}$$

where Z is a normalization constant, chosen to make $\int dP = 1$.

Let us illustrate the implication of this result for the linear harmonic oscillation of an atom in the crystal model introduced in Section 4.2.A. Notice that in contrast to what we did there, we now view this case as a problem of classical statistical physics.

The phase space of this oscillation is two-dimensional, since $f = 1$, and the energy of the oscillating atom at a point (x, p) in phase space is

$$\varepsilon(x, p) = \varepsilon_{\text{kin}}(p) + \varepsilon_{\text{pot}}(x) = \frac{p^2}{2m} + \frac{m\omega^2}{2}x^2, \tag{A3-4}$$

m being the mass, and ω the natural frequency of oscillation of the atom. Let us use this expression for the energy and the probability distribution (A3-3) to calculate the mean energy $\bar{\varepsilon}$ of the oscillator, keeping in mind that in this present case we have

$$d\Omega = dx\, dp$$

for the phase-space volume element. This gives us the following expression for $\bar{\varepsilon}$:

$$\bar{\varepsilon} = \int dx \int dp\, \varepsilon(x, p) e^{-\varepsilon(x,p)/kT} / Z. \tag{A3-5}$$

Now, because ε is the sum of the two terms

$$\varepsilon = \varepsilon_{\text{kin}} + \varepsilon_{\text{pot}},$$

we can write the Boltzmann factor as a product

$$e^{-\varepsilon/kT} = e^{-\varepsilon_{\text{kin}}/kT} \cdot e^{\varepsilon_{\text{pot}}/kT},$$

and the normalization integral, too, becomes a product

$$Z = \int dx\, dp\, e^{-\varepsilon/kT} = \int dp\, e^{-\varepsilon_{\text{kin}}/kT} \int dx\, e^{-\varepsilon_{\text{pot}}/kT}.$$

It is now easily seen that the mean energy $\bar{\varepsilon}$ can be written as the sum of two terms

$$\bar{\varepsilon} = \int dp\, \varepsilon_{\text{kin}} e^{-\varepsilon_{\text{kin}}/kT} \bigg/ \int dp\, e^{-\varepsilon_{\text{kin}}/kT} + \int dx\, \varepsilon_{\text{pot}} e^{-\varepsilon_{\text{pot}}/kT} \bigg/ \int dx\, e^{-\varepsilon_{\text{pot}}/kT}. \tag{A3-6}$$

Notice that this expression for $\bar{\varepsilon}$ is possible because the energy ε is the sum of two terms, the first of which depends only on p, the second only on x.

We now evaluate these integrals. ε_{kin} is proportional to p^2, ε_{pot} to x^2. Both terms in (A3-6), $\overline{\varepsilon_{\text{kin}}}$ and $\overline{\varepsilon_{\text{pot}}}$, are the ratio of two integrals of the same type. In the expression for $\overline{\varepsilon_{\text{kin}}}$, call

$$\frac{1}{2mkT} \equiv \alpha.$$

In the expression for $\overline{\varepsilon_{\text{pot}}}$, call

$$\frac{m\omega^2}{2kT} \equiv \gamma.$$

Then we have

$$\overline{\varepsilon_{\text{kin}}} = kT \int dp (\alpha p^2) e^{-\alpha p^2} \Big/ \int dp\, e^{-\alpha p^2},$$

$$\overline{\varepsilon_{\text{pot}}} = kT \int dx (\gamma x^2) e^{-\gamma x^2} \Big/ \int dx\, e^{-\gamma x^2}. \tag{A3-7}$$

In Appendix 2.A3 to Chapter 2, we showed that the two integrals occurring above have the values

$$\int dp\, e^{-\alpha p^2} = \sqrt{\frac{\pi}{\alpha}},$$

$$\int dp\, p^2 e^{-\alpha p^2} = \frac{1}{2\alpha} \sqrt{\frac{\pi}{\alpha}}. \tag{A3-8}$$

This shows that both $\overline{\varepsilon_{\text{kin}}}$ and $\overline{\varepsilon_{\text{pot}}}$ have the same value

$$\overline{\varepsilon_{\text{kin}}} = \overline{\varepsilon_{\text{pot}}} = \tfrac{1}{2}kT. \tag{A3-9}$$

Notice that this result for the mean kinetic and potential energy of the linear oscillator is independent of the mass of the oscillating particle, and independent of the frequency. It is entirely due to the fact that both kinetic and potential energy are a *quadratic function* of the variables p and x, respectively.

The result [(A3-9)] is an illustrative example of the general *equipartition theorem* of classical statistical mechanics.

This theorem, which we shall state presently, is based on the following premises:

(a) For any element of the system, that has f degrees of freedom, the equilibrium probability distribution in its $2f$-dimensional phase space is of the form (A3-3):

$$dP = d\Omega\, e^{-\varepsilon/kT}/Z. \tag{A3-10}$$

(b) The kinetic energy is the *sum* of f terms depending on only *one* of the f momentum variables p_i $(i = 1, \ldots, f)$:

$$\varepsilon_{\text{kin}} = \sum_{i=1}^{f} p_i^2/2m_i. \tag{A3-11}$$

(c) The elements of the system have an equilibrium configuration of minimum potential energy. For a configuration near equilibrium, the potential energy can be written as the sum

$$\varepsilon_{\text{pot}} = \sum_{s=1}^{f} \tfrac{1}{2}k_s q_s^2(x), \tag{A3-12}$$

where the k_s are constants, and the q_s are f suitable *generalized coordinates*. (These generalized coordinates may be distances between atoms, or angles, etc., replacing the original Cartesian vector components x_i, y_i, z_i used to locate atoms in space.) It is always possible to write ε_{pot} in the form of (A3-12), if an equilibrium configuration exists. We see at once that, when the premises stated above apply, the calculation of the mean kinetic and potential energies leads to expressions identical to those encountered in (A3-7) for the one-dimensional oscillation of an atom. The mean kinetic energy is a sum of terms

$$\overline{(p_i^2/2m)} = \int dp_i\,(p_i^2/2m_i)e^{-p_i^2/2m_ikT} \bigg/ \int dp_i\, e^{-p_i^2/2m_ikT},$$

and the mean potential energy, a sum of terms

$$\overline{\frac{k_s}{2}q_s^2} = \int dq_s \left(\frac{k_s}{2}q_s^2\right) e^{-k_s q_s^2/2kT} \bigg/ \int dq_s\, e^{-k_s q_s^2/2kT}.$$

The integrals encountered in these two expressions are of type (A3-9) again, and we find

$$\overline{\frac{p_i^2}{2m_i}} = \tfrac{1}{2}kT \qquad \text{all } i,$$

$$\overline{\frac{k_s q_s^2}{2}} = \tfrac{1}{2}kT \qquad \text{all } s, \text{ for which } k_s \neq 0.$$

This result is the substance of the equipartition theorem. We can state it as follows: If a system consists of independent elements (quasi-particles) with f degrees of freedom each, then the mean kinetic energy of this element is

$$\overline{\varepsilon_{\text{kin}}} = \frac{f}{2}kT.$$

In addition, every term of type

$$\tfrac{1}{2}k_s q_s^2, \qquad k_s \neq 0,$$

in the potential energy of the element contributes an amount $\tfrac{1}{2}kT$ toward the mean potential energy.

As an illustration, we may mention the diatomic molecule. Here the kinetic energy is the sum of six terms of type $p_i^2/2m_i$; there is only one potential energy term

$$\sim \frac{k_1}{2}(r - r_0)^2,$$

where the generalized coordinate r is the distance between the two atoms in the molecule. The mean energy of the molecule is therefore

$$\tfrac{6}{2}kT + \tfrac{1}{2}kT = \tfrac{7}{2}kT.$$

The value of the mean energy $\bar{\varepsilon}$ as predicted by the equipartition theorem does agree with observation in cases where a classical description is adequate. A necessary condition for this is that the value of kT be large compared to the spacing of energy levels of the elements of the system. At room temperature we can describe the translational and rotational motion of the oxygen molecule in oxygen-gas classically, but *not* the vibration of the two atoms in oxygen.

In the case where a classical treatment is adequate, it can be shown generally that the results of a classical calculation of the mean energy, based on the expression

$$\bar{\varepsilon} = \int d\Omega \, \varepsilon e^{-\varepsilon/kT}/Z \tag{A3-13}$$

and quantum mechanical calculation, based on the expression

$$\bar{\varepsilon} = \sum_n \varepsilon_n e^{-\varepsilon_n/kT}/Z \tag{A3-14}$$

do give the same value for $\bar{\varepsilon}$.

4.5 References and Supplementary Reading

1. Optical Mixing Spectroscopy with Applications to Problems in Physics, Chemistry, Biology and Engineering. In *Polarization, Matter and Radiation. A Jubilee Volume in honor of Alfred Kastler*. Presses Universitaires de France, Paris (1969).

2. Observations of the Spectrum of Light Scattered by Solutions of Biological Macromolecules, by S. B. Dubin, J. H. Lunacek, and G. Benedek. *Proc. Nat. Acad. Sci. USA* **57**, 1164 (1967).

3. Fluctuation Spectroscopy: Determination of Chemical Reaction Kinetics from the Frequency Spectrum of Fluctuations, by G. Feher and M. Weissman. *Proc. Nat. Acad. Sci. USA* **70**, 870 (1973).

4. Vapor Pressure of Water. In *Handbook of Chemistry and Physics*, 75th ed. (1997). A new edition of this handbook appears every year. Each of them contains the required information. The handbook is published by The Chemical Rubber Publishing Company, Cleveland, OH.

5. Gleichgewichte Dampf-Kondensat und Osmotische Phänomene. In *Landolt–Börnstein: Eigenschaften der Materie in ihren Aggregatszuständen*, 6th ed., Vol. II, Part 2a, pp. 19 and 926. Springer-Verlag, Berlin (1964).

6. The Oxygen Equilibrium in Hemoglobin and its Structural Interpretation, by L. Pauling. *Proc. Nat. Acad. Sci. USA.* **21**, 186 (1935).

7. On the Nature of Allosteric Transitions: A Plausible Model, by J. Monod, J. Wyman, and J. P. Changeaux. *Journal of Molecular Biology* **12**, 88–118 (1965).

8. Das Gleichgewicht zwischen Hämoglobin und Sauerstoff, by F. Haurowitz. Hoppe-Seyler *Z. Physiol. Chem* **254**, 266–274 (1938).

9. *Mechanisms of Cooperativity and Allosteric Regulation in Proteins*, by M. Perutz. Cambridge University Press, New York (1990).

10. Application of a Theory of Enzyme Specificity to Protein Synthesis, by D. E. Koshland. *Proc. Nat. Acad. Sci. USA* **44**, 98–104 (1958).

11. Comparison of Experimental Binding Data and Theoretical Models in Proteins Containing Subunits, by D. E. Koshland, G. Nemethy, and D. Filmer. *Biochemistry* **5**, 365–385 (1966).

12. The Hemoglobins, by G. Braunitzer, K. Hilse, V. Rudolf, and N. Hilshmann. In *Advances in Protein Chemistry*. Academic Press, New York (1964).

13. Hemoglobin and Myoglobin, by A. Rossi, E. Fanelli, E. Antonini, and A. Caputo. In *Advances in Protein Chemistry*, Vol. 19. Academic Press, New York (1964).

14. Hemoglobin and Myoglobin in Their Reactions with Ligands, by E. Antonini and M. Brunori. In *Frontiers in Biology*, Vol. 21, edited by A. Neuberger and E. L. Tatum. North-Holland, Amsterdam (1971).

15. Allosteric Regulation. In *Selected Papers in Biochemistry*, Vol. 8, edited by M. Tokushige,, University Park Press, Baltimore (1971).

16. Allosteric Interpretation of the Oxygen-Binding Reaction of Human Hemoglobin Tetramers, by E. Di Cera, C. H. Robert, and S. J. Gill. *Biochemistry* **26**, 4003–4008 (1987).

17. *Allosteric Enzymes*, by G. Herve. CRC Press, Boca Raton, FL (1989).

18. *Functional Chemistry of Biological Macromolecules*, edited by by L. Wyman and S. Gill. University Sciences Books, Mill Valley, CA (1990).

19. *Theory of Site Specific Binding Processes in Biological Macromolecules*, by E. Di Cera. Cambridge University Press, Cambridge, UK (1995).

20. Linkage: Thermodynamics of Macromolecular Interactions. In *Advances in Protein Chemistry*, Vol. 15, edited by E. Di Cera. Academic Press, New York (1998).

21. *Physical Chemistry* (3rd ed.), by F. Daniels and R. A. Alberty. Wiley, New York (1966).

Additional references: Most of the basic material presented in Chapter 4 is covered, in varying degrees of detail, in standard texts on physical chemistry. We have found the following texts particularly useful:

Basic Physical Chemistry for the Life Sciences, by W. R. and H. B. Williams. W. H. Freeman, San Francisco (1967).

Entropy for Biologists: An Introduction to Thermodynamics, by H. J. Morowitz. Academic Press, New York (1970).

Physical Biochemistry, by K. E. van Holde. Prentice Hall, Englewood Cliffs, NJ (1971).

Physical Chemistry for the Life Sciences, by G. M. Barrow. McGraw-Hill, New York (1974).

Physical Chemistry, 6th ed., by R. A. Alberty. Wiley, New York (1983).

Physical Chemistry, 3rd ed., by G. W. Castellan. Benjamin-Cummings, Menlo Park, CA (1983).

Biothermodynamics: The Study of Biochemical Processes at Equilibrium, by J. T. Edsall, and H. Gutfreund. Wiley, New York (1983).

Statistical Thermodynamics for Chemists and Biochemists, by A. Ben-Naim. Plenum Press, New York (1992).

Chemical Thermodynamics: Basic Theory and Methods, 5th ed., by I. M. Klotz and R. M. Rosenberg. Wiley, New York (1994).

Statistical Mechanics and Thermodynamics, by C. Garrod. Oxford University Press, New York (1995).

Thermodynamics and its Applications, 3rd ed., by J. Tester and M. Modell. Prentice Hall, Englewood Cliffs, NJ (1997).

4.6 Problems

(These problems are grouped topically, and in each group they are arranged roughly in order of increasing difficulty.)

Classical Physics Versus Quantum Physics

1. Spacing of energy levels in the harmonic oscillator: Any object, whose energy ε is the sum of a kinetic energy $\varepsilon_{\text{kin}} = (M/2)(dq/dt)^2$ and a potential energy $\varepsilon_{\text{pot}} = \frac{1}{2}Kq^2$, is a harmonic oscillator. The variable q may be a position coordinate, or an angle, or an electric charge, etc. Whatever it is, the variable q

will perform a harmonic oscillation

$$q(t) = a\sin(\omega t + \phi),$$

with $\omega = \sqrt{K/M}$; $\nu = \omega/2\pi$ is the frequency of the oscillation.

(a) Quantum theory shows that the quantized energy levels ε_n are equidistant, and that $\Delta\varepsilon = \varepsilon_{n+1} - \varepsilon_n = h\nu$, h being Planck's constant, and ν the frequency of oscillation. Calculate and compare, for room temperature, the ratio between the mean thermal energy kT and the level spacing of an oscillator with a frequency ν of 500 cycles/sec (tuning fork).

(b) The two atoms of the iodine molecule I_2 oscillate at a frequency $\nu = 6.4 \times 10^{12}\ \text{s}^{-1}$. Find the temperature range for which

$$\Delta\varepsilon = h\nu \ll kT$$

and the range for which $\Delta\varepsilon \gg kT$.

N.B. Use the values

$$h = 6.625 \times 10^{-27}\ \text{erg s},$$
$$k = 1.38 \times 10^{-16}\ \text{erg/}^{\circ}\text{K}.$$

2. Trajectory of the harmonic oscillator in phase space. We may represent the motion of the oscillator by viewing the values of

$$q(t) \qquad \text{and} \qquad p(t) = M\left(\frac{dq}{dt}\right),$$

as the q- and p-coordinates of a moving point in a q–p plane.

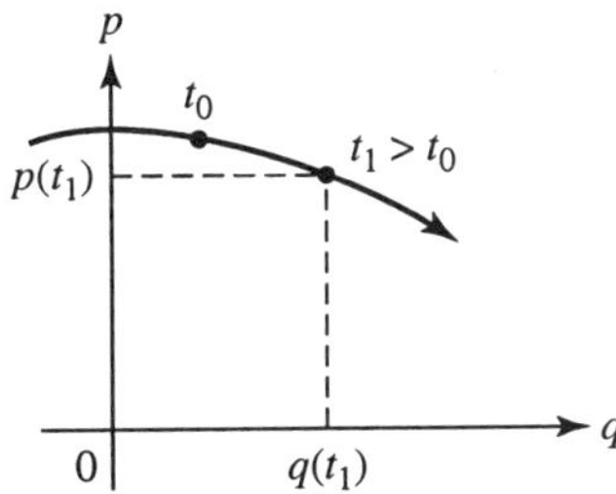

This plane is the *phase space* of the oscillator.

(a) Show that for an oscillator of energy ε, the values of $q(t)$ and $p(t)$ may be written as

$$q(t) = \sqrt{\frac{2\varepsilon}{K}}\,\sin(\omega t), \qquad \omega = \sqrt{\frac{K}{M}}.$$

$$p(t) = \sqrt{2\varepsilon M}\,\cos(\omega t),$$

(b) Show that the integral $\oint p\,dq$, extended over a full cycle of the motion $q(t)$, represents the area of the ellipse:

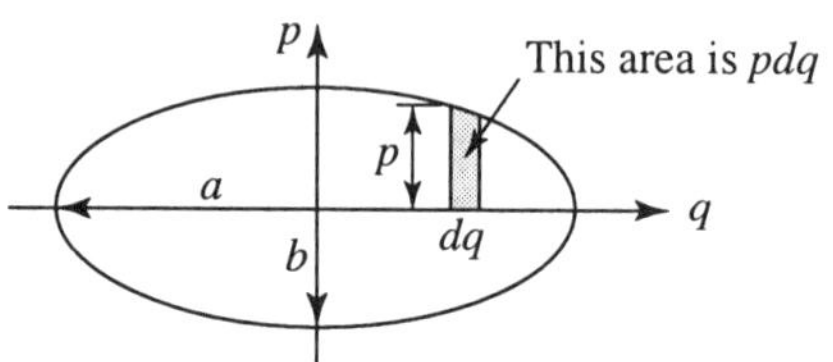

Using the formula for the area of an ellipse $A = \pi ab$, a, b being the principal radii, show that

$$A = \oint p\,dq = \frac{2\pi\varepsilon}{\omega} = \frac{\varepsilon}{\nu}.$$

(c) The quantized values ε_n of the energy can be understood on the basis of De Broglie's view that along the particle's actual spatial trajectory, there must be a standing wave pattern with an *integral number n* of wavelengths. Since the wavelength λ is related to momentum p by De Broglie's relation $p = h/\lambda$ (see Section 4.1.B), one has

$$\oint \frac{dq}{\lambda} = n \qquad \text{or} \qquad \oint \frac{h}{\lambda}dq = \oint p\,dq = nh.$$

Show that combined with the result of (b), this leads to the allowed energy values

$$\varepsilon_n = nh.$$

Microstates and Weight of Macrostates

(Always use Stirling's approximation in the form $\ln N! \simeq N \ln N - N$.)

3. Consider a system of N coins that are tossed again and again. Assume that the probability that a particular coin comes up heads or tails is $\frac{1}{2}$.

(a) Using the notion of microstate and macrostate, explain in words why the probability $P_N(H, T)$ that the N coins will have H heads and T tails, regardless of location in the sequence, is given simply by

$$P_N(H, T) = \left(\frac{1}{2^N}\right) W,$$

with

$$W = \frac{N!}{H!\,T!}.$$

(b) Assume $N, H, T \gg 1$. By maximizing $\ln W$ subject to the constraint $T + H = N$, show that the most probable number of heads $\overline{H}$ is equal to $\overline{H} = (N/2)$.
(*Hint:* Use Stirling's approximation in its weak form, i.e., $\ln N! \simeq N \ln N - N$.)

(c) Show that $\ln W_{\max} \cong N \ln 2$.

(d) Show that *relative error* in the value of $\ln W_{\max}$ arising from the use of the weak form of Stirling's approximation $\ln N! \cong N \ln N - N$, rather than the more exact form

$$\ln N! = \tfrac{1}{2} \ln(2\pi N) + N \ln N - N,$$

is of order $\ln N/(2N \ln 2)$. Show that this represents a small error, if N is a very large number. (A *very large* number is a number, for which $\ln N \gg 1$, i.e., $N > 10^{10}$.)

(e) Show that for a prescribed value of the difference

$$m = H - T,$$

the value of $\ln W$ is given by

$$\ln W = N \ln 2 - \frac{m^2}{2N},$$

as long as $|m| \ll N$.

4. A one-dimensional model of a polymer chain consists of N segments of length l each, that can point either to the right or to the left.

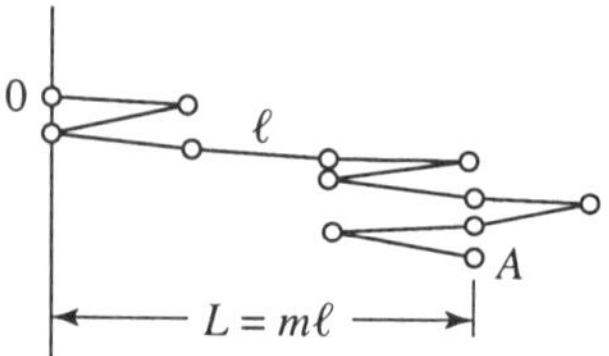

Let a *macrostate* of the chain be defined by the set of all folding patterns, for which the distance $OA = L = ml$ has a fixed value. Each specific folding pattern is a microstate. Find the number $W(N, m)$ of microstates associated with the single macrostate with $L = ml$.

5. Consider a *two*-dimensional model of a chain polymer, as shown below. The chain has N segments of length l_0 each. Each segment can point in any of four directions: right, left, up, down. Call R, L, U, D the corresponding number of segments, so that

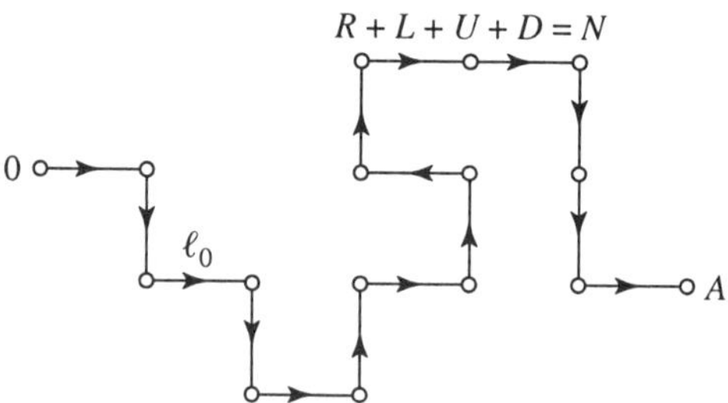

In this sketch, $N = 15$, $R = 7$, $L = 1$, $U = 3$, $D = 4$. Let a macrostate be defined by specified values of R, L, U, D, and a microstate by any particular folding pattern.

(a) Show that the number $W(R, L, U, D)$ of microstates belonging to the macrostate (R, L, U, D) is

$$W = \left(\frac{N!}{N_x! \, N_y!}\right)\left(\frac{N_x!}{R! \, L!}\right)\left(\frac{N_y!}{U! \, D!}\right)$$

where $N_x = R + L$, $N_y = U + D$.

(b) Using the Stirling approximation for $N!$ in the form

$$N! \cong N \ln N - N,$$

show that in this approximation

$$\frac{N!}{\left(\dfrac{N+m}{2}\right)!\left(\dfrac{N-m}{2}\right)!} \cong 2^{N} e^{-m^{2}/2N} \quad \text{for} \quad |m| \ll N.$$

(See also Problem 3, notably parts (d) and (e).)

(c) Using the result of (b) in expression (a), demonstrate that the maximum of W occurs at $R = L = U = D = N/4$.

(d) Find the value of $\ln W_{\text{max}}$ in this case.

(e) Let the end-to-end distance of the polymer chain be constrained to a fixed value l, where

$$l^{2} = l_{0}^{2}\{(R - L)^{2} + (U - D)^{2}\}.$$

Assume that $l \ll N l_0$, and find the expression for W_{max} for this case, as a function of l.

(N.B. The value of $k \ln W_{\text{max}}$, k being the Boltzmann constant, is called the configurational entropy of the polymer chain.)

6. Model of a magnetic material. A simple model of a magnetic material consists of N ($N \gg 1$) elementary magnets, that can either point up or down.

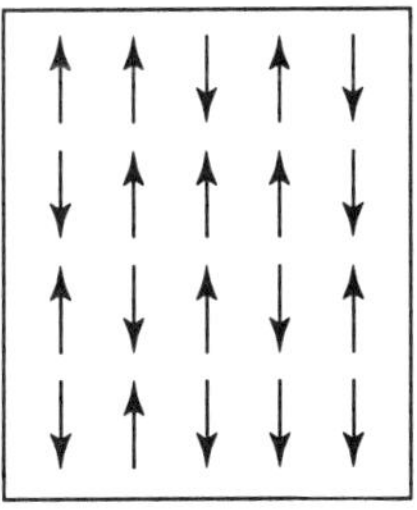

A macrostate of this magnet is defined by the total numbers N_+ and N_- of elementary magnets pointing up or down. A microstate is any specific pattern of orientations of elementary magnets.

(a) Find the number $W(N_+, N_-)$ of microstates associated with a macrostate specified by the values of N_+ and N_-.

(b) Show that if no constraint except $N_+ + N_- = N$ is applied, the weight W is maximum at $N_+ = N_-$.

(c) Suppose the energy ε of each elementary magnet depends on its orientation

$$\varepsilon(\uparrow) = -\Delta\varepsilon$$
$$\varepsilon(\downarrow) = +\Delta\varepsilon$$

so that the total energy E is $-(N_+ - N_-)\Delta\varepsilon$. Find the expressions for N_+ and N_-, for which $\ln W$ is maximum, subject to the constraints that N and E have fixed, prescribed values. (Use the parameters α and β introduced in Section 4.2.)

(d) Find an expression for $\ln W_{\text{max}}$. Show that for $|\beta\Delta\varepsilon| \ll 1$, one has

$$\ln W_{\text{max}} \cong N \ln 2 - \tfrac{1}{2} N (\beta\Delta\varepsilon)^2.$$

(e) Show that in the opposite limit, $\beta\Delta\varepsilon \gg 1$, the value of $\ln W_{\text{max}}$ tends toward zero! What is the physical interpretation of this result?

(f) Find the value of the ratio $(N_+ - N_-)/N$ as a function of $\beta\Delta\varepsilon$. Make a plot of this function on graph paper.

(g) Instead of finding the maximum of $\ln W(N_+, N_-)$ subject to the two constraints

$$N_+ + N_- = N,$$
$$-\Delta\varepsilon(N_+ - N_-) = E,$$

you may solve the constraint equations for N_+ and N_- in terms of N, E, and $\Delta\varepsilon$, and insert the values of N_+, N_- found into the expression for $\ln W(N_+, N_-)$. This gives you $\ln W$ as a function of E rather than of β. You may now verify that

$$\left(\frac{\partial \ln W(E)}{\partial E} \right)_{N=\text{constant}} = \beta.$$

This is, in substance, the equation relating entropy S ($= k \ln W(E)$) to temperature $T = 1/k\beta$:

$$\frac{\partial S(E, N)}{\partial E} = \frac{1}{T}.$$

The Boltzmann Factor

7. The hydrocarbon 2-butene, $CH_3-CH=CH-CH_3$, occurs in two conformations, called the *cis*- and *trans*-conformation:

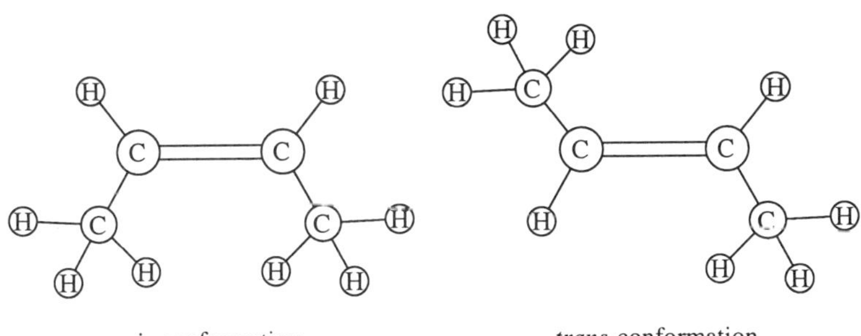

cis-conformation   *trans*-conformation

The energy difference between the two conformations is 1 kcal/mol, *trans* being lower than *cis*. Determine the relative abundance of the two isomers, at $T = 300\,°K$ and $T = 1000\,°K$.
Notice: $R = 2$ cal/mol $°K$.

8. The compound dibromoethane $Br-CH_2-CH_2-Br$ has three stable conformations, schematically represented below (heavy black dots are Br atoms):

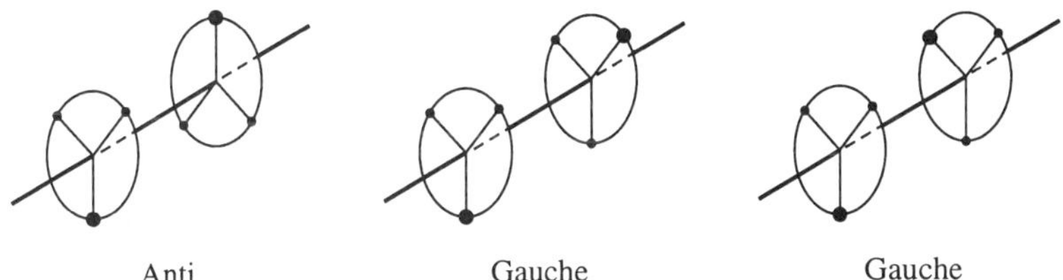

Anti   Gauche   Gauche

The anticonformation is lower by 0.8 kcal/mol than the two gauche-conformations that have equal energies. Notice that these two latter are physically distinct and represent two different quantum states of the molecule. Find the relative abundance of the gauche- and anticonformation at $T = 300\,°K$. Ref: K. Mislow, *Introduction to Stereochemistry*, W. A. Benjamin, New York (1966).

9. The speed at which a chemical reaction proceeds is in general controlled by the height of an energy barrier (the so-called "activation energy") $\Delta\varepsilon$, which the reaction's partners must overcome for the reaction to proceed. The probability that the reaction's partners acquire that needed energy $\Delta\varepsilon$ is given by the Boltzmann factor $e^{-\Delta\varepsilon/kT}$. The diagram below shows the dependence on the rate of photosynthesis in green plants as a function of temperature (at high light intensity). Using the data in this plot, determine the activation energy of the rate limiting step in the process of photosynthesis.

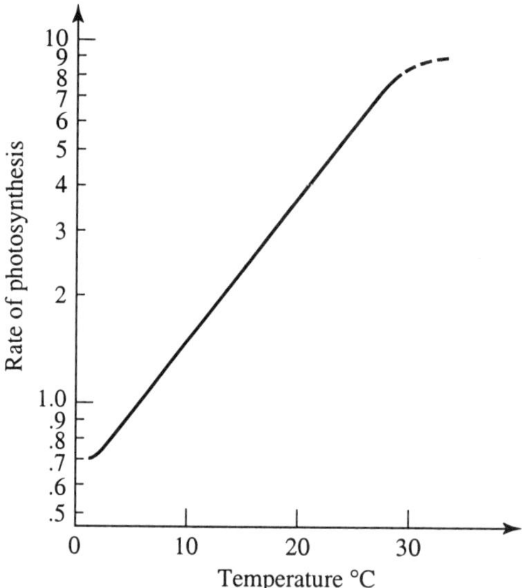

Data from: *Biological Science* (Blue version), Biolog-Sciences Curriculum Study. Houghton Mifflin, Boston (1963).

10. Rotational energy of the hydrogen molecule. The hydrogen molecule has quantized states of rotational excitation, of angular momentum $hJ/2\pi$, and energy

$$\varepsilon_J = \frac{a}{2}J(J+1),$$

a is a constant, and the quantum number J takes the values $0, 1, 2, \ldots$ (integers). The constant a may be written in the form $a = k\theta_r$, where k is Boltzmann's constant, and θ_r a temperature, that has the value 85.4 °K for molecular hydrogen.

(a) Calculate for $T = 300$ °K and 305 °K, the probabilities $P(J)$ that the molecule has rotational energy ε_J, for $J = 0$ to $J = 7$. (See the following table at the end of the problem.) This probability is given by $P(J) = (2J + 1)e^{-\varepsilon_J/kT}/Z$, Z being the normalization constant. The extra factor $(2J + 1)$ arises from the fact that for *each* value of the energy ε_J, there are actually $(2J + 1)$ *distinct* quantum states describing the different orientations of the rotating molecule.

(b) Calculate the *mean* rotational energy of the molecule

$$\overline{\varepsilon}_{\text{rot}} = \sum_J \varepsilon_J P(J)$$

for the two temperatures indicated. (*Hint:* It will be easier to calculate first $(\overline{\varepsilon}_{\text{rot}}/kT)$, which is given by $\overline{\varepsilon}_{\text{rot}}/kT = \sum_J (\varepsilon_J/kT) P(J)$.)

(c) Using the data found in (b), calculate the rotational contribution to the molar specific heat of hydrogen, c_v. c_v is defined by

$$c_v = N_0 \left(\frac{d\overline{\varepsilon}_{\text{rot}}}{dT} \right).$$

Express your results by means of the numerical value of $c_v/R = d\overline{\varepsilon}_{\text{rot}}/d(kT)$.

	$J =$	0	1	2	3	4	5	6	7
$T = 300$ °K	ε_J/kT	0	0.28467	0.8540	1.7080	2.8467	4.2700	5.9781	7.971
	$e^{-\varepsilon_J/kT}$	1	0.7523	0.4257	0.1812	0.0580	0.0140	0.00253	0.00034
$T = 305$ °K	ε_J/kT	0	0.28	0.84	1.68	2.8	4.2	5.88	7.84
	$e^{-\varepsilon_J/kT}$	1	0.7558	0.4317	0.1864	0.0608	0.0150	0.00279	0.00039

11. **Errors in DNA duplication.** The DNA molecule consists of two intertwined linear polymers (the famous "double helix"). The monomers are the so-called nucleotides. Each nucleotide contains one of four different "bases": Adenine (A), Guanine (G), Thymine (T), and Cytosine (C). The sequence of these bases

in the polymer is the genetic message. In the double helix, matching bases from the two strands are hydrogen bonded together. Properly matched pairs are A–T and G–C. They are more tightly bound than improper matches. If the energy difference $\Delta\varepsilon$ between an improper and a proper match is large enough, improper matches become very improbable in the process of DNA duplication.

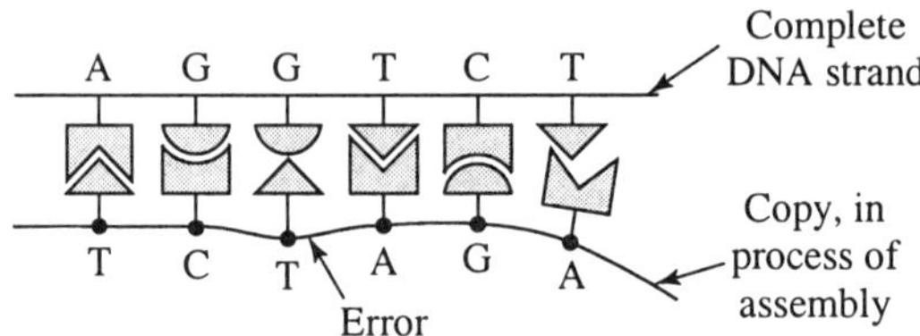

According to an estimate by Watson (*Molecular Biology of the Gene*, 2nd ed., W. A. Benjamin, New York (1970)) the probability of an error may be of order 10^{-9} per nucleotide (at $T = 300\,°K$).

(a) Assuming that the probability of a mismatch is determined by a Boltzmann factor $e^{-\Delta\varepsilon/kT}$, estimate the value of the energy "penalty" $\Delta\varepsilon$.

(b) Assuming that such errors may result in observable mutations, by how much would the mutation rate of an organism (for instance, a bacterium) increase, if the temperature of the environment is raised by $20\,°C$?

Maxwell–Boltzmann Distribution

12. Estimate the translational energy of "the most energetic particle" in a mole of gas, at $T = 300\,°K$. For this estimate, use the expression for the energy distribution function $F(\varepsilon)$, equation (4-93). The question is answered by finding the energy $\varepsilon_{\max}$, beyond which there is just one particle left

$$N_0 \int_{\varepsilon_{\max}}^{\infty} d\varepsilon\, F(\varepsilon) = 1.$$

Hint: You get a good enough approximation for $\varepsilon_{\max}$, if in (4-93) you replace $\sqrt{\varepsilon}$ by $\sqrt{\varepsilon_{\max}}$.

Compare this energy $\varepsilon_{\max}$ with the *mean* translational energy. Express this energy in electron volts ($1\ \text{eV} = 1.6 \times 10^{-12}$ erg). If the gas is hydrogen, with a molecular binding energy of 3.5 eV, show that the bonds of even the most energetic particle are unlikely to get broken in a collision.

13. Using (4-93) for the energy distribution function in a Boltzmann gas, find the root mean square (rms) width of the energy distribution

$$\Delta\varepsilon_{rms} = \sqrt{\overline{(\varepsilon - \bar{\varepsilon})^2}}.$$

14. Gas pressure and number of collisions with a wall. A gas at temperature $T = 300\,°K$ exerts a pressure of 1 atm $= 10^6$ dyn/cm^2 against the walls of the container.

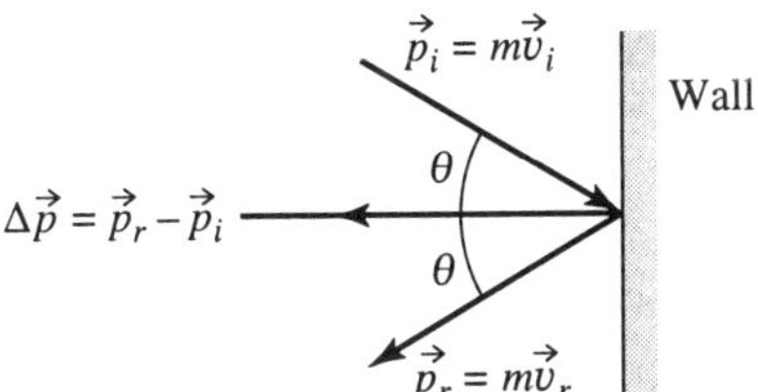

Gas molecules with initial momentum $\vec{p}_i = m\vec{v}_i$ hit the wall and leave with a momentum $\vec{p}_r = m\vec{v}_r$.

(a) Show that one can *estimate* the mean momentum change $\Delta\vec{p}$ in the collision of the gas molecule with the wall to be of order $|\overline{\Delta\vec{p}}| = 2\sqrt{mkT}$, m being the mass of the molecule.

(b) Let n stand for the number of collisions per second and cm^2 of wall area. Show that the gas pressure p is given by

$$p = n\,\overline{|\Delta\vec{p}|}$$

(c) Using $p = 1$ atm, and $m = 6.9 \times 10^{-24}$ g (helium), find the number n of wall collisions/s and cm^2.

Equations of State. Specific Heats. Compressibility. Adiabatic Changes.

15. Real gases (nonideal gases) like CO_2, NH_3, SO_2, are generally described by the so-called van der Waals equation of state. For one mole of gas, this equa-

tion reads

$$\left(p + \frac{a}{V^2}\right)(V - b) = RT.$$

The coefficient b describes the effect of finite molecular size, and the term a/V^2 the effect of the mutual short-range attraction of molecules.

(a) Rewrite this equation in a form so that it applies to n moles of gas.

(b) The van der Waals equation goes over into the ideal gas equation provided the volume V and the temperature T are large enough. Show that the condition for the validity of

$$pV \cong RT$$

is that $b - (a/RT) \ll V$.

16. The heat capacities at constant volume and constant pressure, C_v and C_p, are defined as

$$C_v = \left(\frac{dQ}{dT}\right)_v, \qquad C_p = \left(\frac{dQ}{dT}\right)_p.$$

Using the equation of state of the ideal gas

$$pV = nRT$$

and the first law of thermodynamics

$$dE = dQ - p\,dV,$$

show that $C_p - C_v = nR$.

17. For a system in thermal equilibrium, the total energy E is a state function. It may be viewed, for instance, as a function of T and V: $E(T, V)$. (This means that the given values of T and V uniquely determine the value of E of the (simple) system.)

(a) Show that the specific heat C_v is the partial derivative

$$C_V = \left(\frac{\partial E}{\partial T}\right)_V.$$

(b) Explain why the equation

$$p \overset{?}{=} - \left(\frac{\partial E}{\partial V} \right)_T$$

is *not* true.

(c) The quantity $H = E + pV$ is also a state function (the so-called *Enthalpy*). Show that the change ΔH of H due to heat input ΔQ and to a change of volume and pressure is

$$\Delta H = \Delta Q + V \Delta p.$$

(d) The enthalpy H may be viewed as a function of the variables T and p: $H(T, p)$. Show that it follows from (c) that the specific heat at constant pressure is the partial derivative

$$C_p = \left(\frac{\partial H}{\partial T} \right)_p.$$

18. Equation of state and compressibility. The compressibility κ of a material is defined as the fractional change of volume due to a change of pressure

$$\kappa = -\frac{1}{V} \left(\frac{dV}{dp} \right).$$

Calculate the *isothermal* compressibility κ_{iso}. In this case, you must assume that volume and pressure changes occur while T remains constant. κ has the dimension of a reciprocal pressure. Relate κ to the gas pressure p.

19. Adiabates of the ideal gas. A change of state $(1) \to (2)$ of a system is called *adiabatic*, if along the reversible path from (1) to (2), *no heat* is exchanged with the environment. This implies, according to the first law, that

$$\Delta Q = \Delta E + p \Delta V = 0.$$

For an ideal gas, E depends on T only, and

$$\Delta E = C_V \Delta T,$$

C_V being the heat capacity at constant volume.

(a) Consider 1 mol of ideal gas, and use the equation of state to show that in an adiabatic change, ΔV and ΔT are related by

$$c_v \frac{\Delta T}{T} + R \frac{\Delta V}{V} = 0,$$

when $c_v = C_V/n$ is the *molar* heat capacity.

(b) Integrate the result obtained in (a) and show that for any two states (1) and (2) connected by an adiabatic path, one has

$$T_1 V_1^{R/c_v} = T_2 V_2^{R/c_v}.$$

(c) Using the equation of state, show that the result obtained in (b) can be cast in the alternative forms

$$p_1 V_1^\gamma = p_2 V_2^\gamma,$$
$$T_1 p_1^{(1-\gamma)/\gamma} = T_2 p_2^{(1-\gamma)/\gamma},$$

where $\gamma = c_p/c_v = (c_v + R)/c_v$.

(d) On a p–V diagram (p the vertical, V the horizontal axes) plot the curves $pV^\gamma = $ constant, along with the isotherms $pV = $ constant. Assume for this an ideal diatomic gas like oxygen or nitrogen for which $c_v = \frac{5}{2}R$.

20. Adiabatic compressibility of the ideal gas. Compressibility was defined in Problem 18. To calculate the *adiabatic* compressibility, you must use the relation $Vp^{1/\gamma} = $ constant (see Problem 19).

(a) Show that the adiabatic compressibility κ_{Ad} is smaller than the isothermal one by a factor γ (which is 1.4 for air).

(b) The speed of sound in air is given by the expression

$$v_{\text{sound}} = \frac{1}{\sqrt{\kappa_{Ad}\rho}},$$

ρ being the density of air in g/cm^2. Show that with the result of (a) this can be written as

$$v_{\text{sound}} = \sqrt{\gamma \frac{RT}{M}},$$

M being the mean molecular weight of air.

(c) Calculate the numerical value of the sound velocity in air, at $T = 300\,^\circ$K. Use $M = 29$ g/mol.

(d) The sound velocity is closely related to the root mean square velocity of molecules in a gas. Express v_{rms} in terms of R, T, and M and compare it to v_{sound}.

21. In the atmosphere, the temperature decreases with increasing height z above sea level. When air, heated by the ground, rises, it expands and cools. It also mixes and exhcanges heat with ambient air. After this so-called convective mixing has occurred for a long time, a unique temperature profile $T(z)$ is established, such that any rising air mass expands and cools to the ambient temperature without further heat exchange.

This temperature profile can therefore be calculated from the adiabatic equation

$$T = T_0 \left(\frac{p}{p_0}\right)^{(1-1/\gamma)}, \tag{1}$$

and from the rate of change of the pressure p with height z:

$$\frac{dp}{dz} = -\rho g = -\frac{Mg}{RT(z)}p. \tag{2}$$

(M is the molecular weight, ρ the density: $\rho = Mp/RT$.)

(a) Equation (1) relates p to T. One can, therefore, write equation (2) as

$$\left(\frac{dp}{dT}\right)_{\mathrm{Ad}} \cdot \left(\frac{dT}{dz}\right) = -\frac{Mg}{RT}p$$

or

$$\frac{dT}{dz} = -\frac{Mg}{R}\left(\frac{p}{T}\right)\left(\frac{dT}{dp}\right)_{\mathrm{Ad}}.$$

Find an expression for the right-hand side of this equation, using equation (1). Show that $dT/dz = constant$. That implies that the *temperature decreases linearly with height*.

(b) For air at atmospheric temperature, γ has the value $\frac{7}{5} = 1.4$. Show that

$$\frac{dT}{dz} = -\frac{2}{7}\frac{Mg}{R}.$$

(c) Using a mean molecular weight $M = 29$ g/mol, show that the numerical value of the rate of decrease of temperature with height is

$$\frac{dT}{dz} = 10^{-2}\,^\circ\text{K/m} = 0.54 \times 10^{-2}\,^\circ\text{F/ft}.$$

(d) The mean annual temperature of Bombay, India, is 81 °F, that of Darjeeling, at an altitude of 7340 ft, is 53 °F. Determine how close this temperature change is to an adiabatic decrease with altitude.

Entropy in Thermodynamics

22. Entropy of the ideal gas.

(a) Calculate the entropy difference between the points A and B, $S_B - S_A$, in the diagram below, using the two different paths

$$A \to x \to B,$$
$$A \to y \to B.$$

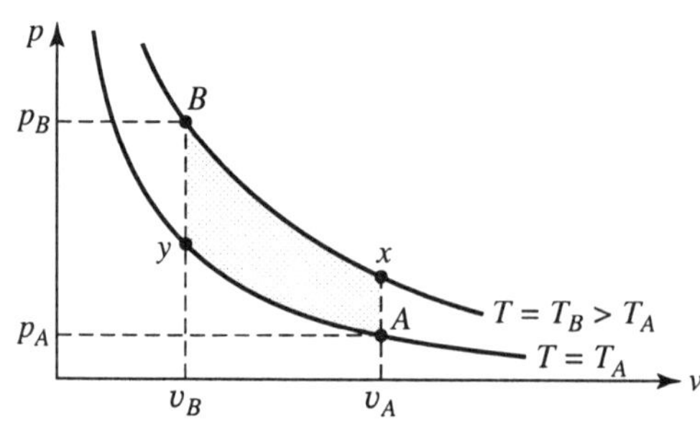

Since you deal with an ideal gas, make use of the information that $E = E(T)$ only.

(b) Show that the shaded area in the diagram measures the *difference* in the total heat absorbed along the paths AXB and AYB.

23. Entropy gain in temperature equalization. A closed composite system consists of two compartments, separated by a rigid, heat conducting wall (see Figure 4-35). Their initial temperatures are T_1^0 and T_2^0, with $T_1^0 > T_2^0$. The heat capacities of the two compartments are C_{1V} and C_{2V}, respectively. $C_{1V} \neq C_{2V}$, but both are temperature-independent constants.

(a) Establish the relation between the temperature changes ΔT_1 and ΔT_2 in compartments (1) and (2), if a finite amount ΔQ is transferred from (1) to (2).

(b) Calculate the entropy change ΔS resulting from a transfer ΔQ of heat from (1) to (2). Express this entropy change in terms of the amount ΔQ of heat transferred. Show that the entropy change is

$$\Delta S = C_1 \ln\left(\frac{T_1 - \Delta Q/C_1}{T_1}\right) + C_2 \ln\left(\frac{T_2 + \Delta Q/C_2}{T_2}\right).$$

(c) The maximum entropy gain obtains when $\Delta Q = \Delta Q_M$. For this value of ΔQ, one has

$$\left.\frac{d(\Delta S)}{d(\Delta Q)}\right|_{\Delta Q = \Delta Q_M} = 0.$$

Show that this condition leads to

$$T_1 - \frac{\Delta Q_M}{C_1} = T_2 + \frac{\Delta Q_M}{C_2}.$$

(d) Using the result of (a), show that the condition you obtain in (c) indicates that the final temperatures of the two compartments are equal.

24. The *Carnot Engine* represents an idealized heat engine in which a sequence of reversible isothermal and adiabatic (isentropic) volume changes of a system are used to do mechanical work, while at the same time absorbing an amount of heat ΔQ_{in}/cycle at a high temperature T_1 and divesting an amount ΔQ_{out} at a low temperature T_2. The cycle is described in the diagram below.

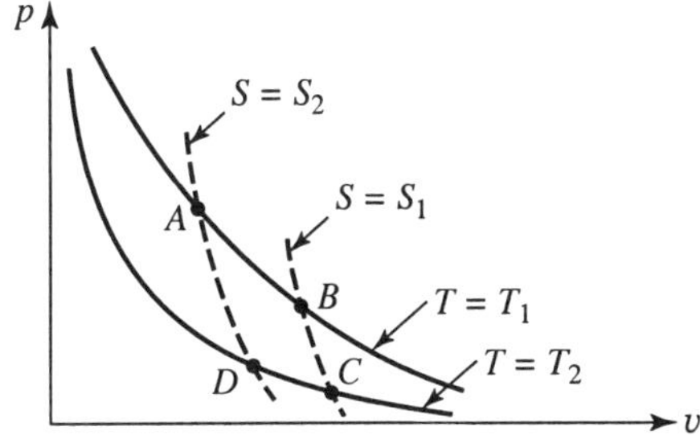

The cycle has four parts:

AB: isothermal volume expansion at T_1 (heat ΔQ_{in} taken in);

BC: isentropic expansion (cooling from $T_1 \to T_2$);

CD: isothermal compression at T_2 (heat ΔQ_{out} expelled);

DA: isentropic compression (heating from $T_2 \to T_1$).

(a) Explain why the integral

$$\oint \frac{dQ}{T} \quad \text{along the } closed \text{ path } ABCDA$$

is zero.

(b) Using this property, establish the relation

$$\frac{\Delta Q_{\text{in}}}{T_1} = \frac{\Delta Q_{\text{out}}}{T_2}.$$

(c) From the first law of thermodynamics, one has

$$\Delta Q_{\text{in}} - \Delta Q_{\text{out}} = \Delta W,$$

ΔW being the total work delivered by the system. Show that the thermal efficiency of the Carnot engine, η, defined by

$$\eta = \frac{\Delta W}{\Delta Q_{\text{in}}} = \frac{\text{work delivered}}{\text{heat intake}}$$

is given by $(T_1 - T_2)/T_1$.

(d) The path in the p–V plane can be gone through in reverse order $ADCBA$. In this case, the engine operates as a heat pump (air conditioner). In this case, one would define its efficiency as

$$\eta' = \frac{\Delta Q_{\text{out}}}{\Delta W} = \frac{\text{heat expelled}}{\text{work needed}}.$$

Show that η' is given by $\eta' = T_2/(T_1 - T_2)$.

(e) A one-room air conditioner is rated at 10,000 Btu/hour. (1 Btu $= 1$ "British Thermal Unit" $= 0.252$ kcal.) Show that in mechanical units, this corresponds to a rate of heat transfer $(dQ_{\text{out}}/dt) = 2.9$ kW. If this air conditioner had the efficiency η' of a Carnot heat pump, operating between temperatures $T_2 = 20\,°\text{C}$ (68 °F) and $T_1 = 40\,°\text{C}$ (104 °F), what would be the electric power requirement (in watts or kilowatts) to run it?

The Helmholtz and Gibbs Free Energies

25. The minimum of the Helmholtz free energy $F = E - TS$ characterizes thermal equilibrium, if the system is subject to externally imposed values of T and V. T and V are therefore the natural systems variables. In Section 4.3.C we showed that the second mixed derivative of F has the two forms

$$\left(\frac{\partial^2 F}{\partial T\, \partial V}\right)_N = \left(\frac{\partial S}{\partial V}\right)_{T,N} = \left(\frac{\partial p}{\partial T}\right)_{V,N}.$$

[equation (4-184)].

(a) Use the above relation to establish the form of the *energy* E viewed as a function of T and V. If one starts from

$$\Delta E = T\Delta S - p\Delta V,$$

one must write ΔS in terms of ΔT and ΔV. Show that this expression is

$$\Delta S = \left(\frac{C_V}{T}\right)\Delta T + \left(\frac{\partial p}{\partial T}\right)_{V,N}\Delta V.$$

(b) Insert this into the expression for ΔE. Show that it follows that

$$\left(\frac{\partial E}{\partial V}\right)_{T,N} = T\left(\frac{\partial p}{\partial T}\right)_{V,N} - p.$$

(c) The result obtained in (b) is completely general, and holds for all simple systems. Consider now the special case of an *ideal gas*, where $p = nRT/V$. Show that it follows from the equation of state *alone* that $(\partial E/\partial V)_{T,N} = 0$.

(d) Calculate $(\partial E/\partial V)_{T,N}$ for the van der Waals gas whose equation of state (for $n = 1$) is

$$\left(p + \frac{a}{V^2}\right)(V - b) = RT.$$

Show that in this case one has

$$\left(\frac{\partial E}{\partial V}\right)_{T,N} = \frac{a}{V^2}.$$

26. **Helmholtz free energy of the ideal, monoatomic gas.** The microscopic, statistical approach gives the following expression for energy and entropy of the ideal monoatomic gas

$$E = \tfrac{3}{2}NkT,$$

$$S = Nk\left\{\tfrac{5}{2} + \ln\left(\frac{V}{NV_F(T)}\right)\right\},$$

where $V_F(T) = (h^2/2\pi mkT)^{3/2}$.

(a) Construct the expression for the Helmholtz free energy $F(T, V, N) = E - TS$.

(b) Verify by calculation that

$$\left(\frac{\partial F}{\partial T}\right)_{V,N} = -S, \qquad \left(\frac{\partial F}{\partial V}\right)_{T,N} = -p.$$

(c) Consider p as a function of T and V: $p = NkT/V$. Verify the "Maxwell relation"

$$\left(\frac{\partial S}{\partial V}\right)_{T,N} = \left(\frac{\partial p}{\partial T}\right)_{V,N}.$$

(d) Calculate the chemical potential μ:

$$\mu = \left(\frac{\partial F}{\partial N}\right)_{T,V}.$$

[See comment preceding (4-199).]

(e) This is more tedious: Verify that μ calculated in (d) is identical with the alternative expression

$$\mu = -T\left(\frac{\partial S}{\partial N}\right)_{E,N}.$$

To calculate this derivative, you must first rewrite S as a function of E, V, N, before taking the N-derivative. In both cases you should get

$$\mu = kT \ln \left(\frac{N V_F(T)}{V} \right).$$

27. The Gibbs free energy G is defined as

$$G = E - TS - pV.$$

The minimum of G characterizes the thermal equilibrium, if the imposed external constraints are the values of p and T.

(a) Verify again that a small change ΔG of G, viewed as a function of the independent variables p, T, N (or n), is given by

$$\Delta G = -S\Delta T + V\Delta p + \mu \Delta N.$$

(b) Explain why it follows from the above equation that the relation

$$\left(\frac{\partial S}{\partial p} \right)_{T,N} = - \left(\frac{\partial V}{\partial T} \right)_{p,N}$$

holds.

(c) Use the relation obtained in (b) to show that if the entropy of a system is viewed as a function T and P, a small change ΔT of T and Δp of p changes S by the amount

$$\Delta S = \frac{C_p}{T} \Delta T - \left(\frac{\partial V}{\partial T} \right)_p \Delta p.$$

(d) Apply this last equation to 1 mol of an ideal, monoatomic gas, for which $C_v = \frac{3}{2}R$, and $V = RT/p$. Calculate for this case the entropy difference $S_2 - S_1$ between two states with values T_1, p_1 and T_2, p_2 of temperature and pressure.

The Clausius–Clapeyron Equation

28. Vapor pressure of water and ice. In the table below, the vapor pressure for water and ice are given for temperatures from $-5\,^\circ$C to $0\,^\circ$C:

Temperature t (°C)	p_V (water) (mmHg)	p_V (ice) (mmHg)
0	4.579	4.579
−1	4.258	4.217
−2	3.965	3.880
−3	3.673	3.568
−4	3.410	3.280
−5	3.163	3.013

(a) Calculate the rate of increase of vapor pressure with temperature t ($t = T - 273°$), (dp_V/dt), using the approximation:

$$(dp_V/dt) \cong \tfrac{1}{2}(p_V(t+1) - p_V(t-1)).$$

Do this, for both water and ice, at $t = -1\,°C$.

(b) Calculate the values of $(1/p_V)(dp_V/dt)$ for both water and ice, at $t = -1\,°C$.

(c) According to the Clausius–Clapeyron equation, the value of $(1/p_V)(dp_V/dt)$ is equal to $\Delta Q_{\mathrm{vap}}/RT^2$, ΔQ_{vap} being the *molar* heat of vaporization. Calculate ΔQ_{vap} (at $t = -1\,°C$) for both water and ice.

(d) You will find in (c) that the two values of ΔQ_{vap} for water and ice are similar, the value for ice being slightly larger. Can you explain this small difference by viewing the evaporation of ice as a two-step process, in which, first the ice melts to form water, and then the water evaporates? The heat of melting of ice is ~ 80 cal/g.

29. In this problem we assess the significance of the numerical values found for ΔQ_{vap}, for both water and ice.

(a) To evaporate 1 mol of ice or water, at an imposed pressure p equal (actually infinitesimally less) to the vapor pressure p_V, we need to overcome the binding energy of the water molecules in ice or water, and also have to supply the energy for the work of expansion. This gives the energy balance

$$\Delta Q_{\mathrm{vap}} = \left(E_{\mathrm{vapor}} - E_{\mathrm{liquid\ or\ solid}}\right) + p_V V_{\mathrm{vapor}}$$
$$\cong \Delta E + RT.$$

Using the data obtained in Problem 28, show that $\Delta Q_{\text{vap}} \gg RT$, and calculate the values of ΔE for both water and ice.

(b) The quantity $\Delta E / N_0$ can be viewed as the energy needed to detach a single water molecule either from the ice lattice or from the fluid water. In ice, the water molecules are viewed as being held in place by hydrogen bonds; two bonds need to be broken to detach a molecule. Show on this basis, that the bond energy of hydrogen bonds in ice is of the order of 5 kcal/mol or 0.2 eV.

(c) What can you infer about the structure of water from the fact that ΔE for water is only slightly less than ΔE for ice? (L. Pauling, *The Chemical Bond*, Cornell University Press, Ithaca (1967), Chap. 12.)

30. The vapor pressure of ethyl ether is plotted in the table below as a function of temperature t (°C):

t (°C)	$-74.3°$	-48.1	-27.7	-11.5	11.7	34.6
p_V (mmHg)	1	10	40	100	400	760

(a) Plot the vapor pressure on logarithmic graph paper, that is, plot $\log_{10} p_V$ versus t (or T). You will see that this graph is not a straight line.

(b) According to the Clausius–Clapeyron equation, one had

$$\frac{1}{p_V}\frac{dp_V}{dT} = \frac{\Delta Q_{\text{vap}}}{RT^2}.$$

The left-hand side of this equation can be written as

$$\frac{d}{dT}(\ln p_V) = 2.303\frac{d}{dT}(\log_{10} p_V).$$

Establish the value of $(d/dT)(\log_{10} p_V)$ at $t = 34.6\,°C$ from the plot of $\log_{10} pV$ versus t.

(c) From the data determined in (b), find the value of ΔQ_{vap} at $t = 34.6\,°C$, in kcal/mol. Compare your result with the measured value $(\Delta Q_{\text{vap}}/M)_{\text{expt}} = 89.3$ cal/g. Ether has the formula $(C_2H_5)_2-O$. Do data and calculated values of ΔQ_{vap} agree? Would measurement of $p_V(t)$ and $(\Delta Q_{\text{vap}}/M)$ be a useful method to determine molecular weights?

Raoult's Law. Boiling Point Elevation. Freezing Point Depression

31. This problem will illustrate how the vapor pressure of a solution may be used to determine the solute molecular weight. Consider a dilute aqueous solution, for which the vapor pressure p_V has been measured to be

$$p_V = 17.30 \text{ mmHg} \quad \text{at } t = 20\,^\circ\text{C}.$$

    Making use of Table 4.5 for the vapor pressure of water, find:

    (a) the relative concentration n_2/n_1 of solute to solvent for this solution;

    (b) the molality m (= moles of solute/kg solvent) of the solution.

    (c) Assuming the solution was obtained by adding 40 g of solute to 1 kg of water, find the molecular weight of the solute.

    (d) Find the boiling point of this solution at $p = 760$ mmHg.

32. Freezing point of an aqueous solution. The freezing point of an aqueous solution is lower than that of pure water at the same pressure. This follows from the equilibrium condition at the freezing temperature T_f:

$$\mu_{\text{ice}}(T_f, p) = \mu_{\text{solution}}(T_f, p)$$
$$= \mu_{\text{water}}(T_f, p) - RT_f \left(\frac{n_2}{n_1}\right),$$

    where (n_2/n_1) is the ratio of mole numbers of solute and solvent. This equation holds for $n_2 \ll n_1$.

    (a) Combining the above equation with the equilibrium condition for the pure water–ice system

$$\mu_{\text{ice}}(T_f^0, p) = \mu_{\text{water}}(T_f^0, p),$$

    and assuming that $\Delta T_f = T_f - T_f^0$ is a small temperature difference, show that the freezing point depression ΔT_f is given by

$$\Delta T_f = -\frac{R(T_f^0)^2}{\Delta Q_m} \left(\frac{n_2}{n_1}\right),$$

    where ΔQ_m is the molar heat of melting.

(b) The heat of melting ΔQ_m of ice is 80 cal/l g at $t = 0\,°C$. The heat of vaporization ΔQ_{vap} of water is much larger, $\Delta Q_{vap} = 539$ cal/g at $t = 100\,°C$. Compare the above equation for ΔT_f with (4-230) for the boiling point elevation. Indicate which of the measurements, ΔT_b or ΔT_f, is likely to give more accurate values of the ratio (n_2/n_1).

(c) Find the freezing point depression ΔT_f resulting from the addition of 1 g mol of solute to 1 kg (10^3 cm^3) of water.

(d) If a *very dilute* solution of sodium chloride (NaCl) is produced, the depression ΔT_f observed is *twice* the value inferred from the above equation for ΔT_f. For a dilute solution of potassium sulfate (K$_2$SO$_4$), the discrepancy is a factor of 3. Give an explanation of this phenomenon.

33. A perpetual motion machine? An inventor presents the device shown in the figure below to the US Patent Office, as a prototype of a perpetual motion machine, designed to extract mechanical work from a heat supply (ocean), without requiring a temperature gradient.

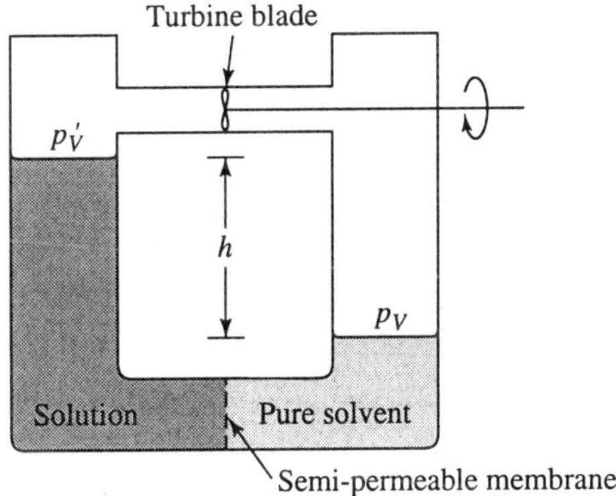

The inventor gives the following description of the manner in which the machine operates:

(a) All parts of the machine are at the *same* temperature T, equal to that of the heat bath surrounding the machine.

(b) The compartment at the left contains a solution, separated from the compartment at the right (containing pure solvent) by a semipermeable membrane, permitting solvent transfer only. A pressure difference $\Delta p = \rho g h$ balances the osmotic pressure difference $\Delta \pi$ across the membrane.

(c) The vapor pressure p_V above pure solvent is higher than the pressure p_V' above the solution (Raoult's law). There is therefore a pressure difference $\Delta p = p_V - p_V'$ in the tube, and the flow of vapor in the upper connecting tube, from right to left, drives a turbine blade.

(d) Vapor moving through the connecting tube condenses at the surface of the solution. It tends to dilute the solution, therefore reducing its osmotic pressure. This means that solvent will flow back through the semipermeable membrane, restoring the original concentration of the solution.

(e) The solvent is therefore set into perpetual motion, flowing as vapor from right to left in the upper part, and from left to right in the lower part, yielding energy.

The US Patent Office rejected the claim. Analyze the operation of this machine carefully, and decide on what grounds you would either accept or reject the inventor's claim.

Diverse Applications

34. Rubber elasticity. This problem is an extension of Problem 5 on the two-dimensional polymer chain. Rubber consists of random polymer chains, forming an irregular network with occasional crosslinks. The elastic properties of rubber can be understood on the basis of the average (in a statistical sense) properties of a single polymer chain. Consider such a chain, of N segments of length l_0 each. It forms a two-dimensional (for our convenience) pattern as shown. R, L, U, D stand for the number of segments pointing to the right, left, or up and down, respectively.

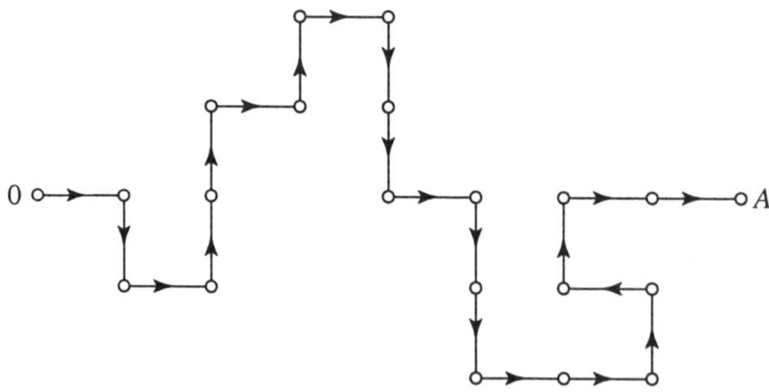

Let a *macrostate* of the chain be defined by the four numbers R, L, U, D, with $R+L+U+D = N$. The weight $W(R, L, U, D)$ is the number of distinct patterns (microstates) that can be found with fixed numbers R, L, U, D.

(a) Find an expression for W and $\ln W$, in the approximation $lN! \cong N \ln N - N$.

(b) The objective is now to find the maximum of $\ln W$, subject to the constraint conditions

$$R + L + U + D = N,$$
$$R - L = l/l_0,$$
$$U - D = O.$$

The two last conditions force the points O and A to be a distance l apart, in the horizontal direction. Show that the condition $\delta \ln W = 0$:

$$\frac{\partial \ln W}{\partial R}\delta R + \cdots + \frac{\partial \ln W}{\partial D}\delta D = 0$$

leads to equations of the form

$$\ln(R/N) = \alpha + \gamma_1, \qquad \ln(U/N) = \alpha + \gamma_2,$$
$$\ln(L/N) = \alpha - \gamma_1, \qquad \ln(D/N) = \alpha - \gamma_2.$$

(c) Verify that the expressions for R, L, U, D found in (b), if inserted into the constraint equations, give the expressions

$$e^\alpha = \tfrac{1}{2}(1 + \cosh \gamma_1),$$
$$\left(\frac{l}{Nl_0}\right) = \frac{\sinh \gamma_1}{1 + \cosh \gamma_1},$$
$$\gamma_2 = 0.$$

In the previously mentioned Problem 5, we considered the case $(l/Nl_0) \ll 1$ only. Show that in this case, $\gamma_1 \cong 2l/Nl_0$, and that

$$\ln W_{\max} \cong N \ln 4 - l^2/Nl_0^2.$$

(d) We now drop the restriction to small values of (l/Nl_0). Using $\cosh x = (e^x + e^{-x})/2$, $\sinh x = (e^x - e^{-x})/2$, show (with some algebra) that

$$e^{\gamma_1} = \frac{1 + l/Nl_0}{1 - l/Nl_0}$$

and that

$$\ln W_{\max} = N \ln 4 - N \ln \left(1 - \left(\frac{l}{Nl_0} \right)^2 \right) - \left(\frac{l}{l_0} \right) \ln \left(\frac{1 + l/Nl_0}{1 - l/Nl_0} \right).$$

(e) In the foregoing, we have assumed no energy difference between the distinct folding patterns of the chain. (In a realistic case, the energy may depend on the number and sequence of bonds in the chain.) The quantity $S_C = k \ln W_{\max}$ represents the configurational entropy of the chain. It contributes a term $-kT \ln W_{\max} = -T S_C$ to the free energy F of the chain. The increase of F at constant T, due to stretching the chain represents the isothermal work done on the chain. It follows that $(\partial F/\partial l)_T$ represents the average force f needed to keep the end points O and A a distance l apart.

(f) Find the expression for this force f, given by

$$f = -T \left(\frac{d S_C}{dl} \right).$$

Notice that this force is purely entropic in origin, in the model discussed here!

(g) Find the expression for f in the limiting case that $(l/Nl_0) \ll 1$. Show, in particular, that this force is of the form

$$f = \text{constant} \cdot T \cdot l.$$

This means that (a) it satisfies Hooke's law, and (b) it increases with T. Both these phenomena apply for rubber.

For further reading, see:

H. M. James and E. Guth, *Journal of Polymer Science* **4**, 153 (1949).

W. Kuhn and H. Kuhn, *Helvetica Chimica Acta* **29**, 1615 (1946).

M. V. Vol'kenstein, Configurational Statistics of High Polymer Chains. In *High Polymers*, Vol. 17. Interscience, New York (1963).

35. The myoglobin oxygen-saturation curve. The degree of saturation s of myoglobin, at a partial oxygen pressure p_{O_2}, is given by equation (4-257):

$$s = \frac{p_{O_2}}{P(T) + p_{O_2}}. \tag{1}$$

In this formula $P(T)$ is a temperature dependent reference pressure, whose value is given by equation (4-256a):

$$P(T) = \left(\frac{kT}{V_F(T)}\right) e^{\Delta E_{10}/RT} = p_F e^{\Delta E_{10}/RT}, \tag{2}$$

where $V_F(T)$ is the "Fermi" volume

$$V_F = \left(\frac{h^2}{2\pi mkT}\right)^{3/2} \tag{3}$$

and ΔE_{10} is the molar energy difference between the oxygenated and deoxygenated Mb. That is, $-\varepsilon_{10} = -\Delta E_{10}/N_0$ is the binding energy of the oxygen molecule to Mb. (ΔE_{10} is negative.) This problem deals with the determination of ΔE_{10} from the data for s versus p_{O_2} for various temperatures (see figure below).

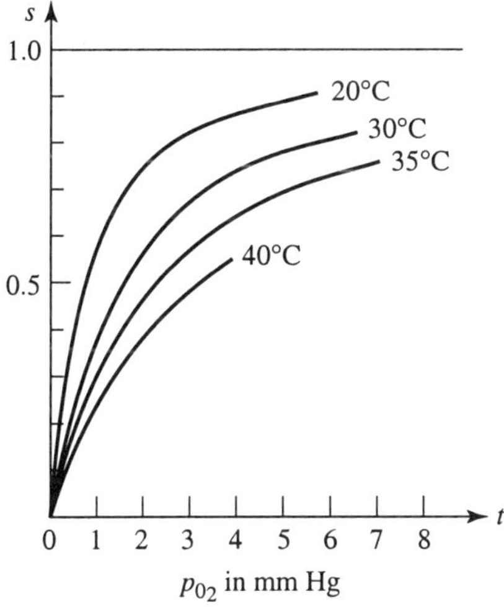

p_{O_2} in mm Hg

Oxygen dissociation curves of human Mb at different temperatures. A. R. Fanelli, E. Antonini, and A. Caputo, Hemoglobin and Myoglobin. In *Advances in Protein Chemistry*, Vol. 19, Academic Press, New York (1964).

You may determine ΔE_{10} in *two* ways.

Method A: ΔE_{10} from equation (2) at a fixed temperature.

(a) Show that $P(T)$ is equal to $p_{1/2}$, the oxygen pressure for which $s = \frac{1}{2}$.
In what follows, use $T = 310°K$ (96.8°F), where $p_{1/2} \cong 2.5$ mmHg.

(b) You must now calculate $V_F(T)$. Show that for oxygen ($m_{O_2} = 5.33 \times 10^{-23}$ g), at $T = 310°K$, $1/V_F(T)$ has the value 1.87×10^{26} cm^{-3}.

(c) Calculate the pressure $p_F \equiv kT/V_F$. (You should get $p_F = 8 \times 10^6$ atm $= 6 \times 10^9$ mmHg.)

(d) You can now determine E_{10} from equation (2):

$$-\Delta E_{10} = RT \ln(p_F/p_{1/2}).$$

Find the value of $-\Delta E_{10}$ in kcal/mol (using $R = 2$ cal/mol °K) and the value of $-\varepsilon_{10} = -E_{10}/N_0$ in electron volts.

Method B: ΔE_{10} from the rate of change of the half-saturation pressure $p_{1/2}$ with temperature: $p_{1/2} = P(T)$ depends on T mainly through the Boltzmann factor $e^{E_{10}/RT}$.

(e) Use the data in the graph above. Show that, at $T = 37°C$ (310°K), the value of $(1/p_{1/2})(dp_{1/2}/dT)$ is approximately

$$\frac{1}{p_{1/2}}\frac{dp_{1/2}}{dT} \cong 0.064/°K.$$

(f) Show from the equality of $p_{1/2}$ with $P(T)$ that

$$\frac{1}{p_{1/2}}\frac{dp_{1/2}}{dT} = \frac{d}{dT}\ln P(T) \cong -\frac{\Delta E_{10}}{RT^2}.$$

(This neglects the T-dependence of p_F.)

(g) Use this relation to calculate ΔE_{10}.

Notice: You should be aware that the two methods to get ΔE_{10} make use of very different information. Method B uses only the *rate of change* of $p_{1/2}$ with temperature T. By contrast, Method A makes use of the *absolute value* of $p_{1/2}$, which brings in Planck's constant h. The two methods should give you values of ΔE_{10} that agree to within about 15%. (As an exercise, one could use the data presented to determine ΔE_{10} by Method B, and by using this value in Method A, determine the value of Planck's constant h!)

36. The effect of 2,3-diphosphoglycerate (2,3-DPG) on the oxygen saturation curve of hemoglobin. If a solution contains both the protein hemoglobin and the organic phosphate ester 2,3-DPG, then the oxygen saturation curve shifts to the right. More specifically, the half-saturation pressure $(p_{O_2})_{1/2}$ in the presence of 2,3-DPG of mole fraction (X_{1p}) is related to its value $(p_{O_2})^0_{1/2}$ in the absence of the ester according to

$$(p_{O_2})_{1/2} = (p_{O_2})^0_{1/2}(1 + aX_{1p})^{1/4}. \tag{1}$$

In this problem you will show that this result is consistent with the simple extension of the MWC model for oxygen binding to hemoglobin. According to this extension:

(i) 2,3-DPG has a single binding site near the center of the Hb tetramer.

(ii) 2,3-DPG binds only to T states of Hb, and if bound, lowers the free energy of each configuration by the same amount $\Delta\mu^0_{TP}$.

(iii) 2,3-DPG does not affect the free-energy advantage of binding oxygen in either the T or the R form.

With these considerations in mind, answer the following questions:

(a) Provide a clear free-energy level diagram of the T states showing the positions and relative occupancies of each configuration, including both oxygen binding and 2,3-DPG binding. Let p be the pressure of oxygen in the overlying gas phase, and let $\Delta\mu^0_T$ be the free energy advantage of binding a single oxygen in the T form, and let (X_{1p}) be the mole fraction of 2,3-DPG in equilibrium in solution.

(b) Provide a clear free-energy level diagram showing the positions and relative occupancies of each state in the R forms. Use the conventional notation for the position of each free-energy level.

(c) Obtain the binding polynomial, using the following symbols:

$$z = \left(\frac{p}{p_F}\right) \exp\left(-\frac{\Delta\mu^0_R}{kT}\right),$$

$$c = \exp\left(\frac{\Delta\mu^0_R - \Delta\mu^0_T}{kT}\right), \tag{2}$$

$$L = \exp\left(\frac{\Delta\mu^0_{RT}}{kT}\right),$$

$$y = X_{1p}\exp\left(-\frac{\Delta\mu^0_{TP}}{kT}\right).$$

(d) Obtain the oxygen saturation function $Y(z)$ in the presence of 2,3-DPG.

(e) Show that the saturation function in the presence of 2,3-DPG has the same form as in its absence except for the fact that the allosteric constant L is replaced by a new constant L'. Obtain the relation between L' and L.

(f) Show that $(p_{O_2})_{1/2}$ varies with the mole fraction of 2,3-DPG according to (1); provide a theoretical expression for, and interpret the meaning of, the coefficient a.

(g) Give a physical interpretation of the features of the model above, that explains the experimental finding that $(p_{O_2})_{1/2}$ increases as found experimentally.

37. Oxygen saturation of mutant hemoglobin. Consider a mutant form of hemoglobin in which one of the four heme groups is unable to bind oxygen. This mutant hemoglobin can exist in either a T form or an R form. In the absence of any oxygen binding the R form lies higher in free energy than the T form by $\Delta\mu^0_{RT}$. Assume a symmetric MWC model for the binding of oxygen to the three available sites. The free-energy change on binding to the T and R forms being $\Delta\mu^0_T$ and $\Delta\mu^0_R$, respectively, with $|\Delta\mu^0_R| > |\Delta\mu^0_T|$.

(a) Draw a free-energy level diagram listing: each configuration, the corresponding position of the free-energy level, the degeneracy. Also tabulate separately the relative probability of each configuration.

(b) Obtain the form of the binding polynomial $B(y)$, where

$$y = \left(\frac{p_{O_2}}{p_F}\right)e^{-\Delta\mu^0_R/kT}.$$

Use the following notation for the free-energy parameters in the binding polynomial

$$c = \exp\left(\frac{\Delta\mu^0_R - \Delta\mu^0_T}{kT}\right), \qquad c < 1, \qquad L = \exp\left(\frac{\Delta\mu^0_{RT}}{kT}\right),$$

(c) Obtain the form of the oxygen saturation function $Y(y)$ for this mutant hemoglobin in terms of the parameters c and L.

(d) "Switchover" between the T and R forms occurs when the occupancy of all T forms is equal to that of all R forms. Show that the value of y at which switchover occurs is

$$(y)_{\text{switchover}} \equiv \frac{L^{1/3} - 1}{(1 - cL^{1/3})}.$$

(e) The switchover from T to R forms is essential in order for the mutant form to show a sigmoidal saturation function. From the result above, show that for sigmoidicity to occur it is necessary that the two conditions (a) and (b) must apply

$$\Delta\mu^0_{RT} > 0 \tag{a}$$

and

$$3(\Delta\mu^0_R - \Delta\mu^0_T) + \Delta\mu^0_{RT} < 0. \tag{b}$$

(f) Interpret physically the meaning of conditions (a) and (b) in terms of the positions of the levels in the free-energy level diagram.

(g) The "median oxygen pressure" P_m is defined as that pressure for which the unoxygenated T state and the fully oxygenated R state have equal probability. Show that on comparing P_m for the mutant form with that for the normal form (having four sites and the same energy parameters) that

$$\frac{(P_m)_{\text{mutant}}}{(P_m)_{\text{normal}}} = L^{(1/12)}.$$

Thus the median pressure for the mutant hemoglobin is greater than that for the normal hemoglobin.

38. **Binding polynomial and shape of oxygen saturation curve.** In terms of the oxygen pressure p_{O_2}, the binding polynomial $B(p_{O_2})$ may be written in the form

$$B(p_{O_2}) = 1 + 4\frac{p_{O_2}}{p_1} + 6\left(\frac{p_{O_2}}{p_2}\right)^2 + 4\left(\frac{p_{O_2}}{p_3}\right)^3 + \left(\frac{p_{O_2}}{p_4}\right)^4.$$

In an experimental paper by Di Cera et al., the parameters p_1 to p_4 have been determined for a variety of samples (and buffer solutions). Below a typical set of data (sample E):

$$p_1 = 25 \text{ torr}, \quad p_2 = 12.7 \text{ torr}, \quad p_3 > 12.6 \text{ torr}, \quad p_4 = 3.71 \text{ torr}.$$

(N.B. In all their samples, p_3 is poorly determined, and compatible with $p_3 \approx \infty$ to within the accuracy of the data.)

(a) Construct the expression for the oxygen saturation curve $Y(p_{O_2})$, defined in terms of $B(p_{O_2})$ by

$$Y(p_{O_2}) = \tfrac{1}{4} p_{O_2} \frac{\partial}{\partial p_{O_2}} \ln B(p_{O_2}).$$

(b) Obtain the expression for the slope of the saturation curve at zero oxygen pressure

$$\left(\frac{dY}{dp_{O_2}} \right)_{p_{O_2}=0}.$$

(c) Show that for large oxygen pressures ($p_{O_2} \gg p_1$ to p_4), the saturation function Y approaches the value 1 in the following way:

$$1 - Y(p_{O_2}) \to \frac{p_a}{p_{O_2}}, \qquad \text{provided} \quad p_3 \neq \infty$$

but

$$1 - Y(p_{O_2}) \to \left(\frac{p_b}{p_{O_2}} \right)^2 \qquad \text{if} \quad p_3 = \infty,$$

and find the expressions for the parameters p_a and p_b.

(d) In a model of Hb oxygenation that makes use of a *single* free-energy ladder only (i.e., a generalized Pauling model), the expression for the relative probabilities R_ν for the occupancy of ν binding sites is

$$R_\nu = \binom{4}{\nu} \left(\frac{p_{O_2}}{p_F} \right)^\nu \exp\left[-\left(\frac{\Delta \mu_\nu^0}{kT} \right) \right].$$

Use this form to obtain expressions for the parameters p_ν ($\nu = 1, 2, 3, 4$) in the expression for the binding polynomial $B(p_{O_2})$.

(e) The data given above were taken at $T = 25\,^\circ$C, in which case the "Fermi pressure" p_F has the value

$$p_F = 5.4 \times 10^9 \text{ torr.}$$

Use this information to determine the quantities $\Delta\mu_\nu^0/kT$, and construct the free-energy ladder. Indicate clearly the spacing between the rungs of the ladder.

39. **Linkage relations.** Consider a macromolecule M with one binding site each for two distinct ligand species "a" and "b." There will be four states: M, M_a, M_b, M_{ab}, with the following chemical potentials:

$$\mu_M = \mu_M^0 + kT \ln X_M,$$
$$\mu_{M_a} = \mu_M + \Delta\mu_{M_a}^0 + kT \ln R_a, \qquad (R_a \equiv X_{M_a}/X_M,\ \text{etc.}),$$
$$\mu_{M_b} = \mu_M + \Delta\mu_{M_b}^0 + kT \ln R_b,$$
$$\mu_{M_{ab}} = \mu_M + \Delta\mu_{M_{ab}}^0 + kT \ln R_{ab}.$$

In addition, one has the chemical potentials μ_a and μ_b of the unliganded species "a" and "b."

(a) State the equilibrium conditions for the reactions

$$M + a \leftrightarrow M_a, \qquad M + b \leftrightarrow M_b, \qquad M_a + b \leftrightarrow M_{ab} \leftrightarrow M_b + a.$$

(b) Obtain expressions for the relative probabilities R_a, R_b, R_{ab} for M_a, M_b, and M_{ab}, in terms of the quantities $\Delta\mu_{M_a}^0$, $\Delta\mu_{M_b}^0$, $\Delta\mu_{M_{ab}}^0$, and the *fugacities*

$$z_a \equiv \exp\left(\frac{\mu_a}{kT}\right) \qquad \text{and} \qquad z_b \equiv \exp\left(\frac{\mu_b}{kT}\right).$$

(c) Now let species "a" be an oxygen molecule, and species "b" a proton (i.e., an H$^+$ ion). One has then

$$\frac{\mu_a}{kT} = \ln p_{O_2} + \text{constant} \qquad \text{and} \qquad \frac{\mu_b}{kT} = -2.3\,\text{pH} + \text{constant}'.$$

It is then convenient to express the binding polynomial

$$P \equiv 1 + R_a + R_b + R_{ab} = 1 + B_a z_a + B_b z_b + B_{ab} z_a z_b = P(\mu_a, \mu_b)$$

in terms of a "free energy" F: $F(\mu_a, \mu_b) \equiv -kT \ln P$.

Show that the mean oxygenation Y and the mean electric charge Q, given by

$$Y(p_{O_2}) \equiv \langle \nu_a \rangle \equiv \frac{R_a + R_{ab}}{1 + R_a + R_b + R_{ab}}$$

and

$$Q \equiv \langle \nu_b \rangle \equiv \frac{R_b + R_{ab}}{1 + R_a + R_b + R_{ab}},$$

may be expressed as

$$\langle \nu_a \rangle = -\frac{\partial F}{\partial \mu_a} \qquad \text{and} \qquad \langle \nu_b \rangle = -\frac{\partial F}{\partial \mu_b}.$$

(d) Show now how the "Maxwell relation"

$$\frac{\partial \langle \nu_a \rangle}{\partial \mu_b} = \frac{\partial \langle \nu_b \rangle}{\partial \mu_a} = -\left(\frac{\partial^2 F}{\partial \mu_a \, \partial \mu_b} \right)$$

leads to the linkage relation

$$\left(\frac{\partial Y}{\partial \text{pH}} \right)_{p_{O_2}} = -2.3 \left(\frac{\partial Q}{\partial \ln p_{O_2}} \right)_{\text{pH}}.$$

(e) Legendre transformed linkage relations. Given that the differential dF has the structure: $dF = -\langle \nu_a \rangle \, d\mu_a - \langle \nu_b \rangle \, d\mu_b$ show that the functions $G_a \equiv F + \langle \nu_a \rangle \mu_a$ and $G_b \equiv F + \langle \nu_b \rangle \mu_b$ may be viewed as functions of the sets $\{\langle \nu_a \rangle, \mu_b\}$ and $\{\mu_a, \langle \nu_b \rangle\}$, respectively.

This gives two new linkage relations

$$\frac{\partial}{\partial \mu_b} \left(\frac{\partial G_a}{\partial \langle \nu_a \rangle} \right) = \frac{\partial}{\partial \langle \nu_a \rangle} \left(\frac{\partial G_a}{\mu_b} \right)$$

and one with (a, b) interchanged. State the explicit form of these relations, in terms of the quantities Y, Q, p_{O_2}, and pH.

(f) One of the relations obtained in question (e) should be of the form

$$\left(\frac{\partial Y}{\partial Q}\right)_{p_{O_2}} = +2.3 \left(\frac{\partial \text{pH}}{\partial \ln p_{O_2}}\right)_Q.$$

This relation can be integrated to give

$$\Delta Y \equiv \{Y(Q = 1) - Y(Q = 0)\}_{p_{O_2}} = \frac{2.3\partial}{\partial \ln p_{O_2}} \int_0^1 dQ\, \text{pH}(p_{O_2}, Q).$$

To do this integral, express pH in terms of Q, by showing that $Q \equiv \langle \nu_b \rangle$ may be written as

$$Q = \frac{z_b}{z_b + \dfrac{1 + B_a z_a}{B_b + B_{ab} z_a}} = \frac{[\text{H}^+]}{[\text{H}^+] + [\text{H}^+]_{1/2}},$$

with $[\text{H}^+]$ being the hydrogen ion concentration, and $[\text{H}^+]_{1/2}$ that at $Q = \frac{1}{2}$.

Show finally that ΔY is given, in terms of $\text{pH}_{1/2}$, by

$$\Delta Y = +2.3 \frac{\partial \text{pH}_{1/2}}{\partial \ln p_{O_2}}.$$

How could one read this number off a curve that plots Y versus p_{O_2} for various values of pH?

40. A generalized Pauling model. The experimental basis for any model of hemoglobin oxygenation lies in the expression for the binding polynomial. From experimental data on the binding polynomial one may construct a single energy level ladder with suitably spaced rungs.

In this problem, we will use a generalized Pauling model, defined below, and determine the four parameters it contains by fitting it to the data presented in Problem 1. In the generalized Pauling model, there is a free energy $\Delta\mu^0$ of binding an oxygen molecule to any of the four heme sites. In addition, there are free energies of interaction between occupied heme sites that contain three parameters $\delta\mu^0$, defined as follows:

$$\delta\mu_1^0 \equiv (\delta\mu^0)_{\alpha_1\beta_1} = (\delta\mu^0)_{\alpha_2\beta_2},$$
$$\delta\mu_2^0 \equiv (\delta\mu^0)_{\alpha_1\beta_2} = (\delta\mu^0)_{\alpha_2\beta_1},$$
$$\delta\mu_3^0 \equiv (\delta\mu^0)_{\alpha_1\alpha_2} = (\delta\mu^0)_{\beta_1\beta_2}.$$

Here α_1, α_2, β_1, β_2 designate the four heme sites, corners of a rectangle, as shown below. $\alpha_1\beta_1$ and $\alpha_2\beta_2$ stand for the two vertical sides, $\alpha_1\alpha_2$ and $\beta_1\beta_2$ for the two horizontal sides, and $\alpha_1\beta_2$, $\alpha_2\beta_1$ for the two diagonals.

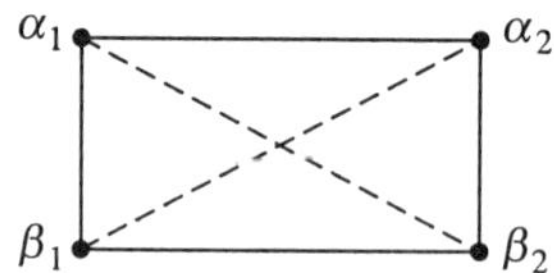

(a) Using the abbreviations

$$z \equiv \left(\frac{p_{O_2}}{p_F}\right) \exp\left(-\frac{\Delta\mu^0}{kT}\right)$$

and $f_1 \equiv \exp(-\delta\mu_1^0/kT)$, $f_2 \equiv \exp(-\delta\mu_2^0/kT)$, and $f_3 \equiv \exp(-\delta\mu_3^0/kT)$, show that the relative probabilities R_ν for ν occupied binding sites are given by

$$R_0 = 1, \qquad R_1 = 4z, \qquad R_2 = 2z^2(f_1 + f_2 + f_3),$$
$$R_3 = 4z^3 f_1 f_2 f_3, \qquad R_4 = z^4 f_1^2 f_2^2 f_3^2.$$

(b) The binding polynomial $B(p_{O_2})$ introduced in Problem 38 contains four parameters p_1 to p_4:

$$B(p_{O_2}) = 1 + 4\frac{p_{O_2}}{p_1} + 6\left(\frac{p_{O_2}}{p_2}\right)^2 + 4\left(\frac{p_{O_2}}{p_3}\right)^3 + \left(\frac{p_{O_2}}{p_4}\right)^4.$$

Use the data given for these parameters to obtain numerical values for the quantities $\frac{\Delta\mu_1^0}{kT}$, $\frac{\Delta\mu_2^0}{kT}$, $\frac{\Delta\mu_3^0}{kT}$, and f_1, f_2, f_3.

(c) Draw an energy ladder representing these values, and indicate the relative spacings.

41. **Oxygen saturation in trimeric hemoglobin.** Consider a form of hemoglobin capable of binding oxygen at any of three equally spaced heme sites. Treat the reversible binding of oxygen using the Pauling model. Let $\Delta\mu^0$ be the attractive energy of binding of each oxygen to a heme group in the absence of neighbors. Let $\delta\mu^0$ be the additional attractive energy of interaction between pairs of oxygen molecules.

(a) Construct a free-energy level diagram that contains: a specification of each possible binding configuration, along with the positions of the corresponding free-energy levels, the degeneracy and the relative occupancies of each configuration.

(b) From the entries in the free energy level diagram, obtain the binding polynomial $B(y)$, where

$$y = \left(\frac{p}{p_F}\right)\exp\left[-\left(\frac{\Delta\mu^0}{kT}\right)\right]\exp\left[-\left(\frac{\delta\mu^0}{kT}\right)\right] \tag{1}$$

and

$$f = \exp\left[-\left(\frac{\delta\mu^0}{kT}\right)\right].$$

(c) Show that the saturation function $Y(y)$ for the binding of oxygen has the form

$$Y(y) = y\frac{(f^{-1} + 2f^{-1}y + y^2)}{(1 + 3yf^{-1} + 3y^2 f^{-1} + y^3)}. \tag{2}$$

(d) Verify that half-saturation occurs for $y = 1$. What experimental information would suffice to determine the magnitude and sign of the two energy parameters $\Delta\mu^0$ and $\delta\mu^0$?

(e) Assume that $\Delta\mu^0$ and $\delta\mu^0$ are independent of the temperature. Obtain an expression for the temperature-dependence of the pressure $(p_{1/2})$ corresponding to half-saturation of this hemoglobin trimer; i.e., obtain an expression for

$$\left(\frac{1}{p_{1/2}}\right)\left(\frac{\partial p_{1/2}}{dT}\right).$$

(f) Obtain expressions for the probability $(P_\nu(y))$ to find this trimer having $\nu = 0, 1, 2,$ or 3 molecules bound, as a function of y.

(g) Make a sketch of each of the species fractions P_0, P_1, P_2, and P_3 as a function not of y, but of $\overline{\nu} = 3Y(y)$. Make no explicit calculations for $P_\nu(\overline{\nu})$; however, explain your reasons for each of the principal features in your sketch.

42. **An exotic hemoglobin.** In remote Lake Baikal, in eastern Siberia, let us suppose that a Russian biologist claimed to have found a hitherto unknown aquatic creature, with a most unusual hemoglobin: It consists of *two* amino acid chains only, with one heme group each, so as to have a total of two binding sites of oxygen per macromolecule.

 You are to establish the possible shapes of the oxygen saturation curve: Saturation Y versus p_{O_2}.

 Notation: There are four states of oxygen binding, that will be called:

M_0: bare macromolecule;

M_{1a}, M_{1b}: one oxygen bound, on site a or b, respectively;

M_2: two O_2 bounds.

The associated chemical potentials will be called

$$\mu_{M_0} = \mu^0_{M_0} + kT \ln X_0,$$
$$\mu_{M_{1a}} = \mu^0_{M_{1a}} + kT \ln X_{1a} \equiv \mu_{M_0} + \Delta\mu^0_1 + kT \ln R_{1a},$$
$$\mu_{M_{1b}} = \mu^0_{M_{1b}} + kT \ln X_{1b} \equiv \mu_{M_0} + \Delta\mu^0_1 + kT \ln R_{1b}.$$

and

$$\mu_{M_2} = \mu^0_{M_2} + kT \ln X_2 \equiv \mu_{M_0} + \Delta\mu^0_2 + kT \ln R_2.$$

Notice that the R's are the mole fractions *relative to the unliganded* macromolecule. Also, sites a and b are assumed to have the *same* free energy $\Delta\mu^0_1$ of binding. Also observe that $R_1 = R_{1a} + R_{1b}$, giving a degeneracy factor of 2. Finally, let $\mu_{O_2} = \mu^0_{O_2} + kT \ln(p_{O_2}/p_F)$ stand for the oxygen chemical potential.

(a) Write the equilibrium conditions for the reactions

$$M_0 + O_2 = M_{1a} \text{ (or } M_{1b}\text{)}, \qquad M_0 + 2O_2 = M_2.$$

(b) Find the expressions for R_1 and R_2.

(c) Introducing the *fugacity* z for oxygen: $z \equiv e^{\mu_{O_2}/kT}$, show that the relative abundances R can be written as

$$R_0 = 1, \qquad R_1 = B_1 z, \qquad R_2 = B_2 z^2,$$

and give expressions for the coefficients B_1 and B_2.

(d) Show that the degree Y of oxygen saturation $(0 \le Y \le 1)$ is given by

$$Y = \frac{1}{2} \frac{R_1 + 2R_2}{(1 + R_1 + R_2)} = \frac{1}{2} z \frac{\partial}{\partial z} \ln B(z), \quad \text{where} \quad B(z) = 1 + B_1 z + B_2 z^2.$$

is the *binding polynomial*.

(e) Show that in the case where the free energy of binding of oxygen to one site does *not* depend on whether the other site is occupied or not, one has the relation $\Delta\mu_2^0 = 2\Delta\mu_1^0$. Also show that, as a consequence, one has the relation between coefficients B_1 and B_2: $\frac{1}{4}B_1^2$. Finally show that in this case, the expression for the saturation Y reduces to

$$Y = \frac{B_1 z/2}{1 + B_1 z/2}.$$

(f) Shape of the saturation curve: $Y = Y(p_{O_2})$.

First it is clear from the definition of the fugacity z, that z is proportional to the oxygen pressure p_{O_2}. This means that the binding polynomial $B(z)$ can be written in the form

$$B(p_{O_2}) = 1 + 2\frac{p_{O_2}}{p_1} + \left(\frac{p_{O_2}}{p_2}\right)^2.$$

Now show that the saturation $Y(p)$ may be written as $Y(p) = \frac{1}{2}p(\partial/\partial p)\ln B(p)$. Then show that the pressures p_1 and p_2 have a simple interpretation

$$\left(\frac{dY}{dp_{O_2}}\right)_{p_{O_2}=0} = \frac{1}{p_1}$$

and that $Y(p_2) = \frac{1}{2}$, so that $p_{1/2} = p_2$.

(g) Now use this information to show that the saturation curve will have the characteristic *sigmoid* shape, provided one has $p_2 < \frac{1}{2}p_1$.
Hint: Make a sketch extending the *tangent* to the saturation curve at $p = 0$: $T(p) \equiv p/p_1$ to and beyond the half-saturation point $p = p_2$, and draw in a hypothetical sigmoid saturation curve.

(h) What relation between the chemical potentials $\Delta\mu_1^0$ and $\Delta\mu_2^0$ is implied by the condition for a sigmoid shape?

Table of Important Constants

Avogadro's Number: $N_0 = 6.023 \times 10^{23}$

Universal Gas Constant: $R = 8.314$ J/$^\circ$K mol

$$= 8.314 \times 10^7 \, (\text{dyn/cm}^2) \cdot \text{cm}^3/^\circ\text{K mol}$$

$$= 1.987 \text{ cal}/^\circ\text{K mol}$$

Boltzmann Constant: $k = 1.381 \times 10^{-16}$ erg/$^\circ$K

$$= 0.863 \times 10^{-4} \text{ eV}/^\circ\text{K}$$

Mass of Hydrogen Atom: $m_{\text{H}} = 1.67 \times 10^{-24}$ g

Planck's Constant: $h = 6.626 \times 10^{-27}$ erg s

Mean Acceleration of Gravity: $g = 9.81$ m/s^2

Melting Point of Ice: $0.00\,^\circ\text{C} = 273.15\,^\circ\text{K}$

Atmospheric Pressure: 1.01×10^6 dyn/cm$^2 = 14.7$ lb/in^2

$$= 760 \text{ mmHg}$$

$$= 1033 \text{ cmH}_2\text{O}$$

Molar Volume of Ideal Gas: 22.41 l at $T = 0\,^\circ\text{C}$, $p = 1$ atm

Table of Units and Conversion Factors

This is a compilation of the main units used in this volume and their conversion to alternate units.

4.6.A.   Units of Length

The basic unit of length in the international mks (meter/kilogram/second) system is the *meter* (m). Subdivision of the meter are the centimeter (cm) and the millimeter (mm):

$$1 \text{ m} = 10^2 \text{ cm} = 10^3 \text{ mm},$$
$$1 \text{ kilometer (km) equals } 10^3 \text{ m}.$$

Other frequently used metric units of length are:

$$1 \text{ micron } (\mu) = 10^{-6} \text{ m},$$
$$1 \text{ Ångstrom unit (Å)} = 10^{-10} \text{ m}.$$

The commonly used British units of length are related to metric units as follows:

$$1 \text{ inch (in)} = 2.54 \text{ cm},$$
$$1 \text{ foot (ft)} = 30.48 \text{ cm},$$
$$1 \text{ yard (yd)} = 0.912 \text{ m},$$
$$1 \text{ US mile (mi)} = 1.609 \text{ km}.$$

4.6.B.   Units of Area and Volume

Metric units of area and volume are derived from the corresponding units of length. A differently named unit is the liter:

$$1 \text{ liter} = 10^3 \text{ cm}^3.$$

Often, 1 cm^3 is referred to as 1 ml. British units of volume are:

$$1 \text{ quart} = \tfrac{1}{4} \text{ gallon} = 32 \text{ fluid ounces},$$
$$= 57.75 \text{ in}^3 = 0.946 \text{ l}.$$

Conversely,

$$1 \text{ in}^3 = 16.39 \text{ cm}^3.$$

4.6.C.   Units of Force

The mks unit of force is the Newton (N):

$$1 \text{ N} = 1 \text{ kg m/s}^2.$$

Often also used is the older cgs (centimeter/gram/second) unit, the dyne:

$$1 \text{ dyn} = 10^{-5} \text{ N} = 1 \text{ g cm/s}^2.$$

The force of 1 pound (lb) is related to (N) by

$$1 \text{ lb} = 4.448 \text{ N}.$$

4.6.D.   Units of Pressure

The mks unit of pressure is Newton per square meter. This is a very small pressure; a larger, metric unit is the bar:

$$1 \text{ bar} = 10^3 \text{ mbar} = 10^5 \text{ N/m}^2 = 10^6 \text{ dyne/cm}^2.$$

The pressure of 1 bar is very close to the standard atmospheric pressure of 1 atm:

$$1 \text{ atm} = 1.013 \text{ bar} = 1013 \text{ mbar}.$$

Derived pressure units use the height of a fluid column to express the exerted pressure. Mercury (Hg) and water (H_2O) columns are most frequently used. One has

$$1 \text{ atm} = 760 \text{ mmHg} = 29.92 \text{ inHg}$$
$$\cong 10^3 \text{ cmH}_2\text{O}.$$

Pressure is also measured in pounds per square inch. One has

$$1 \text{ atm} = 14.70 \text{ lb/in}^2.$$

4.6.E.   Units of Energy

The mks unit of energy is the Joule:

$$1 \text{ J} = 1 \text{ kg m}^2/\text{s}^2 = 1 \text{ N m}.$$

A related metric unit is the erg:

$$1 \text{ erg} = 10^{-7} \text{ J} = 1 \text{ dyn cm}.$$

Thermal units of energy used in this volume are the calorie (cal) and the kilocalorie (kcal):

$$1 \text{ cal} = 4.186 \text{ J}$$
$$= 4.186 \times 10^7 \text{ erg},$$
$$1 \text{ kcal} = 10^3 \text{ cal}.$$

A much used unit of energy in atomic physics is the electron volt (eV). It is the energy imparted to a particle of electronic charge (e) by an electric potential of 1 V. (Energies of chemical bonds are of the order of a few electron volts.) One has

$$1 \text{ eV} = 1.6 \times 10^{-12} \text{ erg}$$

A useful relation is the conversion of one mole (N_0) electron volts into kcals:

$$N_0 \text{ eV} = 2.306 \text{ kcal}$$

Index